Geometry at the Frontier

Symmetries and Moduli Spaces of Algebraic Varieties

CONTEMPORARY MATHEMATICS

766

Geometry at the Frontier

Symmetries and Moduli Spaces of Algebraic Varieties

2016–2018 Workshops
Geometry at the Frontier
November 2016, November 2017, November 2018
Universidad de La Frontera, Pucón, Chile

Paola Comparin
Eduardo Esteves
Herbert Lange
Sebastián Reyes-Carocca
Rubí E. Rodríguez
Editors

AMS
AMERICAN MATHEMATICAL SOCIETY

2020 *Mathematics Subject Classification.* Primary 14H55, 30Fxx, 14Kxx, 14K22, 14H40, 14C30, 14J32, 14H37, 14J50, 14L30.

For additional information and updates on this book, visit
www.ams.org/bookpages/conm-766

Library of Congress Cataloging-in-Publication Data

Names: Comparin, Paola, 1985- editor. | Esteves, Eduardo de Sequeira, 1966- editor. | Lange, H. (Herbert), 1943- editor. | Reyes-Carocca, Sebastián, 1987- editor. | Rodríguez, Rubí E., 1953- editor. | Geometry at the Frontier (1st : 2016 : Pucón, Chile), author. | Geometry at the Frontier (2nd : 2017 : Pucón, Chile), author. | Geometry at the Frontier (3rd : 2018 : Pucón, Chile), author.

Title: Geometry at the Frontier : symmetries and moduli spaces of algebraic varieties : 2016-2018 workshops on geometry at the frontier, Pucón, Chile / Paola Comparin, Eduardo Esteves, Herbert Lange, Sebastián Reyes-Carocca, Rubí E. Rodríguez, editors.

Description: Providence, Rhode Island : American Mathematical Society, [2021] | Series: Contemporary mathematics, 0271-4132 ; volume 766 | Includes bibliographical references.

Identifiers: LCCN 2020042992 | ISBN 9781470453275 (paperback) | 9781470464226 (ebook)

Subjects: LCSH: Geometry, Algebraic–Congresses. | Curves, Algebraic–Congresses. | Moduli theory–Congresses. | Symmetry (Mathematics)–Congresses. | Algebraic varieties–Congresses. | AMS: Algebraic geometry – Curves – Riemann surfaces; Weierstrass points; gap sequences. | Functions of a complex variable – Riemann surfaces. | Algebraic geometry – Abelian varieties and schemes. | Algebraic geometry – Abelian varieties and schemes – Complex multiplication. | Algebraic geometry – Curves – Jacobians, Prym varieties. | Algebraic geometry – Cycles and subschemes – Transcendental methods, Hodge theory, Hodge conjecture. | Algebraic geometry – Surfaces and higher-dimensional varieties – Calabi-Yau manifolds. | Algebraic geometry – Curves – Automorphisms. | Algebraic geometry – Surfaces and higher-dimensional varieties – Automorphisms of surfaces and higher-dimensional varieties. | Algebraic geometry – Algebraic groups – Group actions on varieties or schemes (quotients).

Classification: LCC QA564 .G526 2021 | DDC 516.3/5–dc23

LC record available at https://lccn.loc.gov/2020042992

---

10 9 8 7 6 5 4 3 2 1 26 25 24 23 22 21

Contents

Preface

Geometry is one of the oldest branches of Mathematics, still vigorously alive. It has evolved from the axiomatic approach of Euclid to the study of complex spaces, and it has several applications today.

The Chilean Complex Geometry group has been important for the development of Mathematics in Chile. Its research production has proven to be fruitful, as shown by an important number of high impact publications, a wide network of collaborators, the organization of several international meetings and the fact that numerous former students of the group are now researchers and professionals in universities all over the country and abroad.

The workshops "Geometry at the Frontier" took place in November 2016, 2017 and 2018 in a local campus of the Universidad de La Frontera in the city of Pucón, located in the Chilean Patagonia. These workshops were organized by the members of the Chilean Complex Geometry Group and supported by one of the largest scientific grants funded by the Chilean government in pure mathematics, Anillo en Ciencia y Tecnología 1415 Conicyt-PIA.

The three conferences covered a spectrum of interrelated topics, with emphasis on algebraic varieties, automorphisms and moduli spaces. While algebraic geometry is a classical subject of study, it has recently witnessed a flurry of activity and important progress on many fundamental open questions. The workshops brought together Chilean and foreign researchers working in this and related areas to highlight the recent exciting progress and to explore future avenues of research.

The number of mathematicians working on algebraic geometry, and in particular on moduli spaces, has increased significantly in the last decades. We are truly convinced that this volume will be a useful and solid scientific source of recent developments on algebraic geometry. We believe that this volume will be also very useful for students and young researchers, giving them an overview and an insight on open questions in the area.

We cannot end these words without remembering a constant and very active participant of these meetings, and a permanent collaborator of the Chilean mathematical community, our friend José María Muñoz Porras (1956-2018), from the University of Salamanca. Always ready to discuss new mathematical (and other) ideas, to give new insights with a gentle smile, to stimulate and guide the younger mathematicians, with passion and joy. We miss him sorely.

The Editors

FIGURE 1. The conference venue, thanks to Marian ("Taffy") Griffiths

FIGURE 2. Official Photo, Pucón, 2018. Standing, from right to left: Sebastián Reyes-Carocca, Ara Basmajian, Anita M. Rojas, Rubén A. Hidalgo, Perry Susskind, Abraham Bobadilla, Benjamín Baeza, Felipe Zingali, Ana Victoria Quedo, Eduardo Uribe, Arthur Newlands, Vinicius Marques da Costa, Andrea Tironi, Gian Pietro Pirola, Abel Castorena, Pablo Quezada, José Aburto, Rodrigo Quezada, Xavier Gómez-Mont, Yerko Torres, Juan Desimoni, Ageu Freire, João Paulo Figueredo, Sebastián Rahausen, Eduardo Silva, Giancarlo Lucchini Arteche, Yerika Marín, Claudio Bravo, Francisco Rubilar, Estefanía Bravo, Pietro Beri, Paola Comparin, Luca Ugaglia, Bruno Suzuki, Ana María Porto, Milagros Izquierdo, Juan Carlos Naranjo, Herbert Lange, Elizabeth Gasparim, Matthieu Calvez, Christian Pauly, Ernst Kani, John A. Velling, Rubí E. Rodríguez, Phillip Griffiths, Samuel Boissière, José María Muñoz Porras, Alexis García-Zamora, Saúl Quispe, Víctor González-Aguilera, Alessandra Sarti, Francisco Plaza, Antonio Costa. Sitting, from left to right: Angel Carocca, Mauricio Godoy Molina, Maria Gioia Cifani, Irene Spelta, Ernesto Girondo, Leonardo Opazo, Kelyane Abreu, Danny Flores, María Paz Barría, Cristian Reyes, Yasmina Atarihuana, Juan Carlos García, Carolina Canales, Vicente Chicaiza.

Contemporary Mathematics
Volume **766**, 2021
https://doi.org/10.1090/conm/766/15369

Lower and upper bounds for some generalized arcs

Alexis E. Almendras Valdebenito and Andrea Luigi Tironi

Abstract. Let $\mathbb{F}_q$ be a field with q elements. In this note, we study some generalized arcs, that is, sets of $\mathbb{F}_q$-points in the projective plane $\mathbb{P}^2(\mathbb{F}_q)$ such that no six of them are on a conic. First, we consider the geometric configurations of such generalized arcs for small values of q and then we give some upper and lower bounds for the cardinality of complete generalized arcs, i.e. generalized arcs which are not contained in a bigger one.

Introduction

Let $\mathbb{F}_q$ be a field with q elements, where $q = p^r$ for some prime p and $r \in \mathbb{Z}_{\geq 1}$. Consider the finite projective plane $\mathbb{P}^2(\mathbb{F}_q)$. A k-arc is a set $\mathcal{K}$ of k $\mathbb{F}_q$-points (or, simply, *points*) of $\mathbb{P}^2(\mathbb{F}_q)$ such that no three of them lie on a $\mathbb{F}_q$-line (or, simply, *line*) of $\mathbb{P}^2(\mathbb{F}_q)$. The literature on k-arcs is very vast and many generalizations inherent to this concept have been considered by various authors, even in higher dimensions. In particular, recalling that a k-arc is complete if it is not contained in a $(k+1)$-arc, many authors focused their attention to the study of complete k-arcs and on some upper and lower bounds of their cardinality, obtaining interesting algebraic and geometric results, also in connection with some linear codes of special type. For a survey about all these results we refer, for instance, to **[9]** and the references therein.

In this note, we extend the concept of a k-arc along another direction. More precisely, we consider principally a *generalized k-arc*, that is, a set of k points of $\mathbb{P}^2(\mathbb{F}_q)$ such that no six of them lie on a $\mathbb{F}_q$-conic (or, simply, *conic*) of $\mathbb{P}^2(\mathbb{F}_q)$. Let us observe here that the conics we consider in this definition could be also reducible in two lines. In particular, a generalized k-arc which is also a k-arc is known simply as a *Veronesian k-arc*. In line with the classical results about k-arcs, we start to study here some geometric properties of generalized k-arcs for small values of q (see, e.g., Table 1) and we give some upper and lower bounds about the cardinality of complete generalized and Veronesian k-arcs for sufficiently bigger values of q (see, e.g., Theorems 1, 2, 3).

The paper is organized as follows. After some basic notions and remarks in Section 1, we study in Section 2 the cardinality and the geometry of maximal complete generalized k-arcs in $\mathbb{P}^2(\mathbb{F}_q)$ for $q \leq 9$ (Table 1). Subsequently, for higher

2020 *Mathematics Subject Classification.* Primary 12Y05, 16Z05; Secondary 94B05, 94B35.
Key words and phrases. Finite fields, constacyclic codes, dual codes, skew polynomial rings, semi-linear maps.

values of q, we show some upper bounds for generalized k-arcs, depending on the parity of q (Theorem 1 and Corollary 1). In Section 3, we prove another lower bound for complete k-arcs in $\mathbb{P}^2(\mathbb{F}_q)$ (Proposition 6) and we give principally a lower bound for complete generalized k-arcs $\mathcal{K}_g$, depending on the fact that $\mathcal{K}_g$ is complete as a Veronesian k-arc, or not (Theorems 2 and 3). Finally, in the Appendix, we report the main Magma [**3**] programs used to build all the tables and the corresponding examples.

1. Basic definitions and background material

First of all, let us recall here some basic classical definitions. A $(k,m)_n$-arc is a set of k points of $\mathbb{P}^n(\mathbb{F}_q)$ such that no $m+1$ of them lie on a hyperplane. A $(k,m)_n$-arc is complete if it is not properly contained in a $(h,m)_n$-arc, for some $h>k$. In particular, a $(k,2)_2$-arc is simply a k-arc in $\mathbb{P}^2(\mathbb{F}_q)$ (see, e.g., [**13**], [**14**], [**6**]). Moreover, the size of the largest complete k-arc in $\mathbb{P}^2(\mathbb{F}_q)$ will be denoted by $m(2,q)$, the size of the second largest complete k-arc in $\mathbb{P}^2(\mathbb{F}_q)$ by $m'(2,q)$, and the size of the smallest one by $t(2,q)$.

Now, let us introduce here the two main definitions of this note.

Definition 1. A *generalized k-arc* (or, simply, a k_g-arc) is a set of k points of $\mathbb{P}^2(\mathbb{F}_q)$ such that not six of them are on a (possibly reducible) conic. A k_g-arc is said to be *complete* if it is not contained in a $(k+1)_g$-arc. Moreover, the size of the largest complete k_g-arc is denoted by $m_g(2,q)$, while the size of the smallest one will be denoted by $t_g(2,q)$. Finally, a generalized k-arc is said to be *maximal* if $k=m_g(2,q)$ and *minimal* if $k=t_g(2,q)$.

Definition 2. A *Veronesian k-arc* (or, simply, a k_v-arc) is a k-arc in $\mathbb{P}^2(\mathbb{F}_q)$ such that not six of its points lie on a (necessarily irreducible) conic. A k_v-arc is said to be *complete* if it is not contained in a $(k+1)_v$-arc. Moreover, the size of the largest complete k_v-arc is denoted by $m_v(2,q)$, while the size of the smallest one will be denoted by $t_v(2,q)$. Finally, a generalized k-arc is said to be *maximal* if $k=m_v(2,q)$ and *minimal* if $k=t_v(2,q)$.

Remark 1. From Definitions 1 and 2, one can deduce that a k_g-arc is at most a $(k,3)_2$-arc and that a Veronesian k-arc is a k_g-arc with the additional property of being a $(k,2)_2$-arc. Moreover, since a Veronesian arc is a special case of a generalized arc, we have

$$t_v(2,q) \le t_g(2,q) \quad \text{and} \quad m_v(2,q) \le m_g(2,q)\ .$$

Denote by $|W|$ the cardinality of a set $W \subseteq \mathbb{P}^2(\mathbb{F}_q)$. The following simple technical result will be useful in the sequel.

Lemma 1. *Let $\mathcal{K}_g \subseteq \mathbb{P}^2(\mathbb{F}_q)$ be a generalized k-arc. If l_1 is a line in $\mathbb{P}^2(\mathbb{F}_q)$ such that $|\mathcal{K}_g \cap l_1| = 3$, then for every line l in $\mathbb{P}^2(\mathbb{F}_q)$ we have*

$$|(\mathcal{K}_g \setminus l_1) \cap l| \le 2,$$

i.e. $(\mathcal{K}_g \setminus l_1)$ is a (Veronesian) k-arc in $\mathbb{P}^2(\mathbb{F}_q)$. In particular, a generalized k-arc is at most the disjoint union of a (Veronesian) $(k-3)$-arc and 3 collinear points.

Proof. By Remark 1, we have $|\mathcal{K}_g \cap l| \le 3$ for every line l in $\mathbb{P}^2(\mathbb{F}_q)$. By contradiction, suppose there exists a line l_2 in $\mathbb{P}^2(\mathbb{F}_q)$ such that $|(\mathcal{K}_g \setminus l_1) \cap l_2| = 3$. Then, we get $(l_1 \cap \mathcal{K}_g) \cap (l_2 \cap \mathcal{K}_g) = \varnothing$. Hence, the conic $\Gamma := l_1 \cup l_2$ is such that $|\mathcal{K}_g \cap \Gamma| = 6$, but this is impossible because $\mathcal{K}_g$ is a generalized k-arc. □

2. Some upper bounds for $m_g(2,q)$

Before to study the geometry of generalized k-arcs for small values of q and some upper bounds for $m_g(2,q)$ when q is sufficiently large, let us give here the following result.

PROPOSITION 1. *We have*

$$m_g(2,q) \leq m(5,q)$$

for any $q \geq 2$, *where* $m(5,q)$ *is the largest size of a* $(k,5)_5$*-arc.*

PROOF. Consider the Veronese embedding

$$\begin{aligned} \nu_2 \colon \mathbb{P}^2(\mathbb{F}_q) &\longrightarrow \mathbb{P}^5(\mathbb{F}_q) \\ [x_0 : x_1 : x_2] &\longmapsto [x_0^2 : x_0x_1 : x_0x_2 : x_1^2 : x_1x_2 : x_2^2] \,. \end{aligned} \tag{1}$$

Define $S := \nu_2(\mathbb{P}^2(\mathbb{F}_q)) \subset \mathbb{P}^5(\mathbb{F}_q)$ and denote by $[z_0 : z_1 : z_2 : z_3 : z_4 : z_5]$ the general point of $\mathbb{P}^5(\mathbb{F}_q)$. If H is a hyperplane $\sum_{i=0}^5 \alpha_i z_i = 0$ in $\mathbb{P}^5(\mathbb{F}_q)$, with $\alpha_i \in \mathbb{F}_q$, then $\nu_2^{-1}(H \cap S)$ is a conic in $\mathbb{P}^2(\mathbb{F}_q)$. If $\mathcal{K}_g$ is a generalized k-arc, then $|\nu_2(\mathcal{K}_g) \cap H| \leq 5$. Therefore, the set $\nu_2(\mathcal{K}_g)$ is a $(k,5)_5$-arc and then $m_g(2,q) \leq m(5,q)$. □

First of all, let us compute the exact values of $m_g(2,q)$ for small q's, that is, for $q = 2,3,4,5,7$ and 8.

PROPOSITION 2. $m_g(2,2) = 7$ (*cf.* Table 1).

PROOF. Note that any conic in $\mathbb{P}^2(\mathbb{F}_2)$ has at most 5 points. So, any subset of $\mathbb{P}^2(\mathbb{F}_2)$ is a generalized arc. Therefore, $m_g(2,2) = |\mathbb{P}^2(\mathbb{F}_2)| = 7$. □

EXAMPLE 1. Let $q = 3$ and let $\mathcal{K} = \{p_1, p_2, p_3, p_4\}$ be a complete 4-arc in $\mathbb{P}^2(\mathbb{F}_3)$. Denote by l_{ij} the line passing through the distinct points p_i and p_j. Let q_1 be the intersection of l_{12} and l_{34}, and let q_2 be the intersection of l_{13} and l_{24}. Consider the point q_3 obtained by intersecting the line passing through q_1 and q_2 with l_{14}. So, by construction, the set $\mathcal{K}_g = \{p_1, p_2, p_3, p_4, q_1, q_2, q_3\}$ is a generalized 7-arc. For instance, consider the set $\mathcal{K}_g$ given by the following seven points (see Figure 1):

- $[0:1:2]$
- $[1:1:1]$
- $[1:1:2]$
- $[1:0:1]$
- $[1:2:0]$
- $[1:1:0]$
- $[0:1:0]$

PROPOSITION 3. $m_g(2,3) = 7$ (*see* Figure 1; *cf.* Table 1).

PROOF. Note that any irreducible conic in $\mathbb{P}^2(\mathbb{F}_3)$ has four points. Hence, by Lemma 1 we deduce that $m_g(2,3) \leq m(2,3) + 3 = 7$ and we can conclude by Example 1. □

Consider now, from a combinatorial and geometric point of view, the more intricate cases $q = 4, 5$. For $q = 4$, we have the following result.

PROPOSITION 4. $m_g(2,4) = 7$ (*cf.* Table 1).

PROOF. First of all, we prove a lower bound for $m_g(2,4)$.

CLAIM 1. $m_g(2,4) \geq 7$.

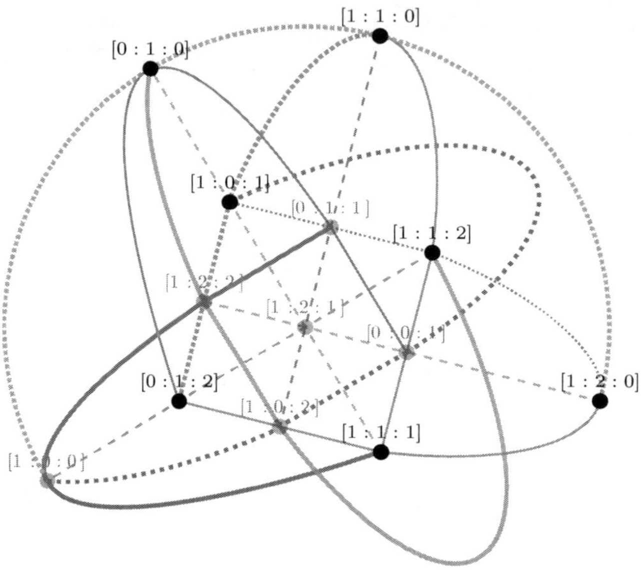

FIGURE 1. A maximal generalized 7-arc in $\mathbb{P}^2(\mathbb{F}_3)$.

Since $q = 4$, consider a complete 6-arc $\mathcal{K} \subset \mathbb{P}^2(\mathbb{F}_4)$, i.e. a conic and its nucleus (that is, the unique intersection point of the $q+1$ tangent $\mathbb{F}_q$-lines to an irreducible conic in $\mathbb{P}^2(\mathbb{F}_q)$ with q even [**6**, p. 143]). Let $p \notin \mathcal{K}$ and write $\mathcal{K}_g := \mathcal{K} \cup \{p\}$. Suppose that $\mathcal{K}_g$ is not a generalized 7-arc. As any irreducible conic has five points, it follows that there exist two lines l_1 and l_2 such that $(l_1 \cap \mathcal{K}_g) \cap (l_2 \cap \mathcal{K}_g) = \varnothing$ and $|\mathcal{K}_g \cap l_i| = 3$, for $i = 1, 2$. Since $|\mathcal{K}| = 6$, after renaming, we can assume that $p \in l_1 \setminus l_2$. Hence $|\mathcal{K} \cap l_2| = 3$, but this is a contradiction because $\mathcal{K}$ is a 6-arc. Thus, $\mathcal{K}_g$ is a generalized 7-arc and $m_g(2,4) \geq 7$. Q.E.D.

By the Claim, we know that there exists a generalized 7-arc in $\mathbb{P}^2(\mathbb{F}_4)$. Now, assume that there exists a generalized 8-arc $\mathcal{K}_g \subset \mathbb{P}^2(\mathbb{F}_4)$. Note that $\mathcal{K}_g$ cannot be a 8-arc because otherwise we would have $|\mathcal{K}_g| \leq q+2 = 6$ (see e.g. [**6**, Chapter 8]). Thus, there is a line $l \subset \mathbb{P}^2(\mathbb{F}_4)$ such that $|\mathcal{K}_g \cap l| = 3$. By Lemma 1, we know that $\mathcal{K}_g \setminus l =: \mathcal{K}$ is a 5-arc in $\mathbb{P}^2(\mathbb{F}_4)$. Write $\mathcal{K}_g := \{p_1, \ldots, p_8\}$ and, after renaming, suppose that $p_1, \ldots, p_5 \in \mathcal{K}$ and $p_6, p_7, p_8 \in l$. Let p be the nucleus of the irreducible conic γ containing $\mathcal{K}$. Since $|\gamma| = |\mathcal{K}| = q+1 = 5$, note that $|\mathcal{K} \cap l| = 0$, otherwise l would contain four distinct points of $\mathcal{K}_g$ and taking any other secant line l' of $\mathcal{K}_g$ such that $(l \cap \mathcal{K}_g) \cap (l' \cap \mathcal{K}_g) = \varnothing$, the reducible conic $l \cup l'$ would have 6 points in common with $\mathcal{K}_g$, a contradiction because $\mathcal{K}_g$ is a generalized 8-arc. In particular, this shows that $p \notin l$ by definition of the nucleus p. Let q_1 and q_2 be the other two points of l distinct from p_6, p_7 and p_8. Considering the pencil of $\mathbb{F}_4$-lines through p, after renaming, we can assume that the line $l_{16} := \langle p_1, p \rangle$ is such that $p_6 \in l_{16}$. Furthermore, considering the pencil of $\mathbb{F}_4$-lines through p_1, after renaming, we can suppose that there exist three lines $l_{1i} := \langle p_1, p_i \rangle$ through p_1 and p_i for $i = 2, 4, 5$ such that $q_2 \in l_{12}, p_7 \in l_{15}$ and $p_8 \in l_{14}$. Now, note that $p_6 \notin l_{12}$ and that the tangent line l' to γ at p_2 cannot contain the point p_6 because $p \in l'$ and $l_{16} = \langle p, p_6 \rangle$. So, consider the pencil of $\mathbb{F}_4$-lines through p_2. Write $l_{2j} := \langle p_2, p_j \rangle$ for $j = 3, 4, 5$. Since $p_6 \notin l' \cup l_{12}$, we see that there exists $j \in \{3, 4, 5\}$ such that $p_6 \in l_{2j}$.

Thus one of the following three cases can occur: (a) $j = 3$, $\{p_2, p_3, p_6\} \subset l_{23}$, (b) $j = 4$, $\{p_2, p_4, p_6\} \subset l_{24}$, (c) $j = 5$, $\{p_2, p_5, p_6\} \subset l_{25}$. In all these cases, having in mind that $\{p_1, p_4, p_8\} \subset l_{14}$ and $\{p_1, p_5, p_7\} \subset l_{15}$, we get a contradiction by taking the reducible conics $l_{23} \cup l_{14}, l_{24} \cup l_{15}, l_{25} \cup l_{14}$, respectively. This shows that actually $m_g(2, 4) = 7$. □

EXAMPLE 2. Let $\mathbb{F}_4 = \{0, 1, a, a+1\}$ be a field with four elements, where $a^2 + a + 1 = 0$. Considering in $\mathbb{P}^2(\mathbb{F}_4)$ the conic $x_0^2 + x_1x_2 = 0$, its nucleus and the point $[1 : 0 : 1]$, from the Claim of Proposition 4 it follows that the set

$$\mathcal{K}_g := \{[1:1:1], [a : a+1 : 1], [a+1 : a : 1], [0:0:1], [0:1:0], [1:0:0], [1:0:1]\}$$

is a maximal generalized 7-arc in $\mathbb{P}^2(\mathbb{F}_4)$ (cf. Table 1).

Now, consider the case $q = 5$.

EXAMPLE 3. In $\mathbb{P}^2(\mathbb{F}_5)$, consider the following set

$$\mathcal{K}' := \{[1:0:0], [0:1:0], [0:0:1], [1:1:1], [1:0:1], [1:1:0], [0:1:1]\}.$$

By the Veronese embedding $\nu_2 \colon \mathbb{P}^2(\mathbb{F}_5) \to \mathbb{P}^5(\mathbb{F}_5)$, we have

$$\begin{aligned}\nu_2(\mathcal{K}') = \{&[1:0:0:0:0:0], [0:1:0:0:0:0], [0:0:1:0:0:0],\\ &[1:1:1:1:1:1], [1:0:1:0:1:0], [1:1:0:1:0:0], [0:1:1:0:0:1]\},\end{aligned}$$

and using the homography given by the matrix

$$\begin{pmatrix} 1 & 0 & 0 & 4 & 4 & 1 \\ 0 & 1 & 0 & 4 & 0 & 0 \\ 0 & 0 & 1 & 0 & 4 & 0 \\ 0 & 0 & 0 & 0 & 0 & 1 \\ 0 & 0 & 0 & 0 & 4 & 1 \\ 0 & 0 & 0 & 4 & 0 & 1 \end{pmatrix},$$

we see that $\nu_2(\mathcal{K}')$ is projectively equivalent to the $(7, 5)_5$-arc

$$\mathcal{K} := \{[1:0:\dots:0], [0:1:0:\dots:0], \dots, [0:\dots:0:1], [1:1:\dots:1]\} .$$

So, $\mathcal{K}'$ is a generalized 7-arc in $\mathbb{P}^2(\mathbb{F}_5)$ (cf. Table 1).

PROPOSITION 5. *$m_g(2, 5) = 7$ (cf. Table 1).*

PROOF. By Proposition 1, we have $m_g(2, 5) \le m(5, 5) = 7$. Thus, we can conclude by Example 3. □

For $q = 7, 8$, from Proposition 1 and [**7**, Theorem 3.17 (iv) and (xiii)], it follows that $m_g(2, q) \le m(5, q) = q + 1$. Using Program 1, we can see that in fact $m_g(2, 7) = 8$ and $m_g(2, 8) = 9$ (see Table 1). In general, this program could theoretically find the exact values of $m_g(2, q)$, but already for $q \ge 9$ the geometry of a generalized k-arc becomes much more intricate and, for this reason, we will give only some upper bounds for $m_g(2, q)$. On the other hand, when $q = 9$, some geometric considerations, up to projectivities, together with slight modifications of Program 1 allowed us to obtain $m_g(2, 9) = 8$ and an example of a complete generalized 8-arc is given in Table 1.

For sufficiently big values of q, we obtain the following upper bounds for $m_g(2, q)$.

q	$m'(2,q)$	$m(2,q)$	$m_g(2,q)$	Examples of complete generalized k-arcs
2	4	4	7	$\mathbb{P}^2(\mathbb{F}_2)$
3	4	4	7	$[0:1:2], [1:1:1], [1:1:2], [1:0:1]$, $[1:2:0], [1:1:0], [0:1:0]$
4	6	6	7	$[1:0:0], [0:1:0], [0:0:1], [1:1:1]$, $[\alpha:\alpha^2:1], [\alpha^2:\alpha:1], [1:0:1]$
5	6	6	7	$[1:0:0], [0:1:0], [0:0:1], [1:1:1]$, $[1:0:1], [1:1:0], [0:1:1]$
7	6	8	8	$[1:0:0], [0:1:0], [0:0:1], [1:1:1]$, $[2:2:1], [5:2:1], [3:5:1], [6:1:1]$
8	6	10	9	$[1:0:0], [0:1:0], [0:0:1], [1:1:1]$, $[1:\alpha^4:1], [\alpha:1:1], [\alpha^5:\alpha^6:1]$, $[\alpha:\alpha^4:1], [\alpha^5:1:0]$
9	8	10	8	$[1:0:0], [0:1:0], [0:0:1], [1:1:1]$, $[\alpha^6:\alpha:1], [\alpha^5:0:1], [\alpha:\alpha:1], [\alpha^7:\alpha^2:1]$

TABLE 1. Maximal generalized k-arcs in $\mathbb{P}^2(\mathbb{F}_q)$ for $q \leq 9$, where $\mathbb{F}_q^* = \langle \alpha \rangle$.

THEOREM 1. *We have*

$$m_g(2,q) \leq \begin{cases} \min\left\{ m(5,q),\ m'(2,q)+3,\ q - \dfrac{\sqrt{q}}{4} + \dfrac{19}{4} \right\} & \text{if } q \geq 7 \text{ is odd} \\ \\ \min\left\{ m(5,q),\ q - \dfrac{\sqrt{q}}{2} + \dfrac{17}{4} \right\} & \text{if } q \geq 16 \text{ is even}\ , \end{cases}$$

where $m(5,q)$ is the largest size of a $(k,5)_5$-arc and $m'(2,q)$ is the second largest size of a complete $(k,2)_2$-arc.

PROOF. First, suppose that $q \geq 7$ is odd. From [**11**, Theorem I], we know that a complete k-arc with $m(2,q)$ points is a conic. Thus, if a k-arc $\mathcal{K}$ has more than $m'(2,q)$ points, then $\mathcal{K}$ can be extended to a conic Γ, so $\mathcal{K}$ must lie on Γ. By Lemma 1, we can assume that a generalized k-arc is contained in the disjoint union of a k-arc and 3 collinear points. Since $m'(2,q) \geq 6$ for any odd integer $q \geq 7$, we conclude that $m_g(2,q) \leq m'(2,q)+3$. Assume now that one of the following conditions is satisfied:

(1) $q \geq 7$ is odd and there is a generalized k-arc $\mathcal{K}_g$ with $k > q - \frac{\sqrt{q}}{4} + \frac{19}{4}$;
(2) $q \geq 16$ is even and there exists a generalized k-arc $\mathcal{K}_g$ with $k > q - \frac{\sqrt{q}}{2} + \frac{17}{4}$.

Consider the Veronese embedding $\nu_2 : \mathbb{P}^2(\mathbb{F}_q) \longrightarrow \mathbb{P}^5(\mathbb{F}_q)$ defined as in (1). Note that the set $\nu_2(\mathcal{K}_g)$ is a $(k,5)_5$-arc. By [**9**, Theorems 4.10 and 4.11], $\nu_2(\mathcal{K}_g)$ lies on a unique normal rational curve Γ of degree 5 in $\mathbb{P}^5(\mathbb{F}_q)$. Moreover, $\nu_2(\mathcal{K}_g) \subseteq \nu_2(\mathbb{P}^2(\mathbb{F}_q)) =: S$, where S is the Veronese surface. Since S is defined by quadrics (see [**5**, Example 2.7]), we see that $\Gamma \not\subseteq S$. Furthermore, there exists a quadric $\mathbb{Q}^4 \subset \mathbb{P}^5(\mathbb{F}_q)$ such that $S \subset \mathbb{Q}^4$ and $\Gamma \not\subset \mathbb{Q}^4$. So, we get

$$|\Gamma \cap S| \leq |\Gamma \cap \mathbb{Q}^4| \leq \deg(\Gamma) \cdot \deg(\mathbb{Q}^4) = 10,$$

where the last inequality is due to the Bézout's Theorem [**10**, Theomem 25.1]. This implies that $k = |\nu_2(\mathcal{K}_g)| = |\nu_2(\mathcal{K}_g) \cap \Gamma \cap S| \leq |\Gamma \cap S| \leq 10$, but this gives a numerical contradiction in both cases (1) and (2). We conclude by Proposition 1. □

REMARK 2. For $q = 2^h$ and $h \geq 4$, there exist irregular hyperovals, that is, $m(2,q)$-arcs in $\mathbb{P}^2(\mathbb{F}_q)$ which are not the union of a conic and its nucleus. For this reason, in these cases we cannot use the same argument as in the first part of the proof of Theorem 1.

REMARK 3. From [**9**, §4.5.2] we know that

$$m'(2,q) \leq \begin{cases} q-1 & q \geq 7 & \text{Segre 1955, Tallini 1957 [\textbf{12}]} \\ q - \frac{1}{4}\sqrt{q} + \frac{25}{16} & q \text{ odd} & \text{Thas 1987 [\textbf{15}]} \\ \frac{44}{45}q + \frac{8}{9} & q \text{ prime} & \text{Voloch 1990 [\textbf{16}]} \\ q - \frac{1}{2}\sqrt{q} + 5 & q = p^h, p \geq 5 & \text{Hirschfeld-Korchmáros 1996 [\textbf{8}]}\ . \end{cases}$$

REMARK 4. For $q \geq 8$ even we have $m(5,q) = q+1$, while for $q \geq 7$ odd we know that $m(5,q) = q+1$, except possibly for $23 \leq q \leq 83$, Hirschfeld 1997 [**7**].

The next result can be easily obtained by comparing directly Theorem 1 and the results recalled in Remarks 3 and 4.

COROLLARY 1. *Let $q \geq 16$ even. Then*

$$m_g(2,q) \leq \begin{cases} q+1 & q = 16, 32 \\ \left\lfloor q - \dfrac{\sqrt{q}}{2} + \dfrac{17}{4} \right\rfloor & q \geq 64\ . \end{cases}$$

Let $q \geq 7$ odd. Then we have one of the following cases:

- *for q prime,*

$$m_g(2,q) \leq \begin{cases} q+1 & 7 \leq q \leq 19 \\ q+2 & 23 \leq q \leq 83 \\ \left\lfloor \dfrac{44}{45}q + \dfrac{8}{9} \right\rfloor + 3 & q \geq 89\ ; \end{cases}$$

- *for $q = p^s$, with p prime greater than or equal to 5 and $s \geq 2$:*

$$m_g(2,q) \leq \begin{cases} q+2 & q \in \{25, 49\} \\ q+1 & q \in \{121, 125\} \\ \left\lfloor q - \dfrac{\sqrt{q}}{2} + 5 \right\rfloor + 3 & q \geq 169\ ; \end{cases}$$

- *for $q = 3^s$ and $s \geq 2$:*

$$m_g(2,q) \leq \begin{cases} 10 & q = 9 \\ q+2 & q \in \{27, 81\} \\ \left\lfloor q - \dfrac{\sqrt{q}}{4} + \dfrac{25}{16} \right\rfloor + 3 & q \geq 243\ . \end{cases}$$

3. Some lower bounds for $t(2,q)$, $t_v(2,q)$ and $t_g(2,q)$

Recall that the size of a smallest complete k-arc in $\mathbb{P}^2(\mathbb{F}_q)$ is denoted by $t(2,q)$. In [**1**], S. Ball showed that

$$(*) \qquad t(2,q) \geq \begin{cases} \lfloor \sqrt{2q} + 2 \rfloor & \text{for any } q \\ \left\lceil \sqrt{3q} + \dfrac{1}{2} \right\rceil & \text{for } q = p^h,\ p \text{ prime},\ h = 1, 2\ . \end{cases}$$

In [**2**], the authors realized a more detailed work on lower bounds when $q \leq 7559$ and they proved that $t(2,q) \leq \lfloor 3\sqrt{q} \rfloor$ for $q \leq 89$.

First of all, let us give here a lower bound for $t(2,q)$ which slightly improves $\lfloor \sqrt{2q} + 2 \rfloor$ only for some values of q (see the below Remark 5).

PROPOSITION 6. *$t(2,q) \geq \left\lceil \sqrt{(2q+\frac{1}{4})} + \frac{3}{2} \right\rceil$ for any $q \geq 2$.*

PROOF. Assume that a set of k points $\{p_1, p_2, \ldots, p_k\}$ is a complete k-arc. Denote by l_{ij} the unique $\mathbb{F}_q$-line passing through p_i and p_j, where $i \neq j$ and $i, j \in \{1, \ldots, k\}$. Write

$$\mathbb{P}^2(\mathbb{F}_q) = \bigcup_{1 \leq i < j \leq k} l_{ij} = \left(\bigcup_{h=1}^{k-1} l_{hk} \right) \cup \left(\bigcup_{1 \leq i < j \leq k-1} l_{ij} \setminus (\cup_{h=1}^{k-1} l_{hk}) \right) .$$

Then we deduce that

$$q^2 + q + 1 = |\mathbb{P}^2(\mathbb{F}_q)| \leq | \cup_{h=1}^{k-1} l_{hk}| + \sum_{1 \leq i < j \leq k-1} |l_{ij} \setminus (\cup_{h=1}^{k-1} l_{hk})\,| =$$

$$= [(k-1)q + 1] + \sum_{1 \leq i < j \leq k-1} [q + 1 - (k-1)] = (k-1)q + 1 + \binom{k-1}{2}(q + 2 - k) ,$$

i.e. $q^2 + q + 1 - [(k-1)q + 1] - \binom{k-1}{2}(q + 2 - k) \leq 0$. Hence we get

$$(q + 2 - k) \cdot \left[q - \binom{k-1}{2} \right] \leq 0 ,$$

that is,

$$[k - (q+2)] \cdot \left[k - \left(\frac{3}{2} - \sqrt{2q + \frac{1}{4}} \right) \right] \cdot \left[k - \left(\frac{3}{2} + \sqrt{2q + \frac{1}{4}} \right) \right] \leq 0 .$$

This leads to the solution $\frac{3}{2} + \sqrt{2q + \frac{1}{4}} \leq k \leq q + 2$ which shows that the inequality of the statement holds for any $q \geq 2$. □

In Tables 2 and 3, we make a comparison between (*) and the lower bound of Proposition 6 when $q \leq 31$. Furthermore, in Table 2 we give also some examples which reach the best known lower bound $t(2,q)$ for a k-arc when $q \leq 11$.

REMARK 5. As shown in Table 2, the lower bound $\left\lceil \sqrt{(2q+\frac{1}{4})} + \frac{3}{2} \right\rceil$ is better than $\lfloor \sqrt{2q} + 2 \rfloor$ for $q = 4, 7, 11$ and it is sharp for $q = 2, 3, 7, 8, 9, 11$. More generally, we have

$$\left\lceil \sqrt{\left(2q + \frac{1}{4}\right)} + \frac{3}{2} \right\rceil = \lfloor \sqrt{2q} + 2 \rfloor + 1 > \lfloor \sqrt{2q} + 2 \rfloor$$

for any q such that $\frac{(h-2)(h-1)}{2} < q < \frac{(h-1)^2}{2}$ for some suitable integer $h \geq 4$. Moreover, the lower bound of Proposition 6 is also better than that obtained in [**4**, Lemma 2.5] for $\mathbb{P}^2(\mathbb{F}_q)$.

Coming back to generalized k-arcs, let us note also that if $\mathcal{K}$ is a subset of $\mathbb{P}^2(\mathbb{F}_q)$, then ($\diamond$) $\quad \mathcal{K}$ is a Veronesian k-arc $\iff$ $\mathcal{K}$ is a generalized k-arc such that all the conics passing through any 5 of its points are irreducible.

q	$\lfloor\sqrt{2q}+2\rfloor$	$\lceil\sqrt{3q}+\frac{1}{2}\rceil$	$\lceil\sqrt{\left(2q+\frac{1}{4}\right)}+\frac{3}{2}\rceil$	$t(2,q)$	Examples of arcs in $\mathbb{P}^2(\mathbb{F}_q)$
2	**4**	3	**4**	**4**	$[0:0:1], [0:1:0], [1:0:0], [1:1:1]$
3	**4**	**4**	**4**	**4**	$[0:0:1], [0:1:0], [1:0:0], [1:1:1]$
4	4	4	5	6	$[0:0:1], [0:1:0], [\alpha:\alpha^2:1], [1:1:1], [1:0:0], [\alpha^2:\alpha:1]$
5	5	5	5	6	$[0:0:1], [0:1:0], [1:0:0], [1:1:1], [3:2:1], [4:3:1]$
7	5	**6**	**6**	**6**	$[0:0:1], [0:1:0], [1:0:0], [1:1:1], [3:2:1], [4:3:1]$
8	**6**	-	**6**	**6**	$[0:0:1], [0:1:0], [\alpha^2:\alpha^3:1], [1:0:0], [1:1:1], [\alpha^4:\alpha^2:1]$
9	**6**	**6**	**6**	**6**	$[0:0:1], [0:1:0], [\alpha^2:\alpha^5:1], [1:0:0], [1:1:1], [\alpha^5:\alpha^2:1]$
11	6	**7**	**7**	**7**	$[0:0:1], [0:1:0], [0:0:1], [1:1:1], [7:3:1], [3:2:1], [10:8:1]$

TABLE 2. Comparison between the known lower bounds for arcs in $\mathbb{P}^2(\mathbb{F}_q)$ for $2 \le q \le 11$, where $\mathbb{F}_q^* = \langle \alpha \rangle$.

q	$\lfloor\sqrt{2q}+2\rfloor$	$\lceil\sqrt{3q}+\frac{1}{2}\rceil$	$\lceil\sqrt{\left(2q+\frac{1}{4}\right)}+\frac{3}{2}\rceil$	$t(2,q)$
13	7	7	7	8
16	7	–	**8**	9
17	7	**8**	**8**	10
19	8	**9**	8	10
23	8	**9**	**9**	10
25	9	**10**	9	12
27	9	–	9	12
29	9	**10**	**10**	13
31	9	**11**	10	14

TABLE 3. Lower bounds for arcs in $\mathbb{P}^2(\mathbb{F}_q)$ for $13 \le q \le 31$.

If $q = 2$, then we have $t_g(2,2) = m_g(2,2) = 7$ and $t_v(2,2) = t(2,2) = 4$, because any conic has at most 5 points.

Let $q = 3$. Using Programs 1 and 2, we obtain that $t_g(2,3) = m_g(2,3) = 7$ and $t_v(2,3) = t(2,3) = 4$, where a minimal complete Veronesian 4-arc is given by $\{\ [1:1:1], [0:0:1], [0:1:0], [1:0:0]\ \}$.

Now, let $q = 4$ and consider a minimal generalized k-arc $\mathcal{K}_g \subset \mathbb{P}^2(\mathbb{F}_4)$. Since $\mathcal{K}_g$ is the smallest complete generalized k-arc and any irreducible conic in $\mathbb{P}^2(\mathbb{F}_4)$ has $q+1 = 5$ points, we deduce that $\mathcal{K}_g$ is not a k-arc. Therefore there are three collinear points in $\mathcal{K}_g$, say p_1, p_2, p_3. Let L be the line through these three points. Since $\mathcal{K}_g$ is a complete generalized k-arc, we deduce that any point $p \notin \mathcal{K}_g$ is contained in a reducible conic of type $L \cup l$, where l is a secant line of $\mathcal{K}_g \setminus \{p_1, p_2, p_3\}$. This implies that

$$2 + 2\binom{k-3}{2} + k \ge q^2 + q + 1 = 21\ .$$

Hence $k \geq 7$ and since $k \leq m_g(2,4) = 7$ by Proposition 4, we conclude that $t_g(2,4) = m_g(2,4) = 7$. By Program 1, we obtain that a minimal generalized 7-arc is given by

$$\{[1:0:0],[0:1:0],[0:0:1],[1:1:1],[\alpha:\alpha^2:1],[\alpha^2:\alpha:1],[1:0:1]\}\ ,$$

where $\mathbb{F}_4^* = \langle\alpha\rangle$. Moreover, by Program 2, one can see that $t_v(2,4) = t(2,4) = 6$ and a minimal complete Veronesian 6-arc is given by

$$\{[1:1:1],[0:0:1],[0:1:0],[1:0:0],[\alpha:\alpha^2:1],[\alpha^2:\alpha:1]\}\ .$$

For $q \geq 5$, as to $t_v(2,q)$, we get the following lower bound.

THEOREM 2. *For $q \geq 5$, we have $t_v(2,q) \geq \lceil t_0 \rceil$, where t_0 is the smallest real positive solution of the following inequality:*

$$\binom{t_0}{5}(q-4) + \binom{t_0}{2}(q-1) + t_0 \geq q^2+q+1\ .$$

PROOF. Let $\mathcal{K}_v \subseteq \mathbb{P}^2(\mathbb{F}_q)$ be a complete Veronesian k-arc. Since any irreducible conic of $\mathbb{P}^2(\mathbb{F}_q)$ has $q+1$ points, from ($\diamond$) we deduce that the maximal number of points covered by conics passing through 5 of k points of $\mathcal{K}_v$ is given by

$$\binom{k}{5}(q+1-5) + k = \binom{k}{5}(q-4) + k\ .$$

Moreover, the maximal number of points covered by lines passing through 2 of k points of $\mathcal{K}_v$ is

$$\binom{k}{2}(q+1-2) + k = \binom{k}{2}(q-1) + k\ .$$

Since the k points of $\mathcal{K}_v$ are common to all the above conics and lines, to cover the projective plane we get the following inequality

$$\binom{k}{5}(q-4) + \binom{k}{2}(q-1) + k \geq q^2+q+1\ .$$

□

REMARK 6. In Table 4, we compare the values of $t(2,q)$ for complete k-arcs with the values of $t_v(2,q)$ obtained by Program 2 and the lower bounds $\lceil t_0 \rceil$ of Theorem 2 for complete Veronesian k-arcs. We would like to stress the fact that a Veronesian k-arc is a particular case of a generalized k-arc and that if a set is complete as a Veronesian k-arc, could be not complete as a generalized k-arc. Furthermore, by using Program 2, we found examples which show that the lower bound $\lceil t_0 \rceil$ is sharp for $q = 5, 8, 9, 11$.

q	$\lceil t_0 \rceil$	$t_v(2,q)$	$t(2,q)$	Examples of complete Veronesian k-arcs
5	**5**	5	6	$[1:0:0],[0:1:0],[0:0:1],[1:1:1],[3:4:1]$
7	5	6	6	$[0:0:1],[0:1:0],[1:0:0],[1:1:1],[3:2:1],[4,3,1]$
8	**6**	6	6	$[0:0:1],[0:1:0],[1:0:0],[1:1:1],[\alpha^3:\alpha^2:1],[\alpha^6:\alpha^4:1]$
9	**6**	6	6	$[0:0:1],[0:1:0],[1:0:0],[1:1:1],[\alpha^6:\alpha^7:1],[\alpha^7:\alpha^3:1]$
11	**6**	6	7	$[0:0:1],[0:1:0],[0:0:1],[1:1:1],[4:3:1],[5:9:1]$

TABLE 4. Minimal Veronesian k-arcs in $\mathbb{P}^2(\mathbb{F}_q)$ for $5 \leq q \leq 11$.

In a similar way as in the proof of Theorem 2, we get the following result.

THEOREM 3. *Let $\mathcal{K}_g \subseteq \mathbb{P}^2(\mathbb{F}_q)$ be a generalized k-arc with $q \geq 5$. If $\mathcal{K}_g$ is complete as a generalized arc, then*

$$k \geq \begin{cases} \lceil t_1 \rceil & \text{if } \mathcal{K}_g \text{ is also a } k\text{-arc} \\ \lceil t_2 \rceil & \text{if } \mathcal{K}_g \text{ is not a } k\text{-arc}, \end{cases}$$

where t_1 and t_2 are the minimum real positive solutions of the following inequalities:

$$\binom{t_1}{5}(q-4) + t_1 \geq q^2 + q + 1 , \qquad \text{and}$$

$$\binom{t_2-3}{5}(q-4) + \binom{t_2-2}{4}(6q-12) + \binom{t_2-3}{2}(q-2) + t_2 \geq q^2 + 3 .$$

In particular, we have $t_g(2,q) \geq \min\{\lceil t_1 \rceil, \lceil t_2 \rceil\}$.

PROOF. Let $\mathcal{K}_g \subseteq \mathbb{P}^2(\mathbb{F}_q)$ be a complete generalized k-arc. If $\mathcal{K}_g$ is also a k-arc, then $\mathcal{K}_g$ is a Veronesian k-arc complete by (irreducible) conics. Thus, in this situation, by arguing in a similar way as in Theorem 2, we get $k \geq \lceil t_1 \rceil$, where $\binom{t_1}{5}(q-4) + t_1 \geq q^2 + q + 1$.

Suppose now that $\mathcal{K}_g$ is not a k-arc. From Lemma 1 it follows that we can write $\mathcal{K}_g = \mathcal{K} \cup \mathcal{P}$, where $\mathcal{K}$ is a (Veronesian) $(k-3)$-arc and $\mathcal{P}$ are 3 distinct collinear points. The conics γ_0 passing through any 5 distinct points of $\mathcal{K}$ are irreducible because $\mathcal{K}$ is a $(k-3)$-arc and they cover at most

$$\binom{3}{0}\binom{k-3}{5}(q+1-5) + k \tag{2}$$

points of the projective plane. On the other hand, the conics γ_1 passing through any 4 distinct points of $\mathcal{K}$ and one point of $\mathcal{P}$, can be either reducible or irreducible. Since the reducible conics cover $2q+1 > q+1$ points of the plane, we can assume that all the γ_1's are reducible. Thus they cover at most

$$\binom{3}{1}\binom{k-3}{4}(2q+1-5) + k \tag{3}$$

points of the plane. Similarly, the conics γ_2 passing through any 3 distinct points of $\mathcal{K}$ and 2 distinct points of $\mathcal{P}$, can be either reducible or irreducible. As in the previous case, we can assume that all the γ_2's are reducible conics. Therefore, they cover at most

$$\binom{3}{2}\binom{k-3}{3}(2q+1-5) + k \tag{4}$$

points of $\mathbb{P}^2(\mathbb{F}_q)$. Finally, the conics γ_3 passing through any 2 distinct points of $\mathcal{K}$ and the 3 points of $\mathcal{P}$ are all reducible and they cover at most

$$\binom{3}{3}\binom{k-3}{2}(q+1-3) + (q+1-3) + k \tag{5}$$

points of the projective plane. Since the k points of $\mathcal{K}_g$ are common to all the above conics γ_i for $i = 0, 1, 2, 3$, using the expressions from (2) to (5), we get

$$S(k) := \binom{k-3}{5}(q-4) + \binom{k-2}{4}(6q-12) + \left[\binom{k-3}{2} + 1\right](q-2) + k,$$

which is greater than or equal to the number of points covered by the span of all the conics passing through any 5 distinct points of $\mathcal{K}_g$. So, if t_2 is the minimum real positive solution of $S(t_2) \geq q^2 + q + 1$, then we have $k \geq \lceil t_2 \rceil$. □

REMARK 7. In Tables 5 and 6, we compare the two lower bounds of Theorem 3 for a generalized k-arc in $\mathbb{P}^2(\mathbb{F}_q)$. Moreover, for $5 \leq q \leq 11$, in Table 5 we find the exact values of $t_g(2,q)$ by using Program 1, and we give their respective examples showing that the lower bound $\lceil t_2 \rceil$ is sharp for $q = 7, 8, 9, 11$.

q	$\lceil t_1 \rceil$	$\lceil t_2 \rceil$	$t_g(2,q)$	Examples of complete generalized k-arcs
5	8	6	7	$[1:0:0], [0:1:0], [0:0:1], [1:1:1]$, $[1:0:1], [1:1:0], [0:1:1]$
7	7	7	**7**	$[0:0:1], [0:1:0], [1:0:0], [1:1:1]$, $[3:0:1], [5:5:1], [2:4:1]$
8	7	7	**7**	$[0:0:1], [0:1:0], [1:0:0], [1:1:1]$, $[\alpha:\alpha^2:1], [\alpha^6:1:0], [\alpha^3:\alpha^6:1]$
9	7	7	**7**	$[0:0:1], [0:1:0], [1:0:0], [1:1:1]$, $[\alpha^3:\alpha^7:1], [1:\alpha^7:1], [\alpha^6:\alpha^3:1]$
11	7	7	**7**	$[0:0:1], [0:1:0], [0:0:1], [1:1:1]$ $[4:5:1], [1:5:1], [5:3:1]$

TABLE 5. Minimal generalized k-arcs in $\mathbb{P}^2(\mathbb{F}_q)$ for $5 \leq q \leq 11$.

q	13	16	17	19	23	25	27	29	31
$\lceil t_1 \rceil$	7	8	8	8	8	8	8	8	8
$\lceil t_2 \rceil$	7	7	7	7	7	7	7	7	7

TABLE 6. Lower bounds for complete generalized k-arcs in $\mathbb{P}^2(\mathbb{F}_q)$ for $13 \leq q \leq 31$.

Finally, looking at the number of 3-secant lines of a generalized k-arc, we can also obtain the following result.

PROPOSITION 7. *Let $\mathcal{K}_g \subseteq \mathbb{P}^2(\mathbb{F}_q)$ be a complete generalized k-arc with $q \geq 5$. Let $T \in \mathbb{Z}_{\geq 0}$ be the number of distinct 3-secant lines of $\mathcal{K}_g$. Then $k \geq \lceil t_3 \rceil$, where t_3 is the minimum real positive solution of the following inequality*

$$\binom{t_3}{5}(q-4) + \frac{T}{2}\Big[(t_3-3)(t_3-4)+1-T\Big]q + t_3 \geq q^2+q+1 \ .$$

PROOF. Write $\mathcal{K}_g := \{p_1, \dots, p_k\}$ and let $I := \{\ (p,\gamma) \mid p \in \mathbb{P}^2(\mathbb{F}_q)\ ,\ \gamma \in \Gamma\ ,\ p \in \gamma\}$, where Γ is the set of all the conics in $\mathbb{P}^2(\mathbb{F}_q)$ passing through any 5 distinct points of $\mathcal{K}_g$. Since any reducible conic of Γ contains at least a 3-secant line, we deduce that the number of distinct reducible conics in Γ is δ given by

$$\delta := \binom{T}{2} + T\left[\binom{k-3}{2} - (T-1)\right] ,$$

while the number of distinct irreducible conics in Γ is $\binom{k}{5} - \delta$, because $|\Gamma| = \binom{k}{5}$. So, the cardinality of I is

$$\left[\binom{k}{5} - \delta\right](q+1) + \delta(2q+1) = \binom{k}{5}(q+1) + \delta q \ ,$$

that is,

$$|I| = \binom{k}{5}(q+1) + \left\{\binom{T}{2} + T\left[\binom{k-3}{2} - (T-1)\right]\right\} q \,.$$

Now, observe that the number of distinct conics in Γ passing through a fixed point $p_j \in \mathcal{K}_g$ is given by $\binom{k-1}{4}$. Note that I can be written as a union $I_1 \cup I_2$, where $I_1 := \{(p_j, \gamma) \in I \mid p_j \in \mathcal{K}_g\}$ and $I_2 := \{(q, \gamma) \in I \mid q \notin \mathcal{K}_g\}$. Since $|I_1| = k\binom{k-1}{4}$ and $|\mathbb{P}^2(\mathbb{F}_q)| - k \leq |I_2|$, we conclude that

$$|I| - k\left[\binom{k-1}{4} - 1\right] \geq |\mathbb{P}^2(\mathbb{F}_q)| \,,$$

i.e. $|I| - 5\binom{k}{5} + k \geq q^2 + q + 1$, which leads to the inequality of the statement. □

Appendix: Magma programs

Let us give here the two main Magma [**3**] programs we used to construct the tables and the examples of the previous sections.

PROGRAM 1. Function to find all the complete generalized k-arcs in $\mathbb{P}^2(\mathbb{F}_q)$.

```
function GArc(q,k)
P<[x]>:=ProjectivePlane(GF(q));
pts:={P![1,0,0],P![0,1,0],P![0,0,1],P![1,1,1]};
S1:={@ e : e in Subsets({p : p in Points(P)} diff pts,k-4) @};
N:=0;
 for j in [1..#S1] do
 ptss:=pts join S1[j];
 pplane:={};
 T:={@ C : C in Set(&cat[[f : f in Basis(&meet[Ideal(Cluster(p)):
 p in S]) | Degree(f) eq 2] : S in Subsets(ptss,5)]) @};
  if #T ge 1 then
  g:=0;
  repeat
  g:=g+1;
  pplane:=pplane join { b : b in Points(Curve(P,[T[g]])) };
  pc:={ b : b in Points(Curve(P,[T[g]])) };
  card:=#(pc meet ptss);
  until g eq #T or card ge 6;
   if card le 5 and #pplane eq q^2+q+1 then
   N:=N+1;
   "Set",N,"=",ptss;
   end if;
  end if;
 end for;
return "end";
end function;
```

PROGRAM 2. Function to find all the complete Veronesian k-arcs in $\mathbb{P}^2(\mathbb{F}_q)$.

```
function VArc(q,k)
P<[x]>:=ProjectivePlane(GF(q));
pts:={P![1,0,0],P![0,1,0],P![0,0,1],P![1,1,1]};
S1:={@ e : e in Subsets({p : p in Points(P)} diff pts,k-4) @};
```

```
N:=0;
 for j in [1..#S1] do
 ptss := pts join S1[j];
 pplane:={};
 T1:={@ C : C in Set(&cat[[f : f in Basis(&meet[Ideal(Cluster(p))
 : p in S]) | Degree(f) eq 1] : S in Subsets(ptss,2)]) @};
 T2:={@ C : C in Set(&cat[[f : f in Basis(&meet[Ideal(Cluster(p))
 : p in S]) | Degree(f) eq 2] : S in Subsets(ptss,5)]) @};
 g:=0;
 repeat
 g:=g+1;
 pplane:=pplane join { b : b in Points(Curve(P,[T1[g]])) };
 pt:={ b : b in Points(Curve(P,[T1[g]])) };
 card1:=#(pt meet ptss);
 until g eq #T1 or card1 ge 3;
  if card1 le 2 then
   if #T2 ge 1 then
   g:=0;
   repeat
   g:=g+1;
   pplane:=pplane join { b : b in Points(Curve(P,[T2[g]])) };
   pc:={ b : b in Points(Curve(P,[T2[g]])) };
   card2:=#(pc meet ptss);
   until g eq #T2 or card2 ge 6;
   else
   card2:=0;
   end if;
   if #pplane eq q^2+q+1 and card2 le 5 then
   N:=N+1;
   "Set",N,"=",ptss;
   end if;
  end if;
 end for;
return "end";
end function;
```

Acknowledgments

Both authors would like to thank the referee for reading the paper carefully and for many useful suggestions which improved the presentation of some results of the first draft. During the preparation of this paper in the framework of the Project Anillo ACT 1415 PIA CONICYT, the authors were partially supported by Proyectos VRID N. 214.013.039-1.OIN and N. 219.015.023-INV, and the first author was also partially supported by CONICYT-PCHA/Magíster Nacional año 2013 - Folio: 221320380.

References

[1] Simeon Ball, *On small complete arcs in a finite plane*, Discrete Math. **174** (1997), no. 1-3, 29–34, DOI 10.1016/S0012-365X(96)00315-9. Combinatorics (Rome and Montesilvano, 1994). MR1477221

[2] Daniele Bartoli, Alexander A. Davydov, Giorgio Faina, Stefano Marcugini, and Fernanda Pambianco, *New upper bounds on the smallest size of a complete arc in a finite Desarguesian projective plane*, J. Geom. **104** (2013), no. 1, 11–43, DOI 10.1007/s00022-013-0154-6. MR3047446

[3] Wieb Bosma, John Cannon, and Catherine Playoust, *The Magma algebra system. I. The user language*, J. Symbolic Comput. **24** (1997), no. 3-4, 235–265, DOI 10.1006/jsco.1996.0125. Computational algebra and number theory (London, 1993). MR1484478

[4] Giorgio Faina, Stefano Marcugini, Alfredo Milani, and Fernanda Pambianco, *The sizes k of the complete k-caps in* $\mathrm{PG}(n, q)$, *for small q and* $3 \leq n \leq 5$, Ars Combin. **50** (1998), 235–243. MR1670593

[5] Joe Harris, *Algebraic geometry*, Graduate Texts in Mathematics, vol. 133, Springer-Verlag, New York, 1992. A first course. MR1182558

[6] J. W. P. Hirschfeld, *Projective geometries over finite fields*, 2nd ed., Oxford Mathematical Monographs, The Clarendon Press, Oxford University Press, New York, 1998. MR1612570

[7] J. W. P. Hirschfeld, *Complete arcs*, Discrete Math. **174** (1997), no. 1-3, 177–184, DOI 10.1016/S0012-365X(96)00330-5. Combinatorics (Rome and Montesilvano, 1994). MR1477235

[8] J. W. P. Hirschfeld and G. Korchmáros, *On the embedding of an arc into a conic in a finite plane*, Finite Fields Appl. **2** (1996), no. 3, 274–292, DOI 10.1006/ffta.1996.0017. MR1398077

[9] J. W. P. Hirschfeld and J. A. Thas, *Open problems in finite projective spaces*, Finite Fields Appl. **32** (2015), 44–81, DOI 10.1016/j.ffa.2014.10.006. MR3293405

[10] Solomon Lefschetz, *Algebraic geometry*, Princeton University Press, Princeton, N. J., 1953. MR0056950

[11] Beniamino Segre, *Ovals in a finite projective plane*, Canadian J. Math. **7** (1955), 414–416, DOI 10.4153/CJM-1955-045-x. MR71034

[12] Beniamino Segre, *Curve razionali normali e k-archi negli spazi finiti* (Italian), Ann. Mat. Pura Appl. (4) **39** (1955), 357–379, DOI 10.1007/BF02410779. MR75608

[13] Beniamino Segre, *Le geometrie di Galois* (Italian), Ann. Mat. Pura Appl. (4) **48** (1959), 1–96, DOI 10.1007/BF02410658. MR116259

[14] Beniamino Segre, *Introduction to Galois geometries* (English, with Italian summary), Atti Accad. Naz. Lincei Mem. Cl. Sci. Fis. Mat. Natur. Sez. Ia (8) **8** (1967), 133–236. MR238846

[15] J. A. Thas, *Complete arcs and algebraic curves in* $\mathrm{PG}(2, q)$, J. Algebra **106** (1987), no. 2, 451–464, DOI 10.1016/0021-8693(87)90007-X. MR880968

[16] José Felipe Voloch, *Arcs in projective planes over prime fields*, J. Geom. **38** (1990), no. 1-2, 198–200, DOI 10.1007/BF01222904. MR1061069

Departamento de Ciencias Básicas, Universidad de Concepción, Los Ángeles, Chile
Email address: `alexisalmendras@udec.cl`

Departamento de Matemática, Universidad de Concepción, Casilla 160-C, Concepción, Chile
Email address: `atironi@udec.cl`

Contemporary Mathematics
Volume **766**, 2021
https://doi.org/10.1090/conm/766/15370

Symmetries of order eight on K3 surfaces without high genus curves in the fixed locus

Dima Al Tabbaa, Annalisa Grossi, and Alessandra Sarti

Abstract. In this paper we classify non-symplectic automorphisms of order eight on complex K3 surfaces in case that the fourth power of the automorphism has only rational curves in its fixed locus. We show that the fixed locus is the disjoint union of a rational curve and ten isolated points or it consists in four isolated fixed points. We give examples corresponding to the case with a rational curve in the fixed locus and to the case with only isolated points in the fixed locus.

Introduction

In this paper we investigate purely non-symplectic automorphisms of order eight on a complex $K3$ surface under certain assumptions on the fixed locus. These automorphisms act non-trivially on $H^{2,0}(X) \simeq \mathbb{C} \cdot \omega_X$ i.e. they multiply the non-degenerate holomorphic 2-form by a primitive 8th root of unity. The study of non-symplectic automorphisms of prime order was completed by several authors: Nikulin [**14**], Artebani, Sarti and Taki[**3**], [**5**] and [**18**]. If the automorphism is not of prime order the setting is more complicated. Indeed, in this situation the purely non-symplectic automorphism of order eight does not admit a trivial action on the Néron-Severi group of the generic K3 surface as it does in the case of the prime order [**7**, Section 11]. In the paper [**19**], Taki studies the case when the order of the automorphism is a prime power and the action is trivial on the Néron-Severi group. If we consider non-symplectic, non-trivial automorphisms of order 2^b, then by results of Nikulin we have $1 \leq b \leq 5$. Further results can be found in a paper by Schütt [**17**] in the case of automorphisms of a two-power order and in a paper by Artebani and Sarti [**5**] in the case of order four. Recently in [**2**] Al Tabbaa, Sarti and Taki completed the study for purely non-symplectic automorphisms of order 16. In [**1**] Al Tabbaa and Sarti studied the case of order eight automorphisms under the assumption that their fourth power σ^4 acts as the identity on the Néron-Severi group and $\mathrm{Fix}(\sigma^4)$ contains an elliptic curve.

This paper deals with purely non-symplectic automorphisms of order eight on K3 surfaces under the assumption that their fourth power σ^4 is the identity on the Néron-Severi group. This corresponds to the situation for the generic K3 surface

Key words and phrases. Non-symplectic automorphisms, K3 surfaces.

The second author was supported by the Laboratoire International LIA LYSM to visit the University of Poitiers in May 2019.

in the moduli space of K3 surfaces with a purely non-symplectic automorphism of order eight and fixed action on the second cohomology group with integer coefficients, see [**7**, Section 10]. The fixed locus $\mathrm{Fix}(\sigma)$ of such an automorphism σ is the disjoint union of smooth curves and points. We give a complete classification of the fixed locus and of some lattice invariants related to σ in the case that $\mathrm{Fix}(\sigma^4)$ is not empty and contains only rational curves. Let X be a K3 surface, ω_X a generator of $H^{2,0}(X)$ and $\sigma \in \mathrm{Aut}(X)$ such that $\sigma^*(\omega_X) = \zeta_8\omega_X$, where ζ_8 denotes a primitive 8th root of unity. We denote by k_σ the number of smooth rational curves fixed by σ and by N_σ the numbers of isolated points in $\mathrm{Fix}(\sigma)$. We denote furthermore by $S(\sigma^4)$ the invariant lattice for the action of σ^4 on the second cohomology group with integer coefficients. We prove the following result:

THEOREM 0.1. *Let σ be a purely non-symplectic automorphism of order 8 on a K3 surface X with $Pic(X) = S(\sigma^4)$. Suppose that* $\mathrm{Fix}(\sigma^4)$ *is not empty and it is the union of smooth rational curves. Then $k_\sigma \in \{0, 1\}$ moreover*

- *if $k_\sigma = 1$ then $N_\sigma = 10$,*
- *if $k_\sigma = 0$ then $N_\sigma = 4$.*

The paper is organized as follows. In Section 1 we recall basic facts on purely non-symplectic automorphisms acting on K3 surfaces. In Section 2 we give several properties of the fixed locus of such automorphisms in particular in the case that the automorphism has even order. The Section 3 is devoted to the proof of our main theorem 0.1. In Table 2 we list in detail all the possibilities for the fixed locus and for some invariants characterizing the action of the automorphism on the second cohomology group with integer coefficients. We have in total four possibilities: one for $k_\sigma = 1$ and three for $k_\sigma = 0$. By using elliptic fibrations we give examples for $k_\sigma = 1$ and for $k_\sigma = 0$ in the case that the number of fixed curves by σ^2 is three, this is done in Section 5 after recalling in Section 4 some basic notions on elliptic K3 surfaces. We do not know if the remaining two cases in Table 2 exist. In any case by using the invariants of the Table 2 one computes that the rank of the Néron-Severi group is 18 or 14 so that by e.g. [**10**, Lemma 1.5] all these K3 surfaces admits an elliptic fibration. This could certainly be helpful in finding the remaining examples.

1. Basic facts

Let X be a K3 surface and $\sigma \in \mathrm{Aut}(X)$ a non-symplectic automorphism of order 8. We assume that $\sigma^*(\omega_X) = \zeta_8\omega_X$ where ζ_8 is a primitive 8th root of unity. Such a σ is called *purely non-symplectic*, for simplicity we just call it *non-symplectic*, always meaning that the action on the holomorphic two-form is by a primitive 8th root of unity.

We denote by $r_{\sigma^j}, l_{\sigma^j}, m_{\sigma^j}$ and m_1 for $j = 1, 2, 4$ the rank of the eigenspace of $(\sigma^j)^*$ in $H^2(X, \mathbb{C})$ relative to the eigenvalues 1, -1, i and ζ_8 respectively (clearly $m_{\sigma^4} = 0$). We recall the invariant lattice:

$$S(\sigma^j) = \{x \in H^2(X, \mathbb{Z}) | (\sigma^j)^*(x) = x\},$$

and its orthogonal complement

$$T(\sigma^j) = S(\sigma^j)^\perp \cap H^2(X, \mathbb{Z}).$$

Since the automorphism acts purely non-symplectically, X is projective, see [**12**, Theorem 3.1], so that if we denote $\mathrm{rk}\, S(\sigma^j) = r_{\sigma^j}$, we have that $r_{\sigma^j} > 0$ for all

$j = 1, 2, 4$, in fact one can always find an invariant ample class. On the other hand, one can easily show that $S(\sigma^j) \subset \mathrm{Pic}(X)$ for $j = 1, 2, 4$ so that the transcendental lattice satisfies $T_X \subseteq T(\sigma^j)$ for $j = 1, 2, 4$.

REMARK 1.1. It is a straightforward computation that the invariants $r_{\sigma^j}, l_{\sigma^j}, m_{\sigma^j}$ and m_1 with $j = 1, 2, 4$ satisfy the following relations:

$$\begin{aligned} r_{\sigma^2} &= r_\sigma + l_\sigma; & r_{\sigma^4} &= r_\sigma + l_\sigma + 2m_\sigma; \\ l_{\sigma^2} &= 2m_\sigma; & l_{\sigma^4} &= 4m_1; \\ m_{\sigma^2} &= 2m_1. \end{aligned}$$

We remark that the invariants l_{σ^2} and m_{σ^2} are even numbers.

The moduli space for $K3$ surfaces carrying a non-symplectic automorphism of even order $n > 2$, with a given action on the $K3$ lattice is known to be a complex ball quotient of dimension $q - 1$ where q is the dimension of the eigenspace V of σ^* in $H^2(X, \mathbb{C})$ relative to the eigenvalues $\zeta_n = e^{\frac{2\pi i}{n}}$, see [**7**, §11]. The complex ball is given by:

$$B = \{[w] \in \mathbb{P}(V) : (w, \overline{w}) > 0\}.$$

If n is even V is the ζ_n eigenspace of σ^* in $T(\sigma^{n/2}) \otimes \mathbb{C}$. This implies that the Néron-Severi group of a $K3$ surface corresponding to the generic point in the moduli space equals $S(\sigma^{n/2})$ see [**7**, Theorem 11.2].

2. The fixed locus

We denote by $\mathrm{Fix}(\sigma^j)$, $j = 1, 2, 4$ the fixed locus of the automorphism σ^j :

$$\mathrm{Fix}(\sigma^j) = \{x \in X \mid \sigma^j(x) = x\}.$$

Clearly $\mathrm{Fix}(\sigma) \subset \mathrm{Fix}(\sigma^2) \subset \mathrm{Fix}(\sigma^4)$. To describe the fixed locus of order 8 non-symplectic automorphisms we start recalling the following result about non-symplectic involutions, see [**14**, Theorem 4.2.2].

THEOREM 2.1. *Let τ be a non-symplectic involution on a K3 surface X. The fixed locus of τ is either empty, the disjoint union of two elliptic curves or the disjoint union of a smooth curve of genus $g \geq 0$ and k smooth rational curves. Moreover, its fixed lattice $S(\tau) \subset \mathrm{Pic}(X)$ is a 2-elementary lattice with determinant 2^a such that:*

- *$S(\tau) \cong U(2) \oplus E_8(2)$ iff the fixed locus of τ is empty;*
- *$S(\tau) \cong U \oplus E_8(2)$ iff τ fixes two elliptic curves;*
- *$2g = 22 - rkS(\tau) - a$ and $2k = rkS(\tau) - a$ otherwise.*

Since $S(\tau)$ is 2-elementary its discriminant group $A_{S(\tau)} = S(\tau)^\vee / S(\tau) \simeq (\mathbb{Z}/2\mathbb{Z})^{\oplus a}$, $a \in \mathbb{Z}_{>0}$. We introduce the invariant $\delta_{S(\tau)}$ of $S(\tau)$ by putting $\delta_{S(\tau)} = 0$ if $x^2 \in \mathbb{Z}$ for any $x \in A_{S(\tau)}$ and $\delta_{S(\tau)} = 1$ otherwise. By [**13**, Theorem 3.6.2], and [**16**, §1] $S(\tau)$ is uniquely determined by the invariant $\delta_{S(\tau)}$, by the rank, the signature and the invariant a. The situation is summarized in Figure 1 from [**15**, §4].

We recall a result about non-symplectic automorphisms of order four on a K3 surface. These results are discussed in [**4**], see also the Appendix of [**2**].

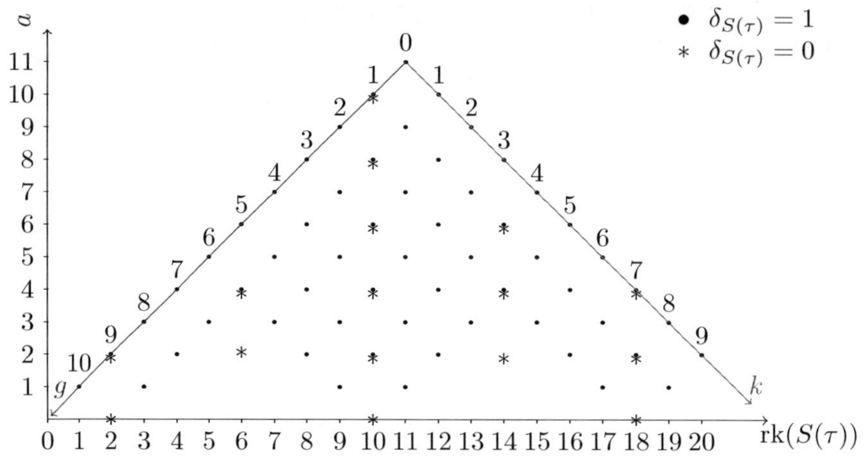

FIGURE 1. Order 2

THEOREM 2.2. *Let X be a K3 surface and γ be a purely non-symplectic automorphism of order four acting on it, with* $\mathrm{Pic}(X) = S(\gamma^2)$. *If* $\mathrm{Fix}(\gamma)$ *contains a smooth rational curve and all curves fixed by* γ^2 *are rational, then the invariants associated to* γ *are as in Table* 1. *All cases in the table do exist. Here m denotes the multiplicity of the eigenvalue i; r denotes the multiplicity of the eigenvalue* 1; *l the multiplicity of the eigenvalue* -1 *and* a_γ *the number of rational curves exchanged by γ and fixed by* γ^2. *Moreover N denotes the number of isolated fixed points and k denotes the number of fixed rational curves.*

m	r	l	N	k	a_γ
4	10	4	6	1	0
3	13	3	8	2	0
	11	5	6	1	1
2	16	2	10	3	0
	14	4	8	2	1
	12	6	6	1	2
1	19	1	12	4	0
	13	7	6	1	3

TABLE 1. The case $g = 0$

REMARK 2.3. For each $p \in \mathrm{Fix}(\sigma)$, there exists a σ^4-fixed smooth rational curve R, such that $p \in R$.

With the notation and the assumptions of the previous remark we have

LEMMA 2.4. *The curve R is σ-invariant.*

PROOF. First of all we notice that since R is fixed by σ^4 then also $\sigma(R)$ is fixed by σ^4. If R was not σ-invariant this means that R is sent to another rational curve

$\sigma(R)$ which meets R in p. This is impossible since these two curves are fixed by σ^4 and the fixed locus of an involution is smooth by Theorem 2.1. □

Observe now that in a neighborhood of a fixed point p of σ the action can be linearized, see e.g. [**12**, Section 5]. We can assume $p = (0,0)$ and we can find z_1 and z_2 local coordinates in a neighborhood of p such that the everywhere non-degenerate holomorphic 2-form can be written as $dz_1 \wedge dz_2$. Since the automorphism is of finite order locally it can be diagonalized and we know that $\sigma^*(dz_1 \wedge dz_2) = \zeta_8(dz_1 \wedge dz_2)$, for this reason the product of the eigenvalues with respect to z_1 and z_2 is equal to ζ_8. We get the following possibilities up to permutation of the coordinates (but this does not play any role in the classification):

$$A_{1,0} = \begin{pmatrix} \zeta_8 & 0 \\ 0 & 1 \end{pmatrix}, \quad A_{2,7} = \begin{pmatrix} i & 0 \\ 0 & \zeta_8^7 \end{pmatrix}, \quad A_{3,6} = \begin{pmatrix} \zeta_8^3 & 0 \\ 0 & \zeta_8^6 \end{pmatrix}, \quad A_{4,5} = \begin{pmatrix} -1 & 0 \\ 0 & \zeta_8^5 \end{pmatrix}.$$

In the first case the point belongs to a smooth fixed curve, since we have an eigenvalue which is equal to 1. In the other three cases it is an isolated fixed point. We say that an isolated point $p \in \mathrm{Fix}(\sigma)$ is of type (t,s) if the local action at p is given by $A_{t,s}$. We denote by $n_{t,s}$ the number of isolated fixed points by σ with matrix $A_{t,s}$.

We further denote by N_{σ^j}, k_{σ^j}, $j = 1,2,4$ the number of isolated points and smooth rational curves in $\mathrm{Fix}(\sigma^j)$. We observe that $N_{\sigma^4} = 0$ since σ^4 only fixes curves or it is empty as explained in Theorem 2.1. We recall [**1**, Proposition 2.2] :

PROPOSITION 2.5. *Let σ be a non-symplectic automorphism of order 8 acting on a K3 surface X. Then* $\mathrm{Fix}(\sigma)$ *is never empty nor it can be the union of two smooth elliptic curves. It is the disjoint union of smooth curves and $N_\sigma \geq 2$ isolated points. We denote $\alpha = \sum_{K \subset \mathrm{Fix}(\sigma)}(1 - g(K))$ and the following relations hold:*

$$n_{2,7} + n_{3,6} = 2 + 4\alpha, \; n_{4,5} + n_{2,7} - n_{3,6} = 2 + 2\alpha, \; N_\sigma = 2 + r_\sigma - l_\sigma - 2\alpha.$$

The fixed locus of such an automorphism σ is then

$$\mathrm{Fix}(\sigma) = C \cup R_1 \cup \cdots \cup R_k \cup \{p_1, \ldots, p_N\} \tag{1}$$

where C is a smooth curve of genus $g \geq 0$ and R_j, $j = 1, \ldots, k$ are disjoint smooth rational curves.

We recall the following remarks and lemmas which are important in the study of the fixed locus of σ.

REMARK 2.6. A non-symplectic automorphism σ of order 8 acts on a set of smooth rational curves of X which are fixed by σ^4 either trivially, i.e. each smooth rational curve is σ-invariant, possibly pointwise fixed by σ, or it exchanges smooth rational curves two by two, or finally σ permutes four rational curves.

LEMMA 2.7. *Four cyclically permuted smooth rational curves by a non-symplectic automorphism σ of order 8 on a K3 surface X are either σ^4-invariant (not pointwise fixed), or pointwise fixed by σ^4.*

PROOF. We can prove it simply as follows. Let C_i, $i \in \{1, \ldots, 4\}$ be four smooth rational curves such that $\sigma(C_i) = C_{i+1}$, $i = 1,2,3$ and $\sigma(C_4) = C_1$, and assume that C_1 is invariant by σ^4, then $\sigma^4(C_2) = \sigma^4(\sigma(C_1)) = \sigma(\sigma^4(C_1)) = \sigma(C_1) = C_2$. In particular if C_1 is pointwise fixed, then one proves in a similar way that

C_2 is pointwise fixed. A similar proof holds also for C_3 and C_4, so we get the statement. □

We denote by $2a_\sigma$ the number of smooth rational curves exchanged by σ and fixed by σ^2, and by $4s_\sigma$ the number of smooth rational curves cyclic permuted by σ and pointwise fixed by σ^4 (and clearly they are interchanged by σ^2 two by two). Let now a_{σ^2} be the number of the pairs of rational curves interchanged by σ^2 and pointwisely fixed by σ^4, then observe that $2a_{\sigma^2} = 4s_\sigma$ and so $a_{\sigma^2} \in 2\mathbb{Z}$.

An important remark about the local behaviour of σ^2 in a neighborhood of a fixed point is the following:

REMARK 2.8. The isolated fixed points by a non-symplectic automorphism σ of type $(2,7)$ and $(3,6)$ are also isolated fixed points in $\mathrm{Fix}(\sigma^2)$. The points of type $(4,5)$ in $\mathrm{Fix}(\sigma)$ are contained in a smooth fixed curve by σ^2. In fact the action of σ^2 at such points is given by the matrix $\begin{pmatrix} 1 & 0 \\ 0 & \zeta_8^2 \end{pmatrix}$ which implies that these points belong to a smooth curve in $\mathrm{Fix}(\sigma^2)$. For this reason we can say that if there exist points of $(4,5)$-type then $k_{\sigma^2} > k_\sigma$.

The following is a result about local actions of σ at a point on a rational curve. With the same notation of Remark 2.3 if R is σ-invariant (not pointwise fixed) each action of σ on R has two fixed points. There are some restrictions about the possible actions of σ, in particular if we have an action on one of the fixed points then the action on the other point is determined. We recall the following result which is stated in a similar way in [**6**, Lemma 8.1].

PROPOSITION 2.9. *Let σ be a non-symplectic automorphism of order 8 on a K3 surface X and suppose that the fixed locus of the involution σ^4 is the union of smooth rational curves. Then if p_1 is an isolated fixed point for σ and it is contained on a rational curve R fixed by σ^4, there exists another fixed point p_2 for σ on R. If the local action in p_1 is of $(2,7)$-type then the local action in p_2 is of $(3,6)$-type and vice-versa. If the action in p_1 is of $(4,5)$-type then the action in p_2 is of $(4,5)$-type.*

PROOF. We know that the action of a non-trivial finite order automorphism on a rational curve has two fixed points. Let p_1 and p_2 be the two fixed points on R. We can assume that the two fixed points are $p_1 = (1:0)$ and $p_2 = (0:1)$. Hence if z is a local coordinate for p_1 on the rational curve $R \simeq \mathbb{P}^1$ then $\frac{1}{z}$ is a local coordinate for p_2 on R. So that if σ acts by $z \mapsto \zeta_8^j z$, $0 \le j \le 7$ at p_1 then at p_2 it acts as $\frac{1}{z} \mapsto \frac{1}{\zeta_8^j z} = \zeta_8^{8-j}\frac{1}{z}$. Observe that $j = 0$ is not possible otherwise R would be pointwise fixed by σ. Also $j = 1,3,5,7$ are not possible. In fact R is fixed by σ^4 so that we must have $\zeta_8^{4j} = (-1)^j = 1$ which implies $j = 2,4,6$. For these j's one can compute that the local action at p_1 and p_2 (as points on X) can be diagonalized with (t,s) as in the statement (up to permutation of the coordinates). □

3. The classification

Our goal is to give a classification of non-symplectic automorphisms of order eight under the assumption that their fourth power fixes only rational curves. Let σ be such an automorphism, then

$$\mathrm{Fix}(\sigma^4) = R'_1 \cup \cdots \cup R'_T$$

where the R_i''s are smooth rational disjoint curves. This implies that:

$$\mathrm{Fix}(\sigma) = R_1 \cup \cdots \cup R_{k_\sigma} \cup \{p_1, \ldots, p_N\}$$

which means that in the description of equation (1) we have $g(C) = 0$.

THEOREM 3.1. *Let σ be a non-symplectic automorphism of order eight acting on a K3 surface X with $\mathrm{Pic}(X) = S(\sigma^4)$. Suppose that $\mathrm{Fix}(\sigma^4)$ is not empty and it is the union of smooth rational curves. Then $k_\sigma \in \{0, 1\}$ and the invariants of σ are as in Table 2.*

PROOF. Consider p_1 a fixed isolated point for σ. From Remark 2.3 there exists a smooth rational curve in $\mathrm{Fix}(\sigma^4)$ such that $p_1 \in R_i'$. From Remark 2.4, R_i' is σ-invariant. Since a finite order automorphism of a rational curve has two fixed points, there exists another fixed point for σ on R_i' : we call it p_2. We conclude that N_σ is even. From what we know about the local behaviour of σ at a fixed point, see Proposition 2.9, we know that if a fixed point is of $(2, 7)$-type then the other fixed point on the same rational curve R_i' is of $(3, 6)$-type. If there is an action of $(4, 5)$-type on p_1 then there is an action of $(4, 5)$-type on p_2. By these considerations we obtain:

$$n_{2,7} = n_{3,6}$$

$$n_{4,5} \in 2\mathbb{Z}.$$

Using Proposition 2.5 we obtain:

$$\alpha = k_\sigma$$

$$n_{3,6} = n_{2,7} = 1 + 2k_\sigma \geq 1$$

$$n_{4,5} = 2 + 2k_\sigma \geq 2$$

Thanks to Remark 2.8 the number of curves in $\mathrm{Fix}(\sigma^2)$ is:

$$k_{\sigma^2} = k_\sigma + \frac{n_{4,5}}{2} + 2a_\sigma.$$

To conclude the proof we consider two cases, $k_\sigma \geq 1$ and $k_\sigma = 0$.

$k_\sigma \geq 1$: In this case $k_{\sigma^2} \geq 3$ so the possible cases are $k_{\sigma^2} = 3$ and $k_{\sigma^2} = 4$ from Table 1. But from Remark 1.1 we know that m_{σ^2} has to be even, so checking again in Table 1, the unique case is $k_{\sigma^2} = 3$ which has $m_{\sigma^2} = 2$. By using the previous equalities we get then that $n_{4,5} = 4$, $k_\sigma = 1$, $a_\sigma = 0$ and $n_{4,5} = 4$, $n_{2,7} = n_{3,6} = 3$ so that $N_\sigma = n_{4,5} + n_{3,6} + n_{2,7} = 10$. We can use Remark 1.1 and Proposition 2.5 to conclude that $r_\sigma = 13$, $l_\sigma = 3$, $m_\sigma = 1$, $m_1 = 1$.

$k_\sigma = 0$: By using the equalities stated at the beginning of the proof we obtain:

$$k_\sigma = 0,$$

$$n_{3,6} = n_{7,2} = 1,$$

$$n_{4,5} = 2.$$

So that in this case $N_\sigma = 4$. Observe that σ^2 is an automorphism of order 4 and it contains a rational curve in the fixed locus since two points of $(4, 5)$-type in $\mathrm{Fix}(\sigma)$ are contained on a fixed curve for σ^2 so the number of curves in $\mathrm{Fix}(\sigma^2)$ is:

$$k_{\sigma^2} = k_\sigma + \frac{n_{4,5}}{2} + 2a_\sigma \geq 1.$$

From Remark 1.1 we know that m_{σ^2} has to be even, so using Table 1 we conclude that there are four possible cases.

If $\boldsymbol{m_{\sigma^2} = 4}$ then $\boldsymbol{k_\sigma^2 = 1}$. From the previous equation $a_\sigma = 0$. So we have $(r_{\sigma^2}, m_{\sigma^2}, l_{\sigma^2}, N_{\sigma^2}, k_{\sigma^2}, a_{\sigma^2}) = (10, 4, 4, 6, 1, 0)$.

If $\boldsymbol{m_{\sigma^2} = 2}$ then $k_\sigma^2 \in \{3, 2, 1\}$. If $\boldsymbol{k_{\sigma^2} = 3}$ from the previous equation $a_\sigma = 1$, so that $(r_{\sigma^2}, m_{\sigma^2}, l_{\sigma^2}, N_{\sigma^2}, k_{\sigma^2}, a_{\sigma^2}) = (16, 2, 2, 10, 3, 0)$.

If $\boldsymbol{k_{\sigma^2} = 2}$ from the previous equation $a_\sigma = \frac{1}{2}$, which is not possible.

If $\boldsymbol{k_{\sigma^2} = 1}$ from the previous equation $a_\sigma = 0$. In this case for σ^2 the invariants are $(r_{\sigma^2}, m_{\sigma^2}, l_{\sigma^2}, N_{\sigma^2}, k_{\sigma^2}, a_{\sigma^2}) = (12, 2, 6, 6, 1, 2)$.

We use then Remark 1.1 and Proposition 2.5 to compute the remaining invariants for σ given in the Table 2.

m_1	m_σ	r_σ	l_σ	N_σ	k_σ	a_σ	Examples
1	1	13	3	10	1	0	5.1
2	2	6	4	4	0	0	
1	1	9	7	4	0	1	5.2
1	3	7	5	4	0	0	

TABLE 2. Invariants of the automorphism

□

4. Elliptic fibrations

DEFINITION 4.1. Let X be a complex surface. An **elliptic fibration** is a holomorphic map $f : X \longrightarrow B$ to a smooth curve B such that the generic fiber is a smooth connected curve of genus one. A **jacobian elliptic fibration** is an elliptic fibration admitting a section $\pi : B \longrightarrow X$ such that $f \circ \pi = Id_B$. The surface X is called an **elliptic surface** if it admits an elliptic fibration (not necessarily jacobian). We call F_v the fiber $f^{-1}(v)$ over a point $v \in B$. The **Mordell-Weil group** is the group of sections of the elliptic fibration.

The **zero section** of an elliptic fibration is a chosen section $s : B \longrightarrow X$ and we identify the map s with the curve $s(B)$ on X. The point of intersection between the zero section and a fiber is the zero of the group law on the fiber.

For $K3$ surfaces we have that $B = \mathbb{P}^1$ (see [**11**]) and, if the fibration is jacobian, it admits a Weierstrass equation:

$$y^2 = x^3 + A(t)x + B(t), \tag{2}$$

where $A(t)$ and $B(t)$ are two polynomials with complex coefficients with $t \in \mathbb{P}^1$ such that $\deg(A(t)) = 8$ and $\deg(B(t)) = 12$. Here the zero section is $t \mapsto (0 : 1 : 0)$.

The discriminant of the fibration is a degree 24 polynomial:

$$\Delta(t) = 4A(t)^3 + 27B(t)^2. \tag{3}$$

The equation (2) defines an elliptic fibration if and only if $\Delta(t)$ does not vanish identically. Each zero of $\Delta(t)$ corresponds to a point v of the base $\mathbb{P}^1$ such that F_v is a singular fiber of the fibration. There are at most finitely many singular fibers. Let δ be the order of vanishing of Δ at a point corresponding to a singular fiber, by the *Kodaira classification* the possible singular fibers are recalled in Table 3 where we denoted by Θ_0 the component of a fiber meeting the zero section. The first column in the Table 3 contains the name of the reducible fiber according to Kodaira classification, the second the Dynkin diagram associated to the fiber, the last column contains the order of vanishing of Δ at the point corresponding to the singular fiber.

Name	Dynkin diagram	Description	δ
II		a cuspidal rational curve	2
III	$\tilde{A}_1$	two rational curves meeting in a point of order 2	3
IV	$\tilde{A}_2$	three rational curves meeting at one point	4
I_1		a nodal rational curve	1
I_2	$\tilde{A}_1$	$\Theta_0 = \Theta_1$	2
I_n	$\tilde{A}_{n-1}$	$\Theta_0 - \Theta_1 - \dots \Theta_i$ / $\Theta_{n-1} - \Theta_{n-2} - \dots \Theta_{i+1}$ ($\Theta_0 \mid \Theta_{n-1}$, $\Theta_i \mid \Theta_{i+1}$)	n
I_k^*	$\tilde{D}_{k+4}$	$\Theta_0, \Theta_1 \setminus / \Theta_2 \dots \Theta_i - \Theta_{i+1} \dots \Theta_{k+2} / \setminus \Theta_{k+3}, \Theta_{k+4}$	$k+6$
IV^*	$\tilde{E}_6$	$\Theta_0 - \Theta_1 - \Theta_2 - \Theta_3 - \Theta_4$; $\Theta_2 \mid \Theta_5 \mid \Theta_6$	8
III^*	$\tilde{E}_7$	$\Theta_0 - \Theta_2 - \Theta_3 - \Theta_4 - \Theta_5 - \Theta_6 - \Theta_7$; $\Theta_4 \mid \Theta_1$	9
II^*	$\tilde{E}_8$	$\Theta_0 - \Theta_1 - \Theta_2 - \Theta_3 - \Theta_4 - \Theta_5 - \Theta_6 - \Theta_7$; $\Theta_2 \mid \Theta_8$	10

TABLE 3. Kodaira classification

A *simple component* of a fiber is a component with multiplicity one. In Table 4 we describe the singular fibers of an elliptic fibration with the multiplicities of the vertices of the extended Dynkin diagrams and we list the components with their multiplicities. The Néron-Severi group of a surface admitting a jacobian elliptic fibration contains the class of a fiber F and the class of the zero section s. Since the fibers are all algebraic equivalent, we have $F \cdot F = 0$. The zero section intersects each fiber in one point, so that $F \cdot s = 1$. The sections of an elliptic fibration on a $K3$ surface are smooth rational curves and this implies that their self-intersection is -2. Moreover, if X is a $K3$ surface that admits a jacobian elliptic fibration, then there is an embedding of $\overline{U}$ in $\mathrm{NS}(X)$, where $\overline{U}$ is the two dimensional lattice

$$\overline{U} = \left\{ \mathbb{Z}^2, \begin{pmatrix} 0 & 1 \\ 1 & -2 \end{pmatrix} \right\}.$$

Observe that by taking the generators F and $F+s$ the lattice $\overline{U}$ is isometric to the hyperbolic two dimensional lattice

$$U = \left\{ \mathbb{Z}^2, \begin{pmatrix} 0 & 1 \\ 1 & 0 \end{pmatrix} \right\}.$$

If $f : X \longrightarrow \mathbb{P}^1$ admits an n-torsion section s_n of order n in the Mordell-Weil group then it induces an automorphism of the same order on X. This acts as the identity on the base of the fibration and as a translation by the section on each fiber, hence it is a symplectic automorphism [**8**, Chapter 15, Lemma 4.4])

5. Examples

We give here examples corresponding to the cases discussed in Theorem 3.1. These are constructed by using jacobian elliptic fibrations on $K3$ surfaces.

Name	simple components	Dynkin diagram
$\tilde{A}_n$	$\Theta_i,\ i=0,\dots,n-1$	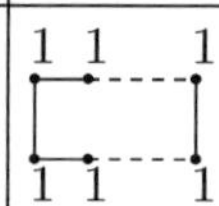
$\tilde{D}_{k+4}$	$\Theta_i,\ i=0,1,k+3,k+4$	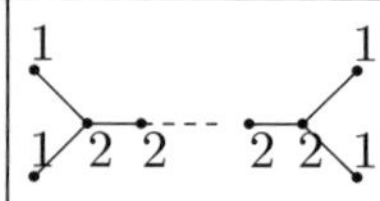
$\tilde{E}_6$	$\Theta_i,\ i=0,4,6$	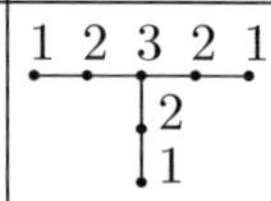
$\tilde{E}_7$	$\Theta_i,\ i=0,7$	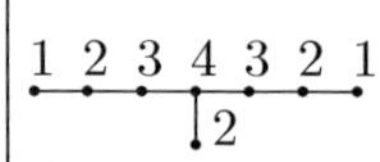
$\tilde{E}_8$	Θ_7	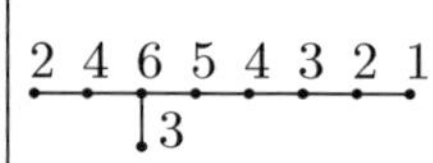

TABLE 4. Dynkin diagrams with the multiplicities of the components

5.1. Example $k_\sigma = 1$. The case $k_\sigma = 1$ in Theorem 3.1 occurs, this means that we can find an example of a non-symplectic automorphism of order 8 on a K3 surface X such that its fixed locus consists of a smooth rational curve and of 10 isolated points and the fixed locus of σ^4 consists of eight smooth rational disjoint curves. Consider the elliptic fibration on X given by:

$$y^2 = x(x^2 + tp_6(t))$$

with $p_6(t) := (a_6t^6 + a_4t^4 + a_2t^2 + a_0) = (t^2 - \alpha_1)(t^2 - \alpha_2)(t^2 - \alpha_3)$, where $a_6, a_4, a_2, a_0, \alpha_1, \alpha_2, \alpha_3 \in \mathbb{C}$, and the order 8 automorphism acting on it:

$$\sigma : (x, y, t) \mapsto (-ix, \zeta_8 y, -t).$$

By [**9**, Section 3] a holomorphic 2-form can be written as $\omega_X = \frac{dt\wedge dx}{2y}$ so that one computes

$$\sigma(\omega_X) = \zeta_8\omega_X,$$

hence σ acts purely non-symplectically. Moreover σ acts as an involution on the base $\mathbb{P}^1$ and it has order four on each fiber of the fibration.

The discriminant is $\Delta(t) = 4t^3(t^2-\alpha_1)^3(t^2-\alpha_2)^3(t^2-\alpha_3)^3$. Recall that $t \in \mathbb{P}^1$ so if we consider the homogenization of the polynomial in coordinates $(t : u)$, we obtain $\Delta(t, u) = 4t^3(t^2 - \alpha_1u^2)^3(t^2 - \alpha_2u^2)^3(t^2 - \alpha_3u^2)^3u^3$.

We take now $\alpha_1 = 0$ and $\alpha_2 = \alpha_3$. Under this assumtion the equation of the elliptic fibration becomes

$$y^2 = x(x^2 + t(t^2 - \alpha_2)^2)$$

and $\Delta(t) = 4t^9(t^2 - \alpha_2)^6$ which in homogeneous coordinates $(t : u)$ is equal to $\Delta(t, u) = 4t^9(t^2 - \alpha_2 u^2)^6 u^3$. For generic choice of the coefficient α_2 the fibration has 4 singular fibers which correspond to the four zeros of $\Delta(t, u)$. To be more precise the fibration has a fiber of type III^* over 0, which corresponds to the $\tilde{E}_7$ Dynkin diagram, a fiber of type III over ∞ which consists in two rational curves tangent in a point and two fibers of type I_0^* over $\pm\sqrt{\alpha_2}$, see Table 3. The action of σ on the base fixes two points: 0 and ∞, and so it preserves the fibers over these two points, the fibers III and III^*, and it exchanges the two fibers of type I_0^*. If $f : X \longrightarrow \mathbb{P}^1$ is the elliptic fibration then $\mathrm{Fix}(\sigma) \subset f^{-1}(0) \cup f^{-1}(\infty)$. Observe that the fibration has a two torsion section s_2, given by $t \mapsto (0 : 0 : 1) = (x : y : z)$ and the zero section s, given by $t \mapsto (0 : 1 : 0) = (x : y : z)$. These sections are preserved by the action of σ and they have two fixed points corresponding to the intersection with the fibers over 0 and ∞. These two sections are pointwise fixed by the action of σ^2. The sections s_2 and s are simple sections which means that they meet one of the components of III^* of multiplicity one (not the same), they meet one of the two components of III (not the same) in a non-singular point and they meet the fibers of type I_0^* in one of the components of multiplicity 1 (not the same). We know that the components are not the same since the sections s and s_2 are contained in the fixed locus of σ^4 and the above mentioned components of the singular fibers are not fixed pointwise by σ^4 and cannot contain more than two fixed points. Now we can see that the two fixed points on each section are contained in III^* and III. The fiber of type III consists in two tangent rational curves. Each of these two rational curves has a fixed point which is not the tangency point, but a non-trivial finite order automorphism acting on a rational curve has two fixed points, so we conclude that the double point on III is fixed by σ. Observe that the sections s and s_2 are preserved so that all the components of the fiber III are preserved. On the fiber of type III^* we have the two fixed points given by the intersections with the two sections s ans s_2. Since the sections s and s_2 are not exchanged then all components in III^* are preserved. The unique component of this fiber of multiplicity four is preserved by σ and it contains three fixed points so it is pointwise fixed. The component of multiplicity two which intersects the component of multiplicity four contains then a fixed point and we know that there is another fixed point on it. In conclusion $N_\sigma = 10$ and $k_\sigma = 1$.

The square of the automorphism $\sigma^2 : (x, y, t) \mapsto (-x, iy, t)$ preserves each fiber and acts as an automorphism of order four on it. Moreover σ^2 fixes two points on the generic smooth fiber, these two fixed points are contained in the two sections s and s_2. This gives that $k_{\sigma^2} \geq 2$.

Since $l_{\sigma^2} = 2m_\sigma = 2$ and $m_{\sigma^2} = 2m_1 = 2$ using Table 1 for the classification of non-symplectic automorphisms of order 4, we know that the invariants for σ^2 are $(r_{\sigma^2}, m_{\sigma^2}, l_{\sigma^2}, N_{\sigma^2}, k_{\sigma^2}, a_{\sigma_2}) = (16, 2, 2, 10, 3, 0)$. The curves fixed by σ^2 are the curves fixed by σ and the two sections s and s_2. The points fixed by σ^2 are 10 but they are not the same fixed points by σ, in fact the 4 fixed points for σ on s and s_2 now lie on fixed curves (namely the curves s and s_2) and they do not give any contribution, but we add exactly 4 other points on the two fibers I_0^*. For this reason the number of fixed points remains the same.

Consider the curve defined by $y = 0$. From the equation of the elliptic fibration we obtain $x = 0$, which gives the zero section, and the curve C: $x^2+t^2(t^2-\alpha_2)^2 = 0$ which has a $2:1$ morphism to $\mathbb{P}^1$ and has possibly ramification points where $t^2(t^2-\alpha_2)^2 = 0$. These points lie on the four singular fibers: the fiber III, the two fibers of type I_0^* and the fiber III^*. One has to study carefully the intersection of C with the Dynkin diagram of the fibers to understand the ramification. The curve C meets III in the double point, I_0^* in the two components of multiplicity one (so it does not ramify here) and III^* in the component of multiplicity two.

Recall that the Riemann-Hurwitz formula applied to a $2:1$ morphism from a curve C to the projective line $\mathbb{P}^1$ is given by:

$$2g(C) - 2 = 2(0-2) + \sum_{p\in C}(e_p - 1)$$

where the sum runs over the ramification points which are two points in this case: the point on the fiber over 0, i.e. on III^* and the point on the fiber over ∞ i.e on III and e_p is the ramification index at a ramification point p. In this case $e_p = 2$. By using the formula we can compute in an easy way the genus of the curve $C \subset \mathrm{Fix}(\sigma^4)$ which is $g(C) = 0$. Hence the fixed curves by σ^4 are all rational and we have eight of them : three components of the fiber III^*, two sections s and s_2, the curve C and two other rational curves which are fixed on the two fibers of type I_0^*.

5.2. Example $k_\sigma = 0$. The case $k_\sigma = 0$ in Theorem 3.1 occurs when $k_{\sigma^2} = 3$. We can consider the same elliptic fibration of the previous example, and we fix $\alpha_1 = 0$ and $\alpha_2 = \alpha_3$ as before. As we have already observed the fibration has a 2-torsion section given by $t \mapsto (0:0:1) = (x:y:z)$. Denote by τ the symplectic involution associated to this 2-torsion section. As we have observed before, this involution is symplectic. The involution exchanges the zero section s and the 2-torsion section s_2. We cannot find fixed points for τ on the generic fiber since it acts as a translation, but we know (see [**12**, Section 5]) that a symplectic involution has eight fixed points on a K3 surface. Consequently the 8 fixed points for τ are on the singular fibers of the elliptic fibration, recall that these are a fiber of type III^*, two fibers of type I_0^* and a fiber of type III.

On each of the two fibers of type I_0^* we have two fixed points on the component of multiplicity two since the sections s and s_2 are exchanged. On the fiber of type III^* we have three fixed points, two of them are on the component of maximal multiplicity and the other is on the component of multiplicity two which intersects the component of maximal multiplicity. On the fiber of type III the two rational curves are exchanged and so we have a fixed point for τ which is the double point.

We consider now $\sigma \circ \tau$. By a direct computation one computes $\sigma \circ \tau = \tau \circ \sigma$ so that we have an automorphism of order eight. Since τ is symplectic and σ acts purely non-symplectically we have that $\sigma \circ \tau$ acts purely non-symplectically too on the elliptic K3 surface. Now the fixed point for τ on III is also a fixed point for σ, so it is a fixed point for $\sigma \circ \tau$. The fibers I_0^* are exchanged by σ so we cannot find fixed points for $\sigma \circ \tau$ on them and two of the three fixed points on III^* are on a fixed curve for σ so they contribute to $\mathrm{Fix}(\sigma \circ \tau)$. Finally we conclude that $\mathrm{Fix}(\sigma \circ \tau)$ consists of four points. Where one of them is the double point of III and three are on the fiber III^*. Finally since σ and τ commute $(\sigma \circ \tau)^2 = \sigma^2$, so that the behaviour of the order four automorphism is the same as it is described in the previous example.

Acknowledgments

We thank Paola Comparin for useful comments and the anonymous referee for a very careful reading of the paper and for the several useful remarks, which improved the paper.

References

[1] Dima Al Tabbaa and Alessandra Sarti, *Order eight non-symplectic automorphisms on elliptic K3 surfaces*, Phenomenological approach to algebraic geometry, Banach Center Publ., vol. 116, Polish Acad. Sci. Inst. Math., Warsaw, 2018, pp. 11–24. MR3889148

[2] Dima Al Tabbaa, Alessandra Sarti, and Shingo Taki, *Classification of order sixteen non-symplectic automorphisms on K3 surfaces*, J. Korean Math. Soc. **53** (2016), no. 6, 1237–1260, DOI 10.4134/JKMS.j150339. MR3570971

[3] Michela Artebani and Alessandra Sarti, *Non-symplectic automorphisms of order 3 on K3 surfaces*, Math. Ann. **342** (2008), no. 4, 903–921, DOI 10.1007/s00208-008-0260-1. MR2443767

[4] Michela Artebani and Alessandra Sarti, *Symmetries of order four on K3 surfaces*, J. Math. Soc. Japan **67** (2015), no. 2, 503–533, DOI 10.2969/jmsj/06720503. MR3340184

[5] Michela Artebani, Alessandra Sarti, and Shingo Taki, *K3 surfaces with non-symplectic automorphisms of prime order*, Math. Z. **268** (2011), no. 1-2, 507–533, DOI 10.1007/s00209-010-0681-x. With an appendix by Shigeyuki Kondō. MR2805445

[6] Jimmy Dillies, *On some order 6 non-symplectic automorphisms of elliptic K3 surfaces*, Albanian J. Math. **6** (2012), no. 2, 103–114. MR3009163

[7] Igor V. Dolgachev and Shigeyuki Kondō, *Moduli of K3 surfaces and complex ball quotients*, Arithmetic and geometry around hypergeometric functions, Progr. Math., vol. 260, Birkhäuser, Basel, 2007, pp. 43–100, DOI 10.1007/978-3-7643-8284-1_3. MR2306149

[8] Daniel Huybrechts, *Lectures on K3 surfaces*, Cambridge Studies in Advanced Mathematics, vol. 158, Cambridge University Press, Cambridge, 2016. MR3586372

[9] Shigeyuki Kondō, *Automorphisms of algebraic K3 surfaces which act trivially on Picard groups*, J. Math. Soc. Japan **44** (1992), no. 1, 75–98, DOI 10.2969/jmsj/04410075. MR1139659

[10] Natsumi Machida and Keiji Oguiso, *On K3 surfaces admitting finite non-symplectic group actions*, J. Math. Sci. Univ. Tokyo **5** (1998), no. 2, 273–297. MR1633933

[11] Rick Miranda, *The basic theory of elliptic surfaces*, Dottorato di Ricerca in Matematica. [Doctorate in Mathematical Research], ETS Editrice, Pisa, 1989. MR1078016

[12] V. V. Nikulin, *Finite groups of automorphisms of Kählerian surfaces of type K3* (Russian), Uspehi Mat. Nauk **31** (1976), no. 2(188), 223–224. MR0409904

[13] V. V. Nikulin, *Integer symmetric bilinear forms and some of their geometric applications* (Russian), Izv. Akad. Nauk SSSR Ser. Mat. **43** (1979), no. 1, 111–177, 238. MR525944

[14] V. V. Nikulin, *Quotient-groups of groups of automorphisms of hyperbolic forms by subgroups generated by 2-reflections. Algebro-geometric applications* (Russian), Current problems in mathematics, Vol. 18, Akad. Nauk SSSR, Vsesoyuz. Inst. Nauchn. i Tekhn. Informatsii, Moscow, 1981, pp. 3–114. MR633160

[15] V. V. Nikulin, *Discrete reflection groups in Lobachevsky spaces and algebraic surfaces*, Proceedings of the International Congress of Mathematicians, Vol. 1, 2 (Berkeley, Calif., 1986), Amer. Math. Soc., Providence, RI, 1987, pp. 654–671. MR934268

[16] A. N. Rudakov and I. R. Shafarevich, *Surfaces of type K3 over fields of finite characteristic* (Russian), Current problems in mathematics, Vol. 18, Akad. Nauk SSSR, Vsesoyuz. Inst. Nauchn. i Tekhn. Informatsii, Moscow, 1981, pp. 115–207. MR633161

[17] Matthias Schütt, *K3 surfaces with non-symplectic automorphisms of 2-power order*, J. Algebra **323** (2010), no. 1, 206–223, DOI 10.1016/j.jalgebra.2009.06.021. MR2564835

[18] Shingo Taki, *Classification of non-symplectic automorphisms of order 3 on K3 surfaces*, Math. Nachr. **284** (2011), no. 1, 124–135, DOI 10.1002/mana.200810070. MR2752672

[19] Shingo Taki, *Classification of non-symplectic automorphisms on $K3$ surfaces which act trivially on the Néron-Severi lattice*, J. Algebra **358** (2012), 16–26, DOI 10.1016/j.jalgebra.2012.02.021. MR2905014

Université de Poitiers, Laboratoire de Mathématiques et Applications, UMR 7348 du CNRS, Bât. H3 - Site du Futuroscope, TSA 61125, 11 bd Marie et Pierre Curie, 86073 POITIERS Cedex 9, France

Email address: dima.al.tabbaa@math.univ-poitiers.fr

Dipartimento di Matematica, Università di Bologna, Piazza di Porta S. Donato, 5, 40126 Bologna, Italy and Universität Augsburg, Institut für Mathematik, Universitätsstr. 14, 86159 Augsburg, Germany

Email address: annalisa.grossi3@unibo.it, annalisa.grossi@math.uni-augsburg.de

URL: https://www.unibo.it/sitoweb/annalisa.grossi3

Université de Poitiers, Laboratoire de Mathématiques et Applications, UMR 7348 du CNRS, Bât. H3 - Site du Futuroscope, TSA 61125, 11 bd Marie et Pierre Curie, 86073 POITIERS Cedex 9, France

Email address: sarti@math.univ-poitiers.fr

URL: http://www-math.sp2mi.univ-poitiers.fr/~sarti/

Contemporary Mathematics
Volume **766**, 2021
https://doi.org/10.1090/conm/766/15371

A decomposition of the Jacobian of a Humbert-Edge curve

Robert Auffarth, Giancarlo Lucchini Arteche, and Anita M. Rojas

Abstract. A *Humbert-Edge curve of type* $n \geq 2$ is a non-degenerate smooth complete intersection of $n-1$ diagonal quadrics in $\mathbb{P}^n$. Such a curve has an interesting geometry since it has a natural action of the group $(\mathbb{Z}/2\mathbb{Z})^n$. We present here a decomposition of its Jacobian variety as a product of Prym-Tyurin varieties, and we compute the kernel of the corresponding isogeny.

To Herbert Lange on his 75th birthday.

1. Introduction

An irreducible non-singular curve $X_n \subseteq \mathbb{P}^n$ over an algebraically closed field of characteristic different from 2 is a *Humbert-Edge curve of type* $n \geq 2$ if it is non-degenerate (i.e. not contained in a hyperplane) and is the complete intersection of $n-1$ diagonal quadrics. Multiplication by -1 on the ith coordinate of $\mathbb{P}^n$ restricts to an automorphism σ_i of X_n, and the group generated by these involutions is a subgroup E_n of $\mathrm{Aut}(X_n)$ isomorphic to $(\mathbb{Z}/2\mathbb{Z})^n$.

Humbert-Edge curves of type 4 were first studied by Humbert [**8**] and later generalized to arbitrary type by Edge in [**5**]. They were then studied more in depth in [**2**] from a complex-analytic point of view and in [**7**] from an algebraic point of view. In particular in this last article a coarse decomposition of the Jacobian variety of a Humbert-Edge curve was given. The purpose of this paper is to give a more refined decomposition of the Jacobian variety of a Humbert-Edge curve by using the automorphism group E_n. Our results imply the following (see Theorems 2.1 and 4.1 for the specific statements):

Theorem 1.1. *The Jacobian of a Humbert-Edge curve of type n splits isogenously as the product of lower-dimensional abelian subvarieties where the dimension of the largest factor grows logarithmically with respect to the genus of the curve. Moreover, each of these factors is a Prym-Tyurin variety of exponent* 2^{n-3} *for the curve.*

We recall that a *Prym-Tyurin variety of exponent e for a curve X* is a principally polarized abelian variety (A, Θ) such that there exists an embedding

2020 *Mathematics Subject Classification.* Primary 14H40; Secondary 14H37, 14K02.

The first author was partially supported by Conicyt via Fondecyt Grant 11180965.

The second author was partially supported by Conicyt via Fondecyt Grant 11170016 and PAI Grant 79170034.

The third author was partially supported by Conicyt via Fondecyt Grant 1180073.

$i : A \hookrightarrow JX$ that satisfies $i^*\Theta_X \equiv e\Theta$, where Θ_X is the theta divisor of JX and $\equiv$ stands for numerical equivalence.

In a sense, our result means that the Jacobian of a Humbert-Edge curve has a fine decomposition. Isogeny decompositions of Jacobians of curves have been a subject of interest ever since Ekedahl and Serre [**6**] found examples of curves of many genera, the largest being 1297, whose Jacobian variety splits isogenously as the product of elliptic curves. They asked if the genera of completely decomposable Jacobians are bounded, and this is a question that remains open in characteristic 0.

We obtain here that the Jacobian variety of a Humbert-Edge curve X_n is decomposed as a product of subvarieties which are Prym-Tyurin varieties for X_n, and are actually isogenous to Jacobian varieties. We were previously unaware of any examples that showed this type of behavior.

REMARK 1.2. Note that Humbert-Edge curves correspond to generalized Fermat curves of type $(2, n)$. When submitting this article, we found out that in [**4**, §5], a decomposition of the Jacobian of a generalized Fermat curve X of type (p, n) for p prime is given in terms of pullbacks of Jacobians of quotients X/H for subgroups $H \leq \mathrm{Aut}(X) \simeq (\mathbb{Z}/p\mathbb{Z})^n$ of index p (i.e. hyperplanes of the $\mathbb{F}_p$-vector space). In the case $p = 2$, this decomposition turns out to correspond to the one we obtain here. However, our approach is different: we follow an inductive procedure instead of cutting out by hyperplanes. Our point of view allows us to describe the decomposition more precisely: we give the exact number of factors and their corresponding dimensions. Moreover, we compute the kernel of the decomposition and the induced polarization on each factor, finding that all of them are Prym-Tyurin varieties. We intend to generalize our results to arbitrary generalized Fermat curves in the future.

2. Main theorem

By [**7**, Lemma 3.2], a Humbert-Edge curve X_n is of genus $2^{n-2}(n-3)+1$ and the quotient of X_n by each σ_i is a Humbert-Edge curve of type $n-1$. In particular, if we denote by S_n the set $\{\sigma_0, \ldots, \sigma_n\} \subset \mathrm{Aut}(X_n)$, then for each subset $T \subset S_n$ we can consider the curve $X_T := X_n/\langle T\rangle$. It is a Humbert-Edge curve of type $n - |T|$ whose corresponding subgroup of automorphisms $E_T \leq \mathrm{Aut}(X_T)$ is naturally isomorphic to $E_n/\langle T\rangle$ and generated by the images of the $\sigma_i \in S_n \smallsetminus T$, as can easily be proved by induction.

By [**7**, Proposition 3.5], the Jacobian of X_n has the following decomposition:

$$JX_n \sim A \oplus JX_n^-, \tag{2.1}$$

where $A = \sum_{i=0}^n \pi_i^* J(X/\sigma_i)$ (here $\pi_i : X_n \to X_n/\sigma_i$ is the natural projection) and JX_n^- is the neutral connected component of

$$\{x \in JX_n : \sigma_i(x) = -x \text{ for all } i\}.$$

Note that in [**7**] the connectedness of this set is assumed, yet not proved.

Our goal is to push this decomposition further by using the previous remark on the quotients X_T for $T \subset S_n$. Our main result states the following:

THEOREM 2.1. *Let X_n be a Humbert-Edge curve of type $n \geq 3$. Then we have the following decomposition of JX_n:*

$$JX_n \sim \bigoplus_{\substack{T \subset S_n \\ |T| \leq n-3}} \pi_T^* JX_T^-,$$

where $\pi_T : X_n \to X_T$ is the natural projection. Moreover, if $n - |T|$ is even, then JX_T^- is trivial; and if $n - |T| = 2m + 1$, then $\dim JX_T^- = m$. In particular, for $1 \leq m \leq \lfloor \frac{n-1}{2} \rfloor$, there are exactly $\binom{n+1}{2m+2}$ summands of dimension m.

We observe that in particular, we can always find $\binom{n+1}{4}$ elliptic curves in the decomposition and the largest factor above is of dimension $\lfloor \frac{n-1}{2} \rfloor$. What is nice about this is that, while the genus of X_n grows exponentially in n, the dimension of the largest factor in its Jacobian decomposition grows linearly and is therefore relatively small.

3. Proof of the main theorem

Note that the last statement of Theorem 2.1 follows immediately from combinatorial considerations. Let us start then by analyzing the dimension of JX_n^-.

LEMMA 3.1. *Let X_n and JX_n^- as above. If n is even, then JX_n^- is trivial. If n is odd, the dimension of JX_n^- is $\frac{n-1}{2}$.*

PROOF. Since we have the action of $E_n \cong (\mathbb{Z}/2\mathbb{Z})^n$ on X_n, we know [**10**] that JX_n decomposes, in general, as

$$JX_n \sim \bigoplus_{\rho \in \mathrm{Irr}_{\mathbb{Q}}(E_n)} B_\rho,$$

where $\mathrm{Irr}_{\mathbb{Q}}(E_n)$ is the set of rational irreducible representations of E_n, up to equivalence. Moreover, each subvariety B_ρ corresponds to the image of the central idempotent e_ρ defined by ρ in the group algebra $\mathbb{Q}[E_n]$.

In our case, we know from [**7**, Thm. 3.4] and the Riemann-Hurwitz formula that the action of E_n on X_n is of signature $(0; 2^{n+1})$. That is, the total quotient X_n/E_n is of genus 0, and there are $n+1$ orbits of fixed points, each with a stabilizer of order 2. In fact, the stabilizers are the $\langle \sigma_i \rangle$ for $i = 0, \ldots, n$. Furthermore, the subvariety JX_n^- corresponds to the representation ρ of E_n given by $\rho(\sigma_i) = -1$ for $i = 0, \ldots, n$. Notice that since $\prod_i \sigma_i = 1$, ρ is defined only in the case of an odd n. It follows immediately that for an even n, the dimension of JX_n^- is 0 (see also [**7**, Remark 3.6]).

To find the dimension of JX_n^- for odd n, we apply the formula in [**11**, Thm. 5.12] (see also [**9**]). Following the notations therein, we have $k_\rho = 1$, $\dim \rho = 1$, $\gamma = 0$, $t = n + 1$, and $\dim \mathrm{Fix}_{G_k} \rho = 0$ since $G_k = \langle \sigma_k \rangle$. This gives immediately $\dim B_\rho = \frac{n-1}{2}$. □

We are ready now to prove Theorem 2.1.

PROOF OF THEOREM 2.1. The statement on the dimension of JX_n^- is Lemma 3.1. The proof of the decomposition follows easily by induction: For $n = 3$, we have that X_n is an elliptic curve and hence the decomposition (2.1) gives us $JX_3 = JX_3^-$ since a Humbert-Edge curve of type 2 is simply a quadric in $\mathbb{P}^2$, hence of genus 0. This gives the result in this case.

Assume now that the result holds for $n-1$ and let us prove it for n. Using once again the decomposition (2.1) for X_n we get

$$JX_n \sim JX_n^- \oplus \sum_{i=0}^{n} \pi_i^* J(X/\sigma_i).$$

Now, using the induction hypothesis for $n-1$ and the fact that the projection $\pi_T : X_n \to X_T$ factors naturally through X/σ_i if $\sigma_i \in T$, we get

$$JX_n \sim JX_n^- \oplus \sum_{i=0}^{n} \bigoplus_{\substack{\sigma_i \in T \subset S_n \\ |T| \leq n-3}} \pi_T^* JX_T^- .$$

Note that $\pi_T^* JX_T^-$ appears in the sum for *every non-empty* T with $|T| \leq n-3$. Considering then, for $i \neq j$ and $T \supset \{\sigma_i, \sigma_j\}$, the commutative diagrams

$$\begin{array}{ccccc} & & X_n & & \\ & \overset{\pi_i}{\swarrow} & \Big\downarrow{\scriptstyle \pi_T} & \overset{\pi_j}{\searrow} & \\ X/\sigma_i & & & & X/\sigma_j \\ & \searrow & & \swarrow & \\ & & X_T & & \end{array} \qquad \begin{array}{ccccc} & & JX_n & & \\ & \overset{\pi_i^*}{\nearrow} & \Big\uparrow{\scriptstyle \pi_T^*} & \overset{\pi_j^*}{\nwarrow} & \\ X/\sigma_i & & & & X/\sigma_j \\ & \nwarrow & & \nearrow & \\ & & JX_T & & \end{array}$$

we see that the sum from $i=0$ to n is simply making up $|T|$ copies of each $\pi_T^* JX_T^-$, which we may clearly erase then in order to get

$$JX_n \sim JX_n^- \oplus \bigoplus_{\substack{\emptyset \neq T \subset S_n \\ |T| \leq n-3}} \pi_T^* JX_T^-,$$

which is the desired decomposition since $JX_n^- = JX_T^-$ for $T = \emptyset$. □

4. Geometric description of the decomposition of JX_n

In this section, we describe the kernel of the decomposition displayed in Theorem 2.1 and the induced polarization on each factor. We prove that each factor in this decomposition for JX_n is a Prym-Tyurin variety of exponent 2^{n-3} with respect to X_n. This description of the induced polarization on each factor allows us to compute the kernel of the isogeny, as we show in Lemma 4.10. The main results in this section are collected in the following theorem:

THEOREM 4.1. *Let X_n be a Humbert-Edge curve of type $n \geq 3$ and genus g_n decomposed as*

$$\varphi : \bigoplus_{\substack{T \subset S_n \\ |T| \leq n-3}} \pi_T^* JX_T^- \to JX_n$$

Then

(1) *Each factor is a Prym-Tyurin variety of exponent 2^{n-3} with respect to X_n.*
(2) *The kernel of the isogeny φ is of order $(2^{n-3})^{g_n}$.*

We prove this theorem in several steps, keeping the notations we have defined in the previous sections. Let us start by the following remark concerning JX_n^-:

LEMMA 4.2. *Let X_n be a Humbert-Edge curve of type n. Let $H_n \leq E_n$ be the subgroup generated by the double products $\sigma_i \sigma_j$ for $i, j \in \{0, \ldots, n\}$. Then $JX_n^- = \pi_{H_n}^* J(X_n/H_n)$, where π_{H_n} denotes the canonical projection $X_n \to X_n/H_n$.*

PROOF. Recall [**3**, Prop. 5.2] that $\pi_{H_n}^* J(X_n/H_n)$ is the subvariety defined as the image of the idempotent $p_{H_n} = \frac{1}{|H|}\sum_{h\in H_n} h$, which is the neutral connected component of $(JX_n)^{H_n}$. The result then follows immediately from [**7**, Prop. 3.5(ii)]. Here once again there is an issue with connectedness, but it does not affect our result since we are taking the neutral connected component on both sides of the equality. □

Using Lemma 3.1 and the Riemann-Hurwitz formula, we immediately get the following result:

COROLLARY 4.3. *Let X_n and H_n be as above and assume that n is odd. Then the quotient morphism $\pi_H : X_n \to X_n/H_n$ is étale.*

REMARK 4.4. Lemma 4.2 may sound troubling in the case of even n since then JX_n^- is trivial. However, note that in this case H_n is the whole E_n and thus $X_n/H_n = X_n/E_n$. And as we saw in the proof of Lemma 3.1, X_n/E_n has genus 0, hence its Jacobian is trivial.

In the odd case, it is easy to see that H_n has index 2 in E_n, hence $|H_n| = 2^{n-1}$.

REMARK 4.5. Note that Lemma 4.2 and Theorem 4.1 imply that the Jacobian of a Humbert-Edge curve decomposes as a product of Jacobians.

We continue now with a lemma on abelian étale covers of curves, such as the one in the last corollary.

LEMMA 4.6. *Let $f : X \to Y$ be a Galois morphism between smooth projective curves with Galois group G and let $f^* : JY \to JX$ be the corresponding morphism on the Jacobian varieties. Let g be the dimension of Y and assume that f is of degree p^{2g} for some prime p. Then the following are equivalent:*

(1) *f is étale and G is isomorphic to $(\mathbb{Z}/p\mathbb{Z})^{2g}$.*
(2) *f factors through every cyclic étale cover $Z \to Y$ of degree p.*
(3) *$\ker f^* = JY[p]$.*

PROOF. From [**1**, Prop. 11.4.3] and its proof, one easily sees that a non-trivial element $\xi \in JY[p]$ is in $\ker f^*$ if and only if f factors through a certain cyclic étale cover $X_\xi \to Y$ of degree p that is uniquely defined by $\langle\xi\rangle$ up to Y-isomorphism. Moreover, cyclic subgroups of $JY[p]$ classify all cyclic étale covers of degree p in this way (see the comment right after *loc. cit.*). Then (3) implies (2) and (2) implies $\ker f^* \supset JY[p]$. However, from *loc. cit.* one deduces that $|\ker f^*|$ is smaller or equal than the degree of f, which is p^{2g}, hence (3).

Assume (1). Then there are precisely $(p^{2g}-1)/(p-1)$ non-trivial quotients of G isomorphic to $\mathbb{Z}/p\mathbb{Z}$. It is easy to see that each one of these induces a different cyclic étale cover of degree p (up to Y-isomorphism). Then f factors through all of these covers and we have (2) since this is exactly the number of cyclic subgroups of $JY[p]$.

Finally, assume (2). As we just said, there are $(p^{2g}-1)/(p-1)$ different cyclic étale covers of degree p and f factors through all of them. Hence, there are at least $(p^{2g}-1)/(p-1)$ non-trivial quotients of G isomorphic to $\mathbb{Z}/p\mathbb{Z}$. Now, the only group of order p^{2g} having this property is $(\mathbb{Z}/p\mathbb{Z})^{2g}$. It is easy then to see that f must be étale: Take for instance the fiber product of $2g$ cyclic covers coming from a generating set of $G = (\mathbb{Z}/p\mathbb{Z})^{2g}$. Then X is isomorphic to this product, which is étale over Y. □

These results allow us to understand the polarization induced on JX_n^-. Indeed:

PROPOSITION 4.7. *Let X_n be a Humbert-Edge curve of type n for odd n and let Θ be the principal polarization of JX_n. Then the polarization induced by Θ on JX_n^- is of type $(2^{n-3}, \dots, 2^{n-3})$.*

PROOF. Using Lemma 4.2, we see JX_n^- as $\pi_{H_n}^* J(X_n/H_n)$. Then, since $|H_n| = 2^{n-1}$, we know by [**1**, Lem. 12.3.1] that the polarization $(\pi_{H_n}^*)^*\Theta$ is 2^{n-1} times the principal polarization of $J(X_n/H_n)$. By Lemma 4.6 we have $\ker(\pi_{H_n}^*) = J(X_n/H_n)[2]$ and hence we have the following commutative diagram

$$\begin{array}{ccc} JX_n^- & \overset{i}{\hookrightarrow} & JX_n \\ & \nwarrow^{\times 2} & \uparrow{\scriptstyle \pi_{H_n}^*} \\ & & J(X_n/H_n) \end{array}$$

This immediately gives

$$2^{n-1}\Theta_{X_n/H_n} \equiv \pi_{H_n}^*\Theta \equiv 2^* i^*\Theta \equiv 4 i^*\Theta$$

and since the Néron-Severi group of an abelian variety is torsion-free, we have that $i^*\Theta \equiv 2^{n-3}\Theta_{X_n/H_n}$, and the result follows. □

REMARK 4.8. From this proof one easily deduces that under the hypothesis (3) of Lemma 4.6, even without assumptions on the order of f or whether it is Galois, the subvariety $f^*JY \subset JX$ is a Prym-Tyurin variety of exponent p^{2g-2}.

We are now ready to prove statement (1) in Theorem 4.1.

PROPOSITION 4.9. *Let $n \geq 3$ be an integer, X_n a Humbert-Edge curve of type n and $T \subset S_n$ with $|S_n \setminus T| > 3$. Then, the induced polarization on the factor $\pi_T^* JX_T^-$ is of type $(2^{n-3}, \dots, 2^{n-3})$.*

PROOF. This is easily proved by induction using Proposition 4.7. For $n = 3$, $JX_3 = JX_3^-$ and it is an elliptic curve with polarization of type $(1) = (2^{3-3})$.

For $n \geq 4$, assume the result for $n-1$. If $T = \emptyset$, then the result corresponds to Proposition 4.7 when n is odd and JX_T^- is trivial when n is even, so we need not consider this case. If $T \neq \emptyset$, consider an element $\sigma_i \in T$. Then π_T factors as

$$X_n \xrightarrow{\pi_i} X_n/\sigma_i \xrightarrow{\pi'} X_T,$$

and hence π_T^* factors as

$$JX_T \xrightarrow{\pi'^*} J(X_n/\sigma_i) \xrightarrow{\pi_i^*} JX_n.$$

By the induction hypothesis, $\pi'^* JX_T^-$ is a Prym-Tyurin subvariety of $J(X_n/\sigma_i)$ of exponent 2^{n-4}. Recalling then that π_i is of degree 2 and ramified, it follows from [**1**, Prop. 12.3.1] and [**1**, Prop. 11.4.3] that the polarization on $\pi_T^* JX_n$ is of type $(2^{n-3}, \dots, 2^{n-3})$. □

In order to prove statement (2) in Theorem 4.1, we prove the following lemma. The result follows immediately then from statement (1).

LEMMA 4.10. *Let (A, Θ) be a principally polarized abelian variety. Assume that there exist abelian subvarieties $A_1, \ldots, A_r$ of A such that $i_j^* \Theta \equiv d\theta_j$ for a principal polarization θ_j on A_j, for $j = 1, \ldots, r$. Denote by $\varphi : \bigoplus_{j=1}^r A_i \to A$ the isogeny given by the sum. Then*

$$|\ker \varphi| = d^{\dim A}.$$

PROOF. We know that $\varphi^* \Theta = d(\theta_1 \boxtimes \cdots \boxtimes \theta_r)$, hence

$$(\varphi^* \Theta)^{\dim A} = d^{\dim A} (\theta_1 \boxtimes \cdots \boxtimes \theta_r)^{\dim A} = d^{\dim A} (\dim A)!,$$

where the last equality follows from Riemann-Roch for abelian varieties. On the other hand,

$$(\varphi^* \Theta)^{\dim A} = \deg \varphi \cdot \Theta^{\dim A}.$$

Once again by Riemann-Roch, we have $\Theta^{\dim A} = (\dim A)!$ and the result follows. □

REMARK 4.11. Using statement (1) in Theorem 4.1 and [**1**, Prop. 12.1.9], we finally deduce that JX_n^-, as defined in [**7**], is connected.

Acknowledgements

We were inspired to work on the Jacobian decomposition of Humbert-Edge curves by a talk given by Alexis Zamora at the Workshop *Geometry at the Frontier III*, which took place in Pucón, Chile, November 12-16, 2018. We would like to thank the organizers for a wonderful conference. We would also like to thank the anonymous referee for his or her useful comments.

References

[1] Christina Birkenhake and Herbert Lange, *Complex abelian varieties*, 2nd ed., Grundlehren der Mathematischen Wissenschaften [Fundamental Principles of Mathematical Sciences], vol. 302, Springer-Verlag, Berlin, 2004. MR2062673

[2] Angel Carocca, Víctor González-Aguilera, Rubén A. Hidalgo, and Rubí E. Rodríguez, *Generalized Humbert curves*, Israel J. Math. **164** (2008), 165–192, DOI 10.1007/s11856-008-0025-2. MR2391145

[3] Angel Carocca and Rubí E. Rodríguez, *Jacobians with group actions and rational idempotents*, J. Algebra **306** (2006), no. 2, 322–343, DOI 10.1016/j.jalgebra.2006.07.027. MR2271338

[4] Mariela Carvacho, Rubén A. Hidalgo, and Saúl Quispe, *Jacobian variety of generalized Fermat curves*, Q. J. Math. **67** (2016), no. 2, 261–284, DOI 10.1093/qmath/haw009. MR3509993

[5] W. L. Edge, *The common curve of quadrics sharing a self-polar simplex*, Ann. Mat. Pura Appl. (4) **114** (1977), 241–270, DOI 10.1007/BF02413789. MR469926

[6] Torsten Ekedahl and Jean-Pierre Serre, *Exemples de courbes algébriques à jacobienne complètement décomposable* (French, with English and French summaries), C. R. Acad. Sci. Paris Sér. I Math. **317** (1993), no. 5, 509–513. MR1239039

[7] Juan B. Frías-Medina and Alexis G. Zamora, *Some remarks on Humbert-Edge's curves*, Eur. J. Math. **4** (2018), no. 3, 988–999, DOI 10.1007/s40879-018-0232-2. MR3851126

[8] G. Humbert. *Sur un complexe remarquable de coniques et sur la surface de troisième ordre.* J. de l'Ecole Poly. 64 (1894), 123-149.

[9] Amy E. Ksir, *Dimensions of Prym varieties*, Int. J. Math. Math. Sci. **26** (2001), no. 2, 107–116, DOI 10.1155/S016117120101153X. MR1836786

[10] H. Lange and S. Recillas, *Abelian varieties with group action*, J. Reine Angew. Math. **575** (2004), 135–155, DOI 10.1515/crll.2004.076. MR2097550

[11] Anita M. Rojas, *Group actions on Jacobian varieties*, Rev. Mat. Iberoam. **23** (2007), no. 2, 397–420, DOI 10.4171/RMI/500. MR2371432

Departamento de Matemáticas, Facultad de Ciencias, Universidad de Chile, Las Palmeras 3425, 7800024 Ñuñoa, Santiago, Chile
Email address: rfauffar@uchile.cl

Departamento de Matemáticas, Facultad de Ciencias, Universidad de Chile, Las Palmeras 3425, 7800024 Ñuñoa, Santiago, Chile
Email address: luco@uchile.cl

Departamento de Matemáticas, Facultad de Ciencias, Universidad de Chile, Las Palmeras 3425, 7800024 Ñuñoa, Santiago, Chile
Email address: anirojas@uchile.cl

Contemporary Mathematics
Volume **766**, 2021
https://doi.org/10.1090/conm/766/15372

Connections on central extensions, lifting gerbes, and finite-dimensional obstruction vanishing

Indranil Biswas and Markus Upmeier

ABSTRACT. Given a central extension of Lie groups, we study the classification problem of lifting the structure group of a principal bundle together with a given connection. For the case of reductive structure groups we introduce a connective structure on the lifting gerbe associated to this problem. Our main result classifies all connections on the extension of structure of a given principal bundle to the central extension. In particular, we find that admissible connections are in one-to-one correspondence with parallel trivializations of the lifting gerbe. Moreover, we prove a vanishing result for Neeb's obstruction classes for finite-dimensional Lie groups.

CONTENTS

Introduction

Consider a central extension of Lie groups

$$1 \longrightarrow A \longrightarrow B \longrightarrow C \longrightarrow 1\,.$$

Given a principal C-bundle

$$\pi : P \longrightarrow X\,, \tag{1}$$

where X is a manifold, it is a classical problem to seek principal B-bundles $\rho : R \longrightarrow X$ equipped with an isomorphism of principal C-bundles

$$\sigma : R/A \longrightarrow P.$$

2020 *Mathematics Subject Classification.* Primary 53C08, 53B05; Secondary 57S15.
Key words and phrases. Connections, gerbes, central extensions.
The first author is partially supported by a J. C. Bose Fellowship.
The second author thanks the TIFR, Mumbai, for hospitality during 03/2017 and was partially funded by DFG grant UP 85/2-1 of the priority program SPP 2026 Geometry at Infinity.

In other words, the principal C-bundle obtained by extending the structure group of R using the projection $B \longrightarrow C$ in (1) is identified with P; see Grothendieck [**4**, **5**] for a solution of the obstruction and classification problem in terms of sheaf cohomology. From a more geometric point of view, one associates a lifting gerbe $\mathcal{G}_\pi$ to this setup, see Murray [**8**] or Brylinski [**2**]; the trivializations of $\mathcal{G}_\pi$ (if they exist) are in bijective correspondence with the central extensions $\rho : R \longrightarrow X$ of the bundle $\pi : P \longrightarrow X$ up to an isomorphism.

Now suppose that the principal C-bundle P is equipped also with a connection ω^π. There being no further topological obstructions, our main result describes all connections on the central extension $\rho : R \longrightarrow X$ in Theorem 3.15, under the assumption that B is a reductive Lie group. The A-bundle $\sigma : R \longrightarrow P$ is naturally B-equivariant and before proving Theorem 3.15 we therefore need to develop some basic theory for equivariant connections in Section 2. In particular we introduce the *distortion* of a G-connection and establish some elementary results for the distortion and for quotient connections; see Propositions 2.6 and 3.1. This is later applied to $G = B$, though it comes at hardly any extra cost to establish Theorem 3.15 in a fully equivariant context, with a Lie group G acting on the base manifold X.

Restricting to admissible connections (19), we then apply Theorem 3.15 to solve the extension of the lifting problem involving connections. Using ω^π we introduce in Proposition 4.4 a connection on the lifting gerbe $\mathcal{G}_\pi$ and explain in Theorem 4.7 that parallel trivializations of $\mathcal{G}_\pi$ correspond to central extensions equipped with admissible connections. This result is equivalent to Theorem 3.12, stated without reference to gerbes.

Using crossed modules, Neeb associates in [**9**] a cohomology class $H^3(X, \mathbb{C})$ to every central extension with $A = \mathbb{C}^*$. In the final section we shall address the question, raised by Neeb [**9**, Problem VI], that asks which obstruction classes may occur. We find in Theorem 5.1 that for finite-dimensional Lie groups C these obstruction classes always vanish. This is proven using classical algebraic topology for Lie groups, mostly due to Borel [**1**] and gives a different approach to that of Brylinski [**2**, Chapter 6]. In infinite dimensions, the situation is drastically different; see Remark 5.5.

1. Preliminaries

1.1. Infinitesimal actions. Let A be a Lie group with Lie algebra $\mathfrak{a}$. Let $P \times A \longrightarrow P$ be a smooth action of A on a manifold P. Then we denote the linearized action by juxtaposition

$$\cdot : TP \times TA \longrightarrow TP\,.$$

If we regard $A \subset TA$ and $P \subset TP$ embedded as the zero section, then we may write

$$X \cdot \xi = X \cdot a + p \cdot \xi \qquad \forall\, X \in T_pP,\ \xi \in T_aA\,.$$

Differentiating the condition for an action gives

$$(X \cdot \xi) \cdot \eta = X \cdot (\xi \cdot \eta) \qquad \forall\, X \in T_pP, \xi \in T_aA,\ \eta \in T_bA\,. \tag{2}$$

In particular, this notation applies to $P = A$ acting on itself by right multiplication and similarly to left actions of A.

1.2. Connections on principal bundles. By convention, all our principal bundles are smooth and have their structure group acting from the right; sometimes the term "principal bundle" will be shortened to "bundle".

DEFINITION 1.1. Let $\pi : P \longrightarrow X$ be a principal A-bundle. A *connection* H^π on π is an A-equivariant complementary subbundle $H^\pi \subset TP$ of the vertical bundle:

$$H^\pi \oplus \ker(d\pi) = TP, \qquad H^\pi_{pa} = H^\pi_p a \qquad \forall\, a \in A,\ p \in P.$$

Equivalently, a connection is a 1-form $\omega^\pi \in \Omega^1(P, \mathfrak{a})$ satisfying

$$\omega^\pi(p \cdot \xi) = \xi, \quad \omega^\pi(v \cdot a) = a^{-1}\omega(v)a \tag{3}$$

for all $p \in P$, $v \in TP$, $\xi \in \mathfrak{a}$ and $a \in A$.

The *horizontal subspace* H^π is simply the kernel of ω^π, the *vertical subspace*

$$T^{\mathrm{vert}}_p P = \ker \pi_{*,p} = p \cdot \mathfrak{a}, \tag{4}$$

giving an identification of vector bundles $T^{\mathrm{vert}} P = P \times \mathfrak{a}$. The adjoint bundle

$$\mathrm{ad}(P) = T^{\mathrm{vert}} P/A = P \times^A \mathfrak{a},$$

is associated to P using the adjoint action of A on $\mathfrak{a}$.

Here we recall that the bundle $P \times^A V \longrightarrow X$ *associated* to an A-space V (also called the *Borel construction*) is the quotient of $P \times V$ by the twisted diagonal action of A; in particular, the points of $\{(pa, a^{-1}v)\}_{a\in A}$ are identified in the quotient $P \times^A V$. When $V = B$ is a group and the action is the left translation using a homomorphism $A \longrightarrow B$, then $P \times^A B$ becomes a principal B-bundle.

PROPOSITION 1.2. *Given a homomorphism (F, f) of A-bundles as in*

$$\begin{array}{ccc} P & \xrightarrow{F} & Q \\ {\scriptstyle\pi}\downarrow & & \downarrow{\scriptstyle\sigma} \\ X & \xrightarrow[f]{} & Y \end{array} \tag{5}$$

we may pull back connections from σ to π by $H^\pi := F_*^{-1} H^\sigma$ or, equivalently, by $\omega^\pi := F^*\omega^\sigma$.

1.3. Equivariant bundles and connections. Let G be another Lie group acting smoothly on a manifold X. The space of G-invariant differential forms on X is denoted $\Omega^*_G(X)$.

DEFINITION 1.3. A *G-equivariant* principal A-bundle is an A-bundle $\pi\colon P \to X$ together with a G-action $G \times P \longrightarrow P$ satisfying

$$g(pa) = (gp)a, \qquad \pi(gp) = g\pi(p)$$

for all $g \in G$, $p \in P$ and $a \in A$.

DEFINITION 1.4. A connection $H^\pi \subset TP$ on a G-equivariant principal A-bundle $\pi : P \longrightarrow X$ is a *G-connection* if for all $g \in G$ we have

$$H^\pi_{gp} = gH^\pi_p\,.$$

Equivalently the connection 1-form ω^π is G-invariant.

Then the horizontal lifting map $P \times_X TX \longrightarrow TP$ is G-equivariant. We write $\overline{v}|_p$ for the *horizontal lift* of $v \in TX$ to $p \in P$. From the definitions it follows that

$$\overline{gv}|_{gp} = g \cdot \overline{v}|_p, \qquad \overline{v}|_{pa} = \overline{v}|_p \cdot a \tag{6}$$

for all $p \in P$, $g \in G$ and $a \in A$.

DEFINITION 1.5. $\mathscr{A}^G(\pi)$ is the *space of G-connections* on π. When $G = \{1\}$ we simply write $\mathscr{A}(\pi)$ for the space of connections on π.

REMARK 1.6. The space $\mathscr{A}^G(\pi)$ may be empty. For example, if A is a Lie subgroup of a Lie group G and π is the principal A-bundle $G \longrightarrow G/A$, consider the left-translation actions of G on G and G/A. Then there is no G-connection if the short exact sequence

$$0 \longrightarrow \mathfrak{a} \longrightarrow \mathrm{Lie}(G) \longrightarrow \mathrm{Lie}(G)/\mathfrak{a} \longrightarrow 0$$

does not admit an A-equivariant splitting. For example, when when $G = \mathrm{SL}(2, \mathbb{C})$ and A is the Borel subgroup of it defined by lower triangular matrices, then there is no G-connection on the principal A-bundle $G \longrightarrow G/A$.

If $\mathscr{A}^G(\pi)$ is non-empty, then it is an affine space modelled on the vector space $\Omega^1_G(P; \mathfrak{a})^A = \Omega^1_G(X; \mathrm{ad}(P))$, the space of G-invariant 1-forms on X with values in the adjoint bundle.

2. Distortion of equivariant connections

In this section, assume that G acts freely and properly on X, so that the quotient map

$$\sigma : X \longrightarrow \underline{X} := G\backslash X$$

is a principal G-bundle over the manifold $\underline{X}$. Let $\pi : P \longrightarrow X$ be a G-equivariant principal A-bundle, and let $\underline{P} = G\backslash P$. Then the quotient map $\underline{\pi} : \underline{P} \longrightarrow \underline{X}$ is a principal A-bundle and we have a commutative diagram

$$\begin{array}{ccc} P & \xrightarrow{\ \pi\ } & X \\ {\scriptstyle \overline{\sigma}}\big\downarrow & & \big\downarrow{\scriptstyle \sigma} \\ \underline{P} & \xrightarrow[\ \underline{\pi}\ ]{} & \underline{X}. \end{array} \tag{7}$$

2.1. Quotients of connections.

PROPOSITION 2.1. *In the situation of* (7), *given a connection ω^σ on the principal G-bundle $\sigma : X \longrightarrow \underline{X}$ and a G-connection ω^π on the principal A-bundle $P \longrightarrow X$, we get a* quotient connection *$\omega^{\underline{\pi}}$ on the quotient bundle $\underline{\pi} : \underline{P} \longrightarrow \underline{X}$ by setting*

$$H^{\underline{\pi}} := \overline{\sigma}_* \left(H^\pi \cap \pi_*^{-1} H^\sigma \right). \tag{8}$$

PROOF. Since G acts freely on $H^\pi \cap \pi_*^{-1} H^\sigma \subset TP$, it follows that (8) is a subbundle of $T\underline{P}$. It is easy to check that it is both G-equivariant and A-equivariant.

Also

$$\begin{aligned}
&\overline{\sigma}_*^{-1}\left(\underline{\pi}_*^{-1}(0)\cap H^{\underline{\pi}}\right)\\
&=\pi_*^{-1}\sigma_*^{-1}(0)\cap\overline{\sigma}_*^{-1}\overline{\sigma}_*(H^{\pi}\cap\pi_*^{-1}H^{\sigma}) && \text{see (7), (8)}\\
&\subset\pi_*^{-1}\sigma_*^{-1}(0)\cap H^{\pi}\cap\pi_*^{-1}H^{\sigma}\\
&=\pi_*^{-1}(\sigma_*^{-1}(0)\cap H^{\sigma})\cap H^{\pi}\\
&=\pi_*^{-1}(0)\cap H^{\pi}=0 && \text{(horizontal)}
\end{aligned}$$

This shows that $H^{\underline{\pi}}$ has trivial intersection with $\underline{\pi}^{-1}(0)$. The rank is

$$\operatorname{rk}H^{\underline{\pi}}=\operatorname{rk}\underline{\pi}_*\overline{\sigma}_*(H^{\pi}\cap\pi_*^{-1}H^{\sigma})=\operatorname{rk}\sigma_*\pi_*(H^{\pi}\cap\pi_*^{-1}H^{\sigma})=\operatorname{rk}\sigma_*H^{\sigma}=\operatorname{rk}T\underline{X},$$

using the fiberwise injectivity of $\underline{\pi}_*$ on $H^{\underline{\pi}}$ just shown; similarly for π_*,σ_*. This proves that (8) is a complementary subbundle of the vertical bundle $\underline{\pi}_*^{-1}(0)$. □

REMARK 2.2. The quotient connection (8) clearly depends on ω^{σ}, a point that is sometimes problematic in the literature.

Using $\overline{\sigma}: P\longrightarrow\underline{P}$ and the pullback connections of Proposition 1.2 we obtain from a connection on $\underline{\pi}$ a G-connection on π. This determines a map

$$r:\mathscr{A}(\underline{\pi})\longrightarrow\mathscr{A}^{G}(\pi)\,. \tag{9}$$

In particular, the set of G-connections on π is non-empty. Also, for each fixed choice of connection on σ, we get from Proposition 2.1 a quotient connection map

$$q(\omega^{\sigma}):\mathscr{A}^{G}(\pi)\longrightarrow\mathscr{A}(\underline{\pi})\,. \tag{10}$$

PROPOSITION 2.3. *Taking pullback connections* (9) *is a right inverse to taking quotients connection* (10), *so* $q(\omega^{\sigma})\circ r=\mathrm{Id}_{\mathscr{A}(\underline{\pi})}$. *In particular,* $q(\omega^{\sigma})$ *is surjective and* r *is injective.*

PROOF. Putting $H^{\pi}=\overline{\sigma}_*^{-1}H^{\underline{\pi}}$ from Proposition 1.2 into (8) gives:

$$\begin{aligned}
&\overline{\sigma}_*\left(\overline{\sigma}_*^{-1}H^{\underline{\pi}}\cap\pi_*^{-1}H^{\sigma}\right)\\
&=\overline{\sigma}_*\left(\overline{\sigma}_*^{-1}H^{\underline{\pi}}\cap\overline{\sigma}_*^{-1}\overline{\sigma}_*\pi_*^{-1}H^{\sigma}\right) && (\overline{\sigma}_*^{-1}\overline{\sigma}_*A=A+\ker\overline{\sigma}_*)\\
&=H^{\underline{\pi}}\cap\overline{\sigma}_*\pi_*^{-1}H^{\sigma} && (\overline{\sigma}_*\ \text{surjective})\\
&\subset H^{\underline{\pi}}\cap\underline{\pi}_*^{-1}\underline{\pi}_*\overline{\sigma}_*\pi_*^{-1}H^{\sigma}\\
&=H^{\underline{\pi}}\cap\underline{\pi}_*^{-1}\sigma_*\pi_*\pi_*^{-1}H^{\sigma} && \text{see (7)}\\
&=H^{\underline{\pi}}\cap\underline{\pi}_*^{-1}\sigma_*H^{\sigma} && (\pi_*\ \text{surjective})\\
&=H^{\underline{\pi}}\cap\underline{\pi}_*^{-1}T\underline{X} && \text{(horizontal)}\\
&=H^{\underline{\pi}}. && (\underline{\pi}_*\ \text{surjective})
\end{aligned}$$

This implies $q(\omega^{\sigma})\circ r=\mathrm{Id}_{\mathscr{A}(\underline{\pi})}$. □

2.2. Distortion. We denote the Lie algebra of G by $\mathfrak{g}$.

Let ω^{π} be a G-connection on π. If we restrict P to a G-orbit of X we have a canonical connection given by the G-action on P, because the action of G on X is free. This gives us a canonical family of connections parametrized by the orbit space $\underline{X}$, and so determines a gauge potential with respect to ω^{π}, which we call the distortion of the G-connection.

DEFINITION 2.4. Let G act freely and properly on X and let $\pi : P \longrightarrow X$ be a G-equivariant principal A-bundle. The *G-distortion* of a G-connection ω^π on P is the following G-equivariant vector bundle homomorphism over X:

$$\tau(\omega^\pi)\colon \mathfrak{g} \times X \longrightarrow \mathrm{ad}(P), \qquad \tau(\xi, x) := \overline{\xi \cdot x}|_p - \xi \cdot p \tag{11}$$

for all $x \in X$ and $p \in \pi^{-1}(x)$. Here the trivial vector bundle $\mathfrak{g} \times X$ gets the G-action

$$g(\xi, x) = (g\xi g^{-1}, gx)$$

and the adjoint bundle $\mathrm{ad}(P) = (P \times \mathfrak{a})/A$ is equipped with the G-action $g(p, \eta) := (gp, \eta)$.

To explain the distortion is well-defined, note that the right hand side of (11) is vertical. If we replace p by pa we get

$$\overline{\xi \cdot x}|_{pa} - \xi \cdot p \cdot a = \left(\overline{\xi \cdot x}|_p - \xi \cdot p\right) \cdot a,$$

which represents the same vector in $\mathrm{ad}(P) = T^{\mathrm{vert}}P/A$. The G-equivariance property is verified by the calculation $\overline{g\xi g^{-1} \cdot gx}|_{gp} - g\xi g^{-1}gp = g\overline{\xi x}|_p - g\xi \cdot p$.

REMARK 2.5. Alternatively, the G-distortion can be identified with the composition

$$\mathfrak{g} \times P \longrightarrow TP \xrightarrow{-\omega^\pi} \mathfrak{a}\,,$$

where the first map is the infinitesimal action. Thus $-\omega^\pi(\xi \cdot p) = \tau(\omega^\pi)(\xi, x)|_p$. For A abelian, $\mathrm{ad}(P) = X \times \mathfrak{a}$, so the distortion may then be regarded as a homomorphism $\mathfrak{g} \times_G X \longrightarrow \mathfrak{a}$.

The distortion measures how far a G-equivariant connection is from being a pullback:

PROPOSITION 2.6. *Let G act freely and properly on X and let $\pi : P \longrightarrow X$ be a G-equivariant principal A-bundle. Then the pullback map (9) is part of an exact sequence*

$$0 \longrightarrow \mathscr{A}(\underline{\pi}) \xrightarrow{r} \mathscr{A}^G(\pi) \xrightarrow{\tau} \mathrm{Hom}_G\left(T^{\mathrm{vert}}X, \mathrm{ad}(P)\right) \longrightarrow 0. \tag{12}$$

A connection ω^σ on $\sigma : X \longrightarrow \underline{X}$ together with a base-point in $\mathscr{A}(\underline{\pi})$ determines a split of (12), in which case we have a dual split short exact sequence

$$0 \longrightarrow \mathrm{Hom}_G(T^{\mathrm{vert}}X, \mathrm{ad}(P)) \longrightarrow \mathscr{A}^G(\pi) \xrightarrow{q(\omega^\sigma)} \mathscr{A}(\underline{\pi}) \longrightarrow 0\,, \tag{13}$$

using the quotient connection map (10) associated to ω^σ.

Recall here that a sequence of affine spaces

$$0 \longrightarrow U \xrightarrow{f} V \xrightarrow{g} W \longrightarrow 0$$

is short exact if $g \circ f$ is a constant map w_0 and if for any base-point $u_0 \in U$ and $v_0 = f(u_0)$ the corresponding sequence of vector spaces is exact (the image of $g \circ f$ is taken as the base-point in W). This is equivalent to $g \circ f = \mathrm{const}_{w_0}$, f injective, g surjective, and $f(U) = g^{-1}(w_0)$. In particular, the definition is independent of the choice of u_0.

PROOF OF PROPOSITION 2.6. Clearly $\tau \circ r = 0$ and r is injective by Proposition 2.3. It is enough to construct an affine linear map

$$r' : \operatorname{Hom}_G(X \times \mathfrak{g}, \operatorname{ad}(P)) \longrightarrow \mathscr{A}^G(\pi)$$

satisfying $\tau \circ r' = 1$. For this we pick a connection on $X \longrightarrow \underline{X}$ and fix an element $\omega^{\underline{\pi}} \in \mathscr{A}(\underline{\pi})$. Then $\mathscr{A}^G(\pi) = \Omega^1_G(X; \operatorname{ad}(P)) + \omega^\pi$ for $\omega^\pi := r(\omega^{\underline{\pi}})$. Using the connection on X we may extend maps on $T^{\mathrm{vert}}X$ by zero on H^σ:

$$r' \colon \operatorname{Hom}_G(X \times \mathfrak{g}, \operatorname{ad}(P)) \xrightarrow{\text{extend}} \operatorname{Hom}_G(TX, \operatorname{ad}(P)) \xrightarrow{+\omega^\pi} \mathscr{A}^G(\pi)\,.$$

Using $\tau(\omega^\pi) = \tau(r(\omega^{\underline{\pi}})) = 0$ we get $\tau \circ r' = 1$. □

3. Central extensions

3.1. Connections on reductions. Consider a Lie group central extension

$$1 \longrightarrow A \xrightarrow{\alpha} B \xrightarrow{\beta} C \longrightarrow 1. \tag{14}$$

The Lie algebras of B and C will be denoted by $\mathfrak{b}$ and $\mathfrak{c}$ respectively. Let G be a Lie group and $\gamma : G \longrightarrow B$ a group homomorphism (mostly the trivial homomorphism). By composing with the adjoint representation we get an induced G-action on $\mathfrak{b}$. Similarly, using $\beta \circ \gamma$ we get an action of G on $\mathfrak{c}$ for which the homomorphism $\beta_* : \mathfrak{b} \longrightarrow \mathfrak{c}$ becomes G-equivariant.

PROPOSITION 3.1. *Let G be a Lie group acting on a manifold P, $\gamma : G \longrightarrow B$ a homomorphism, and $\sigma : R \longrightarrow P$ a G-equivariant A-bundle. Define a G-action on $R \times^A B$ by $g[r, b] := [gr, \gamma(g)b]$ for $[r, b] \in R \times^A B$ and $g \in G$. Then we have an exact sequence*

$$0 \longrightarrow \mathscr{A}^G(\sigma) \longrightarrow \mathscr{A}^G(R \times^A B) \xrightarrow{q} \Omega^1_G(P; \mathfrak{c}), \tag{15}$$

using G-equivariant 1-forms $TP \longrightarrow \mathfrak{c}$ and where the map q is constructed in (18) *below. If β_* admits a G-equivariant section $\mathfrak{c} \longrightarrow \mathfrak{b}$, then q is surjective and a choice of section canonically determines a split of the sequence* (15).

In particular, the sequence always splits when G is compact (for example $G = 1$), or when $G = B$, $\gamma = \mathrm{id}_B$ and B is a reductive Lie group, by which we mean that every real B-representation is completely reducible.

PROOF OF PROPOSITION 3.1. On the level of tangent bundles, the principal A-bundle $\hat{\sigma} \colon R \times B \longrightarrow R \times^A B$ induces an exact sequence

$$0 \longrightarrow \operatorname{ad}(\hat{\sigma}) \longrightarrow \frac{TR \times TB}{A} \longrightarrow T(R \times^A B) \longrightarrow 0,$$

we see that connection 1-forms ω on $R \times^A B$ correspond via

$$\omega(v + \xi) = b^{-1}\omega_0(v)b + b^{-1}\xi \qquad \forall\, v \in T_rR,\ \xi \in T_bB \tag{16}$$

to 1-forms $\omega_0 \in \Omega^1(R; \mathfrak{b})$ satisfying

$$\omega_0(r\eta) = \eta, \qquad\qquad \omega_0(va) = a^{-1}\omega_0(v)a \tag{17}$$

for all $r \in R$, $v \in T_rR$, $a \in A$ and $\eta \in \mathfrak{a}$. Moreover, ω is G-invariant precisely when

$$\omega_0 : TR \longrightarrow \mathfrak{b}$$

is G-equivariant. Hence $\mathscr{A}^G(\sigma) \subset \mathscr{A}^G(R \times^A B)$ is the subspace of $\mathfrak{a}$-valued forms. We may define

$$\sigma^* q(\omega) := \omega_0 \mod \mathfrak{a}, \tag{18}$$

since by (17) the form ω_0 is horizontal and A-invariant, and $\mathscr{A}^G(\sigma)$ is the kernel of q.

Finally, suppose we have a G-equivariant section of β_*. Then we find a G-equivariant bundle homomorphism $R \times^A \mathfrak{c} \longrightarrow R \times^A \mathfrak{b}$ that is a section of $R \times^A \beta_*$. Let $\omega^\sigma \in \mathscr{A}^G(\sigma) \subset \mathscr{A}^G(R \times^A B)$. Then we may identify $\mathscr{A}^G(R \times^A B)$ with $\Omega^1_G(P; \mathfrak{b})$ and a section of q is provided by

$$\mathrm{Hom}_G(TP, \mathfrak{c}) \xrightarrow{s_*} \mathrm{Hom}_G(TP, \mathfrak{b}) \xrightarrow{+\omega^\sigma} \mathscr{A}^G(R \times^A B).$$

This also shows surjectivity of q. □

3.2. Central lifting problems. Throughout, let G be a Lie group acting on a manifold X and fix a central extension of Lie groups as in (14). Given a C-bundle P, we are interested in the lifting problem of whether the structure group may be reduced along β as follows:

DEFINITION 3.2. Let $\pi \colon P \longrightarrow X$ be a G-equivariant principal C-bundle. An *equivariant central extension* of $\pi \colon P \longrightarrow X$ is a pair (ρ, σ) of a G-equivariant principal B-bundle

$$\rho \colon R \longrightarrow X$$

and a G-equivariant B-homomorphism

$$\sigma \colon R \longrightarrow P.$$

(here we use β to convert the C-action on P into a B-action.)

In particular, the principal C-bundles R/A and P are isomorphic via σ. The data in Definition 3.2 will be conveniently denoted by $\rho : R \xrightarrow{\sigma} P \xrightarrow{\pi} X$.

REMARK 3.3. This terminology also makes sense in the topological category for a central extension of topological groups (14) in which β is an A-bundle (automatic when B, C are Lie groups). When C is a Lie group, there is no difference between the topological and the smooth category; this follows from the two facts that B inherits a unique smooth structure and every principal bundle for a Lie group has a unique smooth structure (see Müller–Wockel [**7**]). Similarly, in the holomorphic category we can suppose (14) is a central extension of complex Lie groups, meaning that β is a holomorphic A-bundle.

REMARK 3.4. There is also the weaker equivariance assumption that for each $g \in G$ we have $g^* P \cong P$. It may be of interest to study the relationship between these two versions of the problem, both in the smooth and the holomorphic category.

We are interested in putting connections on central extensions. In principle, connections on R and on P are independent, so we impose the following compatibility condition for a connection on R to qualify as a central extension of a given connection on P:

DEFINITION 3.5. Let $\rho \colon R \xrightarrow{\sigma} P \xrightarrow{\pi} X$ be a central extension as in Definition 3.2, so that we are given a G-equivariant principal B-bundle $\rho \colon R \longrightarrow X$ and a G-equivariant B-homomorphism $\sigma\colon R \longrightarrow P$. Suppose $\pi\colon P \longrightarrow X$ is equipped with a connection ω^π. Then a connection ω^ρ on ρ is *admissible* if

$$\beta_*(\omega^\rho) = \sigma^*\omega^\pi. \tag{19}$$

3.3. Special case of free actions. In this section we briefly discuss the simple case when the action of G on X is free. Then one may understand central extensions from the point of view of sheaf cohomology. The freeness implies that the equivariant central extensions correspond bijectively to central extensions of the quotient principal C-bundle $G\backslash P \longrightarrow G\backslash X$. The pullback of the quotient bundle along the projection $X \longrightarrow G\backslash X$ is isomorphic to P, and similarly for R. The central extension induces an exact sequence

$$\cdots \longrightarrow H^1(G\backslash X,\, B) \longrightarrow H^1(G\backslash X,\, C) \xrightarrow{\ \delta\ } H^2(G\backslash X,\, A). \tag{20}$$

Of course, isomorphism classes of C-bundles on $G\backslash X$ correspond bijectively to classes in $H^1(G\backslash X,\, C)$. We have thus shown:

PROPOSITION 3.6. *Let G act freely on X, and let $P \longrightarrow X$ be a G-equivariant C-bundle. There exists an equivariant central extension of P if and only if $\delta[G\backslash P] = 0$, where δ is the coboundary homomorphism in* (20).

If A, B, C are abelian then the space of equivariant central extensions up to an isomorphism, if non-empty, is an affine space modelled on the quotient group $H^1(G\backslash X, A)/H^0(G\backslash X, C)$.

When B is not abelian, the problem of determining which obstruction classes can occur was raised in [**9**, Problem VI].

REMARK 3.7. One may follow a similar approach in the equivariant case using Grothendieck equivariant sheaves. The corresponding sheaf cohomology $H^1_G(X,\, C)$ is related through acyclic G-covers of X to C-bundles (see [**4**, p. 211]). Here a G-cover $\{U_i\}_{i\in I}$ is an open cover with a fixed-point free G-action on I so that $gU_i = U_{gi}$. It is acyclic when $H^k(U_{i_1\cdots i_n},\, C) = 0$, $k > 0$, for ordinary sheaf cohomology.

3.4. Topological classification. We first explain how to classify central extensions of Definition 3.2 in the smooth category, without connections. Continue to assume that G is a Lie group acting on X and fix a central extension of Lie groups as in (14).

DEFINITION 3.8. Let $\pi\colon P \longrightarrow X$ be a G-equivariant C-bundle. The *difference map* of P is

$$\delta_P : P \times_X P \longrightarrow C, \quad z_1\delta_P(z_1, z_2) := z_2.$$

The map δ_P is G-invariant for the action of G by $g(z_1, z_2) = (gz_1, gz_2)$ on $P \times_X P$. Pullback of β along δ_p defines a G-equivariant A-bundle Q_P, called the *difference bundle* of P:

$$\begin{array}{ccc} Q_P & \longrightarrow & B \\ \big\downarrow & & \big\downarrow \beta \\ P \times_X P & \xrightarrow[\delta_P]{} & C \end{array} \tag{21}$$

The product in B determines the *multiplication map*

$$m\colon Q_P \times_P Q_P \longrightarrow Q_P, \quad m\big((p_1, p_2, b), (p_2, p_3, b')\big) = (p_1, p_3, bb'). \tag{22}$$

The data of Definition 3.8 is a G-equivariant version of the lifting A-gerbe in [**8**], see also Proposition 4.4, which is trivial (in the sense of the following proposition) if and only if a solution to the central lifting problem of Definition 3.2 may be found. This result is well known, but we wish to fix notation and give proof that is particularly convenient for studying connections.

PROPOSITION 3.9. *Let $\pi\colon P \longrightarrow X$ be a G-equivariant C-bundle. Let Q_P be the difference bundle of P with its multiplication map m. Up to an isomorphism, G-equivariant central extensions $\rho\colon R \xrightarrow{\sigma} P \xrightarrow{\pi} X$ of π correspond to pairs (σ, ψ) consisting of a G-equivariant A-bundle $\sigma\colon R \longrightarrow P$ and a G-equivariant A-homomorphism*

$$\psi\colon R \times^A R \longrightarrow Q_P \tag{23}$$

over $P \times_X P$ satisfying the cocycle identity

$$m(\psi[r_1, r_2], \psi[r_2, r_3]) = \psi[r_1, r_3] \qquad \forall (r_1, r_2, r_3) \in R \times_X R \times_X R. \tag{24}$$

REMARK 3.10.

(i) In the Borel construction in (23) the right factor of R is equipped with the left A-action through inversion. So in right hand notation $[r_1, r_2] = [r_1 a, r_2 a]$ for $a \in A$ and the fiber $(R \times^A R)_{(p_1, p_2)}$ is the set of A-maps $R_{p_1} \longrightarrow R_{p_2}$.

(ii) Composing ψ with the map $Q_P \longrightarrow B$ from (21) we may equivalently view ψ as a G-invariant map $R \times_X R \longrightarrow B$ (satisfying some additional properties) for which equation (24) means $\psi(r_1, r_2) \cdot \psi(r_2, r_3) = \psi(r_1, r_3)$.

PROOF OF PROPOSITION 3.9. Suppose first that $\rho\colon R \xrightarrow{\sigma} P \xrightarrow{\pi} X$ is a solution of the extension problem. Then we get the difference map $\psi : R \times_X R \longrightarrow B$ with $r_1 \psi(r_1, r_2) = r_2$. Since A is central, ψ factors through the diagonal action of A giving (23) satisfying (24).

Conversely, suppose an A-bundle $\sigma : R \longrightarrow P$ and ψ as in (23) are given satisfying (24). We must construct a B-action on R. For this we first define a C-action on the bundle $\xi : R \times^A B \longrightarrow P$, where $\xi[r, b] = \sigma(r)$. Given $[r_1, b_1] \in R \times^A B$ and $c \in C$, choose $b_c \in B$ with $\beta(b_c) = c$. Setting

$$p_1 := \sigma(r_1)\,, \qquad p_2 := p_1 c\,,$$

the element $(p_1, p_2, b_c) \in Q_P$ corresponds under ψ to a pair (r_1, r_2), normalized under the A-action by the choice of r_1. Define the action of $c \in C$ by

$$[r_1, b_1] \cdot c := [r_2, b_c^{-1} b_1]. \tag{25}$$

This definition is independent of the choice of lift b_c. We have

$$\xi([r_1, b_1] \cdot c) = \xi[r_1, b_1] \cdot c,$$

since $p_1 \cdot c = p_2$, and the right C-action on $R \times^A B$ commutes with the obvious right B-action. Hence ξ is a C-equivariant B-bundle. The quotient

$$\begin{array}{ccc} R \times^A B & \longrightarrow & (R \times^A B)/C \\ {\scriptstyle \xi}\downarrow & & \downarrow{\scriptstyle \xi/C} \\ P & \xrightarrow{\ \pi\ } & X = P/C. \end{array} \tag{26}$$

is a B-bundle ξ/C that can be identified with $\rho\colon R \longrightarrow X$ via the A-map

$$R \longrightarrow (R \times^A B)/C, \quad r \longmapsto [r, 1]. \tag{27}$$

This allows us to transport the B-operation from $(R \times^A B)/C$ to R. Under the identification $\sigma(r) = \xi[r, 1]$ and so σ is a B-homomorphism, as required in Definition 3.2. □

REMARK 3.11. When A is central and R has a B-action, there is an additional B-action on $R \times_A B$ given by $[r_1, b_1] \star b := [r_1 b, b^{-1} b_1]$ which is trivial on A. In this case $R \times^A B$ becomes a C-equivariant B-bundle.

3.5. Connections on central extensions. With the topological classification out of the way, we now study connections on central extensions of principal C-bundles. For this we shall assume the existence of an $\mathrm{Ad}\,|_B$-equivariant section s of $\beta_*\colon \mathfrak{b} \longrightarrow \mathfrak{c}$. Let

$$0 \longrightarrow \mathfrak{c} \xrightarrow{\ s\ } \mathfrak{b} \xrightarrow{\ t\ } \mathfrak{a} \longrightarrow 0 \tag{28}$$

be the dual short exact sequence, characterized by the formula $1 = \alpha t + s\beta$. Then t is also $\mathrm{Ad}\,|_B$-invariant. We may regard s as a connection on the bundle $\beta\colon B \longrightarrow C$. This yields pullback connections ω^{Q_P} in (21) on all of the difference bundles Q_P of Definition 3.8.

THEOREM 3.12. *Assume an* $\mathrm{Ad}\,|_B$*-equivariant splitting* (28) *of a central extension of Lie groups as in* (14)*. Let* G *be a Lie group acting on a manifold* X *and let* $\pi : P \longrightarrow X$ *be a* G*-equivariant* C*-bundle equipped with* G*-equivariant connection* ω^π*. Let* Q_P *denote the difference bundle of* P *with its pullback connection* ω^{Q_P}*. Up to an isomorphism, every* G*-equivariant central extension* $\rho : R \xrightarrow{\ \sigma\ } P \xrightarrow{\ \pi\ } X$ *of* π *with admissible* G*-equivariant connection* ω^ρ *corresponds to a triple* $(\sigma, \omega^\sigma, \psi)$ *consisting of a* G*-equivariant* A*-bundle*

$$\sigma\colon R \longrightarrow P$$

with G*-equivariant connection* ω^σ *and a* G*-equivariant* A*-homomorphism*

$$\psi\colon R \times^A R \longrightarrow Q_P$$

over $P \times_X P$ *satisfying the cocycle identity* (24) *for* ψ *and also the equation*

$$\mathrm{pr}_2^* \,\omega^\sigma - \mathrm{pr}_1^* \,\omega^\sigma = \psi^* \omega^{Q_P}. \tag{29}$$

Theorem 3.12 will be a direct consequence of the more general Theorem 3.15 below. For greater clarity, we shall restrict to $G = \{1\}$, but the arguments apply in general. Given a central extension $\rho\colon R \xrightarrow{\sigma} P \xrightarrow{\pi} X$, consider from (26) and (27) the diagram

$$\begin{array}{ccc} R \times^A B & \xrightarrow{\ \xi\ } & P \\ {\scriptstyle \overline{\pi}}\big\downarrow & & \big\downarrow{\scriptstyle \pi} \\ R & \xrightarrow[\ \rho\ ]{} & X. \end{array}$$

This exhibits ρ as the C-quotient of ξ. Here $\overline{\pi}[p,b] = pb$. We may therefore apply Propositions 2.6 and 3.1 (for $\gamma = \mathrm{id}_B$) to get the following exact sequences:

$$\begin{array}{ccccccccc} 0 & \longrightarrow & \mathscr{A}(R \to X) & \longrightarrow & \mathscr{A}^C(R \times^A B \to P) & \xrightarrow{\tau} & \mathrm{Hom}_C(P \times \mathfrak{c}, \mathfrak{b}) & \longrightarrow & 0 \\ & & & & \| & & & & \\ 0 & \longrightarrow & \mathscr{A}^B(R \to P) & \longrightarrow & \mathscr{A}^B(R \times^A B \to P) & \xrightarrow{q} & \Omega^1_B(P;\mathfrak{c}) & \longrightarrow & 0 \end{array}$$

Once we fix a connection on $P \longrightarrow X$, both sequences split (but recall that the split depends on a choice of a base-point, in $\mathscr{A}(R \to X)$ and $\mathscr{A}^B(R \to P)$ respectively). We describe the following compositions of maps in the diagram:

$$\mathscr{A}(R \to X) \longrightarrow \Omega^1_B(P;\mathfrak{c}),$$
$$\mathscr{A}^B(R \to P) \longrightarrow \mathrm{Hom}_C(P \times \mathfrak{c}, \mathfrak{b}).$$

The first map assigns to ω^ρ the form κ defined by

$$\kappa(v) := \omega^\rho(\hat{v}) \mod \mathfrak{a} \qquad \forall\, v \in T_pP,\ \sigma_*\hat{v} = v\,.$$

For the second map, the image of $\omega^\sigma \in \mathscr{A}^B(R \to P)$ can be computed from its distortion by

$$(p,\xi) \mapsto \tau(\omega^\sigma)(p,\hat{\xi}) + \hat{\xi} \qquad \forall\, \xi \in \mathfrak{c},\ \beta_*\hat{\xi} = \xi\,.$$

Moreover, the connection H^π on $P \longrightarrow X$ determines an obvious map

$$\mathscr{A}^B(R \to P) \longrightarrow \mathscr{A}(R \to X), \qquad H^\rho_r := \sigma^{-1}_{*,r}\left(H^\pi_{\sigma(r)}\right) \cap H^\sigma_r,$$

which is also obtained from the diagram by using the split. Finally, use the $\mathrm{Ad}\,|_B$-equivariant section s of β_* (this is simply a B-biequivariant connection on the A-bundle $\beta : B \longrightarrow C$, where B acts by multiplication on either side and similarly on C via β) to get

$$\mathscr{A}(R \to X) \cong \ker\left(\mathscr{A}^B(R \to P) \oplus \Omega^1_B(P;\mathfrak{c}) \longrightarrow \mathrm{Hom}_C(P \times \mathfrak{c}, \mathfrak{b})\right).$$

It is useful to make this isomorphism explicit:

Proposition 3.13. *There is a bijection*

$$(30) \quad \mathscr{A}(R \to X) \xrightarrow{\cong} \ker\left(\mathscr{A}^B(R \to P) \xrightarrow{T} \mathrm{Hom}(P \times_C \mathfrak{b}, \mathfrak{a})\right) \oplus \mathscr{A}(P \to X)$$

Here $T(\omega_A) = \tau + t \circ \mathrm{pr}_2$ for $\omega^\sigma \in \mathscr{A}^B(R \to P)$ is essentially the distortion of ω^σ, namely

$$T(\omega_A)(p,\xi_B) = t(\xi_B) - \omega_A(r \cdot \xi_B), \qquad \forall\, r \in R : \sigma(r) = p, \xi_B \in \mathfrak{b}.$$

Proof. $\mathscr{A}(R \to X)$ consists of connection 1-forms $\omega^\rho \in \Omega^1(R;\mathfrak{b})$ satisfying

i) $\omega^\rho\colon TR \longrightarrow \mathfrak{b}$ is B-equivariant,
ii) $\omega^\rho(r \cdot \xi_B) = \xi_B$ for all $\xi_B \in \mathfrak{b}$.

The right hand side of (30) is bijective to all pairs $(\omega^\sigma, \hat{\omega}^\rho) \in \Omega^1(R;\mathfrak{a}) \times \Omega^1(R;\mathfrak{c})$ satisfying:

i) ω^σ B-invariant
ii) $\omega^\sigma(r \cdot \xi_A) = \xi_A$ for all $\xi_A \in \mathfrak{a}$
iii) $T(\omega^\sigma) = 0$
iv) $\hat{\omega}^\rho\colon TR \longrightarrow \mathfrak{c}$ is B-equivariant
v) $\hat{\omega}^\rho(r \cdot \xi_A) = 0$ for all $\xi_A \in \mathfrak{a}$

vi) $\hat{\omega}^\rho(p\cdot\xi_C)=\xi_C$ for all $\xi_C\in\mathfrak{c}$.

This is because from iv) and v) we see that $\hat{\omega}^\rho=\sigma^*\omega^\pi$ is the pullback of a unique form ω^π in $\Omega^1(P;\mathfrak{c})$. In this notation, the bijection (30) is then given by

$$\omega^\sigma := t(\omega^\rho), \qquad \hat{\omega}^\rho := \beta(\omega^\rho),$$
$$\omega^\rho := \alpha\circ\omega^\sigma + s\circ\hat{\omega}^\rho.$$

It is easy to check that these formulas define maps that are inverse to each other. Also properties i), ii) correspond to properties i)–vi) under these bijections. □

PROPOSITION 3.14. *The kernel of* $\mathscr{A}^B(R\to P)\xrightarrow{T}\mathrm{Hom}(P\times_C\mathfrak{b},\mathfrak{a})$ *can be identified with the set of connections* ω^σ *on* $\sigma\colon R\longrightarrow P$ *with the additional property*

$$\mathrm{pr}_2^*\,\omega^\sigma-\mathrm{pr}_1^*\,\omega^\sigma=\psi^*t(\theta_B)\in\Omega^1(R\times_X R;\mathfrak{a}),$$

where θ_B *denotes the Maurer–Cartan form on* B.

Recall here that $\psi\colon\ R\times_X R\longrightarrow B$ encodes the B-action on R, see Remark 3.10.

PROOF. The map (id_R,ψ) is an inverse diffeomorphism to the action

$$R\times B\longrightarrow R\times_X R,\quad (r,b)\longmapsto(r,rb)\,.$$

Taking derivatives, this means that any $(v_{r_1},w_{r_2})\in T_{(r_1,r_2)}(R_1\times_X R_2)$ may be expressed as

$$w_{r_2}=v_{r_1}\cdot b+r_1\cdot\xi_b\quad\text{for}\quad b\in B,\xi_b\in T_bB. \tag{31}$$

For a B-invariant ω^σ in the kernel of T we then compute at (v_{r_1},w_{r_2})

$$\begin{aligned}\omega^\sigma(w_{r_2})-\omega^\sigma(v_{r_1})&=\omega^\sigma(v_{r_1}\cdot b)+\omega^\sigma(r_1\cdot\xi_b)-\omega^\sigma(v_{r_1})\\&=\omega^\sigma(r_1b\cdot b^{-1}\xi_b)=b^{-1}t(\xi_b).\end{aligned}$$

The converse follows from the same computation

$$\omega^\sigma(v_{r_1}\cdot b)+\omega^\sigma(r_1\cdot\xi_b)-\omega^\sigma(v_{r_1})=b^{-1}t(\xi_b)$$

by setting $v_{r_1}=0$ it follows that $\omega^\sigma\in\ker(T)$ and $\xi_b=0$ proving B-invariance. □

Combining Propositions 3.13 and 3.14 yields:

THEOREM 3.15. *Assume an* $\mathrm{Ad}|_B$*-equivariant splitting* (s,r) *of a central extension of Lie groups* (14) *as in* (28). *Let* G *be a Lie group acting on a manifold* X *and let*

$$\rho\colon R\xrightarrow{\ \sigma\ }P\xrightarrow{\ \pi\ }X$$

be a G*-equivariant central extension. The space* $\mathscr{A}^G(R\to X)$ *of* G*-equivariant connections on* ρ *is parameterized by pairs* $(\omega^\sigma,\omega^\pi)$ *of* G*-equivariant connections on* σ *and* π *satisfying*

$$\mathrm{pr}_2^*\,\omega^\sigma-\mathrm{pr}_1^*\,\omega^\sigma=\psi^*t(\theta_B)\in\Omega^1(R\times_X R;\mathfrak{a}),$$

where θ_B *denotes the Maurer–Cartan form on* B *and where, as in Remark* 3.10, *the map* $\psi\colon\ R\times_X R\longrightarrow B$ *encodes the* B*-action on* R.

4. Lifting gerbes and connective structures

The goal of this section is to interpret the conditions appearing in Theorem 3.12 in the terminology of gerbes. Continue to assume an $\mathrm{Ad}|_B$-equivariant splitting (28) of a central extension of Lie groups as in (14). Let G be a Lie group acting on a manifold X and let $\pi\colon P \longrightarrow X$ be a G-equivariant C-bundle equipped with G-equivariant connection ω^π.

4.1. Background on gerbes.

DEFINITION 4.1. Let A be an abelian Lie group. Let $\pi_i : P_i \longrightarrow X$ for $i = 1, 2$ be principal A-bundles. The *tensor product* $P_1 \otimes_A P_2 := (P_1 \times_X P_2)/A$ is another A-bundle over X, where the group A acts by $(p_1, p_2)a = (p_1 a, p_2 a^{-1})$. Equivalently, form the $(A \times A)$-bundle $P_1 \times_X P_2$ and take the associated A-bundle using the multiplication homomorphism $A \times A \longrightarrow A$. Hence $P_1 \otimes_A P_2 = \Delta^*(P_1 \times^A P_2)$ is the pullback of the Borel construction along the diagonal. Given connections ω^{π_i} on P_i, the *tensor product connection* is

$$\omega[v_{p_1}, w_{p_2}] := \omega^{\pi_1}(v_{p_1}) + \omega^{\pi_2}(v_{p_2}), \quad \forall\ [v_{p_1}, w_{p_2}] \in T_{[p_1,p_2]} P_1 \otimes_A P_2.$$

DEFINITION 4.2. An *A-gerbe* $\mathcal{G}$ on X is given by a fiber bundle $Y \longrightarrow X$ together with a principal A-bundle $Q \longrightarrow Y \times_X Y$ and a 'multiplication' A-bundle homomorphism

$$m\colon \mathrm{pr}_{12}^* Q \otimes_A \mathrm{pr}_{23}^* Q \longrightarrow \mathrm{pr}_{13}^* Q, \quad (f, g) \mapsto m(f, g) = g \circ f \tag{32}$$

which is required to be associative $(h \circ g) \circ f = h \circ (g \circ f)$. A gerbe is *$G$-equivariant* if X, Y, Q are equipped with G-actions and all of the involved maps are G-equivariant. A *connection* or *connective structure* on a gerbe $\mathcal{G}$ is a connection on $Q \longrightarrow Y \times_X Y$ for which (32) is parallel. It is called a *G-equivariant connection* if the G-action on Q is parallel.

REMARK 4.3. One pictures a gerbe $\mathcal{G}$ as a "bundle of groupoids", parameterized by $x \in X$. The category $\mathcal{G}_x$ has objects $\pi^{-1}(x)$. The morphism set in $\mathcal{G}_x$ from y_1 to y_2 is the fiber $\mathrm{Hom}_{\mathcal{G}_x}(y_1, y_2) = Q_{(y_1,y_2)}$. Composition, also sometimes called the multiplication in the groupoid, is induced by m, hence the terminology. The existence of identity morphisms and of inverses are a consequence of the invertibility of m, see Murray [**8**].

4.2. Lifting gerbes.

PROPOSITION 4.4. *Assume an* $\mathrm{Ad}|_B$*-equivariant splitting* (s, r) *of a central extension of Lie groups* (14) *as in* (28). *Every central lifting problem as in Definition* 3.2 *with fixed G-equivariant principal C-bundle* $\pi\colon P \longrightarrow X$ *defines a G-equivariant* lifting A-gerbe $\mathcal{G}_\pi$ *which is naturally equipped with a connective structure.*

PROOF. Using the notation of Definition 3.8, take $Y = P$ and take for Q the difference bundle Q_P defined in (21). The multiplication homomorphism is given by (22). The pullback connection ω^{Q_P} on the difference bundle determines a G-equivariant connection on $\mathcal{G}_\pi$. □

REMARK 4.5. For $A = \mathbb{C}^*$ it has been shown that conversely every gerbe is 'stably' isomorphic to one given by a central lifting problem. Here we recall that as gerbes are a two-categorical notion, the usual notion of strict isomorphism should be weakened. The correct notion is that of stable isomorphism, for which we refer to Murray [**8**, Def. 5.3]. For the proof one first shows that a gerbe is classified by its Dixmier–Douady class in $H^2(X, \underline{\mathbb{C}}^*_X) = H^3(X; \mathbb{Z})$.

DEFINITION 4.6. Let $\mathcal{G} = (Y \to X, Q, m)$ be a gerbe with connection. A *parallel trivialization* of $\mathcal{G}$ is an A-bundle $\rho\colon R \longrightarrow X$ together with a connection-preserving isomorphism $R \times^A R \longrightarrow Q$ over $Y \times_X Y$, compatible with (32) in the obvious way, see Murray [**8**]. When $\mathcal{G}$ is G-equivariant, we require R to have a G-action and the isomorphism to be G-equivariant.

With this terminology, we may reformulate Theorem 3.12 as follows:

THEOREM 4.7. *Assume an* $\mathrm{Ad}\,|_B$*-equivariant splitting* (s, r) *of a central extension of Lie groups* (14) *as in* (28). *Let* $\pi\colon P \longrightarrow X$ *be a* G*-equivariant principal* C*-bundle. Parallel* G*-equivariant trivializations of the lifting gerbe* $\mathcal{G}_\pi$ *of* P *are in one-to-one correspondence with central extensions* $\rho\colon R \xrightarrow{\sigma} P \xrightarrow{\pi} X$ *with admissible* G*-equivariant connections.*

5. Vanishing of Neeb's obstructions

Consider a C-bundle $P \longrightarrow X$ and a central extension of groups (14) with $A = \mathbb{C}^*$. Using crossed modules, Neeb associates in [**9**] a cohomology class $H^3(X, \mathbb{C})$ and shows that its vanishing is necessary for the existence of a central extension of P. It was shown in [**3**] that up to torsion this is the full obstruction. In this section we shall address the question, raised by Neeb [**9**, Problem VI], which obstruction classes may occur, in finite dimensions. This gives a different approach to an argument by Brylinski, see [**2**, Chapter 6].

THEOREM 5.1. *Let* C *be a finite-dimensional, connected Lie group. Then all of Neeb's obstruction classes for smooth principal* C*-bundles vanish.*

For the proof we need two lemmas. Recall that by [**4**] the central extension (14) induces an exact sequence in sheaf cohomology

$$\cdots \longrightarrow H^1(X, B) \longrightarrow H^1(X, C) \xrightarrow{\delta} H^2(X, \mathbb{C}^*).$$

The C-bundle determines a class $[P] \in H^1(X, C)$. We will show that δ takes values in the torsion subgroup of $H^2(X, \mathbb{C}^*)$. This suffices, since:

LEMMA 5.2. *Neeb's obstruction class is the image of* $\delta[P]$ *in*

$$H^2(X, \mathbb{C}^*) \cong H^3(X; \mathbb{Z}) \longrightarrow H^3(X; \mathbb{C})\,.$$

All topological spaces are assumed to be paracompact.

DEFINITION 5.3. [**10**, §14.4] Let G be a topological group. A topological G-principal bundle $EG \longrightarrow BG$ is *universal* if every free G-space E admits a G-map $E \longrightarrow EG$, unique up to G-homotopy. The base space BG is the *classifying space* of G.

For any G there exists a universal G-bundle (e.g. [**10**, §14.4]). It is unique up to G-homotopy equivalence. Applying the definition to the trivial bundle $S^n \times G$ we see that all $[S^n, EG]$ are singletons. Hence EG is (weakly) contractible.

The homotopy of BG can be approximated be the following method. Suppose $E \longrightarrow B$ is a G-bundle with n-connected total space E. Then the long exact sequence of homotopy groups for a fibration shows that $B \longrightarrow BG$ is

$(n+1)$-connected. In particular, $H_*(B;\mathbb{Z}) \cong H_*(BG;\mathbb{Z})$ for $* \le n$. Recall here that a topological space X is *n-connected* if $\pi_0(X) = \pi_1(X) = \cdots = \pi_n(X) = \{0\}$. More generally, a continuous map $f\colon X \longrightarrow Y$ is said to be *n-connected* if $\pi_0(f), \dots, \pi_{n-1}(f)$ are bijective and $\pi_n(f)$ is surjective.

LEMMA 5.4. *For all finite-dimensional connected Lie groups C, the cohomology $H^3(BC;\mathbb{Z})$ is a torsion group.*

PROOF. By [**6**, Theorem 6] the group C deformation retracts onto its maximal compact connected subgroup G. Using the long exact sequence of homotopy groups for fibrations, we see that $BG \longrightarrow BC$ is a weak equivalence, so we may assume $C = G$ is compact. We then have an embedding $\rho : G \longrightarrow O(n)$.

For any $N \in \mathbb{N}$ consider the Stiefel manifold $E = V_n(\mathbb{R}^N)$ with its free G-action by ρ. The long exact sequence of homotopy groups for the fibration $O(k) \hookrightarrow O(k+1) \twoheadrightarrow S^k$ combined with $\pi_0(S^k) = \cdots = \pi_{k-1}(S^k)$ shows that $O(k) \to O(k+1)$ is $(k-1)$-connected. Iterating, we find $O(N-n) \to O(N)$ to be $(N-n-1)$-connected. Applying the long exact sequence of homotopy groups to the fibration $O(N-n) \hookrightarrow O(N) \twoheadrightarrow E$ of the Stiefel manifold implies that E is $(N-n-1)$-connected. Then $E \longrightarrow E/G = B$ is a G-bundle with $N-n-1$-connected total space, so we may apply the remarks preceding the lemma. Since B is a compact manifold and N is arbitrary we conclude that all homology groups $H_n(BG;\mathbb{Z}) \cong H_n(B;\mathbb{Z})$ are finitely generated.

According to Borel [**1**], the rational cohomology $H^*(BG;\mathbb{Q})$ is a polynomial ring on even degree generators. Hence $H^3(BG;\mathbb{Q}) = 0$ and so $H_3(BG;\mathbb{Z})$ is a torsion group by the homological universal coefficient theorem. Now apply the cohomological universal coefficient theorem

$$0 \longrightarrow \operatorname{Ext}^1_{\mathbb{Z}}(H_2(BG;\mathbb{Z}),\mathbb{Z}) \longrightarrow H^3(BG;\mathbb{Z}) \longrightarrow \operatorname{Hom}_{\mathbb{Z}}(H_3(BG;\mathbb{Z}),\mathbb{Z}) \longrightarrow 0.$$

Putting $H_2(BG;\mathbb{Z}) = \mathbb{Z}^a \oplus T$ and $H_3(BG;\mathbb{Z}) = T'$ for the torsion subgroups T, T' we see $H^3(BG;\mathbb{Z}) \cong T$ using the standard properties of $\operatorname{Ext}^1_{\mathbb{Z}}$. □

PROOF OF THEOREM 5.1. We claim that each $v = \delta(u)$ is torsion, where $u = [P] \in H^1(X, C)$ corresponds to some C-bundle P. There is a map $f : X \longrightarrow BC$ with $P \cong f^*[EC]$. By using the property of being natural we have a commutative diagram

$$\begin{array}{ccc} H^1(X, C) & \xrightarrow{\ \delta\ } & H^2(X, \mathbb{C}^*) \\ {\scriptstyle f^*}\uparrow & & \uparrow{\scriptstyle f^*} \\ H^1(BC, C) & \xrightarrow[\ \delta\ ]{} & H^2(BC, \mathbb{C}^*) \end{array}$$

It follows that $v = \delta(u) = \delta f^*[EC] = f^*\delta[EC]$ factors through the torsion group $H^2(BC, \mathbb{C}^*) \cong H^3(BC;\mathbb{Z})$ (exponential sequence and Lemma 5.4). □

REMARK 5.5. When we allow infinite-dimensional Lie groups, the situation is drastically different. For a separable complex Hilbert space $\mathcal{H}$, Kuiper's theorem asserts that the bounded invertible operators $GL(\mathcal{H})$ are contractible. Consider

$$1 \longrightarrow \mathbb{C}^* \longrightarrow GL(\mathcal{H}) \longrightarrow PGL(\mathcal{H}) \longrightarrow 1.$$

Then the long exact sequence of homotopy groups shows that $PGL(\mathcal{H}) = K(\mathbb{Z}, 2)$ and $BPGL(\mathcal{H}) = K(\mathbb{Z}, 3)$. By the exponential sequence we have a diagram

$$\begin{array}{ccc} H^1(X, PGL(\mathcal{H})) & \xrightarrow{\delta} & H^2(X, \mathbb{C}^*) \\ \downarrow & & \downarrow \cong \\ [X, BPGL(\mathcal{H})] & \longrightarrow & H^3(X; \mathbb{Z}) \end{array}$$

with lower horizontal isomorphism. Hence δ is surjective.

References

[1] Armand Borel, *Sur la cohomologie des espaces fibrés principaux et des espaces homogènes de groupes de Lie compacts* (French), Ann. of Math. (2) **57** (1953), 115–207, DOI 10.2307/1969728. MR51508

[2] Jean-Luc Brylinski, *Loop spaces, characteristic classes and geometric quantization*, Modern Birkhäuser Classics, Birkhäuser Boston, Inc., Boston, MA, 2008. Reprint of the 1993 edition. MR2362847

[3] Camille Laurent-Gengoux and Friedrich Wagemann, *Obstruction classes of crossed modules of Lie algebroids and Lie groupoids linked to existence of principal bundles*, Ann. Global Anal. Geom. **34** (2008), no. 1, 21–37, DOI 10.1007/s10455-007-9098-0. MR2415176

[4] Alexander Grothendieck, *Sur quelques points d'algèbre homologique* (French), Tohoku Math. J. (2) **9** (1957), 119–221, DOI 10.2748/tmj/1178244839. MR102537

[5] A. Grothendieck, *A general theory of fibre spaces with structure sheaf*, Number 4, University of Kansas, Department of Mathematics, 1958.

[6] Kenkichi Iwasawa, *On some types of topological groups*, Ann. of Math. (2) **50** (1949), 507–558, DOI 10.2307/1969548. MR29911

[7] Christoph Müller and Christoph Wockel, *Equivalences of smooth and continuous principal bundles with infinite-dimensional structure group*, Adv. Geom. **9** (2009), no. 4, 605–626, DOI 10.1515/ADVGEOM.2009.032. MR2574141

[8] Michael K. Murray, *An introduction to bundle gerbes*, The many facets of geometry, Oxford Univ. Press, Oxford, 2010, pp. 237–260, DOI 10.1093/acprof:oso/9780199534920.003.0012. MR2681698

[9] Karl-Hermann Neeb, *Non-abelian extensions of topological Lie algebras*, Comm. Algebra **34** (2006), no. 3, 991–1041, DOI 10.1080/00927870500441973. MR2208114

[10] Tammo tom Dieck, *Algebraic topology*, EMS Textbooks in Mathematics, European Mathematical Society (EMS), Zürich, 2008. MR2456045

School of Mathematics, Tata Institute of Fundamental Research, Homi Bhabha Road, Mumbai 400005

Email address: `indranil@math.tifr.res.in`

The Mathematical Institute, Radcliffe Observatory Quarter, Woodstock Road, Oxford, OX2 6GG, U.K.

Email address: `upmeier@maths.ox.ac.uk`

Contemporary Mathematics
Volume **766**, 2021
https://doi.org/10.1090/conm/766/15373

Some algebraic and geometric properties of hyper-Kähler fourfold symmetries

Samuel Boissière, Marc Nieper-Wißkirchen, and Kévin Tari

ABSTRACT. We complete the classification of order 5 nonsymplectic automorphisms on hyper-Kähler fourfolds deformation equivalent to the Hilbert square of a K3 surface. We then compute the topological Lefschetz number of natural automorphisms of generalized Kummer fourfolds and we describe the geometry of their fix loci.

1. Introduction

Hyper-Kähler manifolds, or equivalently irreducible holomorphic symplectic (IHS) manifolds, are traditionally labelled by their deformation type, which encodes most of the significant features of the manifold. Currently, four deformation types have been exhibited, named after a representative of each class: the Hilbert scheme of n points on a K3 surface, for any $n \geq 1$; the n-th generalized Kummer variety of an abelian surface, for any $n \geq 2$; the O'Grady variety OG6; the O'Grady variety OG10. The present paper focuses on hyper-Kähler fourfolds of Hilbert and of Kummer type, and the topic of interest concerns the classification of the biholomorphic automorphisms of these manifolds.

As for K3 surfaces, the approach for the classification of automorphisms of IHS manifolds is twofold: first a lattice-theoretical classification of the invariant sublattice and of its orthogonal complement inside the second cohomology space with integer coefficients, endowed with the Beauville–Bogomolov–Fujiki quadratic form; then a description of the fix locus of the automorphism which, whenever it is nonempty, is smooth and has usually several connected components of different dimensions.

In the last years, many authors have contributed to the classification of prime order automorphisms on the two families of IHS manifolds in issue here. We refer to [**7**, §1],[**17**, §1] and references therein for a more detailed introduction of the contribution of each author. So far the classification of order 5 nonsymplectic automorphisms on the family of the Hilbert square of a K3 surface had still remained incomplete due to the lack of an algebraic ingredient (Theorem 2.2). Using this result, in Theorem 3.1 we complete the classification of order 5 nonsymplectic automorphisms on the deformation class of the Hilbert square of a K3 surface.

2020 *Mathematics Subject Classification.* 14C05; 14J50.
Key words and phrases. Integral lattice, isometry, irreducible holomorphic symplectic manifold, generalized Kummer variety, abelian surface, automorphism.

In the second part of this note, we describe the fix loci of *natural* automorphisms on generalized Kummer fourfolds, which are the automorphisms coming from the underlying abelian surface. We first give in Proposition 4.1 a generating formula for the topological Lefschetz number of these automorphisms. A similar formula already appeared in [**9**] but the formula given there is not correct when the automorphism contains a nontrivial translation. We correct the formula here and we give a complete proof, which requires substantial work to take care of the missing factor. As an application, we discuss in Section 4.3 the fix loci of natural automorphisms on generalized Kummer fourfolds, whose action on cohomology has prime order.

2. Prime order isometries of integral lattices

A *lattice* L is a free $\mathbb{Z}$-module equipped with a nondegenerate symmetric bilinear form $\langle\cdot,\cdot\rangle_L$ with integer values. Its *dual lattice* is $L^\vee := \operatorname{Hom}_{\mathbb{Z}}(L,\mathbb{Z})$. Clearly L is a sublattice of $L^\vee$ of equal rank, so the *discriminant group* $D_L := L^\vee/L$ is a finite abelian group, whose order $\operatorname{discr}(L)$ is called the *discriminant* of L. The lattice L is called *unimodular* if $\operatorname{discr}(L) = 1$. A sublattice $M \subset L$ is called *primitive* if the quotient L/M is a free $\mathbb{Z}$-module; it is called *p-elementary* for some prime number p if $D_M \cong (\mathbb{Z}/p\mathbb{Z})^{\oplus a}$ for some positive integer a.

In the sequel, we will use the lattices $E_8(-1)$ and $A_4(-1)$, which are the opposites of the usual positive definite root lattices, the hyperbolic plane U, the rank two lattice H_5 with Gram matrix $\begin{pmatrix} 2 & 1 \\ 1 & -2 \end{pmatrix}$, the lattice $A_4^*(-5)$ which is the dual lattice of A_4 with intersection form multiplied by -5, and the rank one lattice $\langle -2\rangle$ generated by an element of square -2. Notation $\frac{\mathbb{Z}}{p\mathbb{Z}}(q)$, with $q \in \mathbb{Q}/2\mathbb{Z}$, means a finite quadratic form generated by an element of square q; we refer to Nikulin [**21**] for deeper reading.

Let $\varphi \in \mathrm{O}(L)$ be an isometry of L, of prime order p. The invariant subspace:

$$T_\varphi := \ker(\varphi - \mathrm{id}_L),$$

is a primitive sublattice of L, since the restriction of the bilinear form to T_φ is nondegenerate. Its orthogonal complement:

$$S_\varphi := T^\perp,$$

is also a primitive sublattice of L. Denote by ξ_p a primitive p-th root of unity, by $K := \mathbb{Q}(\xi_p)$ the p-th cyclotomic field and by $\mathcal{O}_K = \mathbb{Z}[\xi_p]$ its ring of algebraic integers. It is easy to check that $S_\varphi = \ker(\Phi_p(\varphi))$, where Φ_p is the p-th cyclotomic polynomial, so $S_\varphi \otimes_{\mathbb{Z}} \mathbb{Q}$ is endowed with the structure of a finite-dimensional K-vector space. As a consequence, there exists a nonnegative integer m_φ such that:

$$\operatorname{rank}_{\mathbb{Z}} S_\varphi = \dim_{\mathbb{Q}} S_\varphi \otimes_{\mathbb{Z}} \mathbb{Q} = (p-1)m_\varphi.$$

PROPOSITION 2.1. *Let L be an integral lattice and $\varphi \in \mathrm{O}(L)$ an isometry of prime order p. There exists a nonnegative integer a_φ such that $\frac{L}{T_\varphi \oplus S_\varphi} \cong \left(\frac{\mathbb{Z}}{p\mathbb{Z}}\right)^{\oplus a_\varphi}$ and we have $a_\varphi \leq m_\varphi$.*

It is not difficult to see that $\frac{L}{T_\varphi \oplus S_\varphi}$ is a p-torsion module (see [**6**, Lemma 3.1], [**24**, Lemme 2.9]). The property $a_\varphi \leq m_\varphi$ was first proved in [**1**, Theorem 2.1(c)] in the context of isometries of K3 lattices and then in [**7**, Corollary 3.7] in the context of isometries of hyper-Kähler manifolds of $\mathrm{K3}^{[2]}$-type, but only for $p \notin \{5, 23\}$. The above statement is a very useful generalization of these properties, whose proof first appeared in the third author's PhD thesis [**24**, Théorème 2.18], see also [**17**, Lemma 1.8].

THEOREM 2.2. *Let L be an integral lattice and $\varphi \in \mathrm{O}(L)$ an isometry of prime order $p \neq 2$. The integer $p^{m_\varphi}\operatorname{discr}(S_\varphi)$ is a square.*

This result first appeared in this context in [**6**, Equation (2)], in the special case $m_\varphi = 1$, based on results of Bayer–Flückiger [**3**]. It was generalized in the third author's PhD thesis [**24**, Théorème 2.23] by an induction argument. A more general statement, with a shorter proof, is given by Bayer–Flückiger [**4**, Proposition 5.1]: to apply it to our situation, simply observe that the characteristic polynomial of the restriction of φ to $S_\varphi \otimes_{\mathbb{Z}} \mathbb{Q}$ is $\Phi_p^{m_\varphi}$.

COROLLARY 2.3. *Under the assumptions of Theorem 2.2, if furthermore the lattice L is unimodular, then $\operatorname{discr}(S_\varphi) = p^{a_\varphi}$ and $a_\varphi \equiv m_\varphi \mod (2)$.*

PROOF. If L is unimodular, then $D_{S_\varphi} \cong D_{T_\varphi} \cong \left(\frac{\mathbb{Z}}{p\mathbb{Z}}\right)^{\oplus a_\varphi}$ so $\operatorname{discr}(S_\varphi) = p^{a_\varphi}$ (see [**2**, §I.2]), hence by Theorem 2.2, the number $a_\varphi + m_\varphi$ is even. □

3. Order five symmetries of the Hilbert square of a K3 surface

Let X be a projective irreducible holomorphic symplectic manifold, deformation equivalent to the Hilbert square of a K3 surface. We recall that the second cohomology group with integer coefficients $\mathrm{H}^2(X,\mathbb{Z})$ is a rank 23 lattice for the Beauville–Bogomolov–Fujiki quadratic form, which is isometric to the lattice:

$$L := U^{\oplus 3} \oplus E_8(-1)^{\oplus 2} \oplus \langle -2 \rangle.$$

Let $f \in \mathrm{Aut}(X)$ be a nonsymplectic biholomorphic automorphism of prime order p. It is easy to see that $2 \leq p \leq 23$ (see [**10**, §5.4]). We consider the isometry $\varphi := f^*$ induced by f on the lattice $\mathrm{H}^2(X,\mathbb{Z})$. In the sequel we write m, a, S, T instead of $m_\varphi, a_\varphi, S_\varphi, T_\varphi$. Since the representation:

$$\mathrm{Aut}(X) \to \mathrm{O}(\mathrm{H}^2(X,\mathbb{Z})), \quad f \mapsto (f^*)^{-1}$$

is faithful, nonsymplectic prime order automorphisms are classified by the data of the invariant lattice T of φ and its orthogonal complement S. For $2 \leq p \leq 19$ and $p \neq 5$ the classification is given in [**7**], for $p = 23$ it is given in [**6**]. The study of the case $p = 5$ was not finished in [**7**, Table 2] since it was not proven that the list was complete, see [**7**, Remark 6.1]. The goal of our first main result is to fill this gap: it requires the tools developed in the previous section.

THEOREM 3.1. *Let f be an order five nonsymplectic automorphism acting on an irreducible holomorphic symplectic manifold, deformation equivalent to the Hilbert* *square of a K3 surface. Then its invariant lattice T and its orthogonal complement S are one of the following:*

m	a	S	T
1	1	$U \oplus H_5$	$E_8(-1)^{\oplus 2} \oplus H_5 \oplus \langle -2 \rangle$
2	2	$U \oplus H_5 \oplus A_4(-1)$	$E_8(-1) \oplus H_5 \oplus A_4(-1) \oplus \langle -2 \rangle$
3	1	$U \oplus E_8(-1) \oplus H_5$	$E_8(-1) \oplus H_5 \oplus \langle -2 \rangle$
3	3	$U \oplus H_5 \oplus A_4(-1)^{\oplus 2}$	$H_5 \oplus A_4(-1)^{\oplus 2} \oplus \langle -2 \rangle$
4	2	$U \oplus E_8(-1) \oplus H_5 \oplus A_4(-1)$	$H_5 \oplus A_4(-1) \oplus \langle -2 \rangle$
4	4	$U(5) \oplus E_8(-1) \oplus H_5 \oplus A_4(-1)$	$H_5 \oplus A_4^*(-5) \oplus \langle -2 \rangle$
5	1	$U \oplus E_8(-1)^{\oplus 2} \oplus H_5$	$H_5 \oplus \langle -2 \rangle$
5	3	$U \oplus E_8(-1) \oplus H_5 \oplus A_4(-1)^{\oplus 2}$	$U(5) \oplus \langle -10 \rangle$

This result first appeared in the third author's PhD thesis [**24**, Théorème 3.24] (note that there is a typo there on case $(4, 2)$, that case $(5, 3)$ was missing and that in case $(4, 4)$ we have $U(5) \oplus A_4(-1) \cong U \oplus A_4(-5)^*$, showing that our result matches what is written there; the lattices reproduced here are those of [**7**, Table 2]).

PROOF. We start by dressing a list of all possible values of the pairs of invariants (m, a). For this, we use several results from various other sources to narrow down the list of possibilities. By [**10**, Lemma 5.5], the discriminant group of the lattice S is necessarily $D_S \cong (\mathbb{Z}/5\mathbb{Z})^{\oplus a}$, so Theorem 2.2 gives the relation:

$$a \equiv m \mod (2).$$

By Proposition 2.1, we have $a \leq m$ and $4m = \operatorname{rank}_{\mathbb{Z}} S \leq 23$, so $1 \leq m \leq 5$ and moreover, by [**10**, Proposition 4.13] we have also $a \leq 23 - 4m$. This gives a first list of possible pairs (m, a):

$$(1, 1), (2, 0), (2, 2), (3, 1), (3, 3), (4, 0), (4, 2), (4, 4), (5, 1), (5, 3).$$

Again by [**10**, Lemma 5.5], the lattice S has signature $(2, 4m - 2)$, the lattice T is hyperbolic and $D_T \cong \mathbb{Z}/2\mathbb{Z} \oplus D_S$. Since:

$$\operatorname{rank}_{\mathbb{Z}} S = 4m \geq 3 + m \geq 3 + a,$$

we can apply Nikulin's results on orthogonal decomposition of lattices [**21**, Corollary 1.13.5]: the lattice S decomposes as $S = U \oplus S'$ for some 5-elementary lattice S' of signature $(1, 4m - 3)$ and discriminant group $(\mathbb{Z}/5\mathbb{Z})^{\oplus a}$. Applying the classification theorem of Rudakov–Shafarevich [**23**, Section 1], we get that such a lattice S' exists for each value of (m, a) listed aboved, except $(2, 0)$ and $(4, 0)$, and that this lattice is unique with these invariants if $m \geq 2$. For $m = 1$, uniqueness comes from the classification of binary quadratic forms [**15**, Table 15.2a, p. 362]. The genus of the lattice T, which is the orthogonal complement of S for its unique embedding in $\mathrm{H}^2(X, \mathbb{Z})$, is characterized by [**21**, Proposition 1.15.1]. In each case but one, the isometry class of T is determined by [**15**, Corollary 22, p. 395].

The only missing case is $(m, a) = (5, 3)$. The lattice S is isometric to:

$$U \oplus E_8(-1) \oplus H_5 \oplus A_4(-1)^{\oplus 2},$$

so the discriminant form of S is isomorphic to:

$$\frac{\mathbb{Z}}{5\mathbb{Z}}\left(\frac{2}{5}\right) \oplus \frac{\mathbb{Z}}{5\mathbb{Z}}\left(\frac{-4}{5}\right) \oplus \frac{\mathbb{Z}}{5\mathbb{Z}}\left(\frac{-4}{5}\right).$$

To prove that this lattice admits a primitive embedding in $L \cong \mathrm{H}^2(X,\mathbb{Z})$, we apply Nikulin's results on primitive embeddings [**21**, Proposition 1.15.1], here in a quite easy situation. By [**7**, Proposition 2.7 and its proof], the existence of a primitive embedding of S in L is equivalent to the existence of a lattice whose signature and discriminant form are those of an orthogonal complement of S in L, here a lattice of signature $(1,2)$ and discriminant form isomorphic to:

$$\frac{\mathbb{Z}}{5\mathbb{Z}}\left(\frac{-2}{5}\right)\oplus\frac{\mathbb{Z}}{5\mathbb{Z}}\left(\frac{4}{5}\right)\oplus\frac{\mathbb{Z}}{5\mathbb{Z}}\left(\frac{4}{5}\right)\oplus\frac{\mathbb{Z}}{2\mathbb{Z}}\left(\frac{-1}{2}\right).$$

An easy computation shows that the lattice $U(5)\oplus\langle -10\rangle$ satisfies these requirements, so S admits a primitive embedding in L and, again by [**7**, Proposition 2.7], the embedding is unique up to an isometry of L. □

Using [**1**, Theorem 5.3] we see that each case of the classification but the case $(m,a)=(5,3)$ is geometrically realized by an automorphism of the Hilbert square of a K3 surface, induced by an order five nonsymplectic automorphism of the underlying surface. We refer to [**7**, §6.1] for a geometric description of these automorphisms. Following the terminology of [**7**, Definition 4.1], for X deformation equivalent to the Hilbert square $\Sigma^{[2]}$ of a K3 surface Σ, an automorphism f of X is called *natural* if the pair (X,f) deforms to a pair $(\Sigma^{[2]},\sigma^{[2]})$, where σ is an automorphism of Σ. One of the main questions in the study of automorphisms of this kind of varieties is to find out whether they admit nonnatural automorphisms, meaning that they have nonexpected symmetries beyond those of the geometric objects they are built on. Natural automorphisms are easy to detect from the isometry class of their invariant lattice, whose orthogonal decomposition has a $\langle -2\rangle$ factor. In a similar spirit, following the terminology of [**14, 18**], an automorphism is called *(twisted) induced* if it deforms to an automorphism of a (twisted) moduli space of stable sheaves over a K3 surface, with fixed Mukai vector, where the automorphism comes from the underlying K3 surface. This construction permits to realize geometrically more examples whose existence is proven using lattice theory.

EXAMPLE 3.2. The following geometric realization of the case $(m,a)=(5,3)$ has been communicated to us by Chiara Camere. We refer to [**14**, §2.3] for an overview on moduli spaces of twisted sheaves; we follow the presentation given there. Similar examples are given in [**13**]. By [**1**, §5], there exists a projective K3 surface S with an order five nonsymplectic automorphism σ with invariant lattice isometric to the lattice H_5. Furthermore, denoting by H the generator with square $H^2=2$ of this lattice, we may assume that H is ample. Generically in the moduli space of such pairs (S,σ), the Picard group of S is equal to the invariant lattice of σ, hence isometric to the lattice H_5, and its transcendental lattice $\mathrm{T}(S)$ is thus isometric to:

$$H_5\oplus U\oplus E_8(-1)^{\oplus 2}.$$

Denote by e the first generator of the summand U in this decomposition. The surjective morphism $\alpha\colon \mathrm{T}(S)\to\mathbb{Z}/5\mathbb{Z}$ given by intersecting with e defines an order 5 element in the Brauer group $\mathrm{H}^2(\mathcal{O}_S^*)$ of S, that we still denote α. Consider the Mukai lattice:

$$\mathrm{H}^*(S,\mathbb{Z}):=\mathrm{H}^0(S,\mathbb{Z})\oplus\mathrm{H}^2(S,\mathbb{Z})\oplus\mathrm{H}^4(S,\mathbb{Z}),$$

equipped with the pairing $(r,H,s)\cdot(r',H',s')=H\cdot H'-rs'-r's$. We take the B-field $B:=\frac{e}{5}$ and the twisted Mukai vector $v_B:=(0,H,0)\in\mathrm{H}^*(S,\mathbb{Z})$. The moduli space

$M_{v_B}(S,\alpha)$ of stable α-twisted coherent sheaves on S, with twisted Mukai vector v_B is an IHS manifold deformation equivalent to the Hilbert square of a K3 surface, and σ induces an order five nonsymplectic automorphism f on it. We want to check that this automorphism corresponds to the case $(m,a)=(5,3)$. For this we compute its invariant lattice, which is generically the intersection of $v_B^\perp$ with the sublattice of the Mukai lattice generated by $\operatorname{Pic}(S)$ and the classes $(0,0,1)$ and $(5,e,0)$. An easy computation shows that these last two classes are orthogonal to $\operatorname{Pic}(S)$ and generate a sublattice isometric to $U(5)$ and contained in $v_B^\perp$, so the computation reduces to computing the orthogonal complement of the class H in $\operatorname{Pic}(S)$, which is a lattice isometric to $\langle -10\rangle$. As a consequence, the invariant lattice of the action of σ on $M_{v_B}(S,\alpha)$ is isometric to $U(5)\oplus\langle -10\rangle$ and thus corresponds to the case $(m,a)=(5,3)$.

As a conclusion, the following result shows that not only each case of the classification can be geometrically realized as above, but any variety with such an automorphism deforms equivariantly to those. This answers the open question [**7**, Remark 6.1]:

COROLLARY 3.3. *Every order five nonsymplectic automorphism, acting on an irreducible holomorphic symplectic manifold deformation equivalent to the Hilbert square of a K3 surface is natural if* $(m,a)\neq(5,3)$ *and is twisted induced otherwise.*

PROOF. Assume first that $(m,a)\neq(5,3)$. By [**7**, Corollary 5.7], the action $\varphi := f^*$ of the automorphism on $\mathrm{H}^2(X,\mathbb{Z})$, hence f itself, is uniquely determined by the lattices S and T. By [**8**, Theorem 4.5, Section 5.3, Theorem 5.6, Section 7.1] the moduli spaces of lattice polarized nonsymplectic automorphisms considered here are connected (in the case $(m,a)=(1,1)$ it is a point), and since each case is already realized by an automorphism coming from a K3 surface, this proves the statement.

A similar argument shows that, in the case $(m,a)=(5,3)$, every automorphism deforms to a twisted induced one. The tricky point consists in proving that the lattice S given in the table determines the action of φ, up to conjugation by an isometry of S. For this we use a recent work of Brandhorst–Cattaneo [**11**]. The order five isometry $\varphi=f^*$ acts fixed point free on the lattice S, so its minimal polynomial is Φ_5. The minimal polynomial of $\varphi+\varphi^{-1}$ is thus X^2+X-1, so its action on $S_{\mathbb{R}} := S\otimes_{\mathbb{Z}}\mathbb{R}$ admits an orthogonal decomposition:

$$S_{\mathbb{R}} = S_{\mathbb{R}}(\alpha_1)\oplus S_{\mathbb{R}}(\alpha_2),$$

where $S_{\mathbb{R}}(\alpha_i)$ is the eigenspace of $\varphi+\varphi^{-1}$ for the real eigenvalue $\xi_5^i+\xi_5^{-i}$. Both spaces have dimension 10. Since $S_{\mathbb{R}}$ has signature $(2,18)$, the spaces $S_{\mathbb{R}}(\alpha_i)$ have necessarily signatures $(2,8)$ and $(0,10)$. The negative parts $(8,10)$ define the *signatures* of the $\mathbb{Z}[\xi_5]$-lattice associated to the action of φ on S. Taking φ^2 instead of φ simply permutes the signatures. As explained in [**11**, Proposition 2.10] (here the Steinitz class is trivial), it follows that there exists a unique conjugacy class of subgroups generated by a fixed point free isometry of order five on the lattice S. □

4. Natural automorphisms of Kummer fourfolds

4.1. Generalized Kummer varieties. Let A be a two-dimensional complex torus with origin $0\in A$. Let $A^{(n)}$ be the n-th symmetric power of A, let $s\colon A^{(n)}\to A$ the summation morphism and $\pi\colon A^n\to A^{(n)}$ the quotient morphism. Let $A^{[n]}$ be the Hilbert scheme of n points on A, $\rho\colon A^{[n]} \to A^{(n)}$ the Hilbert–Chow morphism and put $\sigma := s \circ \rho$.

The *n-th generalized Kummer variety of A* is the fibre $A^{[\![n]\!]} := \sigma^{-1}(0)$. It fits into the following cartesian diagram:

$$
\begin{array}{ccc}
A \times A^{[\![n]\!]} & \xrightarrow{\ \nu\ } & A^{[n]} \\
{\scriptstyle p}\downarrow & & \downarrow{\scriptstyle \sigma} \\
A & \xrightarrow[\ \mathfrak{n}\ ]{} & A
\end{array}
\tag{4.1}
$$

where p is the projection onto the first factor, $\mathfrak{n}\colon A \to A$ is multiplication by n and $\nu\colon A \times A^{[\![n]\!]} \to A^{[n]}, (a, \xi) \mapsto a + \xi$ is a Galois covering with Galois group $G_n = \mathrm{T}_n(A)$, the finite abelian group of n-torsion points on A. Here, G_n acts on $A \times A^{[\![n]\!]}$ by $g \cdot (a, \xi) = (a - g, g + \xi)$. The variety $A^{[\![n]\!]}$ is a projective irreducible holomorphic symplectic manifold of dimension $2n - 2$.

4.2. Topological Lefschetz numbers. Let $\psi\colon A \to A$ be an automorphism of the complex manifold A. It decomposes as $\psi(x) = b + h(x) = (t_b \circ h)(x)$ for some $b \in A$, where t_b denotes the translation by b and $h\colon A \to A$ is an isomorphism of complex Lie groups, both uniquely determined. It naturally induces an isomorphism $\psi^{[n]}\colon A^{[n]} \to A^{[n]}$. If we further assume that $b \in \mathrm{T}_n(A)$, the restriction of $\psi^{[n]}$ induces an automorphism $\psi^{[\![n]\!]}\colon A^{[\![n]\!]} \to A^{[\![n]\!]}$.

We are interested in the topological Lefschetz number:

$$L(\psi^{[\![n]\!]}) = \sum_{k \geq 0} (-1)^k \operatorname{tr}\left((\psi^{[\![n]\!]})^* |_{\mathrm{H}^k(A^{[\![n]\!]}, \mathbb{C})} \right).$$

The rough idea (see [**19**] for a similar idea) is to compute it by the relation:

$$L(\psi) L(\psi^{[\![n]\!]}) = L(\psi \times \psi^{[\![n]\!]}).$$

Since $L(\psi)$ may be zero, we replace the Lefschetz number $L(f)$ of an automorphism $f\colon X \to X$ of a projective variety X by the polynomial trace:

$$L(f, q) := \sum_{k \geq 0} (-1)^k \operatorname{tr}\left(f^* |_{\mathrm{H}^k(X, \mathbb{C})} \right) q^k,$$

which is always invertible in $\mathbb{C}[\![q]\!]$. Evaluating at $q = 1$, we recover the usual topological Lefschetz number.

In our situation, since translations on A are homotopic to the identity, they act trivially on the cohomology groups of A, hence $L(\psi, q) = L(h, q)$. But for $n \geq 2$, in general $L(\psi^{[\![n]\!]}) \neq L(h^{[\![n]\!]})$, since although translations by n-torsion points do act trivially on $\mathrm{H}^2(A^{[\![n]\!]}, \mathbb{C})$ (see [**9**, Corollary 5]) they act nontrivially on the whole cohomology space of $A^{[\![n]\!]}$, as observed in [**22**, Theorem 1.3].

The third author noticed that there is a mistake in the proof of [**9**, Proposition 7] giving a formula for the Lefschetz number of natural automorphisms of generalized Kummer varieties. Actually, the formula itself is not always true when the automorphism contains a nontrivial translation. In the sequel we state the correct formula and we give a complete proof.

PROPOSITION 4.1. *Let $\psi \in \mathrm{Aut}(A)$, decomposed as $\psi = t_b \circ h$ with $b \in \mathrm{T}_n(A)$, and $\Psi := \psi^*\colon \mathrm{H}^1(A, \mathbb{C}) \to \mathrm{H}^1(A, \mathbb{C})$. Then:*

$$\sum_{n\geq 0} L(\psi^{[\![n]\!]},q)q^{-2n}t^n = \frac{1}{L(\psi,q)} \sum_{\chi\in(G_n^\vee)^h} \chi(b) \prod_{v\geq 1}\prod_{i=0}^{4} \left(\det(1-(\wedge^i\Psi)q^{i-2}t^{v|\chi|})\right)^{(-1)^{i+1}}.$$

In particular,

$$L(\psi^{[\![n]\!]},q) = \frac{q^{2n}}{L(\psi,q)} \left.\frac{\mathrm{d}^n}{n!\,\mathrm{d}t^n}\right|_{t=0} \sum_{\chi\in(G_n^\vee)^h} \chi(b) \prod_{v\geq 1}\prod_{i=0}^{4} \left(\det(1-(\wedge^i\Psi)q^{i-2}t^{v|\chi|})\right)^{(-1)^{i+1}}.$$

PROOF. The morphism ν is not natural with respect to the actions induced by ψ since $\nu\circ(\psi\times\psi^{[\![n]\!]})\neq\psi^{[n]}\circ\nu$, but it commutes with the action of $h\times\psi^{[\![n]\!]}$ on $A\times A^{[\![n]\!]}$ and of $\psi^{[n]}$ on $A^{[n]}$; this is enough for our purpose since $L(\psi,q)=L(h,q)$. Since ν is a finite morphism, its higher direct images vanish, so the Leray–Serre spectral sequence yields a canonical isomorphism:

$$\mathrm{H}^*(A\times A^{[\![n]\!]},\mathbb{C})\cong \mathrm{H}^*(A^{[n]},\nu_*\underline{\mathbb{C}}),$$

which is compatible with the actions of $h\times\psi^{[\![n]\!]}$ and $\psi^{[n]}$, where the action of $\psi^{[n]}$ on $\nu_*\underline{\mathbb{C}}$ is given at each fiber $\zeta\in A^{[n]}$ by:

$$(\psi^{[n]})^*_\zeta\colon (\nu_*\underline{\mathbb{C}})(\zeta)\to(\nu_*\underline{\mathbb{C}})(\psi^{[n]}(\zeta)),\quad f\mapsto f\circ(h\times\psi^{[\![n]\!]}).$$

Since ν is a Galois covering, it is a classical result that $\nu_*\underline{\mathbb{C}}$ decomposes as a direct sum of character sheaves over the character group $G_n^\vee$ of $G_n=\mathrm{T}_n(A)$:

$$\nu_*\underline{\mathbb{C}} = \bigoplus_{\chi\in G_n^\vee} L_{A^{[n]},\chi}$$

where $L_{A^{[n]},\chi}$ is the locally constant sheaf on $A^{[n]}$ with fibre $\mathbb{C}$, associated to the character χ, whose sections over an open set $U\subset A^{[n]}$ are the continuous functions $f\colon\nu^{-1}(U)\to\mathbb{C}$ (where $\mathbb{C}$ is given the discrete topology) such that $f(g\cdot(a,\xi))=\chi(g)f(a,\xi)$ for all $(a,\xi)\in\nu^{-1}(U)$ and $g\in G_n$. It follows (see also [**12**, Proposition 18]) that:

$$\mathrm{H}^*(A\times A^{[\![n]\!]},\mathbb{C})\cong\bigoplus_{\chi\in G_n^\vee}\mathrm{H}^*(A^{[n]},L_{A^{[n]},\chi}).$$

The direct image $\mathfrak{n}_*\underline{\mathbb{C}}$ decomposes similarly as a direct sum of character sheaves over the character group $G_n^\vee$ of $G_n=\mathrm{T}_n(A)$:

$$\mathfrak{n}_*\underline{\mathbb{C}}=\bigoplus_{\chi\in G_n^\vee}L_{A,\chi}$$

where $L_{A,\chi}$ is the locally constant sheaf on A with fibre $\mathbb{C}$, associated to the character χ. Here we make the group G_n act by substraction on A, so the morphism p is G_n-equivariant. With this convention, the sections of $L_{A,\chi}$ over an open set $U\subset A$ are the continuous functions $f\colon\mathfrak{n}^{-1}(U)\to\mathbb{C}$ such that:

$$f(g\cdot x)=f(x-g)=\chi(g)f(x),$$

for all $x\in\mathfrak{n}^{-1}(U)$ and $g\in G_n$.

Following [**20**], we denote by $\mathcal{L}_\chi$ the locally constant sheaf on $A^{(n)}$ such that $\pi^{-1}\mathcal{L}_\chi\cong L_{A,\chi}\boxtimes\cdots\boxtimes L_{A,\chi}$, and we put $L_\chi^{[n]}:=\rho^{-1}\mathcal{L}_\chi$. We construct, for any $\chi\in G_n^\vee$, an isomorphism of local systems $L_\chi^{[n]}\xrightarrow{\sim}L_{A^{[n]},\chi}$ as follows. Given a point $\zeta=(\zeta_1,\ldots,\zeta_n)\in A^{[n]}$ and functions $f_i\in L_{A,\chi}(\zeta_i)$, $i=1,\ldots,n$ defining a section

$f_1 \otimes \cdots \otimes f_n \in L_\chi^{[n]}(\zeta)$, we associate a function $f\colon \nu^{-1}(\zeta) \to \mathbb{C}$ defining a section $f \in L_{A^{[n]},\chi}(\zeta)$. At a point $(a,\xi) \in \nu^{-1}(\zeta)$, with $\xi = (x_1, \ldots, x_n)$, we have $\zeta = a + \xi$, $\zeta_i = a + x_i$ for all i and $\sum_{i=1}^n x_i = 0$. Choose elements $\bar{a}, \bar{x}_1, \ldots, \bar{x}_{n-1} \in A$ such that $n\bar{a} = a$, $n\bar{x}_i = x_i$ for $i = 1, \ldots, n-1$ and put $\bar{x}_n := -\sum_{i=1}^{n-1} \bar{x}_i \in \mathfrak{n}^{-1}(x_n)$. We define:

$$f(a,\xi) := \prod_{i=1}^n f_i(\bar{a} + \bar{x}_i).$$

It is easy to check that this definition does not depend on the choice of $\bar{a}, \bar{x}_1, \ldots, \bar{x}_{n-1}$, that $f \in L_{A^{[n]},\chi}(\zeta)$, *i.e.*:

$$f\left(g \cdot (a,\xi)\right) = \chi(g) f(a,\xi) - f(a-g, g+\xi) \quad \forall g \in G_n,$$

and that this construction defines an isomorphism $L_\chi^{[n]}(\zeta) \xrightarrow{\sim} L_{A^{[n]},\chi}(\zeta)$.

Recall that $\psi^{[n]}$ acts on $\nu_*\underline{\mathbb{C}}$. It is easy to check that if $f \in L_{A^{[n]},\chi}$ then $(\psi^{[n]})^*(f) = f \circ (h \times \psi^{[\![n]\!]}) \in L_{A^{[n]},\chi\circ h}$. Let us compute the action induced on $\bigoplus_{\chi \in G_n^\vee} L_\chi^{[n]}$ by the isomorphisms $L_\chi^{[n]} \xrightarrow{\sim} L_{A^{[n]},\chi}$. At a point $(a,\xi) \in A \times A^{[\![n]\!]}$, using the same notation as above, we compute:

$$\begin{aligned}
((\psi^{[n]})^* f)(a,\xi) &= f\left(h(a), b + h(\xi)\right) \\
&= f\left(h(a), (b + h(x_1), \ldots, b + h(x_n))\right) \\
&= \prod_{i=1}^{n-1} f_i\left(h(\bar{a}) + \bar{b} + h(\bar{x}_i)\right) \cdot f_n\left(h(\bar{a}) + \bar{b} + h(\bar{x}_n) - b\right)
\end{aligned}$$

where the element $\bar{b} + h(\bar{x}_n) - b \in \mathfrak{n}^{-1}(b + h(x_n))$ is chosen according to the construction explained above. Thus:

$$((\psi^{[n]})^* f)(a,\xi) = \chi(b) \prod_{i=1}^n (f_i \circ \psi)(\bar{a} + \bar{x}_i). \tag{4.2}$$

This computation implies the following. Consider the action of the automorphism ψ on $L_{A,\chi}$ defined at any point $x \in A$ and for any element $f \in L_{A,\chi}(x)$, by the formula $\psi^* f = f \circ \psi$. It is easy to check that, if $f \in L_{A,\chi}(x)$, then $\psi^* f \in L_{A,\chi\circ h}(h^{-1}(x))$. Let us denote by $(\psi^*)^{[n]}$ the action of ψ induced on $L_\chi^{[n]}$, such that:

$$(\psi^*)^{[n]}(f_1 \otimes \cdots \otimes f_n) = \psi^* f_1 \otimes \cdots \otimes \psi^* f_n.$$

Equation (4.2) then shows:

$$(\psi^{[n]})^*(f) = \chi(b)(\psi^*)^{[n]}(f) \in L_{A^{[n]},\chi\circ h} \quad \forall f \in L_{A^{[n]},\chi}.$$

In other words, the action $(\psi^{[n]})^*\colon L_{A^{[n]},\chi} \to L_{A^{[n]},\chi\circ h}$ corresponds through the isomorphisms $L_\chi^{[n]} \xrightarrow{\sim} L_{A^{[n]},\chi}$ to the action $\chi(b)(\psi^*)^{[n]}\colon L_\chi^{[n]} \to L_{\chi\circ h}^{[n]}$.

Using [**20**, Theorem 1.2], we get linear isomorphisms:

$$\begin{aligned}
\mathrm{H}^*(A \times A^{[\![n]\!]}, \mathbb{C}[2n]) &\cong \bigoplus_{\chi \in G_n^\vee} \mathrm{H}^*(A^{[n]}, L_{A^{[n]},\chi}[2n]) \\
&\cong \bigoplus_{\chi \in G_n^\vee} \mathrm{H}^*(A^{[n]}, L_\chi^{[n]}[2n]) \\
&\cong \bigoplus_{\chi \in G_n^\vee} \mathrm{S}^* \left(\bigoplus_{\ell \geq 1} \mathrm{H}^*(A, L_\chi^{\otimes \ell}[2]) \right),
\end{aligned}$$

such that the action of $h \times \psi^{[\![n]\!]}$ on the left hand side corresponds to the action of $\chi(b)(\psi^*)^{[n]}$ on the second line, and acts on the last line by ψ^* on each cohomological group, followed by a multiplication by $\chi(b)$ on each super-symmetric tensor.

An element of $\mathrm{H}^k(A \times A^{[\![n]\!]}, \mathbb{C}[2n])$ has *cohomological degree* $k - 2n$ and *conformal weight* n, an element of $\mathrm{H}^k(A, L_\chi^{\otimes \ell}[2])$ has *degree* $k-2$ and *weight* ℓ. Notation S^* denotes the super-symmetric algebra, where the super-structure concerns only the cohomological degree: the weighting does not interfere with the super-structure. The above isomorphisms respect these bigradings.

As explained in [**20**, p. 768], one has $\mathrm{H}^*(A, L_\chi^{\otimes \ell}[2]) = 0$ unless $L_\chi^{\otimes \ell} = \underline{\mathbb{C}}$. In order to keep track of the weighting, we work in the space of formal series in the parameter t encoding the weight, with coefficients in the space of finitely dimensional graded super-vector spaces. We thus have:

$$\bigoplus_{\ell \geq 1} \mathrm{H}^*(A, L_\chi^{\otimes \ell}[2]) t^\ell \cong \bigoplus_{\ell \geq 1} \mathrm{H}^*(A, \mathbb{C}[2]) t^{\ell |\chi|},$$

where $|\chi|$ denotes the order of χ in $G_n^\vee$. We get finally:

$$\bigoplus_{n \geq 0} \mathrm{H}^*(A \times A^{[\![n]\!]}, \mathbb{C}[2n]) t^n \cong \bigoplus_{\chi \in G_n^\vee} \mathrm{S}^* \left(\bigoplus_{\ell \geq 1} \mathrm{H}^*(A, \mathbb{C}[2]) t^{\ell |\chi|} \right).$$

Let us now compute the formal series of polynomial traces $\sum_{n \geq 0} L(h \times \psi^{[\![n]\!]}, q) t^n$ using this isomorphism. First we observe that the action on the right hand side is induced by the natural action of ψ^*, but it sends the block indexed by χ to the block indexed by $\chi \circ h$. So in the computation of the trace, only those characters which are invariant by h contribute to the trace, we denote them $\chi \in (G_n^\vee)^h$. Once the trace of ψ^* is computed, it is multiplied by $\chi(b)$ on each block χ to take into account our previous computation of the action. Now the trace of ψ^* on each super-symmetric block can be computed using standard techniques for traces of natural operators on bigraded super-vector spaces, for which we refer to [**5**, Section 3] and [**9**, Lemma 6]. Denoting by Ψ the action of ψ^* on $\mathrm{H}^1(A, \mathbb{C})$, and using the fact that $\mathrm{H}^*(A, \mathbb{C}) = \bigwedge^* \mathrm{H}^1(A, \mathbb{C})$, we obtain:

$$\sum_{n \geq 0} L(h \times \psi^{[\![n]\!]}, q) q^{-2n} t^n = \sum_{\chi \in (G_n^\vee)^h} \chi(b) \prod_{v \geq 1} \prod_{i=0}^{4} \left(\det(1 - (\wedge^i \Psi) q^{i-2} t^{v|\chi|}) \right)^{(-1)^{i+1}}.$$

Since $L(h \times \psi^{[\![n]\!]}, q) = L(h, q) \cdot L(\psi^{[\![n]\!]}, q) = L(\psi, q) \cdot L(\psi^{[\![n]\!]})$ we get the expected formula. □

REMARK 4.2. The isomorphisms of local systems $L_\chi^{[n]} \xrightarrow{\sim} L_{A^{[n]},\chi}$ constructed in the proof for any $\chi \in G_n^\vee$ can be understood more conceptually as follows. Since diagram (4.1) is G_n-equivariant, cartesian, and since $\mathfrak{n}$ is finite and unramified, we have an isomorphism of G_n-local systems (see also [**12**, Proposition 18]):

$$\nu_*\underline{\mathbb{C}} \cong \nu_* p^{-1}\underline{\mathbb{C}} \cong \sigma^{-1}\mathfrak{n}_*\underline{\mathbb{C}}.$$

Following these isomorphisms, it is easy to see that $L_{A^{[n]},\chi} \cong \sigma^{-1}L_{A,\chi}$ for all $\chi \in G_n^\vee$. We have $\sigma^{-1}L_{A,\chi} = \rho^{-1}s^{-1}L_{A,\chi}$, and it is easy to see that:

$$s^{-1}L_{A,\chi} \cong \mathcal{L}_\chi,$$

so finally $L_{A^{[n]},\chi} \cong \sigma^{-1}L_{A,\chi} \cong \rho^{-1}\mathcal{L}_\chi = L_\chi^{[n]}$.

As a consequence of Proposition 4.1, evaluating at $q = 1$ we recover the formula stated in [**9**, Proposition 7], this time with the missing factors:

COROLLARY 4.3. *Let $\psi \in \operatorname{Aut}(A)$, decomposed as $\psi = t_b \circ h$ with $b \in \mathrm{T}_n(A)$, and $\Psi := \psi^*\colon \mathrm{H}^1(A, \mathbb{C}) \to \mathrm{H}^1(A, \mathbb{C})$. Then:*

$$L(\psi)L(\psi^{[\![n]\!]}) = \left.\frac{\mathrm{d}^n}{n!\,\mathrm{d}t^n}\right|_{t=0} \sum_{\chi\in(G_n^\vee)^h} \chi(b) \prod_{v\geq 1} \exp\left(\sum_{s\geq 1} \frac{\det(1-\Psi^s)}{s} t^{v|\chi|s}\right).$$

4.3. Application. Automorphisms of a two dimensional complex torus A were classified by Fujiki [**16**]. From this we extract a list of those automorphisms whose action on the second cohomology space $\mathrm{H}^2(A, \mathbb{C})$ has prime order, and we consider the automorphisms induced on the generalized Kummer fourfold $A^{[\![3]\!]}$. To get a complete picture, we consider all automorphisms $\psi^{[\![3]\!]}$ with $\psi = t_b \circ (\pm h)$, where h is a group automorphism of prime order on A and t_b is the translation by a 3-torsion point $b \in A$. As an application of our formula in Proposition 4.1, we give in each case the topological Lefschetz number and we briefly discuss the fix locus. Details on the geometric study can be found in [**24**, Section 1.2]. We refer to [**24**, Lemma 1.13] for a discussion of the fix loci of the iterates of the automorphisms listed below. In complement, and in relation to the first part of this note, we refer to [**17**] for a lattice-theoretical classification of these automorphisms.

4.3.1. *Type 0.* Assume that $h = \mathrm{id}$. Clearly $L(\mathrm{id}^{[\![3]\!]}) = 108$. For $b \neq 0$ we get $L(t_b^{[\![3]\!]}) = 27$, the fix locus consists of 27 points.

We have $L((-\mathrm{id})^{[\![3]\!]}) = 60$, the fix locus consists of a copy of $A^{[\![2]\!]}$ and 36 isolated points; the same result holds for $(-t_b)^{[\![3]\!]}$ if $b \neq 0$.

4.3.2. *Type 1.* The torus $A = E \times E'$ is a product of two elliptic curves and $h = \mathrm{id}_E \times (-\mathrm{id}_{E'})$. We have $L(h^{[\![3]\!]}) = 12$, the fix locus consists of a copy of $E \times \mathbb{P}^1$ blown up in 9 points, three copies of $E^{(2)}$, one copy of $E^{[\![3]\!]}$ and one copy of $E \times E$.

Let $b = (u, v)$. If $u = 0$ then $L(\psi^{[\![3]\!]}) = 12$ and the fix locus is the same as above. If $u \neq 0$, we have $L(\psi^{[\![3]\!]}) = 3$, the fix locus consists of 3 points.

4.3.3. *Type 2.* The torus $A = \frac{E\times E'}{\mathbb{Z}/2\mathbb{Z}}$ is the quotient of $E \times E'$ by the translation by a 2-torsion point (a, a') with $a \in \mathrm{T}_2(E)$, $a' \in \mathrm{T}_2(E')$, both nonzero, and $h = \mathrm{id}_E \times (-\mathrm{id}_{E'})$. We have $L(h^{[\![3]\!]}) = 12$, the fix locus consists of a copy of $A/\langle h\rangle$ blown up in 9 points, one copy of $E^{(2)}$ and one copy of $E^{[\![3]\!]}$.

Let $b = (u, v)$. If $u = 0$ then $L(\psi^{[\![3]\!]}) = 12$ and the fix locus is the same as above. If $u \neq 0$, we have $L(\psi^{[\![3]\!]}) = 3$ and the fix locus consists of 3 points.

4.3.4. *Type 3.* The torus is $A = \frac{E\times E'}{(\mathbb{Z}/2\mathbb{Z})^2}$ where $(\mathbb{Z}/2\mathbb{Z})^2 \to E \times E'$ is any group monomorphism, and $h = \mathrm{id}_E \times (-\mathrm{id}_{E'})$. We have $L(h^{[\![3]\!]}) = 12$, the fix locus consists of a copy of $A/\langle h\rangle$ blown up in 9 points and one copy of $E^{[\![3]\!]}$.

Let $b = (u, v)$. If $u = 0$ then $L(\psi^{[\![3]\!]}) = 12$ and the fix locus is the same as above. If $u \neq 0$, we have $L(\psi^{[\![3]\!]}) = 3$ and the fix locus consists of 3 points.

4.3.5. *Type 4.* The torus is $A = E_4 \times E_4$ where $E_4 = \frac{\mathbb{C}}{\mathbb{Z} \oplus \mathrm{i}\mathbb{Z}}$, and $h = \begin{pmatrix} \mathrm{i} & 0 \\ 0 & \mathrm{i} \end{pmatrix}$. We have $L(h^{[\![3]\!]}) = 16$, the fix locus consists of four copies of $\mathbb{P}^1$ and 8 isolated points. The same result holds when h is composed by a nonzero translation.

4.3.6. *Type 5.* The torus is $A = E \times E_6$ where $E_6 = \frac{\mathbb{C}}{\mathbb{Z} \oplus \zeta_6 \mathbb{Z}}$, with ζ_n is a primitive n-th root of unity, and $h = \begin{pmatrix} 1 & 0 \\ 0 & \zeta_3 \end{pmatrix}$. We have $L(h^{[\![3]\!]}) = 27$, the fix locus consists of nine copies of $\mathbb{P}^1$, one copy of E^2, three copies of $E^{[\![3]\!]}$ and three copies of E.

Let $b = (u, v) \in \mathrm{T}_3(A)$. Denote Δ_6 the image in E_6 of the real line $\mathbb{R}\frac{1+\zeta_6}{3}$. If $u = 0$ and $v \in \Delta_6$, then $L(\psi^{[\![3]\!]}) = 27$, the fix locus consists of nine copies of $\mathbb{P}^1$, one copy of E^2, three copies of $E^{[\![3]\!]}$ and three copies of E. Otherwise $L(\psi^{[\![3]\!]}) = 0$ but the fix locus depends on the translation. If $u = 0$ and $v \notin \Delta_6$, it consists of three copies of $E^{(2)}$ and three copies of E; if $u \neq 0$ and $v \in \Delta_6$, it consists of three copies of E_6; if $u \neq 0$ and $v \notin \Delta_6$ the fix locus is empty: this case already appeared in [**9**, Proposition 8] in the context of the construction of Enriques fourfolds.

We have $L((-h)^{[\![3]\!]}) = 9$, the fix locus consists of one copy of E, one copy of $\mathbb{P}^1$ and 7 isolated points. The same result holds when $(-h)$ is composed by a nonzero translation.

4.3.7. *Type 6.* The torus $A = \frac{E \times E_6}{\mathbb{Z}/3\mathbb{Z}}$ is the quotient of $E \times E_6$ by the translation by a 3-torsion point (a, a') with $a \in \mathrm{T}_3(E) \setminus \{0\}$ and $a' = \frac{1+\zeta_6}{3}$, and $h = \begin{pmatrix} 1 & 0 \\ 0 & \zeta_3 \end{pmatrix}$. We have $L(h^{[\![3]\!]}) = 9$, the fix locus consists of three copies of $\mathbb{P}^1$, one copy of $E^{[\![3]\!]}$ and one copy of E.

Let $b \in \mathrm{T}_3(A)$, it can be written as $b = (u, v) + \frac{t}{3}(a, a')$ with $u \in \mathrm{T}_3(E)$, $v \in \mathrm{T}_3(E_6)$ and $t \in \{0, 1, 2\}$. If $t = 0$ and $u \in \mathbb{Z}a$ then $L(\psi^{[\![3]\!]}) = 9$, the fix locus consists of one copy of $E^{[\![3]\!]}$, three copies of $\mathbb{P}^1$ and one copy of E. Otherwise $L(\psi^{[\![3]\!]}) = 0$ but the fix locus depends on the translation. If $t = 0$ and $u \notin \mathbb{Z}a$, it consists of one copy of E_6, if $t \neq 0$ the fix locus is empty.

We have $L((-h)^{[\![3]\!]}) = 9$, the fix locus consists of one copy of $\mathbb{P}^1$ and 7 isolated points. The same result holds when $(-h)$ is composed by a nonzero translation.

4.3.8. *Type 7.* The torus is $A = E_6 \times E_6$ and $h = \begin{pmatrix} \zeta_3 & 0 \\ 0 & \zeta_3 \end{pmatrix}$. Here $L(h^{[\![3]\!]}) = 36$, the fix locus consists of a copy of the minimal resolution of $A/\langle h \rangle$ and 21 isolated points.

Let $b = (u, v)$. If $b \in \Delta_6 \times \Delta_6$, then $L(\psi^{[\![3]\!]}) = 36$ and the fix locus is the same as above, otherwise $L(\psi^{[\![3]\!]}) = 27$ and it consists of nine copies of $\mathbb{P}^1$ and 9 isolated points.

We have $L((-h)^{[\![3]\!]}) = 12$, the fix locus consists of one copy of $\mathbb{P}^1$ and 10 isolated points. The same result holds when $(-h)$ is composed by a nonzero translation.

4.3.9. *Type 8.* The torus is $A = \mathbb{C}^2/\Lambda$ where Λ is the lattice generated by the four vectors:

$$(1, 1), (\zeta_5, \zeta_5^2), (\zeta_5^2, \zeta_5^4), (\zeta_5^3, \zeta_5),$$

and $h = \begin{pmatrix} \zeta_5 & 0 \\ 0 & \zeta_5^2 \end{pmatrix}$. We have $L(h^{[\![3]\!]}) = 13$, the fix locus consists of one copy of $\mathbb{P}^1$ and 11 isolated points. The same result holds when h is composed by a nonzero translation.

We have $L((-h)^{[\![3]\!]}) = 5$, the fix locus consists of 5 points. The same result holds when $(-h)$ is composed by a nonzero translation.

Acknowledgments

The authors warmly thank Simon Brandhorst, Chiara Camere, Paolo Menegatti and Alessandra Sarti for helpful comments, enlightening discussions and precious help during the preparation of this work. We express all our gratitude to the anonymous referee for their very careful reading and valuable observations.

References

[1] Michela Artebani, Alessandra Sarti, and Shingo Taki, *K3 surfaces with non-symplectic automorphisms of prime order*, Math. Z. **268** (2011), no. 1-2, 507–533, DOI 10.1007/s00209-010-0681-x. With an appendix by Shigeyuki Kondō. MR2805445

[2] Thomas Kappeler and Jürgen Pöschel, *KdV & KAM*, Ergebnisse der Mathematik und ihrer Grenzgebiete. 3. Folge. A Series of Modern Surveys in Mathematics [Results in Mathematics and Related Areas. 3rd Series. A Series of Modern Surveys in Mathematics], vol. 45, Springer-Verlag, Berlin, 2003. MR1997070

[3] Eva Bayer-Fluckiger, *Ideal lattices*, A panorama of number theory or the view from Baker's garden (Zürich, 1999), Cambridge Univ. Press, Cambridge, 2002, pp. 168–184, DOI 10.1017/CBO9780511542961.012. MR1975451

[4] Eva Bayer-Fluckiger, *Isometries of quadratic spaces*, J. Eur. Math. Soc. (JEMS) **17** (2015), no. 7, 1629–1656, DOI 10.4171/JEMS/541. MR3361725

[5] Samuel Boissière, *Automorphismes naturels de l'espace de Douady de points sur une surface* (French, with French summary), Canad. J. Math. **64** (2012), no. 1, 3–23, DOI 10.4153/CJM-2011-041-5. MR2932167

[6] Samuel Boissière, Chiara Camere, Giovanni Mongardi, and Alessandra Sarti, *Isometries of ideal lattices and hyperkähler manifolds*, Int. Math. Res. Not. IMRN **4** (2016), 963–977, DOI 10.1093/imrn/rnv137. MR3493439

[7] Samuel Boissière, Chiara Camere, and Alessandra Sarti, *Classification of automorphisms on a deformation family of hyper-Kähler four-folds by p-elementary lattices*, Kyoto J. Math. **56** (2016), no. 3, 465–499, DOI 10.1215/21562261-3600139. MR3542771

[8] Samuel Boissière, Chiara Camere, and Alessandra Sarti, *Complex ball quotients from manifolds of $K3^{[n]}$-type*, J. Pure Appl. Algebra **223** (2019), no. 3, 1123–1138, DOI 10.1016/j.jpaa.2018.05.017. MR3862667

[9] Samuel Boissière, Marc Nieper-Wißkirchen, and Alessandra Sarti, *Higher dimensional Enriques varieties and automorphisms of generalized Kummer varieties* (English, with English and French summaries), J. Math. Pures Appl. (9) **95** (2011), no. 5, 553–563, DOI 10.1016/j.matpur.2010.12.003. MR2786223

[10] Samuel Boissière, Marc Nieper-Wißkirchen, and Alessandra Sarti, *Smith theory and irreducible holomorphic symplectic manifolds*, J. Topol. **6** (2013), no. 2, 361–390, DOI 10.1112/jtopol/jtt002. MR3065180

[11] S. Brandhorst and A. Cattaneo, *Prime order isometries of unimodular lattices and automorphisms of IHS manifolds*, `ArXiv:1912.07119`.

[12] M. Britze, *On the cohomology of generalized Kummer varieties*, Ph.D. thesis, University of Köln, Germany, 2002.

[13] Chiara Camere and Alberto Cattaneo, *Non-symplectic automorphisms of odd prime order on manifolds of $K3^{[n]}$-type*, Manuscripta Math. **163** (2020), no. 3-4, 299–342, DOI 10.1007/s00229-019-01163-4. MR4159799

[14] Chiara Camere, Grzegorz Kapustka, Michał Kapustka, and Giovanni Mongardi, *Verra fourfolds, twisted sheaves, and the last involution*, Int. Math. Res. Not. IMRN **21** (2019), 6661–6710, DOI 10.1093/imrn/rnx327. MR4027563

[15] J. H. Conway and N. J. A. Sloane, *Sphere packings, lattices and groups*, 3rd ed., Grundlehren der Mathematischen Wissenschaften [Fundamental Principles of Mathematical Sciences], vol. 290, Springer-Verlag, New York, 1999. With additional contributions by E. Bannai, R. E. Borcherds, J. Leech, S. P. Norton, A. M. Odlyzko, R. A. Parker, L. Queen and B. B. Venkov. MR1662447

[16] Akira Fujiki, *Finite automorphism groups of complex tori of dimension two*, Publ. Res. Inst. Math. Sci. **24** (1988), no. 1, 1–97, DOI 10.2977/prims/1195175326. MR944867

[17] Giovanni Mongardi, Kévin Tari, and Malte Wandel, *Prime order automorphisms of generalised Kummer fourfolds*, Manuscripta Math. **155** (2018), no. 3-4, 449–469, DOI 10.1007/s00229-017-0942-7. MR3763414

[18] G. Mongardi and M. Wandel, *Induced automorphisms on irreducible symplectic manifolds*, J. Lond. Math. Soc. (2) **92** (2015), no. 1, 123–143, DOI 10.1112/jlms/jdv012. MR3384508

[19] Marc A. Nieper-Wisskirchen, *On the Chern numbers of generalised Kummer varieties*, Math. Res. Lett. **9** (2002), no. 5-6, 597–606, DOI 10.4310/MRL.2002.v9.n5.a3. MR1906063

[20] Marc A. Nieper-Wisskirchen, *Twisted cohomology of the Hilbert schemes of points on surfaces*, Doc. Math. **14** (2009), 749–770. MR2578804

[21] V. V. Nikulin, *Integer symmetric bilinear forms and some of their geometric applications* (Russian), Izv. Akad. Nauk SSSR Ser. Mat. **43** (1979), no. 1, 111–177, 238. MR525944

[22] Keiji Oguiso, *No cohomologically trivial nontrivial automorphism of generalized Kummer manifolds*, Nagoya Math. J. **239** (2020), 110–122, DOI 10.1017/nmj.2018.29. MR4138897

[23] A. N. Rudakov and I. R. Shafarevich, *Surfaces of type $K3$ over fields of finite characteristic* (Russian), Current problems in mathematics, Vol. 18, Akad. Nauk SSSR, Vsesoyuz. Inst. Nauchn. i Tekhn. Informatsii, Moscow, 1981, pp. 115–207. MR633161

[24] K. Tari, *Automorphismes des variétés de Kummer généralisées*, Ph.D. thesis, University of Poitiers, France, 2015.

Laboratoire de Mathématiques et Applications, UMR 7348 CNRS, Bâtiment H3, Boulevard Marie et Pierre Curie, Site du Futuroscope, TSA 61125, 86073 Poitiers Cedex 9, France

Email address: samuel.boissiere@math.univ-poitiers.fr
URL: http://www-math.sp2mi.univ-poitiers.fr/~sboissie/

Lehrstuhl für Algebra und Zahlentheorie, Universitätsstrasse 14, 86159 Augsburg, Germany

Email address: marc.nieper-wisskirchen@math.uni-augsburg.de
URL: http://www.math.uni-augsburg.de/alg/

Kévin Tari, Lycée Henri Moissan, 20 Cours de Verdun, 77100 Meaux
Email address: kevin.tari@free.fr

Contemporary Mathematics
Volume **766**, 2021
https://doi.org/10.1090/conm/766/15374

On the existence of groups of automorphisms of compact Riemann and Klein surfaces

E. Bujalance and F. J. Cirre

ABSTRACT. Each group G of automorphisms of a compact Riemann surface of genus bigger than one can be written as the (finite) quotient $G = \Gamma/\Lambda$ for some Fuchsian groups Γ and Λ where Λ is torsion-free. Amongst the natural questions which arise in this situation, this survey article focuses on the following ones. First, which groups can be realized as the full group of all automorphisms of some compact Riemann surface? Second, does every Fuchsian group Γ contain a torsion-free normal subgroup Λ of finite index? Third, for a fixed finite group G, is it possible to characterize all Fuchsian groups Γ containing a torsion-free normal subgroup Λ such that $G = \Gamma/\Lambda$? And fourth, which conditions assure that a group G of automorphisms of a compact Riemann surface is the full group of all its automorphisms? We will consider these topics not only in the setting of compact Riemann surfaces and Fuchsian groups but also in the setting of compact Klein surfaces and non-euclidean crystallographic groups.

1. Introduction

As a consequence of Riemann Uniformization Theorem, every compact Riemann surface S of genus $g \geq 2$ is isomorphic to the orbit space U/Λ of the hyperbolic plane U under the action of a torsion-free Fuchsian group Λ. The (finite) groups of automorphisms of S can be described as quotients Γ/Λ where Γ is a Fuchsian group containing Λ as a normal subgroup of finite index. Several questions arise in a natural way. The following ones will be considered here:

(1) Which finite groups can be realized as the full group of automorphisms of some compact Riemann surface?
(2) Does every Fuchsian group Γ contain a torsion-free normal subgroup Λ of finite index?
(3) For a fixed a finite group G, is it possible to characterize all Fuchsian groups Γ containing a torsion-free normal subgroup Λ such that G is isomorphic to Γ/Λ?
(4) Given a group $G = \Gamma/\Lambda$ of automorphisms of the compact Riemann surface $S = U/\Lambda$, which conditions on G or Γ assure that G is the full group

2020 *Mathematics Subject Classification.* Primary 30F10, 30F50; Secondary 14H37, 20B25.
Key words and phrases. Riemann and Klein surfaces, groups of automorphisms.
Both authors were supported in part by PGC2018-096454-B-I00.

of automorphisms of S, so that S admits no more automorphisms than those in G?

Answers to these questions have been the inspiration of much research over the past few decades. In this paper we will focus on the combinatorial techniques used to solve the above questions. We will consider these topics not only in the setting of compact Riemann surfaces but also in the setting of compact Klein surfaces. These can be seen as the natural generalization of compact Riemann surfaces when we allow non-orientable surfaces or surfaces with non-empty boundary while keeping some "dianalytic" structure on them. A uniformization theorem for compact Klein surfaces also exists, and it provides a combinatorial approach to their groups of automorphisms.

2. Preliminaries

We begin with some background about compact Riemann and Klein surfaces and their groups of automorphisms, a lot of which can be found in [**18**].

By a *non-euclidean crystallographic group* (shortly, *NEC group*) we mean a discrete cocompact subgroup of the group $\mathrm{Aut}(U) = \mathrm{PGL}(2, \mathbb{R})$ of automorphisms of the upper half-plane U, including those which reverse orientation. The structure of an NEC group Γ is determined by its *signature*, introduced by Macbeath in [**36**] as a collection of symbols and non-negative integers of the form

$$\sigma(\Gamma) = (\gamma; \pm; [m_1, \ldots, m_r]; \{(n_{11}, \ldots, n_{1s_1}), \ldots, (n_{k1}, \ldots, n_{ks_k})\}). \tag{1}$$

The integers $m_1, \ldots, m_r$ are called *proper periods*; each bracket $C_i = (n_{i1}, \ldots, n_{is_i})$ is a *period cycle*, and the integers n_{ij} are called *link periods*. An empty set of proper periods (where $r = 0$) will be denoted by $[-]$, an empty period-cycle (where $s_i = 0$) by $(-)$, and an empty set of period cycles (where $k = 0$) by $\{-\}$.

Some topological features of the compact orbit space U/Γ can be derived from the signature of Γ : (a) γ is the topological genus of the orbit space U/Γ, (b) k is the number of its boundary components, (c) the sign is $+$ or $-$ according to whether or not U/Γ is orientable, (d) the integers $m_i \geq 2$ are the orders of the r branch points in the interior of U/Γ, and (e) the $n_{j\ell}$ are the orders of the s_j branch points on the jth boundary component of U/Γ, for $1 \leq j \leq k$.

Associated with every signature $\sigma(\Gamma)$ is a canonical presentation for the group Γ, and also a formula for the hyperbolic area of a fundamental domain for Γ. If the sign is $+$ then Γ is generated by elliptic elements $x_1, \ldots, x_r$, reflections $c_{10}, \ldots, c_{1s_1}, \ldots, c_{k0}, \ldots, c_{ks_k}$, orientation preserving transformations (mostly hyperbolic ones) $e_1, \ldots, e_k$, and hyperbolic elements $a_1, b_1, \ldots, a_\gamma, b_\gamma$, which satisfy the defining relations

$$\begin{aligned}
&x_i^{m_i} = 1 \quad (\text{for } 1 \leq i \leq r),\\
&c_{ij-1}^2 = c_{ij}^2 = (c_{ij-1}c_{ij})^{n_{ij}} = 1, \;\; c_{is_i} = e_i^{-1} c_{i0} e_i \quad (\text{for } 1 \leq i \leq k, 1 \leq j \leq s_i),\\
&x_1 \ldots x_r \, e_1 \ldots e_k \, a_1 b_1 a_1^{-1} b_1^{-1} \ldots a_\gamma b_\gamma a_\gamma^{-1} b_\gamma^{-1} = 1.
\end{aligned}$$

On the other hand, if the sign is $-$, then $\gamma > 0$ and the generators a_i and b_i are replaced by glide reflections $d_1, \ldots, d_\gamma$, and the final relation replaced by $x_1 \ldots x_r \, e_1 \ldots e_k \, d_1^2 \ldots d_\gamma^2 = 1$. A set of generators satisfying these relations will be called *canonical.*

An NEC group with no orientation reversing elements is a *Fuchsian group.* Observe that it has signature $(\gamma; +; [m_1, \ldots, m_r]; \{-\})$. An NEC group Γ which is not Fuchsian is a *proper* NEC group, and its *canonical Fuchsian subgroup* Γ^+, which consists of its orientation preserving elements, has index two in Γ.

The hyperbolic area of a fundamental region for Γ is given by

$$\mu(\Gamma) = 2\pi\left(\varepsilon\gamma - 2 + k + \sum_{i=1}^{r}\left(1 - \frac{1}{m_i}\right) + \frac{1}{2}\sum_{i=1}^{k}\sum_{j=1}^{s_i}\left(1 - \frac{1}{n_{ij}}\right)\right),$$

where $\varepsilon = 2$ if the sign is $+$, and $\varepsilon = 1$ if the sign is $-$. Furthermore, if Δ is any subgroup of finite index in Γ, then Δ is also an NEC group, and the hyperbolic areas of fundamental regions for Δ and Γ satisfy the Riemann-Hurwitz formula $\mu(\Delta) = |\Gamma : \Delta|\,\mu(\Gamma)$.

The *algebraic genus* g of a compact surface of topological genus γ with k boundary components, is defined as

$$g = \begin{cases} 2\gamma + k - 1 & \text{if the surface is orientable,} \\ \gamma + k - 1 & \text{otherwise.} \end{cases}$$

It follows from the Uniformisation Theorem that any surface of algebraic genus $g \geq 2$ is of the form U/Λ for some NEC group Λ with signature

$$(\gamma; \pm; [-]; \{(-), \overset{k}{\ldots}, (-)\}),$$

where the sign is $+$ if the surface is orientable, and $-$ otherwise. We will call each NEC group with such a signature a *surface NEC group,* and will distinguish between *bordered surface NEC groups* (those with $k > 0$), *unbordered non-orientable surface NEC groups* (those with $k = 0$ and sign $=$ "$-$"), and *surface Fuchsian groups* (those with $k = 0$ and sign $=$ "$+$"). Observe that surface NEC groups are precisely those with no orientation preserving element of finite order (apart from the identity), that is, those whose canonical Fuchsian subgroup is torsion free.

A finite group G acts (faithfully) as a group of automorphisms of a surface represented as U/Λ where Λ is a surface NEC group, if and only if there exist an NEC group Γ and an epimorphism $\theta\colon \Gamma \to G$ whose kernel is Λ. In this case we say (for short) that θ is a *smooth epimorphism*, and that *G acts with signature $\sigma(\Gamma)$*. Any such action can be described by means of a *generating vector*, which consists of the images under the corresponding smooth epimorphism θ of a canonical set of generators for Γ.

It can happen that a group $G \cong \Gamma/\Lambda$ acts on a compact surface $S \cong U/\Lambda$, but is not the full automorphism group of S. When that happens, there exists an NEC group Γ' containing Γ and normalising Λ, so Γ'/Λ is a group of automorphisms of S larger than G. If Γ' and Γ are proper NEC groups with the same Teichmüller dimension then the signatures of Γ' and Γ must appear in the lists of finite index inclusions of NEC groups, given by the first author [**3**] for the case where Γ is normal in Γ', and by Estévez and Izquierdo [**23**] for the case where Γ is not normal in Γ'. These inclusions are analysed in further detail in [**9**]. These lists are based on the previous study of finite index inclusions of Fuchsian groups considered by Greenberg [**27**] and completed by Singerman [**46**].

If a pair (σ, σ') of signatures occurs in any of these lists, then for any NEC group Γ with signature σ there always exists another NEC group Γ' with signature σ' containing Γ with finite index. We will say that σ is a non-maximal signature. In this case, the action of $G = \Gamma/\Lambda$ on $S = U/\Lambda$ might be extendable to the action of the larger group Γ'/Λ. On the other hand, if a signature σ does not occur as the first entry of a pair in any of the above lists, then there exists a *maximal* NEC group Γ with signature σ, so Γ is not contained in any other NEC group with finite index. In this case, if Λ is a surface normal subgroup of Γ then the action of $G = \Gamma/\Lambda$ on $S = U/\Lambda$ cannot be extended, and $G = \mathrm{Aut}(S)$.

3. Finite groups as full automorphism groups

Every finite group acts as a group of automorphisms of some compact surface. This is a well known result which, for Riemann surfaces, goes back to Hurwitz [**32**]. A modern proof in the setting of algebraic function fields can be found in [**38**].

It is a rather difficult task to decide whether a finite group G is the full group of all automorphisms of the surface or, on the contrary, the surface admits more automorphisms than those in G. This is the problem of extendability of group actions. Signatures of Fuchsian and NEC groups prove to be a useful tool in dealing with this problem, as explained at the end of Section 2.

For group actions on Riemann surfaces, we have the following result, which is a consequence of the more general result [**28**, Theorem 6'] by Greenberg.

THEOREM 3.1. *Every finite group is the full group of conformal automorphisms of some compact Riemann surface of genus bigger than one.*

The same result for compact Klein surfaces was obtained in [**5**]. We give a proof inspired by [**28**, Theorem 6'] and based on NEC groups to illustrate their use in this setting.

THEOREM 3.2. *Every finite group is the full automorphism group of some unbordered non-orientable compact Klein surface, of some bordered orientable compact Klein surface, and of some bordered non-orientable compact Klein surface.*

PROOF. Let $g_1, \ldots, g_n$ be generators of the finite group G. We just show the case of bordered orientable surfaces, the others being similar. Let us consider a maximal NEC group Γ with signature $(n+1; +; [-]; \{(-)\})$, and let $a_1, b_1, \ldots, a_{n+1}, b_{n+1}, e, c$ be a canonical set of generators of Γ, where c is a reflection and the other generators are hyperbolic. There does exist such a maximal NEC group (see the end of Section 2) because this signature does not occur as the first entry of a pair in any of the lists in [**3**] and [**23**] (observe that we are considering NEC signatures, and not Fuchsian signatures as in [**27**, **46**]). We define an epimorphism $\theta : \Gamma \to G$ by

- $\theta(a_i) = \theta(b_i) = g_i$ for $i = 1, \ldots, n$;
- $\theta(a_{n+1}) = \theta(b_{n+1}) = g_n$;
- $\theta(e) = \theta(c) = 1$.

This epimorphism is in fact well defined because it preserves the defining relations $c^2 = e\, a_1 b_1 a_1^{-1} b_1^{-1} \ldots a_{n+1} b_{n+1} a_{n+1}^{-1} b_{n+1}^{-1} = 1$. The only orientation reversing canonical generator of Γ is the reflection c and, since $c \in \ker\theta$, it follows that the signature of $\ker\theta$ has sign "+" (see Theorems 2.1.2 and 2.1.3 in [**18**]). Moreover, $\ker\theta$ contains no elliptic element but it contains a reflection. Therefore $\ker\theta$ is a bordered orientable surface NEC group. Therefore $U/\ker\theta$ is a bordered orientable compact Klein surface, on which $G \simeq \Gamma/\ker\theta$ acts as its full automorphism group. □

4. Surface normal subgroups of finite index

In this section we consider the following question: does any NEC group Γ possess a surface normal subgroup Λ of finite index? The importance of this question in the setting of Riemann and Klein surfaces and their groups of automorphisms comes from the Uniformization Theorem, as explained in Section 2. An affirmative answer implies the existence of a compact surface U/Λ on which the finite group $G \simeq \Gamma/\Lambda$ acts.

For Fuchsian groups, this question is Fenchel's conjecture, proven in all its generality in a sequence of papers by J. Nielsen [**41**], S. Bundgaard and J. Nielsen [**20**], and R. H. Fox [**26**], see also [**21**].

THEOREM 4.1. *Every Fuchsian group possesses a torsion-free normal subgroup of finite index.*

An outline of the proof is the following. A finite collection of homomorphisms from the Fuchsian group Γ onto finite groups are constructed in such a way that each homomorphism preserves the order of some torsion canonical generator x of Γ. The kernel of each homomorphism, which is a normal subgroup of finite index, contains no power of x. Therefore, intersecting all these kernels yields a torsion-free normal subgroup Λ. Moreover, Λ has finite index in Γ, by Lemma 4.2 below, a proof of which can be found in [**20**].

LEMMA 4.2. *If $N_1, \ldots, N_s$ are normal subgroups of a group G which have finite index in G, then their intersection $N_1 \cap \cdots \cap N_s$ has also finite index in G.*

The above technique works fairly well except when the Fuchsian group has a small number of generators, especially when it is a triangle group. This particular case is solved with the following Lemma 4.3 by Fox in [**26**], with a correction by Chau in [**21**]. This is the final step in the proof of Fenchel's conjecture for Fuchsian groups.

LEMMA 4.3. *Given three integers $m_1 > 1$, $m_2 > 1$ and $m_3 > 1$ there exists a permutation group containing two permutations u and v such that u has order m_1, v has order m_2 and the product uv has order m_3.*

For proper NEC groups, Fenchel's conjecture splits naturally into three disjoint cases according to whether the surface normal subgroup is Fuchsian, bordered or unbordered non-orientable. The answer is affirmative in the first case, see Theorem 4.5. Its proof is based on Theorem 4.1 and Lemmas 4.2 and 4.4.

LEMMA 4.4. *Let H and K be subgroups of finite index in a group G with $K \lhd H$ and $|G : H| = 2$. Let us write $G = H \cup cH$. Then $K \cap cKc^{-1}$ is a normal subgroup of G which also has finite index in G.*

THEOREM 4.5. *Every proper NEC group possesses a torsion free normal subgroup of finite index consisting of orientation preserving elements.*

PROOF. (Included in [**10**].) Applying Theorem 4.1 to the canonical Fuchsian subgroup Γ^+ of a proper NEC group Γ gives the existence of a torsion-free normal subgroup K of finite index in Γ^+. Let us write $\Gamma = \Gamma^+ \cup c\Gamma^+$, with $c \in \Gamma - \Gamma^+$. Lemma 4.4 yields that the intersection $\Lambda := K \cap cKc^{-1}$ is also a normal subgroup of finite index in Γ. Clearly, Λ consists of orientation preserving elements and is torsion-free. □

In the case of bordered surface normal subgroups, it turns out that the answer to Fenchel's conjecture is not always affirmative, as shown in [**19**].

THEOREM 4.6. *Let* Γ *be an NEC group with signature* (1) *with* $k > 0$. *Then* Γ *possesses a bordered surface normal subgroup of finite index if and only if the signature has either an empty period cycle or a period cycle with two consecutive link periods equal to* 2.

PROOF. Assume first that Γ possesses a bordered surface normal subgroup Λ with finite index N. Let $\theta : \Gamma \to \Gamma/\Lambda$ be the corresponding smooth epimorphism. If N is odd then all canonical generating reflections c_{ij} belong to $\ker\theta = \Lambda$. This prevents Γ from having a pair of consecutive generating reflections c_{ij}, c_{ij+1} since otherwise its product $c_{ij}c_{ij+1}$ would be a non-trivial elliptic element in $\ker\theta$. Therefore, all period cycles in $\sigma(\Gamma)$ are empty. Assume now that N is even. Since $\ker\theta$ is bordered, it contains a canonical generating reflection $c \in \Gamma$. Let $\mathcal{C}$ be its associated period cycle in $\sigma(\Gamma)$ and assume $\mathcal{C}$ contains a link period $n = \text{order}(cc')$, where c' is another canonical generating reflection. (If there is no such n then $\mathcal{C}$ is empty are we are done.) Both c and its conjugate $c'cc'$ belong to $\ker\theta$ and so the same happens to their product $c(c'cc') = (cc')^2$. So $(cc')^2 = 1$, since otherwise $(cc')^2$ would be a non-trivial elliptic element in $\ker\theta$. So $n = 2$. If there were no other link period in $\mathcal{C}$ then c and c' would be conjugate reflections (by means of a boundary transformation e), and so their product cc' would be a non-trivial element in $\ker\theta$. This contradiction shows that there exists another link period in $\mathcal{C}$, and we may assume it is of the form $m = \text{order}(c''c)$ for some canonical generating reflection c''. The same reasoning as above yields $m = 2$. This concludes the proof of the "only if" part.

The proof of the "if" part is inspired by the method of Fox in [**26**] and explained above. The details can be found in [**19**]. □

Fenchel's conjecture for the case of unbordered non-orientable surface normal subgroups is still an open question, although some progress has been achieved, see [**10**].

THEOREM 4.7. *Every NEC group with signature*$(g; -; [m_1, \ldots, m_r]; \{\mathcal{C}_1, \ldots, \mathcal{C}_k\})$ *possesses a torsion free normal subgroup of finite index whose signature has sign "$-$". The same holds true for signatures* $(g; +; [m_1, \ldots, m_r]; \{\mathcal{C}_1, \ldots, \mathcal{C}_k\})$ *whenever* $g > 0$.

When $g = 0$ we also have some partial results: for instance, it is true when the signature is $(0; +; [m_1, \ldots, m_r]; \{(-)\})$. But the answer for the general case is unknown.

5. Necessary and sufficient conditions for the existence of a finite group action

As said in Section 3, every finite group G can be written as Γ/Λ for some NEC group Γ and some surface subgroup Λ. In this section we address the following question: is it possible to describe (the isomorphism class of) all such groups Γ? This is equivalent to finding necessary and sufficient conditions on the signature of Γ for it to admit a smooth epimorphism onto a fixed finite group G.

The first case to be considered is when G is cyclic. The question was completely solved in the setting of Riemann surfaces by Harvey [**31**].

THEOREM 5.1. *Let Γ be a Fuchsian group with signature $(\gamma; +; [m_1, \ldots, m_r]; \{-\})$ and let $M = \text{lcm}\{m_1, \ldots, m_r\}$. There is a smooth epimorphism from Γ onto the cyclic group of order N if and only if the following conditions are satisfied:*

(*i*) $\text{lcm}\{m_1, \ldots, \hat{m}_i, \ldots, m_r\} = M$ *for all i, where $\hat{m}_i$ denotes the omission of m_i;*
(*ii*) *M divides N, and if $\gamma = 0$ then $M = N$;*
(*iii*) *$r \neq 1$, and if $\gamma = 0$ then $r \geq 3$;*
(*iv*) *if M is even then the number of periods m_i such that M/m_i is odd is also even.*

As an easy application of this result, Harvey proved that the maximum order of a cyclic group of automorphisms of a Riemann surface of genus g is $4g + 2$, a result already shown by Wiman [**49**]. This solved the so called maximum order problem for cyclic groups. Harvey also computed the minimum genus of a Riemann surface on which a cyclic group of order N acts. This solved the so called minimum genus problem for cyclic groups. Theorem 5.1 was a seminal result in the study of groups of automorphisms of compact Riemann surfaces. Since then, much research has been conducted to extend Harvey's theorem to other classes of finite groups acting on Riemann surfaces. In particular, the maximum order and the minimum genus problems for different families of finite groups have received a good deal of attention, see [**14**].

Most of this research considers groups whose elements act preserving the orientation of the Riemann surface. It is interesting, however, to consider also groups which contain orientation reversing automorphisms. For instance, in modern terminology two related parameters have been defined taking into account this difference: the *strong symmetric genus* of a group G is the minimum genus of a Riemann surface on which G acts preserving its orientation, but if we allow some elements of G to reverse the orientation of the surface then we are looking for the *symmetric genus* of G. We refer again to [**14**].

Related to these two parameters there exist the notions of the *real genus* and the *symmetric crosscap number* of a group. They arise naturally when we consider Klein surfaces instead of Riemann surfaces. The real genus of a group was introduced by May in [**39**] as the minimum algebraic genus of the bordered surfaces on which the group acts. The symmetric crosscap number is the minimum genus of the unbordered non-orientable surfaces on which the group acts. The systematic study of the symmetric crosscap number was begun by May in [**40**], although previous results from other authors are also to be noted, see for instance [**4**], [**24**], [**29**] and [**30**]. There is a significant amount of research dealing with the real genus and the symmetric crosscap number of a group, and most of it uses the combinatorial theory of NEC groups. The reader interested in these two topics is referred to the surveys [**11**] and [**12**].

Back to the question we are considering in this section, Etayo in [**25**] found the necessary and sufficient conditions on the signature of a proper NEC group for it to admit a smooth epimorphism onto a cyclic group whose kernel is a surface Fuchsian group. In other words, he considered cyclic groups acting on Riemann surfaces and containing elements which reverse the orientation of the surface. As an easy application, Etayo finds the symmetric genus of a cyclic group. The study splits naturally into two cases, depending on whether or not the cyclic group contains an orientation reversing element of order two. This last case has consequences in the study of the so called pseudo-real Riemann surfaces.

A Riemann surface is *pseudo-real* (or *asymmetric*) if it admits orientation reversing automorphisms but no orientation reversing automorphism of order two. The importance of pseudo-real surfaces stems from the fact that in the moduli space of compact Riemann surfaces of given genus, pseudo-real surfaces represent the points that have real moduli but are not definable over the reals. Etayo's result in [**25**] can be extended to determine the necessary and sufficient conditions on the signature of a proper NEC group for it to admit a smooth epimorphism onto a cyclic group whose kernel uniformises a pseudo-real Riemann surface, see [**48**]. It is worth mentioning that the study of pseudo-real surfaces requires the knowledge of the full automorphism group of the surface, in order to assure the non-existence of orientation reversing automorphisms of order two. This leads to the problem of extendability of group actions, as explained at the end of Section 2.

The analogous of Harvey's conditions in Theorem 5.1 for cyclic actions on Riemann surfaces, have already been found for cyclic actions on Klein surfaces: in [**4**] for unbordered non-orientable surfaces, and in [**18**, Chapter 3] for the general case of bordered surfaces (orientable or not). In this last case it is worth mentioning that a refined solution is achieved by considering surfaces with a fixed topological type. For instance, if we want to consider non-orientable surfaces with a fixed number $k > 0$ of boundary components then we have to deal with smooth epimorphisms whose kernel has signature $(\gamma; -; [-]; \{(-), \overset{k}{\ldots}, (-)\})$. In the case of cyclic actions of odd order we have the following, see [**18**, Theorem 3.1.3].

THEOREM 5.2. *Let Γ be an NEC group with signature $(\gamma'; \pm; [m_1, \ldots, m_r]; \{\mathcal{C}_1, \ldots, \mathcal{C}_{k'}\})$ and let $N > 1$ be an odd number. There is a smooth epimorphism θ from Γ onto the cyclic group of order N whose kernel has signature $(\gamma; -; [-]; \{(-), \overset{k}{\ldots}, (-)\})$ if and only if the following conditions are satisfied:*

(*i*) $\mathrm{sign}(\Gamma) =$ "$-$";
(*ii*) $\mathcal{C}_j = (-)$ *for all* j;
(*iii*) m_i *divides* N *for all* i;
(*iv*) $\mu(\ker\theta) = N\mu(\Gamma)$;
(*v*) *there exist* $l_1, \ldots, l_{k'}$ *positive divisors of* N *such that* $k = \sum_{i=1}^{k'} N/l'_i$;
(*vi*) *if* $\gamma' = 1$ *then* $N = \mathrm{lcm}(m_1, \ldots, m_r, l_1, \ldots, l_{k'})$.

Abelian groups constitute, as far as we know, the only other family of groups for which the above problem has been completely solved for both Riemann and Klein surfaces. For Riemann surfaces, the first approach was given by Breuer [**2**], whose conditions for the existence of a smooth epimorphism (with Fuchsian surface kernel) onto an abelian group were refined by Rodríguez, [**42**], expressing such conditions in terms of the signature of the Fuchsian group and the invariant factors defining the abelian group.

Extending this study, Rodríguez himself also considered abelian group actions on bordered Klein surfaces (orientable or not) in [**43**], and on unbordered non-orientable Klein surfaces in [**44**]. However, it is still an open problem to find necessary and sufficient conditions for abelian actions on pseudo-real Riemann surfaces, although some progress has been achieved in [**15**].

We finally mention the family of dihedral groups, which has also been the topic of several papers, both in the setting of Riemann surfaces, as in [**22**], and in the setting of Klein surfaces, as in [**33**]. The main tool used in these papers are Fuchsian and NEC groups respectively, and both papers are focussed on dihedral groups of order $2p$ with p a prime. For dihedral groups of arbitrary order, a complete solution to the problem we are dealing with in this section, has been obtained for Riemann surfaces in [**13**]. For unbordered non-orientable Klein surfaces, there is still some work in progress, [**45**].

It is also worth mentioning some results on the so called genus spectrum problem (or Hurwitz problem) for p-groups. The genus spectrum problem for a finite group G consists of determining the values of $g \geq 2$ for which there exists a compact Riemann surface of genus g on which G acts. The genus spectrum for cyclic p-groups is determined by Kulkarni and Maclachlan in [**35**], and a larger class of p-groups is considered by Maclachlan and Talu in [**37**]. In both papers, the combinatorial theory of Fuchsian groups is a basic tool.

6. Conditions for the existence of a finite full group action

Assume there exists a smooth epimorphism $\theta : \Gamma \to G$ for an NEC group Γ onto a finite group G. Then G acts on the compact surface $S = U/\ker\theta$ as a group of automorphisms. As explained at the end of Section 2, the surface S might admit more automorphisms than those in G, so that G would not be the full automorphism group of S. If the signature σ of Γ is a non-maximal signature (occurring in the lists in [**3**] and [**23**]) then for every group Γ with signature σ there always exists another group Γ' containing Γ with finite index. In this case, the action of G might be extended to the action of a larger group G'. But this is not always the case, see Example 6.1 below. In this section we consider the problem of finding conditions to assure that a finite group G acting with a non-maximal signature is the full automorphism group of a surface.

The conditions for the extendability of the action of G depend on the way in which Γ is embedded in Γ', but can be translated into algebraic conditions of the finite group G. This way, the action of some groups can be extended to a larger action, while other groups (acting with the same signature) cannot. Furthermore, the same abstract group G can act (with non-maximal signature) in different ways, and some actions can be extended while others cannot.

The family of cyclic groups is the first to be dealt with. The conditions for the extendability of a cyclic group acting with non-maximal signature on a compact surface were found in [**16**] for the case of Riemann surfaces, in [**7**] for the case of non-orientable unbordered Klein surfaces, and in [**8**] for the case of bordered Klein surfaces. It is worth mentioning that in the last two cases, every action of a finite cyclic group with non-maximal signature always extends to the action of a larger group. In contrast, this is not so for cyclic actions on Riemann surfaces, where the extendability depends on the signature itself and on the type of action. We illustrate this situation with the following example.

EXAMPLE 6.1. *The cyclic group* C_p *of prime order* p *acts with non-maximal Fuchsian signature* $(0;+;[p,p,p];\{-\})$ *on compact Riemann surfaces of genus* $g = (p-1)/2$. *There are different actions, described by means of different smooth epimorphisms* $\theta : \Gamma \to C_p = \langle u\rangle$. *One of them is given by* $\theta(x_1) = \theta(x_2) = u$, $\theta(x_3) = u^{-2}$, *where* $\{x_1, x_2, x_3\}$ *is a canonical set of generators of* Γ. *According to* [**16**, Theorem 4.1(v)] *this action of* C_p *extends to a larger group. Another action is given by* $\theta(x_1) = u$, $\theta(x_2) = u^2$, $\theta(x_3) = u^{-3}$. *Assume that* $p > 7$. *Then, according to*

[**16**, Theorem 4.2] *this action of* C_p *does not extend to a larger group, so* C_p *is the full automorphism group of the surface. It is also worth mentioning that there exist pairs* (n, g) *such that the cyclic group* C_n *never acts as the full automorphism group of any Riemann surface of genus* g*, see* [**47**].

A detailed study of the extendability of arbitrary group actions (not only cyclic ones) on Riemann surfaces was carried out in [**6**]. The conditions for the extendability obtained there have been applied to different families of groups: for non-cyclic abelian actions in [**6**] itself, for dihedral actions in [**13**], for p-cyclic actions on the so called multiple prime covers in [**50**]. They have also been employed to determine the full automorphism groups of some families of Riemann surfaces, as cyclic covers of the sphere branched over three points in [**34**], cyclic p-gonal surfaces branched over any number of points, for p prime in [**51**], pseudo-real surfaces in [**17**] and [**1**], and surfaces of genus 7 in [**52**].

In the setting of Klein surfaces, the corresponding study of the extendability of group actions in all its generality, is a work in progress, [**9**].

References

[1] Michela Artebani, Saúl Quispe, and Cristian Reyes, *Automorphism groups of pseudo-real Riemann surfaces*, J. Pure Appl. Algebra **221** (2017), no. 9, 2383–2407, DOI 10.1016/j.jpaa.2016.12.039. MR3631722

[2] Thomas Breuer, *Characters and automorphism groups of compact Riemann surfaces*, London Mathematical Society Lecture Note Series, vol. 280, Cambridge University Press, Cambridge, 2000. MR1796706

[3] E. Bujalance, *Normal N.E.C. signatures*, Illinois J. Math. **26** (1982), no. 3, 519–530. MR658461

[4] Emilio Bujalance, *Cyclic groups of automorphisms of compact nonorientable Klein surfaces without boundary*, Pacific J. Math. **109** (1983), no. 2, 279–289. MR721920

[5] E. Bujalance, *A note on the group of automorphisms of a compact Klein surface* (Spanish, with English summary), Rev. Real Acad. Cienc. Exact. Fís. Natur. Madrid **81** (1987), no. 3, 565–569. MR950620

[6] Emilio Bujalance, F. J. Cirre, and Marston Conder, *On extendability of group actions on compact Riemann surfaces*, Trans. Amer. Math. Soc. **355** (2003), no. 4, 1537–1557, DOI 10.1090/S0002-9947-02-03184-7. MR1946404

[7] E. Bujalance, F. J. Cirre, and M. D. E. Conder, *Extensions of finite cyclic group actions on non-orientable surfaces*, Trans. Amer. Math. Soc. **365** (2013), no. 8, 4209–4227, DOI 10.1090/S0002-9947-2013-05776-5. MR3055694

[8] Emilio Bujalance, Francisco-Javier Cirre, and Marston Conder, *Extensions of finite cyclic group actions on bordered surfaces*, Rev. Mat. Iberoam. **31** (2015), no. 1, 349–372, DOI 10.4171/RMI/837. MR3320844

[9] E. Bujalance, F. J. Cirre and M. D. E. Conder: Analysis of NEC group inclusions, *in preparation.*

[10] E. Bujalance, F. J. Cirre, M. D. E. Conder and A. F. Costa: On Fenchel's conjecture for NEC groups, *in progress.*

[11] E. Bujalance, F. J. Cirre, J. J. Etayo, G. Gromadzki, and E. Martínez, *A survey on the minimum genus and maximum order problems for bordered Klein surfaces*, Groups St Andrews 2009 in Bath. Volume 1, London Math. Soc. Lecture Note Ser., vol. 387, Cambridge Univ. Press, Cambridge, 2011, pp. 161–182. MR2858856

[12] E. Bujalance, F. J. Cirre, J. J. Etayo, G. Gromadzki, and E. Martínez, *Automorphism groups of compact non-orientable Riemann surfaces*, Groups St Andrews 2013, London Math. Soc. Lecture Note Ser., vol. 422, Cambridge Univ. Press, Cambridge, 2015, pp. 183–193. MR3495655

[13] E. Bujalance, F. J. Cirre, J. M. Gamboa, and G. Gromadzki, *On compact Riemann surfaces with dihedral groups of automorphisms*, Math. Proc. Cambridge Philos. Soc. **134** (2003), no. 3, 465–477, DOI 10.1017/S030500410200662X. MR1981212

[14] Emilio Bujalance, Francisco-Javier Cirre, and Grzegorz Gromadzki, *A survey of research inspired by Harvey's theorem on cyclic groups of automorphisms*, Geometry of Riemann surfaces, London Math. Soc. Lecture Note Ser., vol. 368, Cambridge Univ. Press, Cambridge, 2010, pp. 15–37, DOI 10.1007/978-3-642-14828-6. MR2665004

[15] E. Bujalance, F. J. Cirre and J. Rodríguez: Abelian actions on pseudo-real Riemann surfaces, *in preparation.*

[16] Emilio Bujalance and Marston Conder, *On cyclic groups of automorphisms of Riemann surfaces*, J. London Math. Soc. (2) **59** (1999), no. 2, 573–584, DOI 10.1112/S0024610799007115. MR1709666

[17] Emilio Bujalance, Marston D. E. Conder, and Antonio F. Costa, *Pseudo-real Riemann surfaces and chiral regular maps*, Trans. Amer. Math. Soc. **362** (2010), no. 7, 3365–3376, DOI 10.1090/S0002-9947-10-05102-0. MR2601593

[18] Emilio Bujalance, José J. Etayo, José M. Gamboa, and Grzegorz Gromadzki, *Automorphism groups of compact bordered Klein surfaces*, Lecture Notes in Mathematics, vol. 1439, Springer-Verlag, Berlin, 1990. A combinatorial approach. MR1075411

[19] E. Bujalance and E. Martínez, *A remark on NEC groups representing surfaces with boundary*, Bull. London Math. Soc. **21** (1989), no. 3, 263–266, DOI 10.1112/blms/21.3.263. MR986369

[20] Svend Bundgaard and Jakob Nielsen, *On normal subgroups with finite index in F-groups*, Mat. Tidsskr. B **1951** (1951), 56–58. MR48447

[21] T. C. Chau, *A note concerning Fox's paper on Fenchel's conjecture*, Proc. Amer. Math. Soc. **88** (1983), no. 4, 584–586, DOI 10.2307/2045442. MR702279

[22] Buddha Prasad Chetiya, Sanjay Kumar Dutta, and Kuntala Patra, *Genera of compact Riemann surfaces with dihedral automorphism groups*, J. Ramanujan Math. Soc. **11** (1996), no. 2, 127–138. MR1429305

[23] José Luis Estévez and Milagros Izquierdo, *Non-normal pairs of non-Euclidean crystallographic groups*, Bull. London Math. Soc. **38** (2006), no. 1, 113–123, DOI 10.1112/S0024609305017984. MR2201610

[24] Jose Javier Etayo Gordejuela, *Sobre grupos de automorfismos de superficies de Klein* (Spanish), ProQuest LLC, Ann Arbor, MI, 1984. Thesis (Ph.D.)–Universidad Complutense de Madrid (Spain). MR3564052

[25] J. J. Etayo Gordejuela, *Nonorientable automorphisms of Riemann surfaces*, Arch. Math. (Basel) **45** (1985), no. 4, 374–384, DOI 10.1007/BF01198242. MR810257

[26] Ralph H. Fox, *On Fenchel's conjecture about F-groups*, Mat. Tidsskr. B **1952** (1952), 61–65. MR53937

[27] Leon Greenberg, *Maximal Fuchsian groups*, Bull. Amer. Math. Soc. **69** (1963), 569–573, DOI 10.1090/S0002-9904-1963-11001-0. MR148620

[28] Leon Greenberg, *Maximal groups and signatures*, Discontinuous groups and Riemann surfaces (Proc. Conf., Univ. Maryland, College Park, Md., 1973), Princeton Univ. Press, Princeton, N.J., 1974, pp. 207–226. Ann. of Math. Studies, No. 79. MR0379835

[29] Grzegorz Gromadzki, *Abelian groups of automorphisms of compact nonorientable Klein surfaces without boundary*, Comment. Math. Prace Mat. **28** (1989), no. 2, 197–217. MR1024938

[30] W. Hall: Automorphisms and coverings of Klein surfaces, *PhD Thesis,* University of Southampton, 1977. `http://eprints.soton.ac.uk/259986/`

[31] W. J. Harvey, *Cyclic groups of automorphisms of a compact Riemann surface*, Quart. J. Math. Oxford Ser. (2) **17** (1966), 86–97, DOI 10.1093/qmath/17.1.86. MR201629

[32] A. Hurwitz, *Ueber algebraische Gebilde mit eindeutigen Transformationen in sich* (German), Math. Ann. **41** (1892), no. 3, 403–442, DOI 10.1007/BF01443420. MR1510753

[33] M. Izquierdo, *On Klein surfaces and dihedral groups*, Math. Scand. **76** (1995), no. 2, 221–232, DOI 10.7146/math.scand.a-12538. MR1354580

[34] Sadok Kallel and Denis Sjerve, *On the group of automorphisms of cyclic covers of the Riemann sphere*, Math. Proc. Cambridge Philos. Soc. **138** (2005), no. 2, 267–287, DOI 10.1017/S0305004104008096. MR2132169

[35] Ravi S. Kulkarni and Colin Maclachlan, *Cyclic p-groups of symmetries of surfaces*, Glasgow Math. J. **33** (1991), no. 2, 213–221, DOI 10.1017/S0017089500008247. MR1108745

[36] A. M. Macbeath, *The classification of non-euclidean plane crystallographic groups*, Canadian J. Math. **19** (1967), 1192–1205, DOI 10.4153/CJM-1967-108-5. MR220838

[37] C. Maclachlan and Y. Talu, *p-groups of symmetries of surfaces*, Michigan Math. J. **45** (1998), no. 2, 315–332, DOI 10.1307/mmj/1030132185. MR1637662

[38] Daniel J. Madden and Robert C. Valentini, *The group of automorphisms of algebraic function fields*, J. Reine Angew. Math. **343** (1983), 162–168, DOI 10.1515/crll.1983.343.162. MR705883

[39] Coy L. May, *Finite groups acting on bordered surfaces and the real genus of a group*, Rocky Mountain J. Math. **23** (1993), no. 2, 707–724, DOI 10.1216/rmjm/1181072586. MR1226198

[40] Coy L. May, *The symmetric crosscap number of a group*, Glasg. Math. J. **43** (2001), no. 3, 399–410, DOI 10.1017/S0017089501030038. MR1878584

[41] Jakob Nielsen, *The commutator group of the free product of cyclic groups* (Danish), Mat. Tidsskr. B **1948** (1948), 49–56. MR29378

[42] Jesús Rodríguez, *Some results on abelian groups of automorphisms of compact Riemann surfaces*, Riemann and Klein surfaces, automorphisms, symmetries and moduli spaces, Contemp. Math., vol. 629, Amer. Math. Soc., Providence, RI, 2014, pp. 283–297, DOI 10.1090/conm/629/12569. MR3289647

[43] Jesús Rodríguez Martín de los Santos, *Abelian actions on compact bordered Klein surfaces*, Rev. R. Acad. Cienc. Exactas Fís. Nat. Ser. A Mat. RACSAM **111** (2017), no. 1, 189–204, DOI 10.1007/s13398-016-0287-1. MR3596048

[44] Jesús Rodríguez: Abelian actions on compact nonorientable Riemann surfaces, *submitted.*

[45] Jesús Rodríguez: Dihedral groups of automorphisms of compact unbordered Klein surfaces, *in preparation.*

[46] David Singerman, *Finitely maximal Fuchsian groups*, J. London Math. Soc. (2) **6** (1972), 29–38, DOI 10.1112/jlms/s2-6.1.29. MR322165

[47] David Singerman and Paul Watson, *Non-maximal cyclic group actions on compact Riemann surfaces*, Rev. Mat. Univ. Complut. Madrid **10** (1997), no. 2, 423–439. MR1605674

[48] Ewa Tyszkowska, *On (q,n)-gonal pseudo-real Riemann surfaces*, Internat. J. Math. **28** (2017), no. 13, 1750095, 14, DOI 10.1142/S0129167X17500951. MR3737073

[49] A. Wiman: Über die hyperelliptischen Curven und diejenigen vom Geschlechte $p = 3$, welche eindeutigen Transformationen in sich zulassen, *Bihang Kongl. Svenska Vetenskaps-Akademiens Handlingar* (Stockholm 1895-96).

[50] Aaron Wootton, *Multiple prime covers of the Riemann sphere*, Cent. Eur. J. Math. **3** (2005), no. 2, 260–272, DOI 10.2478/BF02479202. MR2129913

[51] A. Wootton, *The full automorphism group of a cyclic p-gonal surface*, J. Algebra **312** (2007), no. 1, 377–396, DOI 10.1016/j.jalgebra.2007.01.018. MR2320463

[52] Reza Zomorrodian, *Group action on genus 7 curves and their Weierstrass points*, Q. J. Math. **61** (2010), no. 4, 511–540, DOI 10.1093/qmath/hap015. MR2735837

Departamento de Matemáticas Fundamentales, Facultad de Ciencias, Universidad Nacional de Educación a Distancia, Madrid, Spain

Email address: `eb@mat.uned.es`

Departamento de Matemáticas Fundamentales, Facultad de Ciencias, Universidad Nacional de Educación a Distancia, Madrid, Spain

Email address: `jcirre@mat.uned.es`

Contemporary Mathematics
Volume **766**, 2021
https://doi.org/10.1090/conm/766/15375

Hyperbolic structures for Artin-Tits groups of spherical type

Matthieu Calvez and Bert Wiest

ABSTRACT. In this mostly expository paper we present several candidates for hyperbolic structures on irreducible Artin-Tits groups of spherical type and elucidate some relations between them. Most constructions are algebraic analogues of well-known hyperbolic structures on Artin braid groups coming from curve graphs and arc graphs of punctured disks. These constructions are not known to be hyperbolic structures in general. On the other hand, the purely algebraic (Garside theoretical) construction of the *absorbable elements* yields the only candidate which is known to be a hyperbolic structure, although its geometric meaning remains mysterious.

1. Introduction

Given a group G and a generating set X of G, the word metric d_X turns G into a metric space; this space is $(1,1)$-quasi-isometric to $\Gamma(G,X)$, the Cayley graph of G with respect to X, endowed with the usual graph metric where each edge is identified to an interval of length 1. Hyperbolic structures on groups were recently introduced and studied in [**1**]. A *hyperbolic structure* on a group G is a generating set X of G such that (G, d_X) is Gromov-hyperbolic; note that X must be infinite whenever G is not itself hyperbolic.

In this paper, we are interested in hyperbolic structures on Artin-Tits groups of spherical type. One motivation is trying to prove that irreducible Artin-Tits groups of spherical type are hierarchically hyperbolic [**3, 4**], where the hierarchical structure should come from the hierarchy of parabolic subgroups of these groups.

After [**9, 10**], we know that irreducible Artin-Tits groups of spherical type admit non-trivial hyperbolic structures; i.e. each irreducible Artin-Tits group of spherical type A contains an (infinite) generating set X^A_{abs} such that the corresponding Cayley graph $\Gamma(A, X^A_{abs})$ is a Gromov-hyperbolic metric space with *infinite diameter*. This hyperbolic structure was defined in a purely algebraic manner, using only the Garside structure on A, and it consists of the set of the so-called *absorbable elements* (to which we must add the cyclic subgroup generated by the square of the so-called Garside element if A is of dihedral type). We shall briefly review this construction in Section 4.

2020 *Mathematics Subject Classification.* Primary 20F65, 20F36.

The first author is supported by FONDECYT grant 1180335; he also acknowledges support by Anillo PIA-CONICYT ACT 1415, CONICYT PAI 79160023, MTM2016-76453-C2-1-P and FEDER.

Unfortunately, these absorbable elements are poorly understood – for instance we do not know any polynomial-time algorithm which recognizes whether any given element belongs to X^A_{abs} – and this makes it quite difficult to work with the graph $\Gamma(A, X^A_{abs})$.

In this paper we generalize to any irreducible Artin-Tits group of spherical type some well-known hyperbolic structures on Artin's braid groups with $n+1$ strands $\mathcal{B}_{n+1}$ (a.k.a. Artin-Tits groups of type A_n), $n \geqslant 3$.

Because it can be identified with the mapping class group of a $n+1$ times punctured disk $\mathcal{D}_{n+1}$, Artin's braid group on $n+1$ strands admits nice actions on the curve graph of $\mathcal{D}_{n+1}$ (denoted by $\mathcal{C}(\mathcal{D}_{n+1})$), the arc graph of $\mathcal{D}_{n+1}$ (denoted by $\mathcal{A}(\mathcal{D}_{n+1})$) and the graph of arcs in $\mathcal{D}_{n+1}$ both of whose extremities lie in $\partial\mathcal{D}_{n+1}$ (denoted by $\mathcal{A}_\partial(\mathcal{D}_{n+1})$). All these graphs can be shown to be connected and Gromov-hyperbolic; this was first shown in [**22**] but the circle of ideas around [**18, 27**] provides simpler arguments.

All these actions are cobounded (actually cocompact); according to a standard argument (Lemma 2.5, close in spirit to Svarc-Milnor's lemma [**1**, Section 3.2]), we extract from each of these actions a hyperbolic structure on $\mathcal{B}_{n+1}$, which consists of the union of the stabilizers of a (finite) family of representatives of the orbits of vertices.

Each of these generating sets can be algebraically described in terms of the *parabolic subgroups* of the Artin-Tits group of type A_n, allowing to extend the definitions to any irreducible Artin-Tits group of spherical type. Given an irreducible Artin-Tits group of spherical type A, we define:

- X^A_P is the union of all proper irreducible standard parabolic subgroups of A and the cyclic subgroup generated by the square of the Garside element;
- X^A_{NP} is the union of the normalizers of all proper irreducible standard parabolic subgroups of A;
- X^A_{abs} is the set of absorbable elements (together with the cyclic subgroup generated by the square of the Garside element, if A is of dihedral type: 2 generators).

Note that X^A_{NP} contains the cyclic subgroup generated by the square of the Garside element (which is central). Similarly for X^A_{abs}: any power of the Garside element can be written as a product of at most 3 absorbable elements, provided A is not of dihedral type [**9**, Example 3]. Therefore the center of A has bounded diameter with respect to the word metric on A induced by any of the above generating sets.

We then study the relationships between X^A_P, X^A_{NP} and X^A_{abs}. Following [**1**], given two generating sets X, Y of a group G, we write $X \preccurlyeq Y$ if the identity map from (G, d_Y) to (G, d_X) is Lipschitz (or equivalently, if $\sup_{y\in Y} d_X(1_G, y) < \infty$). The sets X and Y are *equivalent* if both $X \preccurlyeq Y$ and $Y \preccurlyeq X$ hold (or equivalently, if the identity map is a bilipschitz equivalence between (G, d_X) and (G, d_Y)).

Table 1 summarizes the main contents of this paper, for any irreducible Artin-Tits group of spherical type A with at least 3 generators. Vertical arrows indicate the identity of A. The *graph of irreducible parabolic subgroups* and the *additional length graph* (denoted by $\mathcal{C}_{parab}(A)$ and $\mathcal{C}_{AL}(A)$, respectively) were defined in [**13**] and [**9**], respectively. For Artin-Tits groups *of type A*, all the mentioned generating sets are hyperbolic structures. In any case, all spaces under consideration have infinite diameter (Corollary 4.13).

Gen. set	Cayley graph		Hyperbolicity
X_P^A	$\Gamma(A, X_P^A)$	generalized $\mathcal{A}_\partial(\mathcal{D}_{n+1})$	Conjectured
$\curlywedge$	$\downarrow$	Lipschitz. Conj. 4.18: not equivalent	
X_{NP}^A	$\Gamma(A, X_{NP}^A)$	generalized $\mathcal{C}(\mathcal{D}_{n+1})$, q.isom. to $\mathcal{C}_{parab}$ [**13**]	Conjectured
$\curlywedge$	$\downarrow$	Lipschitz [**2, 9**]. Quest. 4.14: equivalent ?	
X_{abs}^A	$\Gamma(A, X_{abs}^A)$	q.isom. to $\mathcal{C}_{AL}(A)$ [**9**]	Proved [**9, 10**]

TABLE 1. Summary of the main results presented in the paper when A has at least 3 generators.

A space similar to $\Gamma(A, X_P^A)$ was defined in [**11**]; that paper gives necessary and sufficient conditions on an Artin-Tits group (not necessarily of spherical type) for the union of its standard parabolic subgroups of spherical type to be a hyperbolic structure. However, when the group is itself of spherical type, this results in a Cayley graph of finite diameter. It is also interesting to note that a construction analogous to $\mathcal{C}_{parab}(A)$ has been recently announced –and conjectured to be hyperbolic– for each Artin-Tits group of FC type [**25**].

When A is of spherical dihedral type, all three sets X_P^A, X_{NP}^A and X_{abs}^A are hyperbolic structures but their orders differ slightly, namely X_P^A and X_{NP}^A are equivalent to each other and smaller than X_{abs}^A – see Proposition 4.11; also, all corresponding Cayley graphs have infinite diameter.

The plan of the paper is as follows. In Section 2 we review necessary definitions and facts about Artin-Tits groups of spherical type; we also prove Lemma 2.5. The latter is used in Section 3 to describe hyperbolic structures on $\mathcal{B}_{n+1}$; at the same time, these structures are reinterpreted algebraically in terms of parabolic subgroups of $\mathcal{B}_{n+1}$. In Section 4, we justify the assertions and present the conjectures from Table 1.

2. Preliminaries

2.1. Preliminaries on Artin-Tits groups of spherical type. Let S be a finite set. A *Coxeter matrix* over S is a symmetric matrix $M = (m_{s,t})_{s,t\in S}$ with $m_{s,s} = 1$ for $s \in S$ and $m_{s,t} \in \mathbb{N}_{\geqslant 2} \cup \{\infty\}$ for $s \neq t \in S$. The *Coxeter graph* $\Gamma = \Gamma_S = \Gamma(M)$ associated to M is a labeled graph whose vertices are the elements of S and in which two distinct vertices are connected by an edge labeled by $m_{s,t}$ whenever $m_{s,t} \geqslant 3$.

For two symbols a, b and an integer m, define

$$\Pi(a, b; m) = \begin{cases} (ab)^k & \text{if } m = 2k, \\ (ab)^k a & \text{if } m = 2k+1. \end{cases}$$

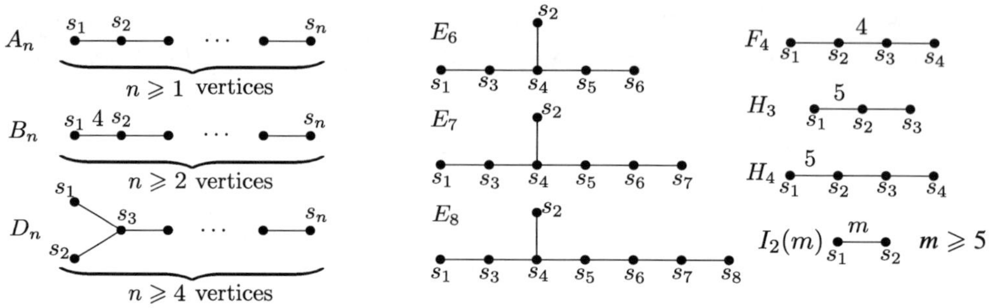

FIGURE 1. Coxeter's classification of irreducible Artin-Tits groups of spherical type

The Artin-Tits system associated to M (or to Γ) is the pair (A, S) where $A = A_S = A_\Gamma$ is the group presented by

$$\langle S \mid \Pi(s,t;m_{s,t}) = \Pi(t,s;m_{s,t}), \forall s \neq t \in S, \text{with } m_{s,t} < \infty \rangle .$$

As relations are homogeneous, there is a well-defined homomorphism $\ell : A_S \longrightarrow \mathbb{Z}$ which sends each $s \in S$ to 1. The *rank* of A_S is the number of elements of S.

Given a subset T of S, denote by Γ_T the subgraph of Γ_S induced by the vertices in T and by A_T the subgroup of A_S generated by T. It is a theorem of Van der Lek [**29**] that (A_T, T) is the Artin-Tits system associated to the graph Γ_T. Such a subgroup A_T ($T \subset S$) is called a *standard parabolic subgroup* of A_S. A *parabolic subgroup* of A_S is a subgroup which is conjugate to some standard parabolic subgroup.

The group A_S is said to be *irreducible* if the graph Γ_S is connected, equivalently if A_S cannot be expressed as the direct product of two standard parabolic subgroups. The group A_S is of *spherical type* if the Coxeter group obtained as the quotient of A_S by the normal closure of the squares of elements in S is finite. The group A_S is said to be of *dihedral type* if it has rank 2 (in this case, the corresponding Coxeter group is exactly the usual dihedral group).

Irreducible Artin-Tits groups of spherical type have been classified [**12**] (see also [**19**] for a modern exposition) into type A_n ($n \geqslant 1$), B_n ($n \geqslant 2$), D_n ($n \geqslant 4$), E_6, E_7, E_8, F_4, H_3, H_4 and $I_2(m)$ ($m \geqslant 5$). Figure 1 summarizes this classification. Note that for $n \geqslant 1$, the Artin-Tits group of type A_n is precisely Artin's braid group $\mathcal{B}_{n+1}$. In this paper, we will be concerned only with irreducible Artin-Tits groups of spherical type. For an integer $m = 3, 4$, we write also $I_2(m)$ for A_2 and B_2 respectively.

We recall that each irreducible Artin-Tits groups of spherical type $A = A_S$ can be equipped with a finite Garside structure – this was proved in [**7**, **14**]. The reader is referred to [**17**, Section 2] for a gentle introduction to Garside structures. We just recall the existence of a Garside element Δ_S; either Δ_S or Δ_S^2 generates the center of A_S and we denote by Ω_S this minimal central element. (It is known that $\Omega_S = \Delta_S$ for groups of type A_1, B_n ($n \geqslant 2$), D_n (n even), E_7, E_8, F_4, H_3, H_4 and $I_2(m)$ (m even), and $\Omega_S = \Delta_S^2$ for groups of type A_n ($n \geqslant 2$), D_n (n odd), E_6, and $I_2(m)$ (m odd)). Similarly, for each irreducible standard parabolic subgroup A_T ($T \subset S$), there is a Garside element Δ_T and $\Omega_T = \Delta_T$ or Δ_T^2 generates the center of A_T. Moreover, if P is any irreducible parabolic subgroup of A_S, then

$P = a^{-1}A_Ta$ for some element $a \in A$ and some $T \subset S$ – note that this expression is by no means unique. However there is a well-defined element $\Omega_P = a^{-1}\Omega_T a$ which does not depend on a nor on T in the above expression of P and generates the center of P (see [**13**]).

The two following results on conjugacy of parabolic subgroups will be useful later.

PROPOSITION 2.1. [**26**, Theorem 6.1] *Let $A = A_S$ be an irreducible Artin-Tits group of spherical type and let $T \subset S$ such that A_T is an irreducible Artin-Tits group. Then*

$$N_{A_S}(A_T) = Z_{A_S}(\Omega_T).$$

PROPOSITION 2.2. [**13**, Section 11] *Let $A = A_S$ be an irreducible Artin-Tits group of spherical type. Let P, Q be two irreducible parabolic subgroups of A_S and suppose that Ω_P and Ω_Q commute. Then there exists $g \in A$ and $T_1, T_2 \subset S$ such that $g^{-1}Pg = A_{T_1}$ and $g^{-1}Qg = A_{T_2}$.*

We conclude this section with a few words on the case of dihedral type Artin-Tits groups.

PROPOSITION 2.3. [**24**, Theorem 6.2 (2)] *Let $m \geqslant 3$ be an integer. Let A be an Artin-Tits group of type $I_2(m)$. Then the quotient $A/\langle \Delta^2 \rangle$ has a finite index subgroup isomorphic to a free group of rank $m-1$.*

COROLLARY 2.4. *Let m and A be as in Proposition 2.3. Let X be any finite generating set of A. Then the generating set $X \cup \langle \Delta^2 \rangle$ is a hyperbolic structure on A. All hyperbolic structures obtained in this way are equivalent.*

PROOF. Let $\overline{X}$ be the set of cosets of elements of X mod $\langle \Delta^2 \rangle$; this generates $A/\langle \Delta^2 \rangle$. Observe that $\Gamma(A, X \cup \langle \Delta^2 \rangle)$ and $\Gamma(A/\langle \Delta^2 \rangle, \overline{X})$ are quasi-isometric spaces. From Proposition 2.3 it follows applying Svarc-Milnor lemma that $\Gamma(A/\langle \Delta^2 \rangle, \overline{X})$ is quasi-isometric to a $m-1$-regular tree, hence hyperbolic. Therefore, for any X, $\Gamma(A, X \cup \langle \Delta^2 \rangle)$ is quasi-isometric to a regular tree, whence the claim. □

2.2. Cocompact actions on connected graphs. The unique result in this subsection will allow us to exhibit a number of hyperbolic structures on Artin's braid group $\mathcal{B}_{n+1}$. Although this is a standard fact, we include a stand-alone proof for the convenience of the reader.

LEMMA 2.5. [**22**, Lemma 3.2][**20**, Lemma 10] *Let X be a connected graph and denote by d_X the graph-metric on X obtained by identifying each edge to an interval of length 1. Let G be a group acting by isometries on X such that*

(i) *there exists a* finite *set A of vertices of X such that every vertex orbit has a (non-necessarily unique) representative in A,*
(ii) *there exists a* finite *set B of edges of X such that every edge orbit has a (non-necessarily unique) representative in B,*
(iii) *the set $T = \bigcup_{a \in A} Stab_G(a)$ generates G.*

Then (G, d_T) is quasi-isometric to (X, d_X).

PROOF. It is known that (X, d_X) is (1,1)-quasi-isometric to $(\mathcal{V}_X, d_X)$, the vertex set of X endowed with the restriction of the graph metric on X; we focus on $(\mathcal{V}_X, d_X)$ rather than X itself.

Let $M_1 = Diam(A)$. Let $A_0 \subset A$ having exactly one representative of each vertex orbit under the action of G; for each $a \in A$, choose $g_a \in G$ such that $g_a{\cdot}a \in A_0$ and let $M_2 = \max\{\|g_a\|_T,\ a \in A\}$ (where we denote $\|g_a\|_T = d_T(1_G, g_a)$).

Let $\{\{u_i, v_i\}, 1 \leqslant i \leqslant m\}$ be an enumeration of the finite set B and fix $\alpha_i, \beta_i \in G$ so that $\alpha_i^{-1} \cdot u_i \in A$ and $\beta_i^{-1} \cdot v_i \in A$. Let

$$M_3 = \max_{1 \leqslant i \leqslant m} \{\|\alpha_i\|_T, \|\beta_i\|_T\}.$$

Fix $a_0 \in A$ and define a map $\psi\colon G \longrightarrow \mathcal{V}_X$ by $g \mapsto g \cdot a_0$. We first prove that the map $\psi\colon (G, d_T) \to (\mathcal{V}_X, d_X)$ is Lipschitz. Let $g, h \in G$, let $n = \|g^{-1}h\|_T$ and write $g^{-1}h = t_1 \dots t_n$, with $t_i \in T$. We have

$$d_X(\psi(g), \psi(h)) = d_X(a_0, g^{-1}h \cdot a_0) = d_X(a_0, t_1 \dots t_n \cdot a_0)$$

and

$$\begin{aligned} d_X(a_0, t_1 \dots t_n \cdot a_0) &\leqslant d_X(a_0, t_1 \cdot a_0) + d_X(t_1 \cdot a_0, t_1 t_2 \cdot a_0) + \dots \\ &\qquad \dots + d_X(t_1 \dots t_{n-1} \cdot a_0, t_1 \dots t_n \cdot a_0) \\ &= \sum_{i=1}^{n} d_X(a_0, t_i \cdot a_0). \end{aligned}$$

But for any $t \in T$ there is at least some $a_t \in A$ so that $t \cdot a_t = a_t$; so we have

$$\begin{aligned} d_X(a_0, t \cdot a_0) &\leqslant d_X(a_0, a_t) + d_X(a_t, t \cdot a_0) \\ &= d_X(a_0, a_t) + d_X(t \cdot a_t, t \cdot a_0) \\ &= 2 d_X(a_0, a_t) \leqslant 2M_1. \end{aligned}$$

In summary, we have

$$d_X(\psi(g), \psi(h)) \leqslant 2M_1 d_T(g, h).$$

On the other hand, given any vertex x of X, there exist $a \in A$ and $g \in G$ so that $x = g \cdot a$. If $a, a' \in A$ and $g, g' \in G$ are such that $g \cdot a = g' \cdot a' = x$, we have $g_a \cdot a = g_{a'} \cdot a' \in A_0$ whence $g'^{-1} g g_a^{-1} g'_a$ fixes a' and so lies in T. It follows that the set of elements $g \in G$ such that $g \cdot a = x$ for some $a \in A$ has diameter at most $1 + 2M_2$ in (G, d_T). We define $\varphi : \mathcal{V}_X \longrightarrow G$ in the following way: to each $x \in \mathcal{V}_X$ we associate some g in G such that $g^{-1} \cdot x \in A$. It is now clear that $\varphi \circ \psi$ is at distance at most $2M_2 + 1$ from Id_G and $\psi \circ \varphi$ is at distance at most M_1 from $Id_{\mathcal{V}_X}$.

We now show that $\varphi\colon (\mathcal{V}_X, d_X) \longrightarrow (G, d_T)$ is also Lipschitz. Let x_1, x_2 be adjacent vertices of X. There is some g in G and $1 \leqslant i \leqslant m$ such that $\{g{\cdot}x_1, g{\cdot}x_2\} = \{u_i, v_i\}$ is an edge belonging to B. By construction, $\alpha_i^{-1} g \cdot x_1 = \alpha_i^{-1} \cdot u_i \in A$ and $\beta_i^{-1} g \cdot x_2 = \beta_i^{-1} \cdot v_i \in A$; it follows that $d_T(\varphi(x_1), g^{-1}\alpha_i) \leqslant 2M_2 + 1$ and $d_T(\varphi(x_2), g^{-1}\beta_i) \leqslant 2M_2 + 1$.

We then have

$$\begin{aligned} d_T(\varphi(x_1), \varphi(x_2)) &\leqslant d_t(\varphi(x_1), g^{-1}\alpha_i) + d_T(g^{-1}\alpha_i, g^{-1}\beta_i) + d_T(g^{-1}\beta_i, \varphi(x_2)) \\ &\leqslant 4M_2 + 2 + d_T(\alpha_i, \beta_i) \\ &\leqslant 4M_2 + 2 + 2M_3. \end{aligned}$$

In summary, if x_1, x_2 are arbitrary vertices of X then

$$d_T(\varphi(x_1), \varphi(x_2)) \leqslant 2(2M_2 + M_3 + 1) \cdot d_X(x_1, x_2). \qquad \square$$

3. Hyperbolic structures for $\mathcal{B}_{n+1}$

In this section, we review some Gromov-hyperbolic graphs on which Artin's braid group $\mathcal{B}_{n+1}$ acts. Lemma 2.5 is put to work in order to derive hyperbolic structures on $\mathcal{B}_{n+1}$ from these actions. These hyperbolic structures are interpreted in an algebraic context.

3.1. The curve graph. Fix an integer $n \geqslant 3$. Artin's braid group on $n+1$ strands is identified with the mapping class group of a closed, $n+1$ times punctured disk $\mathcal{D}_{n+1}$ (with boundary fixed pointwise). We consider the following model for $\mathcal{D}_{n+1}$: the disk of diameter $[0, n+2]$ in $\mathbb{C}$ with the standard orientation, and $n+1$ punctures at $1, \ldots, n+1$ numbered accordingly $p_1, \ldots, p_{n+1}$. For $1 \leqslant i \leqslant n$, write s_i for the half-Dehn twist along an horizontal arc connecting p_i and p_{i+1}. Then $(\mathcal{B}_{n+1}, \{s_1, \ldots, s_n\})$ is exactly the Artin-Tits system associated to the Coxeter graph A_n.

The *curve graph* $\mathcal{C}(\mathcal{D}_{n+1})$ of the $n+1$ times punctured disk $\mathcal{D}_{n+1}$ is the (locally-infinite) graph defined as follows:

- its vertices are isotopy classes of essential simple closed curves in $\mathcal{D}_{n+1}$, that is isotopy classes of simple closed curves in $\mathcal{D}_{n+1}$ surrounding at least 2 and at most n punctures;
- two distinct vertices are connected by an edge if and only if the corresponding isotopy classes can be realized disjointly.

According to [**22**], this graph is connected and Gromov-hyperbolic; a simpler proof can be found in [**18**]. The group $\mathcal{B}_{n+1}$ acts naturally by isometries (simplicial automorphisms) on $\mathcal{C}(\mathcal{D}_{n+1})$.

An essential simple closed curve in $\mathcal{D}_{n+1}$ (a vertex of $\mathcal{C}(\mathcal{D}_{n+1})$) is said to be *round* or *standard* if it is isotopic to a geometric circle. Note that there are exactly $\frac{n(n+1)}{2} - 1$ such vertices: each of them can be described by two integers $1 \leqslant i < j \leqslant n+1$, where p_i (p_j respectively) is the first (the last respectively) puncture surrounded; we denote this curve by c_{ij}. For $1 \leqslant i < j \leqslant n+1$, we denote by $\mathcal{B}_{ij}$ the set of braids whose support is enclosed by c_{ij}; this is the standard irreducible parabolic subgroup generated by $\{s_t, i \leqslant t \leqslant j-1\}$, which is isomorphic to a braid group on $j-i+1$ strands. The square of the Garside element of this group Δ_{ij}^2 is the Dehn twist around the curve c_{ij}.

LEMMA 3.1. *Let $X_\mathcal{C}$ be the union of stabilizers of all standard curves. Then $\mathcal{B}_{n+1}$ equipped with the word metric $d_{X_\mathcal{C}}$ is quasi-isometric to the curve graph of the $n+1$ times punctured disk.*

PROOF. First recall that the curve graph is connected. Also, the action of each generator s_i on the curve graph fixes some standard curve; therefore, $X_\mathcal{C}$ generates $\mathcal{B}_{n+1}$. According to [**15**, Section 1.3.1], the set A of all standard curves contains a representative of each vertex orbit under the action of $\mathcal{B}_{n+1}$. According to [**21**, Proposition 4.4], the *finite* set of edges in $\mathcal{C}(\mathcal{D}_{n+1})$ both of whose extremities are standard curves contains a representative of each edge orbit under the action of $\mathcal{B}_{n+1}$. Hence we see that the isometric action of $\mathcal{B}_{n+1}$ on the curve graph $\mathcal{C}(\mathcal{D}_{n+1})$ satisfies the hypotheses of Lemma 2.5. □

PROPOSITION 3.2. *Let $X_{NP}^{\mathcal{B}_{n+1}}$ be the union of all normalizers of proper irreducible standard parabolic subgroups of $\mathcal{B}_{n+1}$. Then $\mathcal{B}_{n+1}$ equipped with the word* *metric* $d_{X_{NP}^{\mathcal{B}_{n+1}}}$ *is quasi-isometric to the curve graph of the* $n+1$ *times punctured disk. In particular,* $X_{NP}^{\mathcal{B}_{n+1}}$ *is a hyperbolic structure on* $\mathcal{B}_{n+1}$.

PROOF. According to [**15**, Fact 3.8], the stabilizer of any curve c is exactly the centralizer of the Dehn twist around c; if $c = c_{ij}$ is standard, this is the centralizer of the square of the Garside element Δ_{ij} of $\mathcal{B}_{ij}$ (this is the same as the centralizer of s_i if $j = i+1$). In other words, in view of Proposition 2.1, $Stab_{\mathcal{B}_{n+1}}(c_{ij})$ is the same as the normalizer of the standard parabolic subgroup $\mathcal{B}_{ij}$. □

3.2. The boundary connecting arc graph. Again, we fix an integer $n \geqslant 3$; we keep the same model for the punctured disk $\mathcal{D}_{n+1}$ and s_i, $1 \leqslant i \leqslant n$, are the usual Artin generators of $\mathcal{B}_{n+1}$.

The *boundary connecting arc graph* $\mathcal{A}_\partial(\mathcal{D}_{n+1})$ of the $n+1$ times punctured disk $\mathcal{D}_{n+1}$ is the (locally-infinite) graph defined as follows:

- its vertices are isotopy classes of essential (i.e. not homotopic into a subset of $\partial\mathcal{D}_{n+1}$) simple arcs in $\mathcal{D}_{n+1}$ both of whose endpoints lie in the boundary $\partial\mathcal{D}_{n+1}$ and where in addition the endpoints are allowed to slide freely along the boundary during the isotopy.
- two distinct vertices are connected by an edge if and only if the corresponding isotopy classes of arcs can be realized disjointly.

The argument given in [**18**] can be easily adapted to show that $\mathcal{A}_\partial(\mathcal{D}_{n+1})$ is a Gromov-hyperbolic connected graph. The group $\mathcal{B}_{n+1}$ acts by isometries on $\mathcal{A}_\partial(\mathcal{D}_{n+1})$.

Given $1 \leqslant i \leqslant n$, let a_i be the arc consisting simply of the vertical line with real part $i + \frac{1}{2}$; let $A = \{a_i, 1 \leqslant i \leqslant n\}$. Then A exhausts all the orbits of vertices of $\mathcal{A}_\partial(\mathcal{D}_{n+1})$ under the action of $\mathcal{B}_{n+1}$ (the orbit of a boundary connecting arc a is uniquely characterized by the number of punctures in any of the two connected components of $\mathcal{D}_{n+1} \setminus a$). Two different disjoint boundary connecting arcs (forming an edge of $\mathcal{A}_\partial(\mathcal{D}_{n+1})$) can be transformed simultaneously into two arcs a_i, a_j for some $1 \leqslant i \neq j \leqslant n$ by the action of a braid: this says that there is a finite system B of representatives of the orbits of edges. Finally, each half-twist s_i stabilizes some arc in A. (Indeed, s_i fixes a_{i-1} if $i \geqslant 2$, and s_1 fixes a_2.) Therefore, all the hypotheses of Lemma 2.5 are satisfied.

It follows from Lemma 2.5 that the union $X_{\mathcal{A}_\partial}$ of stabilizers of arcs a_i, $1 \leqslant i \leqslant n$, is a hyperbolic structure on $\mathcal{B}_{n+1}$ and that $(\mathcal{B}_{n+1}, d_{X_{\mathcal{A}_\partial}})$ is quasi-isometric to $\mathcal{A}_\partial(\mathcal{D}_{n+1})$.

LEMMA 3.3. *If* n *is odd and* $i = \frac{n+1}{2}$, *the stabilizer of* a_i *is generated by* Δ *and the direct product of standard parabolic subgroups* $\langle s_1, \dots, s_{i-1}\rangle \times \langle s_{i+1} \dots s_n\rangle$. *In any other case, the stabilizer of* a_i *is generated by* Δ^2 *and the direct product of standard parabolic subgroups* $\langle s_1, \dots, s_{i-1}\rangle \times \langle s_{i+1} \dots s_n\rangle$.

PROOF. The arc a_i cuts $\mathcal{D}_{n+1}$ into two connected components which are punctured disks with i and $n+1-i$ punctures respectively. Except if $2i = n+1$, an element g in the stabilizer of a_i must preserve the boundary of each of these two disks. Following the conventions in [**21**], this means that g can be written as a product of an interior braid (which is an element of $\langle s_1, \dots, s_{i-1}\rangle \times \langle s_{i+1} \dots s_n\rangle$) and a tubular braid of the form

$$((s_i \dots s_1)(s_{i+1} \dots s_2) \dots (s_n \dots s_{n-i+1})(s_{n-i+1} \dots s_1) \dots (s_n \dots s_{i+1}))^k$$

for some $k \in \mathbb{Z}$ (a pure braid on 2 fat strands). We observe however that the above braid can be written as

$$\Delta^{2k}\Delta^{-2k}_{\langle s_1,\dots,s_{i-1}\rangle}\Delta^{-2k}_{\langle s_{i+1}\dots s_n\rangle}.$$

If $2i = n+1$, an element of the stabilizer of a_i can be written as the product of an interior braid as above and a tubular braid (on 2 fat strands) which needs not be pure, that is a power of

$$(s_i \dots s_1)(s_{i+1} \dots s_2) \dots (s_n \dots s_{n-i+1}),$$

which we can rewrite as

$$\Delta\Delta^{-1}_{\langle s_1,\dots,s_{i-1}\rangle}\Delta^{-1}_{\langle s_{i+1}\dots s_n\rangle}. \qquad \square$$

PROPOSITION 3.4. *Let $X_P^{\mathcal{B}_{n+1}}$ be the subset of $\mathcal{B}_{n+1}$ consisting of the union of all proper irreducible standard parabolic subgroups together with the center $\langle\Delta^2\rangle$. Then $\mathcal{B}_{n+1}$ equipped with the word metric $d_{X_P^{\mathcal{B}_{n+1}}}$ is quasi-isometric to $\mathcal{A}_\partial(\mathcal{D}_{n+1})$. In particular, $X_P^{\mathcal{B}_{n+1}}$ is a hyperbolic structure on $\mathcal{B}_{n+1}$.*

PROOF. Let $X_{\mathcal{A}_\partial}$ be the union of stabilizers of all arcs a_i, $1 \leqslant i \leqslant n$, as above. Then $X_{\mathcal{A}_\partial}$ contains the generating set $X_P^{\mathcal{B}_{n+1}}$ (every element in a proper irreducible standard parabolic subgroup fixes some arc a_i). On the other hand, each element in $X_{\mathcal{A}_\partial}$ can be written as a product of at most 6 elements of $X_P^{\mathcal{B}_{n+1}}$, according to the above claim (observe that Δ can be decomposed as a product of three elements of proper irreducible standard parabolic subgroups). This shows that both generating sets are equivalent. The remaining part of the statement follows as $X_{\mathcal{A}_\partial}$ is a hyperbolic structure and $(\mathcal{B}_{n+1}, d_{X_{\mathcal{A}_\partial}})$ is quasi-isometric to $\mathcal{A}_\partial(\mathcal{D}_{n+1})$. $\square$

4. Generalizing to Artin-Tits groups of spherical type

Throughout this section, $A = A_S$ is some fixed non-cyclic irreducible Artin-Tits group of spherical type.

4.1. Definitions.

4.1.1. *Generalized boundary connecting arc graph.* Let X_P^A be the union of all proper irreducible standard parabolic subgroups of A and the cyclic subgroup generated by the square of Δ_S. It is clear that X_P^A generates A because each generator in S belongs to some proper irreducible standard parabolic subgroup.

PROPOSITION 4.1. *Suppose that one of the following holds.*

(i) *A is of dihedral type, or*
(ii) *A is of type A_n ($n \geqslant 3$).*

Then X_P^A is a hyperbolic structure on A.

PROOF. We refer to Proposition 4.11 for the proof of (i). For (ii), this is Proposition 3.4: in this case, the Cayley graph $\Gamma(A, X_P^A)$ is quasi-isometric to the graph $\mathcal{A}_\partial(\mathcal{D}_{n+1})$. $\square$

CONJECTURE 4.2. *Let A be any non-cyclic irreducible Artin-Tits group of spherical type. The set X_P^A is a hyperbolic structure on A.*

REMARK 4.3. After this paper was accepted, the first author proved that the set X_P^A is also a hyperbolic structure when A is of type B_n ($n \geqslant 3$) [**8**].

4.1.2. *Generalized curve graph.* Let X^A_{NP} be the union of the normalizers of all proper irreducible standard parabolic subgroups of A. Of course, X^A_{NP} generates A, as each $s \in S$ normalizes any parabolic subgroup containing it and hence belongs to X^A_{NP}. Note also that any power of Δ^2_S normalizes every parabolic subgroup.

Recall from [**13**] the definition of the *graph of parabolic subgroups* of A, denoted $\mathcal{C}_{parab}(A)$:

- its vertices are proper irreducible parabolic subgroups of A,
- two distinct vertices P, Q are connected by an edge if the minimal central elements Ω_P and Ω_Q commute.

This graph was shown to be connected [**25**, Lemma 5.2] as soon as the Coxeter graph defining A has at least 3 vertices. The group A acts naturally (by conjugation on vertices) by isometries on $\mathcal{C}_{parab}(A)$. Observing that the set of *standard* parabolic subgroups is finite and using Proposition 2.2, Lemma 2.5 implies along the same lines as Lemma 3.1:

PROPOSITION 4.4. *Suppose that A is not of dihedral type. The graphs $\Gamma(A, X^A_{NP})$ and $\mathcal{C}_{parab}(A)$ are quasi-isometric.*

In the special case where A_S is of type A_n ($n \geqslant 3$), it is shown in [**13**] that $\mathcal{C}_{parab}(A_S)$ is exactly the same graph (isometric) as the curve graph of a $n+1$ times punctured disk. We have:

PROPOSITION 4.5. *Suppose that one of the following holds.*

(i) *A is of dihedral type, or*
(ii) *A is of type A_n ($n \geqslant 3$).*

Then X^A_{NP} is a hyperbolic structure on A.

PROOF. We refer to Proposition 4.11 for the proof of (i). For (ii), this is Proposition 3.2: in this case, the Cayley graph $\Gamma(A, X^A_{NP})$ is quasi-isometric to the graph $\mathcal{C}(\mathcal{D}_{n+1})$. □

REMARK 4.6. If A is of dihedral type, $\mathcal{C}_{parab}(A)$ is a totally disconnected space. Therefore Lemma 2.5 does not apply to the action of A on $\mathcal{C}_{parab}(A)$ and there is no quasi-isometry between $\Gamma(A, X^A_{NP})$ and $\mathcal{C}_{parab}(A)$. In this special case, X^A_{NP} is a hyperbolic structure but $\mathcal{C}_{parab}(A)$ is not hyperbolic.

CONJECTURE 4.7. *Let A be any non-cyclic irreducible Artin-Tits group of spherical type. The set X^A_{NP} is a hyperbolic structure on A.*

REMARK 4.8. After this paper was accepted, the first author proved this conjecture in the case when A is of type B_n ($n \geqslant 3$) [**8**].

4.1.3. *Absorbable elements and additional length graph.* Let X^A_{abs} be the set of absorbable elements (together with the cyclic subgroup generated by Δ^2 if A is of dihedral type). The definition of absorbable element of a Garside group was proposed in [**9**]. It is not difficult to see that each element in S is absorbable, hence X_{abs} generates A.

Recall from [**9**] the definition of the *additional length graph* $\mathcal{C}_{AL}(A)$: its vertices are right cosets of A modulo the cyclic subgroup generated by Δ; there is an edge between the vertices $a\langle\Delta\rangle$ and $b\langle\Delta\rangle$ if there is some u which is either simple (i.e. a positive prefix of Δ) or absorbable such that au belongs to the coset $b\langle\Delta\rangle$.

LEMMA 4.9. *$\mathcal{C}_{AL}(A)$ and $\Gamma(A, X^A_{abs})$ are quasi-isometric.*

PROOF. It is enough to check the quasi-isometry of the respective vertex sets. Any coset $a\langle\Delta\rangle$ has diameter at most 4 in $\Gamma(A, X^A_{abs})$: any two elements of such a coset differ by a power of Δ; and any power of Δ can be written as a product of at most 4 elements of X^A_{abs} in the dihedral case (because $\Delta^q = \Delta^{2\lfloor\frac{q}{2}\rfloor} \cdot a \cdot b \cdot b^{-1}a^{-1}\Delta$), and of at most 3 absorbable elements in the general case: (the argument of [**9**, Example 3] for braid groups with at least 4 strands can be easily adapted to any Artin-Tits group of rank $\geqslant 3$). For any coset $a\langle\Delta\rangle$, choose an arbitrary representative; this defines a map φ from the vertices of $\mathcal{C}_{AL}$ to A. Conversely, given $a \in A$, we associate to a the coset $\psi(a) = a\langle\Delta\rangle$. It is straightforward to check that these maps are quasi-inverses of each other with respect to the graph metrics on $\mathcal{C}_{AL}(A)$ and $\Gamma(A, X^A_{abs})$. It is also clear that ψ is 1-Lipschitz. On the other hand, as the set of simple elements (positive divisors of Δ) is finite, we can define M as the maximum of $d_{X^A_{abs}}(1, s\Delta^j)$, for s simple and $j \in \mathbb{Z}$. It then follows that φ is M-Lipschitz. □

The graph $\mathcal{C}_{AL}(A)$ was shown to be hyperbolic in [**9**]; therefore Lemma 4.9 shows:

PROPOSITION 4.10. *Let A be any non-cyclic irreducible Artin-Tits group of spherical type. The set X^A_{abs} is a hyperbolic structure on A.*

4.2. Comparisons and further results. In this subsection, we compare the different generating sets for A described above. We will first look at Artin-Tits groups of dihedral type, and then at all other types.

PROPOSITION 4.11. *Assume that A is of type $I_2(m)$ $(m \geqslant 3)$. Then we have*

$$X^A_P \sim X^A_{NP} \preccurlyeq X^A_{abs} \sim S \cup \langle\Delta^2\rangle,$$

where the middle inequality is strict. All the sets mentioned are hyperbolic structures on A and the corresponding Cayley graphs all have infinite diameter.

PROOF. Denote by a, b the generators of A. In order to prove that $X^A_P \sim X^A_{NP}$, we first observe that $X^A_P \subset X^A_{NP}$ whence $X^A_{NP} \preccurlyeq X^A_P$. For the converse inequality, it will be enough to show that the normalizer of $\langle a\rangle$ ($\langle b\rangle$ respectively) is generated by Ω and a (Ω and b respectively). In order to do so, we first observe that for any $g \in N_A(\langle a\rangle)$ we have $g^{-1}ag \in \langle a\rangle$, and using the abelianisation we see that $g^{-1}ag = a$, i.e., the normalizer of $\langle a\rangle$ coincides with $Z_A(a)$, the centraliser of a. This centraliser can be very easily computed using the algorithm in [**16**]. For m even, we get that $Z_A(a) = \langle a, \Delta\rangle$. For m odd, we obtain $\{a, \Pi(b, a; m-2)a^2\Pi(b, a; m-2)\}$ as a generating set; up to right and left multiplying the second generator by a, this gives the generating set $\{a, \Delta^2\}$. In any case, an element of $N_A(\langle a\rangle)$ can be written as $\Omega^p a^q$, which has length at most $(2 + \|\Delta\|_{X^A_P})$ with respect to $d_{X^A_P}$ – and similarly for $N_A(\langle b\rangle)$. It follows that the identity map from $(A, d_{X^A_{NP}})$ to $(A, d_{X^A_P})$ is $(2 + \|\Delta\|_{X^A_P})$-Lipschitz. This shows the first equivalence.

In order to prove the middle inequality, we note that the set of absorbable elements is finite with $4m - 8$ elements (see [**9**, Example 2.2] for $m = 3$). Letting $M = \sup\{\|x\|_{X^A_P},\ x \in X^A_{abs} \setminus \langle\Delta^2\rangle\}$, we see that the identity is an M-Lipschitz map from $(A, d_{X^A_{abs}})$ to $(A, d_{X^A_P})$, whence the middle inequality. This inequality is not an equivalence, because we have $d_{X^A_P}(Id, a^n) = 1$ for all n but $d_{X^A_{abs}}(Id, a^n) \to_{n\to\infty} +\infty$.

For the second equivalence, we use again that there are only finitely many absorbable elements. It follows from Corollary 2.4 that X^A_{abs} and $S \cup \langle \Delta^2 \rangle$ are equivalent hyperbolic structures on A.

Next we prove that X^A_P is a hyperbolic structure. The proof uses the "guessing geodesics lemma" [**23**, Theorem 3.15].

Given g, h vertices of X^A_P, consider the subgraph A_{gh} consisting of the projections to $\Gamma(A, X^A_P)$ of the set of geodesics in $\Gamma(A, S \cup \langle \Delta^2 \rangle)$ between g and h. Because the identity from $\Gamma(A, S \cup \langle \Delta^2 \rangle)$ to $\Gamma(A, X^A_P)$ is Lipschitz and $\Gamma(A, S \cup \langle \Delta^2 \rangle)$ is hyperbolic, the different subgraphs $A_{g,h}$ form thin triangles in $\Gamma(A, X^A_P)$ so they satisfy the second condition of [**23**, Theorem 3.15].

What happens if g, h are 1 apart in $\Gamma(A, X^A_P)$? Observe first that if $g, h \in A$ satisfy $g^{-1}h = a^n$ (or b^n, with $n \in \mathbb{Z}$), then in $\Gamma(A, S \cup \langle \Delta^2 \rangle)$, there is a unique geodesic between g and h, namely $g, ga, ga^2, \ldots, h$. Thus the subgraph A_{gh} has diameter 1 as well; hence the first condition in [**23**, Theorem 3.15] is also satisfied.

We conclude that $\Gamma(A, X^A_P)$ is hyperbolic.

Finally, we have to prove that $\Gamma(A, X^A_P)$ has infinite diameter. Let us assume, for a contradiction, that there exists some number L such that every element x of A can be written in the form

$$x = \Delta^{k_0} a^{k_1} b^{k_2} a^{k_3} b^{k_4} \ldots c^{k_L} \quad \text{or} \quad x = \Delta^{k_0} b^{k_1} a^{k_2} b^{k_3} a^{k_4} \ldots c^{k_L}$$

with $k_0, k_1, \ldots, k_L \in \mathbb{Z}$, and $c = a$ or $c = b$, according to the parity of L. Now, we recall from Proposition 2.3 that $\Gamma(A, S \cup \langle \Delta^2 \rangle)$ is quasi-isometric to an infinite regular $2(m-1)$-valent tree – in particular, as a metric space it has uncountably many ends. Also, powers of a and powers of b represent quasi-geodesics in this quasi-tree. However, the set of points in the tree represented by words as above has only countably many ends, leading to a contradiction. □

For the rest of the section, we now assume that A has rank at least 3.

PROPOSITION 4.12. *Suppose that A is an irreducible Artin-Tits group of spherical type with rank at least 3. Then $X^A_{abs} \preccurlyeq X^A_{NP} \preccurlyeq X^A_P$.*

PROOF. For the second inequality, it is enough to observe that $X^A_P \subset X^A_{NP}$ whence the identity map $(A, d_{X^A_P}) \longrightarrow (A, d_{X^A_{NP}})$ is 1-Lipschitz.

The first inequality was shown for braids in [**9**]: any braid in the stabilizer of a standard curve can be written as a product of 9 absorbable braids [**9**, Lemma 11]. This result has been generalized to the case of a general Artin-Tits group A of spherical type in the paper [**2**], where it is shown that any element which normalizes a proper irreducible standard parabolic subgroup of A is a product of at most 9 absorbable elements. This is to say that the identity map from $(A, d_{X^A_{NP}})$ to $(A, d_{X^A_{abs}})$ is 9-Lipschitz. □

COROLLARY 4.13. *The metric spaces $(A, d_{X^A_{NP}})$ and $(A, d_{X^A_P})$ have infinite diameter.*

PROOF. This is just a combination of Proposition 4.12 and Lemma 4.9 together with [**10**, Theorem 1.1], which asserts that $\mathcal{C}_{AL}(A)$ has infinite diameter. □

4.3. Open problems.

QUESTION 4.14. Is it true that the identity map $\Gamma(A, X^A_{NP}) \to \Gamma(A, X^A_{abs})$ is a quasi-isometry, equivalently, that the inequality $X^A_{abs} \preccurlyeq X^A_{NP}$ is not strict? (In the case of Artin braid group, this is asking whether $\mathcal{C}_{AL}(\mathcal{B}_n)$ is quasi-isometric to the curve graph of the punctured disk $\mathcal{C}(\mathcal{D}_n)$.)

A positive answer to Question 4.14 would of course imply that X^A_{NP} is a hyperbolic structure (and hence hyperbolicity of the graph of irreducible parabolic subgroups $\mathcal{C}_{parab}(A)$). It would also imply that Garside normal forms are unparameterized quasi-geodesics in $\mathcal{C}_{parab}(A)$ (since they are in $\mathcal{C}_{AL}(A)$); this would contrast with the recently announced example in **[28]** of a family of geodesics in a mapping class group whose shadows in the corresponding curve graph are not unparameterized quasi-geodesics.

Note that the answer to Question 4.14 is not even known in the specific case of Artin's braid groups – see [**9**, Conjecture 1]. Indeed, as pointed out to us by Ursula Hamenstädt, it is very much conceivable that $\mathcal{C}_{AL}(\mathcal{B}_n)$ is quasi-isometric to a tree.

QUESTION 4.15. Is every large piece of 2-dimensional quasi-flat in the Cayley graph of a mapping class group sent to a subset of bounded diameter by the projection to the curve complex?

This question is deliberately vague, but to the best of our knowledge, no result along these lines is known, except that *maximal-dimensional* quasi-flats are known to be squashed down to bounded diameter in the curve complex [**5**, **6**]. We will give a more precise version of the question below (Question 4.16), and we will show that a positive answer would imply a positive answer to Question 4.14.

Let us look at the Cayley graph of A with respect to the Garside generators, modulo the Δ-action: for any $z \in A$, all the vertices corresponding to elements of the form $x\Delta^k$ ($k \in \mathbb{Z}$) get identified, and if any two edges in the Cayley graph have the same endpoints after the Δ-action, then they get identified as well. The quotient space is quasi-isometric to the Cayley graph of $A/Z(A)$, and we will denote it $Cay(A)/\langle\Delta\rangle$ – see [**9**] for a detailed account.

We now discuss the connection between the geometry of this graph and the notion of an absorbable element. For completeness, recall that an element $y \in A$ is *absorbed* by $x \in A$ if $\inf(x) = \inf(y) = \inf(xy) = 0$ and $\sup(x) = \sup(y) = \sup(xy) = L$ [**9**, Definition 1, Lemma 3]. In this situation, there is an equilateral triangle with corners 1, x and xy (red in Figure 2), and with sides of length L (blue in the figure) representing the Garside normal form words for x, y, and xy. Let us denote the Garside normal form words by $x = x_1x_2 \dots x_L$, and $y = y_1y_2 \dots y_L$. Now, if $1 \leqslant \ell \leqslant L$, then x also absorbs $y_1 \dots y_\ell$. This means that in $Cay(A)/\langle\Delta\rangle$ we have $d(1, xy_1 \dots y_\ell) = L$: every vertex on the edge between x and xy is at the same distance (namely L) from the opposite corner of the triangle (namely 1). More generally, if $1 \leqslant \ell^y \leqslant \ell^x \leqslant L$, then $x_{L-\ell^x+1} \dots x_L$ absorbs $y_1 \dots y_{\ell^y}$, and $d(x_1 \dots x_{L-\ell^x}, xy_1 \dots y_{\ell^y}) = \ell^x$.

Another crucial observation is that the three sides of the triangle play completely symmetrical roles: if x absorbs y, then y absorbs $(xy)^{-1}\Delta^L$ and $\Delta^L(xy)^{-1}$ absorbs x. In particular, every vertex on every side of the triangle (between 1 and x, or between x and xy, or between xy and 1) is at distance L from the opposite corner. More generally, if two vertices of $Cay(A)/\langle\Delta\rangle$ lie on two different sides of the triangle, and if they are at distance d_1 and d_2, respectively, from the corner of the triangle shared by the two sides, then their distance in $Cay(A)/\langle\Delta\rangle$ is $\max(d_1, d_2)$. Let us call a triangle satisfying this condition "fat" – indeed, in such a triangle all distances are at least as large as in a Euclidean comparison triangle. Notice that

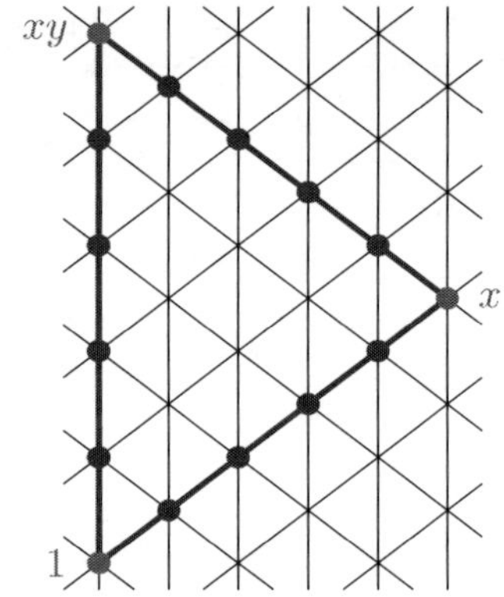

FIGURE 2. A typical fat equilateral triangle in $Cay(A)/\langle\Delta\rangle$, here with $L = 5$. For instance, in $A = \mathcal{B}_4$, one can take $x = \sigma_1^5$, $y = \sigma_3^5$ and $xy = (\sigma_1\sigma_3)^5$.

for every fat triangle induced from an absorbable element, the image in $\mathcal{C}_{AL}(A)$ is of diameter at most 2 (because the image of every edge is of diameter 1).

QUESTION 4.16. We have seen that every absorbable element gives rise to a *fat* equilateral triangle in $Cay(A)/\langle\Delta\rangle$. Are the images of all such triangles in $\Gamma(A, X^A_{NP})$ (or, equivalently, in the complex of irreducible parabolic subgroups) of uniformly bounded diameter? For $A = \mathcal{B}_{n+1}$, do the images in $\mathcal{C}(\mathcal{D}_{n+1})$ of all fat equilateral triangles have uniformly bounded diameter?

OBSERVATION 4.17. A positive answer to Question 4.16 would imply a positive answer to Question 4.14.

CONJECTURE 4.18. *The inequality $X^A_{NP} \preccurlyeq X^A_P$ is strict, i.e., the identity map $\Gamma(A, X^A_P) \to \Gamma(A, X^A_{NP})$ is Lipschitz but* not *a quasi-isometry.*

PROPOSITION 4.19. *Conjecture* 4.18 *does hold when A is a braid group with at least 4 strands: for $n \geqslant 3$, the natural map $\mathcal{A}_\partial(\mathcal{D}_{n+1}) \to \mathcal{C}(\mathcal{D}_{n+1})$ is not a quasi-isometry.*

PROOF. We refer to [**23**, Section 5]. Consider the open disk D of radius 1 centered at $\frac{3}{2}$ containing the first two punctures of $\mathcal{D}_{n+1}$ (see the beginning of Section 3.1); let $X = \mathcal{D}_{n+1} \setminus D$: this subsurface is a hole for $\mathcal{A}_\partial(\mathcal{D}_{n+1})$ according to [**23**, Definition 5.2]. Note also that X is homeomorphic to $\mathcal{D}_n$. Now, let β be a pseudo-Anosov braid on n strands and consider the $n+1$-strand braid $\hat\beta$ obtained from β by doubling the first strand. Fix c greater than the constant C_0 relative to the complex $\mathcal{A}_\partial(\mathcal{D}_{n+1})$ in [**23**, Theorem 5.14]. It is known [**22**, Proposition 4.6] that β acts in a loxodromic way on the curve graph of $\mathcal{D}_n$, hence it acts loxodromically on $\mathcal{C}(X)$, the curve graph of X. In other words, we can find $\alpha > 0$ so that

$$d_{\mathcal{C}(X)}(1, \beta^k) \geqslant \alpha k,$$

for every integer k, and in particular $d_{\mathcal{C}(X)}(1, \beta^k) \geqslant c$, for k big enough. Now, [**23**, Theorem 5.14] implies that there exists a constant $A = A(c) \geqslant 1$ such that for k big enough, $d_{\mathcal{A}_\partial(\mathcal{D}_{n+1})}(1, \hat\beta^k) \geqslant \frac{\alpha k - A}{A}$, whence $\hat\beta$ acts loxodromically on $\mathcal{A}_\partial(\mathcal{D}_{n+1})$.

However, as it preserves the round curve bounding D, $\hat{\beta}$ acts elliptically on $\mathcal{C}(\mathcal{D}_{n+1})$. This finishes the proof of Conjecture 4.18 for braid groups. $\square$

References

[1] Carolyn Abbott, Sahana H. Balasubramanya, and Denis Osin, *Hyperbolic structures on groups*, Algebr. Geom. Topol. **19** (2019), no. 4, 1747–1835, DOI 10.2140/agt.2019.19.1747. MR3995018

[2] **Yago Antolín, María Cumplido**, *Parabolic subgroups acting on the additional length graph*, arXiv:1906.06325

[3] Jason Behrstock, Mark Hagen, and Alessandro Sisto, *Hierarchically hyperbolic spaces II: Combination theorems and the distance formula*, Pacific J. Math. **299** (2019), no. 2, 257–338, DOI 10.2140/pjm.2019.299.257. MR3956144

[4] Jason Behrstock, Mark F. Hagen, and Alessandro Sisto, *Hierarchically hyperbolic spaces, I: Curve complexes for cubical groups*, Geom. Topol. **21** (2017), no. 3, 1731–1804, DOI 10.2140/gt.2017.21.1731. MR3650081

[5] Jason A. Behrstock and Yair N. Minsky, *Dimension and rank for mapping class groups*, Ann. of Math. (2) **167** (2008), no. 3, 1055–1077, DOI 10.4007/annals.2008.167.1055. MR2415393

[6] Jason Behrstock, Bruce Kleiner, Yair Minsky, and Lee Mosher, *Geometry and rigidity of mapping class groups*, Geom. Topol. **16** (2012), no. 2, 781–888, DOI 10.2140/gt.2012.16.781. MR2928983

[7] Egbert Brieskorn and Kyoji Saito, *Artin-Gruppen und Coxeter-Gruppen* (German), Invent. Math. **17** (1972), 245–271, DOI 10.1007/BF01406235. MR323910

[8] **Matthieu Calvez, Bruno Cisneros**, *Curve graphs for Artin-Tits groups of type B, $\widetilde{A}$ and $\widetilde{C}$ are hyperbolic*, arXiv:2003.04796

[9] Matthieu Calvez and Bert Wiest, *Curve graphs and Garside groups*, Geom. Dedicata **188** (2017), 195–213, DOI 10.1007/s10711-016-0213-x. MR3639631

[10] Matthieu Calvez and Bert Wiest, *Acylindrical hyperbolicity and Artin-Tits groups of spherical type*, Geom. Dedicata **191** (2017), 199–215, DOI 10.1007/s10711-017-0252-y. MR3719080

[11] Ruth Charney and John Crisp, *Relative hyperbolicity and Artin groups*, Geom. Dedicata **129** (2007), 1–13, DOI 10.1007/s10711-007-9178-0. MR2353977

[12] *Harold Scott MacDonald Coxeter*, Bull. London Math. Soc. **11** (1979), no. 1, 111–112, DOI 10.1112/blms/11.1.111. MR535802

[13] María Cumplido, Volker Gebhardt, Juan González-Meneses, and Bert Wiest, *On parabolic subgroups of Artin-Tits groups of spherical type*, Adv. Math. **352** (2019), 572–610, DOI 10.1016/j.aim.2019.06.010. MR3964156

[14] Pierre Deligne, *Les immeubles des groupes de tresses généralisés* (French), Invent. Math. **17** (1972), 273–302, DOI 10.1007/BF01406236. MR422673

[15] Benson Farb and Dan Margalit, *A primer on mapping class groups*, Princeton Mathematical Series, vol. 49, Princeton University Press, Princeton, NJ, 2012. MR2850125

[16] Nuno Franco and Juan González-Meneses, *Computation of centralizers in braid groups and Garside groups*, Proceedings of the International Conference on Algebraic Geometry and Singularities (Spanish) (Sevilla, 2001), Rev. Mat. Iberoamericana **19** (2003), no. 2, 367–384, DOI 10.4171/RMI/352. MR2023190

[17] Volker Gebhardt and Juan González-Meneses, *The cyclic sliding operation in Garside groups*, Math. Z. **265** (2010), no. 1, 85–114, DOI 10.1007/s00209-009-0502-2. MR2606950

[18] Sebastian Hensel, Piotr Przytycki, and Richard C. H. Webb, *1-slim triangles and uniform hyperbolicity for arc graphs and curve graphs*, J. Eur. Math. Soc. (JEMS) **17** (2015), no. 4, 755–762, DOI 10.4171/JEMS/517. MR3336835

[19] James E. Humphreys, *Reflection groups and Coxeter groups*, Cambridge Studies in Advanced Mathematics, vol. 29, Cambridge University Press, Cambridge, 1990. MR1066460

[20] Sang-Hyun Kim and Thomas Koberda, *The geometry of the curve graph of a right-angled Artin group*, Internat. J. Algebra Comput. **24** (2014), no. 2, 121–169, DOI 10.1142/S021819671450009X. MR3192368

[21] Eon-Kyung Lee and Sang-Jin Lee, *A Garside-theoretic approach to the reducibility problem in braid groups*, J. Algebra **320** (2008), no. 2, 783–820, DOI 10.1016/j.jalgebra.2008.03.033. MR2422316

[22] Howard A. Masur and Yair N. Minsky, *Geometry of the complex of curves. I. Hyperbolicity*, Invent. Math. **138** (1999), no. 1, 103–149, DOI 10.1007/s002220050343. MR1714338
[23] Howard Masur and Saul Schleimer, *The geometry of the disk complex*, J. Amer. Math. Soc. **26** (2013), no. 1, 1–62, DOI 10.1090/S0894-0347-2012-00742-5. MR2983005
[24] Jon McCammond, *Combinatorial descriptions of multi-vertex 2-complexes*, Illinois J. Math. **54** (2010), no. 1, 137–154. MR2776989
[25] **Rose Morris-Wright**, *Parabolic subgroups of Artin groups of FC type*, arXiv:1906.07058
[26] Luis Paris, *Parabolic subgroups of Artin groups*, J. Algebra **196** (1997), no. 2, 369–399, DOI 10.1006/jabr.1997.7098. MR1475116
Math. Soc. Japan, Tokyo, 2017
[27] Piotr Przytycki and Alessandro Sisto, *A note on acylindrical hyperbolicity of mapping class groups*, Hyperbolic geometry and geometric group theory, Adv. Stud. Pure Math., vol. 73, Math. Soc. Japan, Tokyo, 2017, pp. 255–264, DOI 10.2969/aspm/07310255. MR3728501
[28] **Kasra Rafi, Yvon Verberne**, *Geodesics in the mapping class group*, arXiv:1810.12489
[29] **Harm Van der Lek**, *The Homotopy Type of Complex Hyperplane Complements*, Ph.D. thesis (1983), Nijmegen

Instituto de Matemática de la Universidad de Valparaíso, Gran Bretaña 1111, Valparaíso 2340000, Chile
Email address: calvez.matthieu@gmail.com
URL: https://calvezmatthieu.wixsite.com/matthieucalvez

Bert Wiest, Univ Rennes, CNRS, IRMAR - UMR 6625, F-35000 Rennes, France
Email address: bertold.wiest@univ-rennes1.fr
URL: http://perso.univ-rennes1.fr/bertold.wiest

Contemporary Mathematics
Volume **766**, 2021
https://doi.org/10.1090/conm/766/15376

Linear stability for line bundles over curves

Abel Castorena, E.C. Mistretta, and H. Torres-López

ABSTRACT. Let C be a smooth irreducible projective curve and let $(L, H^0(L))$ be a complete and globally generated linear series on C. Denote by M_L the syzygy bundle, kernel of the evaluation map $H^0(L) \otimes \mathcal{O}_C \to L$. In this work we restrict our attention to the case of globally generated line bundles L over a curve with $h^0(L) = 3$. The purpose of this short note is to connect Mistretta-Stoppino Conjecture on the equivalence between linear (semi)stability of L and slope (semi)stability of M_L with the existence of extensions of line bundles of L by certain quotients Q of M_L. Also, we give numerical conditions to produce examples of line bundles L which are linearly semistables but with syzygy bundle M_L unstable, that is, we find numerical conditions to look for counter-examples to Mistretta-Stoppino Conjecture of rank 2.

1. Introduction

Let C be a smooth irreducible projective curve of genus g. A globally generated g^r_d over C is a pair (L, V), where L is a line bundle of degree d on C and $V \subseteq H^0(L)$ is a linear subspace of dimension $r + 1$ such that the evaluation map $V \otimes \mathcal{O}_C \to L$ is surjective. The rank r kernel $M_{V,L}$ of the evaluation map fits into the following exact sequence

$$(1) \qquad 0 \longrightarrow M_{V,L} \longrightarrow V \otimes \mathcal{O}_C \longrightarrow L \longrightarrow 0.$$

The bundle $M_{V,L}$ is called a syzygy bundle (or dual span bundle). When $V = H^0(L)$, we will denote the bundle $M_{H^0(L),L}$ by M_L. The vector bundle $M_{V,L}$ and its dual $M^\vee_{V,L}$ have been studied from various points of view. The study of the stability of $M_{V,L}$ is related with the study of Brill-Noether varieties and the Minimal Resolution Conjecture (see [**5**]). L. Ein and R. Lazarsfeld showed in [**4**] that M_L is stable for $d > 2g$, and it is semistable for $d = 2g$ (see §3 for the notions of stability for vector bundles). In [**11**], the authors proved that M_{K_C} is semistable,

---

2020 *Mathematics Subject Classification.* Primary 14C20; Secondary 14J26.

Key words and phrases. Slope stability, linear stability, syzygy bundle, linear series.

The first author was supported by PAPIIT IN100419 (Universidad Nacional Autónoma de México) and by CONACyT, A1-S-9029 "Moduli de curvas y curvatura en A_g".

The second author was partially funded by the PRIN research project "Geometria delle Varietà Algebriche" code 2015-EYPTSB-PE1, and partially funded by the research project SID 2016 - MISTRETTA "Vector Bundles, Tropicalization, Fano Manifolds".

The third author was supported by PAPIIT IN100419 (Universidad Nacional Autónoma de México).

where K_C is the canonical line bundle. Recently, the semistability of $M_{V,L}$ was proved for general curves (see [**2**]).

In [**10**], D. Mumford introduced linear semistability for projective varieties $X \subset \mathbb{P}^n$ (cf. Definition 3.3). This implies Chow semistability for curves $C \subset \mathbb{P}^n$ (see [**10**]), and Mumford uses this to construct the moduli space of smooth irreducible projective curves of genus g. For this reason, it is interest to know when a curve $C \subset \mathbb{P}^n$ is linearly semistable.

Later, linear semistability was generalized for a pair (L, V) over a curve C, and linear semistability of the pair (L, V) is equivalent to linear semistability on the image curve induced by the linear system (L, V) (cf. [**12**]).

On the other hand, (semi)stability of the vector bundle $M_{V,L}$ is a stronger condition than linear (semi)stability of the generated linear series (L, V), *i.e.* (semi)stability of $M_{V,L}$ implies linear (semi)stability of the pair (L, V) (cf. Remark 3.4).

It is interesting to know when linear semistability of the pair (L, V) implies semistability of $M_{V,L}$. In this direction, in [**8**, Conjectures 8.6 and 8.7] the authors give two conjectures about this equivalence, and give some conditions under which the equivalence between semistability of $M_{V,L}$ and linear semistability of (L, V) holds, then they used this equivalence to prove semistability of $M_{V,L}$ in some cases see [**8**, Theorem 1.3]. In particular they conjecture that linear semistability of a complete linear system L is equivalent to slope semiastability of the syszygy bundle M_L. Afterwards, in [**3**], the first and third named authors proved this conjecture holds when C is a general Brill-Noether curve and when C is hyperelliptic.

Previously, the second named author of the present work proved in [**7**, Lemma 2.2] that semistability of $M_{V,L}$ is equivalent to linear semistability of (L, V) when $d \geq 2g + 2c$ and $V \subseteq H^0(L)$ is a subspace of codimension $c \leq g$. Using this equivalence he showed that for a general subspace $V \subseteq H^0(L)$ of codimension $c \leq g$, $M_{V,L}$ is semistable (see [**7**, Theorems 2.7 and 2.8]).

The purpose of this work is to investigate further on the relationship between linear semistability of a complete linear series and semistability of M_L in case $\dim H^0(C, L) = 3$. For any such linear series we have the following:

THEOREM 1.1. *Let L be a globally generated line bundle with $h^0(L) = 3$, then L is linearly semistable.*

Therefore any line bundle L with $h^0(L) = 3$ such that M_L is not semistable would provide a counter-example to Mistretta-Stoppino conjecture. This must be produced on a non Brill-Noether general curve, after [**3**].

The important point in the proof of the Theorem and in the following analysis is a characterization of destabilizing quotients of M_L and the extensions they induce:

LEMMA 1.2. *Let Q be a quotient line bundle of M_L with $\deg Q < -\deg L/2$, then there exist an unique globally generated extension*

$$0 \to Q \to F \to L \to 0 \ ,$$

with $h^0(F) = 3$.

Our theorem provides a natural and intrinsic characterization of the kind of singularities of the image curve induced by linear system (see Corollary 3.8):

COROLLARY 1.3. Let L be a globally generated line bundle with $h^0(L) = 3$ inducing a birational map $\phi_L : C \to \mathbb{P}^2$. Denote $\bar{C} \subset \mathbb{P}^2$ its image and d the degree of $\bar{C}$ in $\mathbb{P}^2$. Then, for any point $p \in \bar{C}$, we have multiplicity $m_p \leq \frac{d}{2}$.

The article is organized as follows: In section 2, we construct a globally generated extension (Theorem 1.2). In section 3, we prove that $(L, H^0(L))$ is linearly semistable (Theorem 1.1) and we show Corollary 1.3.

NOTATION. K_C denote the canonical line bundle on C. Given a vector bundle E over C we denote by d_E (or $\deg(E)$) the degree of E, and by n_E the rank (or $\mathrm{rk}(E)$) of E. The slope of E is defined as the rational number $\mu(E) := \frac{d_E}{n_E}$. Given vector bundles M and N on C, the Ext functor $\mathrm{Ext}^1(M, N)$ is canonically isomorphic to the cohomology space $H^1(M^* \otimes N)$ and classifies extensions of M by N.

2. Extensions of line bundles

Let L be a globally generated line bundle over a curve C, and assume that $h^0(L) = r + 1 = 3$. Let M_L be the rank 2 syzygy bundle of L, that is, we have an exact sequence of bundles

$$0 \to M_L \to H^0(L) \otimes \mathcal{O}_C \to L \to 0. \tag{2}$$

Recall that as L is globally generated and non trivial then $\deg L > 0$, so $\mu(M_L) = -\deg L/2 < 0$.

In this section, we are interested in constructing globally generated non-trivial extensions of L by Q, where Q is a quotient line bundle of M_L. This will allow us to analyze possible destabilizations of M_L, or linear destabilizations of $|L|$.

We have the following

LEMMA 2.1. *Let Q be a quotient line bundle of M_L such that $\deg Q < -\deg L/2$, then there exist an unique non trivial extension*

$$0 \to Q \to F \to L \to 0,$$

with $h^0(F) = 3$.

PROOF. Let $\mathrm{Ext}^1(L, Q)$ be the space which parametrizes extensions of the form

$$u : 0 \to Q \to F_u \to L \to 0.$$

For $u \in \mathrm{Ext}^1(L, Q) \simeq H^1(Q \otimes L^\vee) \simeq H^0(K_C \otimes L \otimes Q^\vee)$ we have the coboundary map in cohomology $\partial_u : H^0(L) \to H^1(Q)$.

Consider the line bundle S defined by $S := Ker(M_L \to Q)$, that is, there exists the following exact sequence

$$0 \to S \to M_L \to Q \to 0. \tag{3}$$

Let

$$\partial : \mathrm{Ext}^1(L, Q) = H^1(L^\vee \otimes Q) \to \mathrm{Hom}(H^0(L), H^1(Q)) \simeq H^0(L)^\vee \otimes H^1(Q)$$

be the map which associates u to its coboundary map

$$\partial_u : H^0(L) \to H^1(Q).$$

As $\deg Q < -\deg L/2 < 0$, then $H^0(Q) = 0$. It follows from the long exact sequence in cohomology

$$0 \to H^0(Q) \to H^0(F_u) \to H^0(L) \xrightarrow{\partial_u} H^1(Q)$$

that $u \in \mathrm{Ker}(\partial)$ if and only if $h^0(F_u) = h^0(L) = 3$.

By dualizing the sequence (2) and twisting by Q, we get the exact sequence of vector bundles

$$0 \to L^\vee \otimes Q \to H^0(L)^\vee \otimes Q \to M_L^\vee \otimes Q \to 0.$$

Since $H^0(Q) = 0$, we obtain the following long exact sequence in cohomology:

$$0 \to H^0(M_L^\vee \otimes Q) \to H^1(L^\vee \otimes Q) \xrightarrow{\partial} H^0(L)^\vee \otimes H^1(Q).$$

Therefore it is enough to show that $h^0(M_L^\vee \otimes Q) = 1$ in order to conclude that $\dim \ker(\partial) = 1$. Now, dualizing the exact sequence (3) and twisting by Q we get

$$0 \to \mathcal{O}_C \to M_L^\vee \otimes Q \to S^\vee \otimes Q \to 0$$

and as $\deg Q < -\deg L/2 = \mu(M_L)$ then $\mu(M_L) < \deg S$ (cf. Remark 3.1 below). Therefore $h^0(S^\vee \otimes Q) = 0$, so $h^0(M_L^\vee \otimes Q) = 1$ and $\dim\ \ker(\partial) = 1$. We recall the fact that if $\lambda \in \mathbb{C}^*$ then u and λu determine the same extension up to isomorphism [**9**, Lemma 3.3]. Therefore there is only one non-trivial extension $u : 0 \to Q \to F \to L \to 0$ such that $h^0(F) = 3$. □

Following the notation of the above theorem we have:

LEMMA 2.2. *The vector bundle F defined by the extension u is globally generated.*

PROOF. By contradiction, assume that $0 \neq u \in \ker(\partial)$ is such that the corresponding rank two vector bundle F is not globally generated. Then, the three sections of F generate a subsheaf $\tilde{F} = \mathrm{Im}(\mathrm{ev}\colon H^0(F) \otimes \mathcal{O}_C \to F) \subset F$. As $\tilde{F}$ is a torsion free sheaf on a curve, then $\tilde{F}$ is a globally generated vector bundle.

Since L is globally generated and $H^0(L) = H^0(F)$, we have a surjective map $\tilde{F} \twoheadrightarrow L$, then $\mathrm{rk}(\tilde{F}) = 2$, otherwise $\tilde{F} \cong L$ and the extension u would split. Moreover, since F is not globally generated, there exists an effective divisor D contained in the support $\mathrm{Bs}(F) := \{p \in C \mid \mathrm{ev}\colon H^0(F) \otimes \mathcal{O}_C \to F \text{ not surjective at } p\}$, and F fits into the following commutative diagram

(4)
$$\begin{array}{ccccccccc}
0 & \longrightarrow & Q(-D) & \longrightarrow & \tilde{F} & \longrightarrow & L & \longrightarrow & 0 \\
 & & \big\downarrow & & \big\downarrow & & \| & & \\
0 & \longrightarrow & Q & \longrightarrow & F & \longrightarrow & L & \longrightarrow & 0.
\end{array}$$

Since $h^0(F) = 3$ and $\tilde{F}$ is globally generated of rank 2 and non trivial, it follows that $h^0(\tilde{F}) = 3$.

Then, for any point $p \in D$ there is an exact sequence

$$u_p : 0 \to Q(-p) \to E \to L \to 0,$$

with $h^0(E) = 3$, such that the extension u is induced from u_p in the following sense: the inclusion $Q(-D) \subset Q$ gives rise to a map $\phi\colon \mathrm{Ext}^1(L, Q(-D)) \to \mathrm{Ext}^1(L, Q)$. After diagram (4) the map ϕ associates to the extension $\tilde{u}\colon 0 \to Q(-D) \to \tilde{F} \to L \to 0$ the extension $u\colon 0 \to Q \to F \to L \to 0$. Now, for any $p \in D$, the map ϕ factors through the maps $\mathrm{Ext}^1(L, Q(-D)) \to \mathrm{Ext}^1(L, Q(-p)) \to \mathrm{Ext}^1(L, Q)$, therefore the diagram (4) can be extended to a commutative diagram:

(5)
$$\begin{array}{ccccccccc}
0 & \longrightarrow & Q(-D) & \longrightarrow & \tilde{F} & \longrightarrow & L & \longrightarrow & 0 \\
 & & \downarrow & & \downarrow & & \| & & \\
0 & \longrightarrow & Q(-p) & \longrightarrow & E & \longrightarrow & L & \longrightarrow & 0 \\
 & & \downarrow & & \downarrow & & \| & & \\
0 & \longrightarrow & Q & \longrightarrow & F & \longrightarrow & L & \longrightarrow & 0.
\end{array}$$

As $h^0(\tilde{F}) = h^0(F) = 3$, then $h^0(E) = 3$ as well.

From the fact that $h^0(E) = h^0(F) = h^0(L)$ and $h^0(Q(-p)) = 0$, then $u_p \in \mathrm{Ker}(\mathrm{Ext}^1(L, Q(-p)) \xrightarrow{\tilde{\partial}} \mathrm{Hom}(H^0(L), H^1(Q(-p)))$. Now, we compute the dimension of the space of all extensions $u_p \in \ker(\mathrm{Ext}^1(L, Q(-p)) \xrightarrow{\tilde{\partial}} \mathrm{Hom}(H^0(L), H^1(Q(-p)))$, where

$$\mathrm{Ext}^1(L, Q(-p)) \simeq H^0(K_C \otimes L \otimes Q^\vee(p))^\vee,$$

and

$$\mathrm{Hom}(H^0(L), H^1(Q(-p))) \simeq H^0(L)^\vee \otimes H^0(K_C \otimes Q^\vee(p))^\vee.$$

The map $\tilde{\partial}$ is dual to

$$\mu_1 : H^0(L) \otimes H^0(K_C \otimes Q^\vee(p)) \to H^0(K_C \otimes L \otimes Q^\vee(p)).$$

Then, $\ker(\tilde{\partial}) = \mathrm{cork}(\mu_1)$, where $ker(\tilde{\partial})$ parametrizes non-trivial extensions u inside $\ker(\partial)$, such that the corresponding rank two vector bundle F_u is not globally generated. We know that $\dim(\ker(\partial)) = 1$, then we need to show that $\dim(\ker(\tilde{\partial})) = 0$, that is, we have to show that μ_1 is surjective.

Note that $h^1(K_C \otimes S \otimes Q^\vee(p)) = h^0(S^\vee \otimes Q(-p)) = 0$. Twisting the exact sequence (2) by $K_C \otimes Q^\vee(p)$ and taking cohomology, we obtain that $h^1(M_L \otimes K_C \otimes Q^\vee(p)) = h^1(K_C(p)) = 0$. Hence we have the following exact sequence

$$0 \to H^0(M_L \otimes K_C \otimes Q^\vee(p)) \to H^0(L) \otimes H^0(K_C \otimes Q^\vee(p)) \xrightarrow{\mu_1} H^0(L \otimes K_C \otimes Q^\vee(p)) \to 0.$$

So μ_1 is surjective and $\ker(\tilde{\partial}) = 0$, thus F is a globally generated rank two vector bundle with $h^0(F) = 3$. □

3. Linear stability and stability of Syzygy bundle

In this section we will use the results obtained in the previous section to show that a globally generated line bundle L with 3 sections is linearly semistable. Moreover, we give numerical conditions that can be useful in finding counterexamples to the conjecture proposed by E. C. Mistretta and L. Stoppino [**8**, Conjecture 8.7]. This conjecture affirms the equivalence between the (semi)stability of M_L and the linear (semi)stability of L.

In order to make the exposition self-contained we recall some facts on vector bundles.

We say that a vector bundle E is stable (semistable) if for all non trivial subbundles $F \subset E$

$$\mu(F) < \mu(E) \qquad (\text{resp. } \leq).$$

If E is not semistable, then we say that E is unstable.

REMARK 3.1. Suppose $0 \to F \to E \to Q \to 0$ is an exact sequence of vector bundles. Then

$$\mu(F) < \mu(E) \iff \mu(E) < \mu(Q)$$

and the same holds for $\geq$ and all other inequalities.

Is well known that for any unstable vector bundle E there exists an unique filtration (*Harder-Narasimhan*)

$$0 \subset E_1 \subset E_2 \subset \cdots \subset E_k = E,$$

where the grading $gr_i = E_i/E_{i-1}$ satisfies the following conditions:

(1) the grading gr_i is semistable;
(2) $\mu(gr_i) > \mu(gr_{i+1})$ for $i = 1, \ldots, k-1$.

REMARK 3.2. Let E be an unstable vector bundle of rank 2, then there exists a unique sub-line bundle $S \subset E$ with $\mu(S) > \mu(E)$; indeed suppose that there exists subline bundles S_1 and S_2 with $\mu(S_1), \mu(S_2) > \mu(E)$. Then consider the following filtration

$$0 \subset S_i \subset E \quad \text{for } i = 1, 2. \tag{6}$$

We have S_i and E/S_i satisfies the following conditions

(1) S_i and E/S_i are line bundles;
(2) $\mu(S_i) > \mu(E/S_i)$ for $i = 1, 2$.

Hence for $i = 1, 2$ the filtration $0 \subset S_i \subset E$ is Harder-Narasimhan filtrations and by uniqueness we have that $S_1 = S_2$. The bundle S is called the *maximal destabilizing subbundle of E.*

DEFINITION 3.3. Let (L, V) be a globally generated g^r_d over a curve C, that is, $\deg(L) = d$ and $V \subseteq H^0(L)$ with $r + 1 = \dim(V)$. We say that (L, V) is linearly semistable (respectively linearly stable) if for any linear subspace $W \subset V$ of dimension w,

$$\frac{\deg(\tilde{L})}{w-1} \geq \frac{\deg(L)}{r} \qquad (\text{respectively } >),$$

where $\tilde{L}$ is the line bundle generated by W, namely, there exists the following commutative diagram

$$\begin{array}{ccccccccc} 0 & \longrightarrow & M_{W,\tilde{L}} & \longrightarrow & W \otimes \mathcal{O}_C & \longrightarrow & \tilde{L} & \longrightarrow & 0 \\ & & \downarrow & & \downarrow & & \downarrow & & \\ 0 & \longrightarrow & M_{V,L} & \longrightarrow & V \otimes \mathcal{O}_C & \longrightarrow & L & \longrightarrow & 0. \end{array} \tag{7}$$

REMARK 3.4. Linear (semi)stability of (L, V) is equivalent to the condition that the bundle $M_{V,L}$ can not be destabilized by subbundles of the form $M_{W,\tilde{L}}$, where $(\tilde{L}, W)$ is a generated subseries of (L, V).

CONJECTURE 3.5. [**8**, Conjecture 8.7] Let C be a curve, and let L be a globally generated line bundle on C. The linear (semi)stability of $(L, H^0(L))$ is equivalent to (semi)stability for M_L.

LEMMA 3.6. *Let L be a globally generated line bundle with $h^0(L) = 3$. Let $W \subset H^0(L)$ be a linear subspace with $dim(W) = 2$, then $h^0(\tilde{L}) \leq 3$, where $\tilde{L}$ is the line bundle generated by W. Moreover, $h^0(\tilde{L}) = 3$ if and only if W generates L.*

PROOF. Let $W \subset H^0(L)$ be a subspace of dimension 2, let $\tilde{L}$ be the line bundle generated by W, which fits into the following diagram

$$\begin{array}{ccccccccc} 0 & \longrightarrow & \tilde{L}^\vee & \longrightarrow & W \otimes \mathcal{O}_C & \longrightarrow & \tilde{L} & \longrightarrow & 0 \\ & & \big\downarrow & & \big\downarrow & & \big\downarrow & & \\ 0 & \longrightarrow & M_L & \longrightarrow & H^0(L) \otimes \mathcal{O}_C & \longrightarrow & L & \longrightarrow & 0. \end{array}$$

We have $\tilde{L} = L(-D)$ with D an effective divisor, which can be zero. That is, if W generates L then $\tilde{L} = L$ and if W doesn't generates L then there exists an effective divisor $D \neq 0$ such that $\tilde{L} = L(-D)$. Since L is generated and D is effective, we see that $h^0(\tilde{L}) = h^0(L(-D)) = 2$ and this completes the proof. □

THEOREM 3.7. *Let L be a globally generated line bundle with $h^0(L) = 3$, then L is linearly semistable.*

PROOF. By contradiction, assume that L is not linearly semistable, so there exists a linear subspace $W \subset H^0(L)$ of dimension 2 with $\deg(\tilde{L}) < \frac{d}{2}$, where $\tilde{L}$ is the line bundle generated by W, and $d = \deg L$. The line bundle $\tilde{L}$ fits into the following diagram

$$\begin{array}{ccccccccc} 0 & \longrightarrow & \tilde{L}^\vee & \longrightarrow & W \otimes \mathcal{O}_C & \longrightarrow & \tilde{L} & \longrightarrow & 0 \\ & & \big\downarrow & & \big\downarrow & & \big\downarrow & & \\ 0 & \longrightarrow & M_L & \longrightarrow & H^0(L) \otimes \mathcal{O}_C & \longrightarrow & L & \longrightarrow & 0. \end{array}$$

From Lemma 3.6, we have $h^0(\tilde{L}) \leq 3$ and $h^0(\tilde{L}) = 3$ if and only if W generates L. In this last case, we have $deg(\tilde{L}) = d > \frac{d}{2}$. Thus $h^0(\tilde{L}) = 2$ because $\deg(\tilde{L}) < \frac{d}{2}$. On the other hand, we can show that $h^0(\tilde{L}) \geq 3$.

Note that M_L is unstable and that the line bundle $S := \tilde{L}^\vee$ is a destabilizing subbundle of M_L, in fact $-\deg(\tilde{L}) = \mu(S) > \mu(M_L) = -\frac{d}{2}$ by contradiction hypothesis. □

Let Q be the quotient line bundle of M_L by S.

CLAIM 1. $h^0(\tilde{L}) \geq 3$.

PROOF OF THE CLAIM. first notice that Q satisfies the following:

$$\deg(Q) = \deg(M_L) - \deg(S) = -d - \deg(S) < -d + \frac{d}{2} = -\frac{d}{2} < 0.$$

From Lemmas 2.1 and 2.2, there exists an unique globally generated vector bundle F of rank two which fits into the following exact sequence

$$0 \to Q \to F \xrightarrow{\alpha} L \to 0.$$

Since $H^0(Q)=0$, we complete the diagram

$$\begin{array}{ccccccccc} 0 & \longrightarrow & M_F & \longrightarrow & H^0(F)\otimes\mathcal{O}_C & \longrightarrow & F & \longrightarrow & 0 \\ & & \downarrow{\scriptstyle \alpha_1} & & \downarrow{\scriptstyle \alpha_2} & & \downarrow{\scriptstyle \alpha} & & \\ 0 & \longrightarrow & M_L & \longrightarrow & H^0(L)\otimes\mathcal{O}_C & \longrightarrow & L & \longrightarrow & 0, \end{array}$$

such that α_2 is an isomorphism i.e $H^0(F)\simeq H^0(L)$. By snake lemma, we have an exact sequence

$$\ker(\alpha_1)\to\ker(\alpha_2)\to\ker(\alpha)\to\operatorname{coker}(\alpha_1)\to\operatorname{coker}(\alpha_2)\to\ldots.$$

where $\ker(\alpha_1)=\ker(\alpha_2)=\{0\}$, $\ker(\alpha)=Q$, $\operatorname{coker}(\alpha_2)=\{0\}$ since $\alpha_2:(H^0(F)\to H^0(L))$ is an isomorphism. Thus, $\ker(\alpha)\simeq\operatorname{coker}(\alpha_1)$, $Q=M_L/M_F$ and $M_F\simeq S$, that is, we have the following exact sequence

$$0\to S\to H^0(F)\otimes\mathcal{O}_C\to F\to 0.$$

Dualizing the above exact sequence and taking cohomology, we get

$$0\to H^0(F^\vee)\to H^0(F)^\vee\to H^0(S^\vee)\to\cdots$$

Since F is globally generated without trivial summands, we have $h^0(F^\vee)=0$. So $h^0(S^\vee)\geq h^0(F)=3$. This proves the claim.

So from the first part, we have $h^0(\tilde{L})=h^0(S^\vee)-2$ and from the Claim we have $h^0(\tilde{L})\geq 3$, which is a contradiction. Thus, L is linearly semistable and this complete the proof. □

Corollary 3.8. Let L be globally generated line bundle with $h^0(L)=3$ inducing a birational map $\phi_L:C\to\mathbb{P}^2$. Denote $\bar{C}\subset\mathbb{P}^2$ its image and d the degree of $\bar{C}$ in $\mathbb{P}^2$. Then, for any point $p\in\bar{C}$, we have multiplicity $m_p\leq\frac{d}{2}$.

Proof. The proof follows from Theorem 3.7 and [**8**, Proposition 8.1]. □

Remark 3.9. Let L be a globally generated line bundle over a curve C of degree d such that $h^0(L)=3$. From the exact sequence

$$0\to M_L\to H^0(L)\otimes\mathcal{O}_C\to L\to 0$$

we have for any line bundle B the sequence

$$0\to M_L\otimes B\to H^0(L)\otimes B\to L\otimes B\to 0.$$

So, if there exists a line bundle B on C with $h^0(L\otimes B)<3h^0(B)$, then $h^0(M_L\otimes B)\neq 0$, which implies that $O_C\hookrightarrow M_L\otimes B$, that is, $B^\vee\hookrightarrow M_L$. We have that M_L is unstable if $2\deg(B)<d$. Thus, should be interesting to find some numerical conditions to find a linearly semistable line bundle L with M_L unstable.

References

[1] E. Arbarello, M. Cornalba, P. A. Griffiths, and J. Harris, *Geometry of algebraic curves. Vol. I*, Grundlehren der Mathematischen Wissenschaften [Fundamental Principles of Mathematical Sciences], vol. 267, Springer-Verlag, New York, 1985. MR770932

[2] U. N. Bhosle, L. Brambila-Paz, and P. E. Newstead, *On coherent systems of type $(n,d,n+1)$ on Petri curves*, Manuscripta Math. **126** (2008), no. 4, 409–441, DOI 10.1007/s00229-008-0190-y. MR2425434

[3] Abel Castorena and H. Torres-López, *Linear stability and stability of syzygy bundles*, Internat. J. Math. **29** (2018), no. 11, 1850080, 14, DOI 10.1142/S0129167X18500805. MR3871724

[4] Lawrence Ein and Robert Lazarsfeld, *Stability and restrictions of Picard bundles, with an application to the normal bundles of elliptic curves*, Complex projective geometry (Trieste, 1989/Bergen, 1989), London Math. Soc. Lecture Note Ser., vol. 179, Cambridge Univ. Press, Cambridge, 1992, pp. 149–156, DOI 10.1017/CBO9780511662652.011. MR1201380

[5] Gavril Farkas, Mircea Mustaţă, and Mihnea Popa, *Divisors on $\mathcal{M}_{g,g+1}$ and the minimal resolution conjecture for points on canonical curves* (English, with English and French summaries), Ann. Sci. École Norm. Sup. (4) **36** (2003), no. 4, 553–581, DOI 10.1016/S0012-9593(03)00022-3. MR2013926

[6] Robin Hartshorne, *Geometry: Euclid and beyond*, Undergraduate Texts in Mathematics, Springer-Verlag, New York, 2000. MR1761093

[7] Ernesto C. Mistretta, *Stability of line bundle transforms on curves with respect to low codimensional subspaces*, J. Lond. Math. Soc. (2) **78** (2008), no. 1, 172–182, DOI 10.1112/jlms/jdn016. MR2427058

[8] Ernesto C. Mistretta and Lidia Stoppino, *Linear series on curves: stability and Clifford index*, Internat. J. Math. **23** (2012), no. 12, 1250121, 25, DOI 10.1142/S0129167X12501212. MR3019424

[9] M. S. Narasimhan and S. Ramanan, *Moduli of vector bundles on a compact Riemann surface*, Ann. of Math. (2) **89** (1969), 14–51, DOI 10.2307/1970807. MR242185

[10] David Mumford, *Stability of projective varieties*, Enseign. Math. (2) **23** (1977), no. 1-2, 39–110. MR450272

[11] Kapil Paranjape and S. Ramanan, *On the canonical ring of a curve*, Algebraic geometry and commutative algebra, Vol. II, Kinokuniya, Tokyo, 1988, pp. 503–516. MR977775

[12] Lidia Stoppino, *Slope inequalities for fibred surfaces via GIT*, Osaka J. Math. **45** (2008), no. 4, 1027–1041. MR2493968

Centro de Ciencias Matemáticas (Universidad Nacional Autónoma de México), Campus Morelia, Apdo. Postal 61-3(Xangari). C.P. 58089, Morelia, Michoacán. México.

Email address: abel@matmor.unam.mx

Dipartimento di Matematica, Università di Padova, Via Trieste 63, 35121 Padova, Italy.

Email address: ernesto@math.unipd.it

CONACyT-UAZ. Unidad Académica de Matemáticas, Paseo la Bufa, Calzada Solidaridad, 98060 Zacatecas, Zac. México.

Email address: hugo@cimat.mx

Contemporary Mathematics
Volume **766**, 2021
https://doi.org/10.1090/conm/766/15377

The contribution of W. L. Edge to the study of Humbert's curves

Juan B. Frías-Medina and Alexis G. Zamora

To Herbert Lange, on the occasion of his 75th birthday

ABSTRACT. The aim of this note is to explain the contribution of W. L. Edge to the study of Humbert's curves.

1. Introduction

The French mathematician G. Humbert (1859-1921) studied ([**9**]) the following geometric situation: let $\{p_1, ..., p_5\}$ be a set of 5 points in general position in $\mathbb{P}^3$, and $\{\mathcal{C}_t\}$ the family of twisted cubic curves through the fixed 5 points. Consider another point $p \in \mathbb{P}^3$ general with respect to $p_1, ..., p_5$ and define

$$X := \{q \in \mathbb{P}^3 \mid \exists t : q \in \mathcal{C}_t \text{ and } \overline{pq} \text{ is tangent to } \mathcal{C}_t \text{ at } q\},$$

($\overline{pq}$ standing for the line joining p and q).

This variety turns out to be an irreducible genus five curve. In [**2**] Baker considered the curve of contact of a Weddle surface with one of its tangent cones to a node (for the definition of Weddle surface and its basic properties see [**11**], see also [**3**], specially Remark 8.6.2).

In 1951 ([**4**]) Edge published the first of several papers dedicated to the study of Humbert's curves and different specializations and generalizations. His first observation is that Humbert's and Baker's curves are isomorphic. Furthermore, X is a Humbert's curve if and only if it is the projection of a canonical curve in $\mathbb{P}^4$ with special properties. These properties are: be defined as a complete intersection of three diagonal quadrics or equivalently they admit the action of an automorphisms subgroup isomorphic to $\mathbb{Z}_2^4$ (see the following section).

Starting from the definition of a Humbert's curve as a canonical one defined by intersection of three diagonal quadrics, we explain the contributions of Edge. We concentrated in the paper of 1951 and by the end of this article give a brief comment on some of his others contributions to the issue.

2020 *Mathematics Subject Classification.* Primary 14H37, 14H40, 14H45.

This paper was partially supported by CONACyT Grant 257079. The first author acknowledges the financial support of Fondo Institucional de Fomento Regional para el Desarrollo Científico, Tecnológico y de Innovación, FORDECYT 265667, during 2018 and the support of "Programa de Becas Posdoctorales 2019", DGAPA, UNAM during the revision of this article.

2. Edge's approach to Humbert's curves

We work over the complex field $\mathbb{C}$. In 1951 W. L. Edge published the paper "Humbert's plane sextics of genus 5" ([**4**]).

We start with an irreducible sextic plane curve C with five nodes $n_0, \cdots, n_4$ in general position. The geometric genus of C is 5. The linear system $\mathbb{P}H^0(\mathcal{O}(3) \otimes I_Z)$ of cubic curves passing through the nodes ($Z = n_0 + \cdots + n_4$) induces on C the canonical linear system and defines the canonical embedding of C into $\mathbb{P}^4$. Denote by X its image.

The image of $\mathbb{P}^2$ under this map is a degree 4 Del Pezzo surface $S \subset \mathbb{P}^4$. The surface S can be identified with the blowing-up of $\mathbb{P}^2$ with center at Z. Thus, we have:

$$X \subset S \subset \mathbb{P}^4.$$

It is well known that in suitable system of projective coordinates S can be defined as the complete intersection of two diagonal quadrics Q_0, Q_1. By a diagonal quadric we mean a quadric given by an equation:

$$a_0 x_0^2 + \cdots + a_4 x_4^2 = 0.$$

X is at the same time the normalization and the canonical model of C. It has degree 8. If, moreover, X is non trigonal, then a third quadric Q_2 exists such that X is the complete intersection of Q_0, Q_1 and Q_2 (even if X is trigonal, it is contained in the intersection of three quadrics, but in this case their intersection is the surface S ([**10**])). In this way, if X is non trigonal, then a net $\mathcal{N}$ of quadrics exists such that X is the base locus of $\mathcal{N}$.

The inverse process consists in choosing two points $p, q \in X$ and projecting $\mathbb{P}^4$ onto $\mathbb{P}^2$ with center the chord $\gamma = \overline{pq}$. The linear system $\mathcal{N}(-\gamma)$ of quadrics in $\mathcal{N}$ containing the chord γ has as base locus a degree 4 Del Pezzo surface S. The projection

$$\pi : S \to \mathbb{P}^2,$$

is the contraction of five (-1)-curves on S and X is projected under π to a sextic nodal curve C. The images of the points p, q can be identified as follows: if $n_0, \cdots, n_4$ are the nodes of C, then the conic Γ passing through these five points intersects C in another two points, that are the images of p and q. Also, Γ turns out to be the image of the chord γ under π.

More intrinsically, fixing $p, q \in X$ determines the g_6^2 given by $H^0(K_X(-p-q))$, i.e. a map:

$$\pi : X \to \mathbb{P}^2,$$

whose image is a nodal plane sextic.

A particularly interesting case occurs when the third quadric Q_2 can also be chosen as a diagonal one. This was the starting point of Edge's study of Humbert's curves, his first contribution was the following algebraic characterization:

THEOREM 2.1. *A genus* 5 *curve* X *is isomorphic to a Humbert's curve if and only if its canonical model is the complete intersection of* 3 *diagonal quadrics.*

In this way, a Humbert's curve is given, according to Edge's approach, by the solutions of a system of equations:

$$\{(a_{ij})\begin{pmatrix} x_0^2 \\ \vdots \\ x_4^2 \end{pmatrix} = 0\} \subseteq \mathbb{P}^4, \tag{2.1}$$

with $A = (a_{ij}) \in M_{3\times 5}(\mathbb{C})$. The matrix A is not uniquely determined, it is defined up to the action of $PGL_2(\mathbb{C})$ by the left.

The condition for X not be contained in any hyperplane is that every 3×3 minor of A must be different from zero, this guarantees in particular that X is non-singular. The net of quadrics so defined will be denoted by $\mathcal{N}$. Thus, X is the base locus of $\mathcal{N}$.

In classical language the property of being defined by diagonal quadrics is expressed by saying that the quadrics "have a common self-polar simplex"([**4**], §2, [**3**] §1.3).

Both, Humbert and Baker, noticed that a Humbert's curve admits "five linearly independent elliptic integrals". In Edge's article these appear in a very geometric and natural way. Fix, indeed, the point $e_0 = (1 : 0 : ... : 0)$. Let $\mathcal{N}(-e_0)$ be the linear system of quadrics in $\mathcal{N}$ vanishing at e_0. Using the notation in (2.1), an explicit expression for this pencil is:

$$\lambda(a_{10}Q_0 - a_{00}Q_1) + \mu(a_{20}Q_0 - a_{00}Q_2) = 0.$$

The base locus of this pencil is an intersection of two diagonal quadrics in $\mathbb{P}^4$. Thus, it is a Del Pezzo degree four surface S, and $X \subseteq S$. Projecting onto the hyperplane $H_0 = V(x_0) \simeq \mathbb{P}^3$ we get an intersection of two quadrics in $\mathbb{P}^3$, that is, an elliptic degree 4 curve $E_0 \subset \mathbb{P}^3$. Restricting this projection to X we obtain a $2:1$ covering:

$$\pi_0 : X \to E_0,$$

ramified on the eight points of intersection of $X.H_0$. The pull-back of a generator of $H^0(E_0, \omega_{E_0})$ to X gives rise to one of these "elliptic integrals". The others are obtained considering the different values of i.

It must be noticed that points in the intersection $X.H_i$ are Weierstrass points of X, each one with gap sequence $\{1, 2, 3, 5, 7\}$. Therefore, they form the totality of Weierstrass points of X.

The maps π_i have yet another interpretation. The curve X is invariant under the group of projective transformations generated by the involutions:

$$\sigma_i : \mathbb{P}^4 \to \mathbb{P}^4,$$
$$(x_0 : \ldots : x_i : \ldots : x_4) \mapsto (x_0 : \ldots : -x_i : \ldots : x_4).$$

The map π_i is the quotient of X by the action of σ_i. The group $\mathcal{E} = \langle \sigma_0, \ldots, \sigma_4 \rangle$ is isomorphic to $\mathbb{Z}_2^4$ and the σ_i's satisfy the relation: $\sigma_0 \ldots \sigma_4 = 1$. In Edge's notation σ_i are "central inversions"and "it is by exploiting this property (...) that we obtain our results"([**4**], page 486).

Previous properties can be summarized in the following:

THEOREM 2.2. *Let $X \subset \mathbb{P}^4$ be a canonical curve. The following statements are equivalent:*

i) *X is a Humbert's curve.*
ii) *X admits five involutions $\sigma_0, ..., \sigma_4$ such that $[\sigma_i, \sigma_j] = 1$, $\sigma_0 \cdots \sigma_4 = 1$, and $E_i := X/\langle \sigma_i \rangle$ are elliptic curves.*
iii) *X admits 10 even and effective theta-characteristics.*
iv) *The discriminant curve of Π, $\Delta \subset \mathbb{P}^2$, is a product of 5 lines.*

COROLLARY 2.3. *Let X be a Humbert's curve. The set of Weierstrass points of X is formed by the 40 intersections of X with the hyperplanes H_i. Each of these points have weight 3 and gap sequence $\{1, 2, 3, 5, 7\}$.*

Edge proved some of these characterizations and probably assumed the validity of Theorem 2.2. For proofs of these results see [**11**], [**7**] and [**1**], part III.5.

3. Chords and projections

Let $\mathcal{C}(X)$ denotes the variety of chords of X. The group $\mathcal{E}$ acts on $\mathcal{C}(X)$ in a natural way. Let $\overline{pq} \in \mathcal{C}(X)$, and assume that $\overline{pq} \not\subset H_i$, $i = 0, \ldots 4$. Then the orbit $\mathcal{E}.\overline{pq}$ consists of 16 elements of $\mathcal{C}(X)$. Each element $\overline{p'q'} \in \mathcal{E}.\overline{pq}$ intersects exactly 5 other elements of the orbit, namely $\overline{p'q'} \cap \overline{pq} \neq \emptyset$ if and only if $\overline{p'q'} = \sigma_i \overline{pq}$ for some i. This follows from the fact that no point of X lies in the intersection of H_i and H_j if $i \neq j$.

Consider the linear system $\mathcal{N}(-\overline{pq})$ consisting of quadrics in $\mathcal{N}$ containing the line $\overline{pq}$. Since $p, q \in X$, that is the base locus of $\mathcal{N}$, $\mathcal{N}(-\overline{pq})$ turns out to be a pencil of diagonal quadrics. Therefore, the base locus S of $\mathcal{N}(-\overline{pq})$ is a degree 4 Del Pezzo surface containing X and the orbit $\mathcal{E}.\overline{pq}$. This orbit is the set of (-1)-curves contained in S.

The projection of $\mathbb{P}^4$ onto $\mathbb{P}^2$ with center $\overline{pq}$ can be performed, as explained in the previous section. The five (-1)-curves contracted by

$$\pi_{pq} : S \to \mathbb{P}^2,$$

are $\sigma_i \overline{pq}$ $(i = 0, \ldots 4)$. The image of $\overline{pq}$ is the conic Γ passing through the image n_i of $\sigma_i \overline{pq}$ and the image of $\sigma_i \sigma_j \overline{pq}$ corresponds to the plane line $\overline{n_i n_j}$.

The image of X is then a sextic curve $\mathcal{H} \subset \mathbb{P}^2$ which has nodes in the points n_i. If $\mathcal{H}.\Gamma = t + s + 2(n_0 + \ldots n_4)$, then s and t are the images of p and q under π_{pq}.

Exploiting the fact that $\mathcal{H}$ is the image of a Humbert's curve, a lot of extra projective properties can be deduced. For instance,

PROPOSITION 3.1. *Let $\{l_{ij}\}_{j=1,2}$ be the set of tangent lines to X at n_i. Then, a reordering exits such that $\{l_{i1}\}_{i=0}^4$ pass through s and $\{l_{i2}\}_{i=0}^4$ pass through t.*

A proof based on projective geometry is given in [**4**], §9. The fundamental fact is that any line in $\mathbb{P}^4$ intersecting the chord $\overline{pq}$ is contracted under π_{pq} to a point in $\mathbb{P}^2$.

Also we have:

PROPOSITION 3.2. *([**4**], §14) For each node n_i of $\mathcal{H}$ there pass eight tangents to $\mathcal{H}$, corresponding to Weierstrass points of X, and four bitangents to $\mathcal{H}$.*

PROOF. The image of the chord $\sigma_0 \overline{pq}$ is the node n_0 and a simple explicit computation shows that any pair of points $r, \sigma_0 r \in X$ determines a line $\overline{\pi r \pi \sigma_0 r}$ passing through the node n_0.

Consider the covering:

$$\phi : X \xrightarrow{4:1} \mathbb{P}^1 ,$$

induced by the projection of $\mathcal{H}$ with center n_0. The fact that n_0, r and $\sigma_0 r$ are collinear means that ϕ factorizes through the quotient π_0 by the action of σ_0:

$$\begin{array}{ccc} X & \xrightarrow{\phi} & \mathbb{P}^1 , \\ {\scriptstyle \pi_0}\downarrow & \nearrow {\scriptstyle \varphi} & \\ E & & \end{array}$$

φ being a $2:1$ cover. Therefore, the ramifications of ϕ correspond to the ramification of π_0 (the Weierstrass points corresponding to σ_0) and to the preimage under π_0 of the 4 ramification points of φ.

□

The previous construction can be specialized to the case when the chord $\overline{pq}$ is replaced by a tangent T_pX. In this case, the resulting plane sextic has five cusp corresponding to the images $\sigma_i p$. Analogous projective properties can be deduced in this case. The study of the cuspidal form of Humbert's curves was a constant interest for Edge, for instance in [**6**] he studied the cuspidal form of a particular Humbert's curve.

4. Differentials and equations for $\mathcal{H}$

In this section we identify plane curves with its defining equations. For two homogeneous polynomials, F, G, we denote by $F \cdot G$ the intersection of the curves defined by F and G as a zero-cycle of $\mathbb{P}^2$. On the other hand FG denotes the product of polynomials.

Recall that $X \subset \mathbb{P}^4$ is a canonical curve and the generators of the group $\mathcal{E}$ acts diagonally on the projective coordinates x_i. Therefore, a basis $\{\omega_0, \ldots, \omega_4\}$ of regular differentials on X exists such that:

$$\sigma_i \omega_j = (-1)^{\delta_{ij}} \omega_j,$$

with δ_{ij} denoting Kronecker's delta. Call this the standard basis.

On the plane curve $\mathcal{H}$ differentials are identified with plane cubic curves passing through the nodes n_i of $\mathcal{H}$.

The standard basis of $H^0(X, K_X)$ can be identified with specific plane cubic curves using the fact that their set of zeroes on $\mathcal{H}$ are the projections of the intersections $X.H_i$.

Fix, for instance $i = 0$, then H_0 contains the intersection r_i of the lines $\sigma_i \overline{pq}$ and $\sigma_0 \sigma_i \overline{pq}$ for $i = 1, 2, 3, 4$. Now, the tangent plane $T_{r_i} S$ is generated by the lines $\sigma_i \overline{pq}$ and $\sigma_0 \sigma_i \overline{pq}$. When projecting onto $\mathbb{P}^2$ this means that the resulting cubic ϕ_0 satisfies that:

$$n_0 \in T_{n_i} \phi_0, \quad i = 1, ..., 4.$$

Using similar considerations for the lines $\sigma_i \sigma_j \overline{pq}$, $\sigma_l \sigma_k \overline{pq}$ and pq, $\sigma_0 \overline{pq}$, it can be deduced that:

LEMMA 4.1. *Let ϕ_0 be the plane curve corresponding in the standard basis of $H^0(X, K_X)$ with the hyperplane H_0, then:*

i) *For any $i = 1, 2, 3, 4,\quad n_0 \in T_{n_i}\phi_0$.*
ii) *The cubic ϕ_0 passes through the three diagonal points of the quadrangle $\{n_1, n_2, n_3, n_4\}$.*
iii) $T_{n_0}\phi_0 = T_{n_0}\Gamma_0$.

Now, it is an easy matter to write equations for the plane cubic curve ϕ_0 satisfying the previous geometric conditions. We give here an expression slightly different from the presented by Edge.

Since

$$2n_0 + n_1 + n_2 + n_3 + n_4 = \Gamma.\phi_0 = \Gamma.(l_{12}l_{34}T_{n_0}\Gamma),$$

(with l_{ij} standing for the line joining n_i and n_j), we can apply Noether's $AF + BG$ Theorem, ([**8**]) in order to obtain:

$$\phi_0 = \Gamma L_0 + \lambda l_{12}l_{34}T_{n_0}\Gamma,$$

with L_0 some linear polynomial and $\lambda \in \mathbb{C}^*$.

The indeterminate coefficients in this expression can be computed using the conditions established in Lemma 4.1. Moreover, it follows from the expression for ϕ_0 that the line L_0 must contains the point of intersection $l_{12} \cdot l_{34}$.

In order to explicitly compute the coefficients of the cubic ϕ_i, Edge noticed it could be helpful to choose coordinates in such a way that $\Gamma = V(xz - y^2)$. In this way, we can assume that $n_i = (1 : \theta : \theta^2)$. He later applied this idea to the study of generalizations of Humbert's curves ([**5**]).

Consider in $H^0(\mathcal{O}_{\mathbb{P}^2}(6) \otimes I_{2Z})$ the space of solutions of the systems of linear equations:

$$\sum_{i=0}^{4} \alpha_i \phi_i^2 = 0,$$

this is a 2-dimensional vector space W, that can be identified with the space of diagonal quadrics containing the Del Pezzo surface S in $\mathbb{P}^4$.

Thus, elements of $H^0(\mathcal{O}_{\mathbb{P}^2}(6) \otimes I_{2Z})/W$ correspond to Humbert's plane sextic curves. In terms of the geometry of $\mathbb{P}^4$, we have the exact sequence:

$$0 \to V \otimes I_S \to V \to H^0(\mathcal{O}_S(2)),$$

where $V \subset H^0(\mathcal{O}_{\mathbb{P}^4}(2))$ is the linear system of diagonal quadrics:

$$V = \{\sum a_i x_i^2\}.$$

5. Planar representation of σ_i

As seen before the choice of $p, q \in X$ determines a Del Pezzo surface S. Consider in $\mathbb{P}^4$ the cone with vertex e_0 and base S. This is a rank 4 quadric $\mathcal{Q}$ and has in consequence two complementary rulings $\{\Lambda_\lambda\}$ and $\{\Lambda'_\mu\}$ of planes. They can be distinguished by the chord $\gamma = \overline{pq}$. In fact, the plane spanned by γ and $\sigma_0\gamma$ is contained in $\mathcal{Q}$. Thus, there is a ruling containing $\langle\gamma, \sigma_0\gamma\rangle$ and another complementary to it.

The intersection of Λ_λ and Λ'_μ is a line l passing through e_0. If $r \in X$ then λ and μ exist such that $r \in \Lambda_\lambda.\Lambda'_\mu$. Thus, $\sigma_0(r)$ is determined by the residual intersection of l with X.

From this, a geometric description of the automorphism of $\mathcal{H}$ induced by σ_0 can be deduced. Just note that the projection of the ruling $\{\Lambda_\lambda\}$ that contains $\langle\gamma, \sigma_0\gamma\rangle$ is the linear system $\mathbb{P}H^0(\mathcal{O}_{\mathbb{P}^2}(2) \otimes I_{Z'})$, with $Z' = n_1 + n_2 + n_3 + n_4$ and the projection of $\{\Lambda'_\mu\}$ is the linear system $\mathbb{P}H^0(\mathcal{O}_{\mathbb{P}^2}(1) \otimes I_{n_0})$.

Thus, given $r \in \mathcal{H}$ consider the unique element $C \in \mathbb{P}H^0(\mathcal{O}_{\mathbb{P}^2}(2) \otimes I_{Z'} \otimes I_r)$ and the unique $L \in \mathbb{P}H^0(\mathcal{O}_{\mathbb{P}^2}(1) \otimes I_{n_0} \otimes I_r)$. The intersection of the conic C and the line L is:

$$L.C = r + \sigma_0(r).$$

It must be remarked that the group $\mathcal{E}$ can be represented as a subgroup of the Cremona group of birational transformations of $\mathbb{P}^2$. Edge did not mention this representation. A discussion of this approach will be published in a forthcoming work.

6. Moduli and degenerations

In [**5**] Edge considered a generalization of Humbert's curves to the case when $X \subset \mathbb{P}^n$ is a complete intersection of $n-1$ diagonal quadrics. One of the most remarkable facts of these curves is that any set of equations:

$$A\begin{pmatrix} x_0^2 \\ \vdots \\ x_n^2 \end{pmatrix} = 0, \quad A \in M_{n-1\times n+1}(\mathbb{C})$$

has a "normal" form.

For instance, in the case $n = 4$, that we have hitherto discussed the argument is: considering the columns of A as elements in $\mathbb{P}^2$, namely $p_0, p_1, \ldots, p_4$, we use the fact that given 5 points of $\mathbb{P}^2$ in general position a unique conic exists passing through these points. Denote such conic by Γ. On the other hand, we consider the curve given by the Veronese embedding

$$\begin{aligned} \nu : \mathbb{P}^1 &\to \mathbb{P}^2 \\ (a:b) &\mapsto (a^2 : ab : b^2) \end{aligned}$$

It follows that $\nu(\mathbb{P}^1) = V(y^2 - xz)$. Later, we can construct a change of coordinates which carries Γ to $V(y^2 - xz)$ and such that p_0, p_1, p_2 are sent to $(1:0:0), (1:1:1), (1:-1:1)$ respectively. The images of the points p_3 and p_4 are of the form $(1:\alpha:\alpha^2)$ and $(1:\beta:\beta^2)$. Finally, we use this new points as the columns of the matrix that defines the Humbert's curve.

The normal form of X is in this way given by:

$$\begin{pmatrix} 1 & 1 & 1 & 1 & 1 \\ 0 & 1 & -1 & \alpha & \beta \\ 0 & 1 & 1 & \alpha^2 & \beta^2 \end{pmatrix}\begin{pmatrix} x_0^2 \\ \vdots \\ x_n^2 \end{pmatrix} = 0.$$

Moreover, since $X \subset \mathbb{P}^4$ is a canonical curve, then another Humbert's curve $X' \subset \mathbb{P}^4$ is isomorphic to X if and only if X and X' are projectively equivalents. Thus, the normal form is uniquely determined. Therefore a moduli space of Humbert's curves, and even a compactification of it, can be described.

In the general setting of a complete intersection of $n-1$ quadrics in $\mathbb{P}^n$, a similar argument follows: we consider the columns of the matrix A as points in $\mathbb{P}^{n-2}$, we have $n+1$ of them. There is a unique rational curve C that contains all the points, and using a change of coordinates we can take C to the image of the map

$$\begin{array}{rcl} \nu : \mathbb{P}^1 & \rightarrow & \mathbb{P}^{n-2} \\ (a:b) & \mapsto & (a^{n-2} : a^{n-3}b : \cdots : b^{n-2}) \end{array}$$

So, the image of the point p_{i-1} corresponding to the i-th column has the form $(1 : \alpha_i : \alpha_i^2 : \cdots : \alpha_i^{n-2})$ for all $i = 1, \ldots, n+1$.

By considering particular kinds of degenerated normal forms, Humbert's curves admitting larger automorphism groups can be constructed. Some of this particular cases can be consulted in [**5**] and [**6**], see also the discussion in [**7**].

References

[1] Robert D. M. Accola, *Riemann surfaces, theta functions, and abelian automorphisms groups*, Lecture Notes in Mathematics, Vol. 483, Springer-Verlag, Berlin-New York, 1975. MR0470198

[2] H. F. Baker, *An Introduction to the Theory of Multiply periodic functions*, Cambridge University Press (1907).

[3] Igor V. Dolgachev, *Classical algebraic geometry*, Cambridge University Press, Cambridge, 2012. A modern view. MR2964027

[4] W. L. Edge, *Humbert's plane sextics of genus* 5, Proc. Cambridge Philos. Soc. **47** (1951), 483–495, DOI 10.1017/s030500410002689x. MR42154

[5] W. L. Edge, *The common curve of quadrics sharing a self-polar simplex*, Ann. Mat. Pura Appl. (4) **114** (1977), 241–270, DOI 10.1007/BF02413789. MR469926

[6] W. L. Edge, *A plane sextic and its five cusps*, Proc. Roy. Soc. Edinburgh Sect. A **118** (1991), no. 3-4, 209–223, DOI 10.1017/S030821050002905X. MR1121664

[7] Juan B. Frías-Medina and Alexis G. Zamora, *Some remarks on Humbert-Edge's curves*, Eur. J. Math. **4** (2018), no. 3, 988–999, DOI 10.1007/s40879-018-0232-2. MR3851126

[8] William Fulton, *Algebraic curves*, Advanced Book Classics, Addison-Wesley Publishing Company, Advanced Book Program, Redwood City, CA, 1989. An introduction to algebraic geometry; Notes written with the collaboration of Richard Weiss; Reprint of 1969 original. MR1042981

[9] G. Humbert, *Sur un complexe remarquable de coniques*, Jour. de l'Ecole Poly. 64 (1894), 123-149.

[10] A. N. Tyurin, *On Intersections of Quadrics*, Russ. Math. Surv. 30 (1975), 51-105.

[11] Robert Varley, *Weddle's surfaces, Humbert's curves, and a certain* 4*-dimensional abelian variety*, Amer. J. Math. **108** (1986), no. 4, 931–951, DOI 10.2307/2374519. MR853219

Centro de Ciencias Matemáticas, Universidad Nacional Autónoma de México Campus Morelia, Antigua Carretera a Pátzcuaro # 8701, C.P. 58089, Morelia, Mich. México

Email address: bosco@matmor.unam.mx

U. A. Matemáticas, U. Autónoma de Zacatecas, Calzada Solidaridad entronque Paseo a la Bufa, C.P. 98000, Zacatecas, Zac. México.

Email address: alexiszamora06@gmail.com

Contemporary Mathematics
Volume **766**, 2021
https://doi.org/10.1090/conm/766/15378

Deformations of noncompact Calabi–Yau manifolds, families and diamonds

Elizabeth Gasparim and Francisco Rubilar

ABSTRACT. We discuss a new notion of deformation of complex structure, which we use as an adaptation of Kodaira's theory of deformations, but that is better suited to the study of noncompact manifolds. We present several families of deformations illustrating this new approach. Our examples include toric Calabi–Yau threefolds, cotangent bundles of flag manifolds, and semisimple adjoint orbits. We describe their Hodge theoretical invariants, depicting Hodge diamonds and KKP diamonds.

1. Deformations of complex structures

We present a survey of recent results on deformations of noncompact complex manifolds giving a variety of examples. We explore some new Hodge theoretical invariants of Landau–Ginzburg models, motivated by the KKP conjecture. This conjecture states that three invariants arising from a priori unrelated fundamental ideas in complex geometry actually coincide. It has been recently proved that this conjecture is false in full generality, but as shown in the examples in this paper, there are important families where it does hold. Some families of deformations treated in this survey come from adjoint orbits of certain complex semisimple Lie algebras. We focus on the noncompact case, which turns out to behave radically different from the compact one.

Deformation theory of complex structures for compact manifolds was initially developed by K. Kodaira and D.C. Spencer in the second half of the twentieth century. Their seminal work is summarized in the textbook "Complex manifolds and deformations of complex structures" [**Ko**]. We recall some of the main results of the compact case, explain some difficulties encountered when using the same definition for the noncompact case, and then state the definition we propose Def. 2.3, which seems better suited to the case of noncompact manifolds. We give several examples initially in two and three complex dimensions, and then in further generality using adjoint orbits and cotangent bundles, we give examples in any complex dimension.

2020 *Mathematics Subject Classification.* Primary 32G05.

E. Gasparim was partially supported by Simons Associateship grant of the Abdus Salam International Centre for Theoretical Physics and Network grant NT8 of the Office of External Activities, ICTP.

F. Rubilar was partially supported by Beca Doctorado Nacional Conicyt Folio 21170589 and Network grant NT8 of the Office of External Activities, ICTP.

1.1. The compact case. Let us recall some fundamental results from the theory of classical deformations of complex structures for compact manifolds. The intuitive idea underlying the deformation theory of complex manifolds goes as follows. As Kodaira states in [**Ko**, Sec. 4.1], roughly speaking, a finite dimensional complex manifold $M = M^n$ is obtained gluing domains $U_1, \dots, U_j, \dots$ in $\mathbb{C}^n$, i.e., $M = \bigcup_j U_j$. By [**Ko**, Thm. 2.1], these domains can be considered as polydisks. Hence, if M is a compact complex manifold and $\mathfrak{U} = \{U_j\}$ is an open covering of M, then we can assume $\mathfrak{U}$ finite. Thus a compact complex manifold M is obtained by glueing a finite number of polydisks $U_1, \dots, U_j, \dots, U_l$ by identifying $z_k \in U_k$ and $z_j = f_{jk}(z_k) \in U_j : M = \bigcup_{j=1}^{l} U_j$ where f_{jk} denotes the transition function associated to $U_j, U_k \in \mathfrak{U}$ such that the intersection $U_j \cap U_k$ is not empty. Hence, roughly speaking a deformation of M is considered to be the glueing of the same polydisks U_j via different identifications, i.e., replacing $f_{jk}(z_k)$ by $f_{jk}(z_k, t)$ where t is called the parameter of deformation. Determining *when* and *how* the structure of M as a complex manifold depends on this parameter t introduced in the transition functions is the fundamental idea behind the concept of deformations of complex structures.

Now we recall two basic definitions in deformation theory, namely the concepts of family and of infinitesimal deformation.

DEFINITION 1.1 ([**Ko**, Def. 2.8]). Suppose given a compact complex manifold $M_t = M_t^n$ for each point t of a domain B of $\mathbb{C}^m$. A smooth manifold $\mathcal{M} = \{M_t | t \in B\}$ is called an analytic *family* of compact complex manifolds if there exists a holomorphic surjection $\overline{\omega}\colon \mathcal{M} \to B$ satisfying the following conditions:

ι. For each $t \in B$, $\overline{\omega}^{-1}(t)$ is a compact connected subset of $\mathcal{M}$.
$\iota\iota$. $\overline{\omega}^{-1}(t) = M_t$.
$\iota\iota\iota$. The rank of the Jacobian matrix of $\overline{\omega}$ equals m at every point of $\mathcal{M}$.
$\iota\nu$. There are a locally finite open covering $\{U_j | j = 1, 2, \dots\}$ of $\mathcal{M}$ and complex-valued holomorphic functions $z_j^1(p), \dots, z_j^n(p), j = 1, 2, \dots,$ defined on U_j such that for each t

$$\{p \to z_j^1(p), \dots, z_j^n(p) | U_j \cap \overline{\omega}^{-1}(t) \neq \emptyset\} \tag{1.1}$$

forms a system of local complex coordinates of M_t.

By ι. and $\iota\iota$. each $\overline{\omega}^{-1}(t)$ is a compact differentiable manifold. Condition $\iota\iota\iota$. means that $\overline{\omega}^{-1}(t)$ is the underlying differentiable manifold of M_t. With this notation $t \in B$ is the parameter of the analytic family $\{M_t | t \in B\}$ and B its parameter space or base space. This definition can be extended to the case when B is an arbitrary complex manifold.

NOTATION 1.2. With the same notation as before, in order to make explicit the parameter space it is usual to write down an analytic complex family as $(\mathcal{M}, \overline{\omega}, B)$.

DEFINITION 1.3 ([**Ko**, Def. 2.9]). Let M and N be two compact complex manifolds. N is called a *deformation* of M if M and N belong to the same complex analytic family, that is, if there is a complex analytic family $(\mathcal{M}, \overline{\omega}, B)$ with a complex manifold B as its parameter space such that $M = \overline{\omega}^{-1}(t_0)$ and $N = \overline{\omega}^{-1}(t_1)$ for some $t_0, t_1 \in B$.

With the above definition in mind we get that any two elements of an analytic complex family are diffeomorphic [**Ko**, Thm. 2.3]. Hence, it follows that the differentiable structure of complex manifolds does not change under deformation. Several examples of deformations can be found in [**Ko**, Sec. 2.3].

The fundamental issue when we consider deformations of complex manifolds is the following: given an analytic family $(\mathcal{M}, \overline{\omega}, B)$ of compact complex manifolds, how can we determine whether the complex structure of $M_t = \overline{\omega}^{-1}$ actually depends on t? Essentially the approach goes via differentiating the transition functions with respect to the parameter t. Accordingly, it is possible to show that deformations of a compact complex manifold M are parametrized by $\mathrm{H}^1(M, TM)$, that is, the first cohomology group with coefficient in its tangent sheaf. Thus, if θ is any nonzero class in $\mathrm{H}^1(M, TM)$, we can associate to θ a deformation of M which is called an *infinitesimal deformation*. In general, computing the dimension of the first cohomology group with coefficients in the tangent sheaf is not enough to prove existence of deformation of a compact complex manifold. It is also necessary verify *integrability*, that is, if there exist manifolds realizing such directions of deformations, and it is also necessary to check whether there are *obstructions* which correspond to elements in $\mathrm{H}^2(M, TM)$, [**Ko**, Thm. 5]. We exhibit how this works in some examples.

EXAMPLE 1.4. The complex projective space $\mathbb{P}^n$ has no (classical) deformations of its complex structure. This result follows directly from the fact that $\mathrm{H}^1(\mathbb{P}^n, T\mathbb{P}^n) = 0$ and $\mathbb{P}^n$ is compact. For instance, if we consider $n = 1$, we have that $T\mathbb{P}^1 \cong \mathcal{O}_{\mathbb{P}^1}(2)$. Since $\mathrm{H}^1(\mathbb{P}, \mathcal{O}_{\mathbb{P}^1}(2)) = 0$, the result follows.

EXAMPLE 1.5. If M is a curve of genus g, then $\mathrm{H}^2(M, TM) = 0$. Hence there are no obstructions to deforming its complex structure.

EXAMPLE 1.6. The Hirzebruch surface $\mathbb{F}_k$ defined as

$$\mathbb{F}_k = \mathrm{Proj}(\mathcal{O}_{\mathbb{P}^1}(-k) \oplus \mathcal{O}_{\mathbb{P}^1}), \quad k \geq 0$$

is compact and has $\lfloor \frac{k}{2} \rfloor$ deformations, these are the $\mathbb{F}_{k-2i}$.

2. Deformations of noncompact manifolds

In this section we describe our methods of finding infinitesimal deformations of noncompact manifolds which were used in [**BG1, R**] and [**GKRS**]. When looking for deformations of noncompact manifolds one needs to keep in mind the caveat that cohomology calculations are generally not enough to decide questions of existence of infinitesimal deformations, as the following example illustrates. Suppose we naively extended Definition 1.1 to the noncompact case by simply removing the requirements of compactness, that is, repeating the same definition only with all appearances of the word compact deleted. This would work poorly, as the following example shows.

EXAMPLE 2.1. Edoardo Ballico gave us the following illustration that cohomological rigidity may not imply absence of deformations,

We consider deformations of $X = \mathbb{C}$. Clearly $\mathrm{H}^1(X, TX) = 0$. However, with the naive generalizations from the compact case, there do exist nontrivial deformations of X as the following family shows.

Consider $\pi\colon \mathbb{P}^1 \times D \to D$ with D any smooth manifold (even $\mathbb{P}^1$ or a disc) and a specific $o \in D$. Take $s_\infty\colon D \to \mathbb{P}^1 \times D$ the section of π defined by

$$s_\infty(x) = (\infty, x)$$

then take another section s of π with:

$$s(o) = (\infty, o), \qquad s(x) = (a_x, x)$$

with $a_x \in \mathbb{C} = \mathbb{P}^1 \setminus \{\infty\}$. Take as the total space for our family Y as $\mathbb{P}^1 \times D$ minus the images of the two sections. Then we obtain a deformation of $\mathbb{C}$ in which at all points of $D \setminus \{o\}$ you have $\mathbb{C} \setminus \{0\}$. Thus not a trivial deformation in any reasonable sense.

REMARK 2.2. Example 2.1 shows that vanishing of cohomology may not imply nonexistence of deformations, at least not if we allow the smooth type of the manifold to vary in the family. Nevertheless, as we shall see, cohomology calculations are still useful to find deformations.

In this work by deformation we mean the following:

DEFINITION 2.3. A *deformation family* of a complex manifold X is a holomorphic surjective submersion $\tilde{X} \xrightarrow{\pi} D$, where D is a complex disc centred at 0 (possibly a vector space, possibly infinite dimensional), satisfying:

- $\pi^{-1}(0) = X$,
- $\tilde{X}$ is locally trivial in the C^∞ category.

The fibres $X_t = \pi^{-1}(t)$ are called *deformations* of X.

In further generality, allowing the parameter space to be a variety or a scheme, and requiring that the bundle be locally trivial, we obtain the concept of a *family* of complex structures.

REMARK 2.4. We say that a deformation is trivial if it is isomorphic in the holomorphic category to the original manifold.

REMARK 2.5. Our choice for the dimension of D is $n = h^1(X, TX)$ whenever possible. The case $n = 0$ may be included in the following definition.

DEFINITION 2.6. We call a manifold X *formally rigid* when $\mathrm{H}^1(X, TX) = 0$.

DEFINITION 2.7. We call a manifold X *rigid* if any deformation $\tilde{X} \xrightarrow{\pi} D$ is biholomorphic to the trivial bundle $X \times D \to D$.

EXAMPLE 2.8. Projective spaces are rigid. Indeed, Ex. 1.4 shows it is formally rigid, and in the compact case formally rigid implies rigid.

In general, formally rigid does not imply rigid, as example 2.1 shows, but we improve on the situation with our proposed definition. With Def. 2.3 we do not claim to solve the problem that a manifold X does not deform under the condition $\mathrm{H}^1(X, TX) = 0$, however we certainly eliminate some unwanted pathological cases such as the one in Example 2.1.

Observe that the deformations considered in [**BG1**] satisfy Definition 2.3, hence maintain the C^∞ type of the manifold.

2.1. Surfaces. We give examples of noncompact surfaces which have finite dimensional spaces of deformations, and exhibit their families of deformations. Let Z_k be the noncompact surface defined as

$$Z_k = \mathrm{Tot}(\mathcal{O}_{\mathbb{P}^1}(-k)); \qquad k \geq 1.$$

Barmeier and Gasparim [**BG1**] described the classical deformation theory of the surfaces Z_k, they have also described their noncommutative deformation theory in

[**BG2**], but in this work we will restrict ourselves to classical deformations. Here we recall a definition taken from [**M**], in order to clarify the exposition of the results for the non-compact surfaces Z_k.

These surfaces are total spaces of negative line bundles on the projective line. A preliminary estimate via cohomology calculations [**BG1**, Thm. 5.4] showed that Z_k admits a $(k-1)$-dimensional semiuniversal family of classical deformations. Here by semiuniversal it is meant that the family is versal and the Kodaira–Spencer map is bijective, see [**M**]. Denoting by $\mathcal{Z}_k$ any nontrivial deformation of Z_k for $k \geq 2$, [**BG1**, Thm. 6.6] showed that $\mathcal{Z}_k$ contains no compact complex analytic curves. Furthermore, [**BG1**, Thm. 6.14] showed that any holomorphic vector bundle on $\mathcal{Z}_k$ splits as a direct sum of algebraic line bundles. This is somewhat surprising, given the existence of nontrivial moduli of vector bundles on the original Z_k surfaces as constructed in [**BGK**]. Also in [**BG1**] it is proved that any nontrivial deformation of Z_k is affine (this situation contrasts with the case of the threefolds considered in the next section).

Deformation theory effects on moduli of vector bundles have applications to mathematical physics. The effects for the case of surfaces Z_k can be interpreted in Yang–Mills theory. They imply that moduli of $SU(2)$ instantons on noncompact surfaces are sensitive to the complex structure: the moduli of irreducible instantons of normalized charge (or splitting) j over the noncompact surfaces Z_k are of dimension $2j-k-2$, whereas a nontrivial (commutative) deformation of Z_k admits no instantons [**BG1**, Thm. 7.4]. In comparison, in the context of deformations of compact curves and surfaces and their moduli we note that by Grothendieck's splitting theorem any holomorphic vector bundle on $\mathbb{P}^1$ splits as a direct sum of (algebraic) line bundles. Neither the curve $\mathbb{P}^1$ itself nor its moduli spaces of vector bundles admit any deformations. Curves of higher genus do admit deformations and a celebrated theorem of Narasimhan and Ramanan [**NR, BV**] shows that all classical deformations of the moduli of stable bundles on a smooth curve come from deformations of the curve itself (case $g > 1, \quad (r,d) = 1$). As we saw, deformations of the surfaces Z_k does not produce new moduli of bundles. The situation for Calabi–Yau threefolds turns out a lot more interesting, as we shall now see.

2.2. Calabi–Yau threefolds. We define Calabi–Yau threefolds W_k as total spaces of holomorphic vector bundles on $\mathbb{P}^1$, i.e.,

$$W_k = \mathrm{Tot}(\mathcal{O}(-k) \oplus \mathcal{O}(k-2)); \qquad k \geq 1.$$

We put *canonical coordinates* on W_k by taking the open covering $\mathfrak{U} = \{U, V\}$, where $U = \{(z, u_1, u_2)\} \simeq \mathbb{C}^3 \simeq V = \{(\xi, v_1, v_2)\}$ such that

$$(\xi, u_1, u_2) = (z^{-1}, z^k u_1, z^{2-k} u_2),$$

on the intersection $U \cap V \simeq \mathbb{C}^* \times \mathbb{C}^2$. The behaviour of the deformations of these manifolds is quite different from the one of deformations of the surfaces Z_k. This situation is described by the following results proved by Rubilar in [**R**] and Gasparim and Suzuki in [**GS**]. First, Rubilar proved that while W_1 is formally rigid [**R**, Lem. 5.2.1], W_2 has an infinite dimensional family of deformations [**R**, Thm. 6.3.2]. The latter was accomplished by showing that $\dim \mathrm{H}^1(W_2, TW_2) =$

∞ and then proving that directions of deformations parametrized by such cohomology are integrable, by explicitly constructing the corresponding families. Moreover, he also showed that W_2 has both affine and non-affine deformations, situation which contrasts with the one of the surfaces Z_k. Subsequently, Gasparim and Suzuki showed explicitly that there are infinitely many isomorphism types among the deformations of W_2 [**GS**, Thm. 3.7]. Then, presenting transformations taking deformations of W_k for $k > 2$ to deformations of W_2 and conversely, they obtain also infinitely many isomorphism types of deformations of W_k for $k > 2$.

2.2.1. *Moduli of vector bundles.* The study of deformations is very closely linked to the theory of moduli spaces. In fact, there is a precise sense in which infinitesimal deformations can be used to generate neighborhoods of moduli spaces. For the case of vector bundles this is known as Kuranishi theory. Here we only comment on the results for our surfaces and threefolds, without discussing the general theory.

For deformations of surfaces Z_k, Barmeier and Gasparim showed that moduli spaces of holomorphic vector bundles with fixed topological invariants are trivial [**BG1**, Cor. 6.19], i.e. consist of a single point. For the threefold W_2, Gasparim and Suzuki showed that some nontrivial deformations of W_2 have the effect of decreasing the dimension of the moduli of vector bundles by just one dimension, hence keeping nontrivial moduli [**GS**, Thm. 8.13].

In conclusion, we have observed that the theories of deformations of surfaces Z_k and threefolds W_k are qualitatively different, the latter being very rich in terms of applications to moduli of vector bundles.

3. Hyperkähler families

DEFINITION 3.1. A *hyperkähler manifold* is a Riemannian manifold of real dimension $4k$ and holonomy group contained in $\mathrm{Sp}(k)$.

Here $\mathrm{Sp}(k)$ denotes a compact form of a symplectic group, identified with the group of quaternionic-linear unitary endomorphisms of a k-dimensional quaternionic Hermitian space. Every hyperkähler manifold M has a 2-sphere of complex structures with respect to which the metric is Kähler.

In particular, it is a hypercomplex manifold, meaning that there are three distinct complex structures, I, J, and K, which satisfy the quaternion relations

$$I^2 = J^2 = K^2 = IJK = -1.$$

Any linear combination

$$aI + bJ + cK$$

with a, b, c real numbers such that $a^2 + b^2 + c^2 = 1$ is also a complex structure on M. Hence, hyperkähler manifolds are especially rich examples where to study deformations of complex structures. Each of the complex structures $aI + bJ + cK$ may be regarded as a deformation of I.

REMARK 3.2. To start with I, J, K are almost complex structures, these are then required to be integral, and in such case the complex structures obtained agree with the ones defined by Kodaira using charts.

REMARK 3.3. We have the following general facts:

(1) Hyperkähler manifolds are special classes of Kähler manifolds.
(2) All hyperkähler manifolds are Ricci-flat and are thus Calabi–Yau manifolds.

Compact hyperkähler manifolds have been extensively studied using techniques from algebraic geometry, sometimes under the alternative name of holomorphically symplectic manifolds. Due to Fedor Bogomolov's decomposition theorem (1974,[**Bo**]), the holonomy group of a compact holomorphically symplectic manifold M is exactly $\mathrm{Sp}(k)$ if and only if M is simply connected and any pair of holomorphic symplectic forms on M are scalar multiples of each other.

For applications to the study of Hodge theory of Landau–Ginzburg (LG) models (see Def. 4.1), here we wish to focus on those hyperkähler families that contain the adjoint orbits of semisimple Lie groups. An adjoint orbit of a compact Lie group is called a *flag manifold.*

Families of hyperkähler structures containing semisimple adjoint orbits were studied by Kronheimer [**Kr1**], Biquard [**Bq**], and Kovalev [**Kv**]. Such families include both cotangent bundles of flag manifolds and adjoint orbits of semisimple Lie algebras. In the following section, we will add superpotentials to these families and then study them from the point of view of Landau–Ginzburg models. The approach we propose here goes in a different way, we study the hyperkählerian structures using Lie theory techniques. We first recall some definitions and results.

THEOREM 3.4. [**Kv**, Thm. 1.1] *Let G be a compact semisimple Lie group with $\mathfrak{g}$ its Lie algebra, H a stabilizer of some non-zero element of $\mathfrak{g}$ under the adjoint action of G, and T a maximal torus such that $G \supseteq H \supseteq T$. Let G^c, H^c be the complex forms of G and H. Then the manifold G^c/H^c has a family of hyperkählerian structures with the following properties:*

(1) (a) *the hyperkählerian metrics are complete,*
(b) *the family admits parametrisation as follows: let $\mathfrak{t}$ be a Cartan subalgebra of $\mathfrak{g}$ corresponding to T, then the parameter space consists of those triples (τ_1, τ_2, τ_3) in $\mathfrak{t}$ which each have the stabilizer H;*

(2) *On G^c/H^c there is a complex structure I which is independent of τ_1. If also the complex group H^c is the stabilizer of the element $\tau^c = \tau_2 + i\tau_3$ then the underlying complex manifold determined by I is isomorphic to the complex adjoint orbit $\mathrm{Ad}(G^c)\tau^c$.*

REMARK 3.5. Observe that putting $H^c = T^c$ we obtain the result of Kronheimer [**Kr1**] on regular semisimples orbits, that is, those of type G^c/T^c. In general T^c is a proper subgroup of H^c, therefore in such case the family of hyperkählerian structures is smaller.

To motivate the parametrisation, consider the adjoint action of G^c. Let τ^c be some element of $\mathfrak{g}^c$ and H^c its stabilizer. The orbit $\mathrm{Ad}(G^c)\tau^c$ is diffeomorphic to G^c/H^c. The second de Rham cohomology of G^c/H^c can be identified with a real commutative Lie algebra $\mathfrak{h}_\tau \subset \mathfrak{g}$ determined (up to isomorphism) by *regularity properties* of τ^c: for example, $\mathfrak{h}_\tau = \mathfrak{t}$ if τ^c is a regular element of $\mathfrak{t}^c$. The three Kähler structures on a complex orbit and the cohomology classes of their Kähler forms can be put in correspondence to elements τ_1, τ_2, τ_3 in $\mathfrak{h}_\tau$ such that $\tau_2 + i\tau_3 = \tau^c$. Then τ_1 serves as a parameter in the family of hyperkählerian structures induced on G^c/H^c from the orbit of $\tau_2 + i\tau_3$.

THEOREM 3.6. [**Kv**, Thm. 1.2] *Let G and $\mathfrak{g}$ be as in Thm. 3.4. If $\tau_2, \tau_3 \in \mathfrak{t}$ and σ^c is a nilpotent element of the centralizer $z(\tau_2 + i\tau_3)$ in $\mathfrak{g}^c$ then the complex adjoint orbit $\mathrm{Ad}(G^c)(\tau_2 + i\tau_3 + \sigma^c)$ has a family of hyperkählerian structures. The* *family is parametrised by elements $\tau_1 \in \mathfrak{t}$ such that the centralizers in $\mathfrak{g}$ of the pair (τ_2, τ_3) and the triple (τ_1, τ_2, τ_3) coincide.*

REMARK 3.7. If $\tau_2 = \tau_3 = 0$ then $\tau_1 = 0$ is the only possible value and we obtain the following result due to Kronheimer.

COROLLARY 3.8. [**Kr2**] *A nilpotent adjoint orbit of G^c is a hyperkähler manifold.*

It remains to point out that Theorem 3.6 applies to all adjoint orbits of G^c. This follows immediately from standard properties of complex semisimple Lie algebras, the details can be found e.g. in Humphreys [**H**] or San Martin [**SM**]. Any element of $\mathfrak{g}^c$ can be written uniquely in the form $x = s + n$, with n nilpotent, s semisimple and $[s, n] = 0$. As a semisimple element s is contained in some Cartan subalgebra, that is in $\mathrm{Ad}(g)\mathfrak{t}^c$ for some $g \in G^c$. So any element of $\mathfrak{g}^c$ is G^c-conjugate to one of the form $(\tau_2 + i\tau_3) + \sigma^c$.

We state in a simplified form of a consequence of Thm. 3.4 which we will use to study deformations.

COROLLARY 3.9. *Let $\mathfrak{g}$ be a semisimple complex Lie algebra, and $H_0 \in \mathfrak{g}$ a regular element. Let G be a connected Lie group with Lie algebra $\mathfrak{g}$ and K a compact form of G. Then the adjoint orbit $\mathrm{Ad}\, G(H_0)$ is a deformation of the cotangent bundle of the flag manifold $\mathrm{Ad}\, K(H_0)$.*

4. Adjoint orbits and cotangent bundles

Using the result of Thm. 3.4 we see that adjoint orbits may be regarded as deformations of cotangent bundles of flag manifolds. That is, if G is a complex Lie group and K its compact part, then an adjoint orbit $\mathrm{Ad}\, G(H_0)$ can be regarded as a deformation of the cotangent bundle of the flag manifold $\mathrm{Ad}\, K(H_0)$. We wish to add superpotentials to these families, to look at them from the point of view of Landau–Ginzburg models, and then to study how their Hodge diamonds vary in the families. We first recall the construction given in [**GGSM1**].

DEFINITION 4.1. *A Landau–Ginzburg model* (LG) is a pair (Y, w), where

(1) Y is a smooth complex quasi-projective variety with trivial canonical bundle K_Y;
(2) $w\colon Y \to \mathbb{C}$ is a morphism with a compact critical locus $\mathrm{crit}(w) \subset Y$, called the superpotential.

4.1. LG models on adjoint orbits. Let $\mathfrak{g}$ be a complex semisimple Lie algebra with Cartan subalgebra $\mathfrak{h}$, and $\mathfrak{h}_{\mathbb{R}}$ the real subspace generated by the roots of $\mathfrak{h}$. Let Π denote the set of all roots of $\mathfrak{h}$. An element $H \in \mathfrak{h}$ is called *regular* if $\alpha(H) \neq 0$ for all $\alpha \in \Pi$.

NOTATION 4.2. $\mathcal{O}(H_0) := \mathrm{Ad}\, G(H_0)$.

THEOREM 4.3. [**GGSM1**, Thm. 3.1] *Given $H_0 \in \mathfrak{h}$ and $H \in \mathfrak{h}_{\mathbb{R}}$ with H a regular element, the potential $f_H : \mathcal{O}(H_0) \to \mathbb{C}$ defined by*

$$f_H(x) = \langle H, x \rangle \qquad x \in \mathcal{O}(H_0)$$

has a finite number of isolated singularities and defines a symplectic Lefschetz fibration; that is to say

(1) *the singularities are (Hessian) nondegenerate;*
(2) *if* $c_1, c_2 \in \mathbb{C}$ *are regular values then the level manifolds* $f_H^{-1}(c_1)$ *and* $f_H^{-1}(c_2)$ *are diffeomorphic;*
(3) *there exists a symplectic form* Ω *on* $\mathcal{O}(H_0)$ *such that the regular fibres are symplectic submanifolds;*
(4) *each critical fibre can be written as the disjoint union of affine subspaces contained in* $\mathcal{O}(H_0)$, *each symplectic with respect to* Ω.

Given the Iwasawa decomposition: $\mathfrak{g} = \mathfrak{k} \oplus \mathfrak{a} \oplus \mathfrak{n}$, if Π is a set of roots of $\mathfrak{a}$, with a choice of a set of positive roots Π^+ and simple roots $\Sigma \in \Pi^+$, then the corresponding Weyl chamber is $\mathfrak{a}^+$. A subset $\Theta \in \Sigma$ defines a parabolic subalgebra $\mathfrak{p}_\Theta$ with parabolic subgroup P_Θ and a flag $\mathbb{F}_\Theta = G/P_\Theta$. A element $H_\Theta \in cl\mathfrak{a}^+$ is characteristic for $\Theta \subset \Sigma$ if $\Theta = \{\alpha \in \Sigma : \alpha(H_\Theta) = 0\}$.

THEOREM 4.4. [**GGSM2**, Thm. 2.1] *The adjoint orbit* $\mathcal{O}(H_\Theta) = \mathrm{Ad}(G)\cdot H_\Theta \simeq G/Z_\Theta$ *of the characteristic element* H_Θ *is a* C^∞ *vector bundle over* $\mathbb{F}_\Theta$ *isomorphic as a vector bundle to the cotangent bundle* $T^*\mathbb{F}_\Theta$. *Moreover, we can write down a diffeomorphism* $\iota : \mathrm{Ad}(G)\cdot H_\Theta \to T^*\mathbb{F}_\Theta$ *such that*

(1) ι *is equivariant with respect to the actions of* K, *that is, for all* $k \in K$,

$$\iota \circ \mathrm{Ad}(k) = \widetilde{k} \circ \iota$$

where K *is the compact subgroup in the Iwasawa decomposition* $G = KAN$, *and* $\widetilde{k}$ *is the lifting to* $T^*\mathbb{F}_\Theta$ *(via the differential) of the action of* k *on* $\mathbb{F}_\Theta$.
(2) *The pullback of the canonical symplectic form on* $T^*\mathbb{F}_\Theta$ *by* ι *is the (real) Kirillov–Kostant–Souriau form on the orbit.*

Viewing the orbit as the cotangent bundle of a flag manifold, we can identify the topology of the fibres in terms of the topology of the flag. Denote by $\mathcal{W}$ the Weyl group of G, then we have the following two corollaries.

COROLLARY 4.5. [**GGSM1**, Cor. 3.4] *The homology of a regular fibre coincides with the homology of* $\mathbb{F}_\Theta \setminus \mathcal{W}\cdot H_\Theta$. *In particular the middle Betti number is* $k - 1$ *where* k *is the number of singularities of the fibration (equal to the number of elements in* $\mathcal{W}\cdot H_\Theta$*).*

For the case where singular fibres have only one critical point, we have the following corollary.

COROLLARY 4.6. [**GGSM1**, Cor. 4.2] *The homology of the singular fibre though* wH_Θ, $w \in \mathcal{W}$, *coincides with that of*

$$\mathbb{F}_{H_\Theta} \setminus \{uH_\Theta \in \mathcal{W}\cdot H_\Theta | u \neq w\}.$$

In particular, the middle Betti number of this singular fibre equals $k - 2$, *where* k *is the number of singularities of the fibration* f_H.

These corollaries show that Hodge diamonds for the LG models can have arbitrarily high numbers in their middle cohomology, and that these Lefschetz fibrations may have large quantities of vanishing cycles.

5. Diamonds

5.1. Hodge diamonds. We now wish to consider some examples of how diamonds vary under deformations. Consider first the standard Hodge diamond of a variety:

$$
\begin{array}{ccccccccc}
 & & & & h^{n,n} & & & & \\
 & & & h^{n,n-1} & & h^{n-1,n} & & & \\
 & & h^{n,n-2} & & h^{n-1,n-1} & & h^{n-2,n} & & \\
 & \cdot^{\cdot^{\cdot}} & & & \vdots & & & \ddots & \\
h^{n,0} & & \cdots & & \curvearrowleft & & \cdots & & h^{0,n} \quad \updownarrow \star \\
 & \ddots & & & \vdots & & & \cdot^{\cdot^{\cdot}} & \\
 & & h^{2,0} & & h^{1,1} & & h^{0,2} & & \\
 & & & h^{1,0} & & h^{0,1} & & & \\
 & & & & h^{0,0} & & & & \\
 & & & & \longleftrightarrow & & & &
\end{array}
$$

where the symbols $\curvearrowleft, \longleftrightarrow, \updownarrow \star$ are there to remind us that in the case of smooth projective varieties there are symmetries of the diamond corresponding to Serre duality, conjugation, and Hodge star, respectively.

5.2. KKP diamonds. For the case of Landau–Ginzburg models (Y, w) there are three new Hodge theoretical invariants which were defined by Katzarkov, Kontsevich, and Pantev in [**KKP**]. These invariants take into account not just the variety, but also the potential together with its critical points and vanishing cycles. They are the numbers $f^{p,q}(Y, w)$ which come from sheaf cohomology of logarithmic forms, the numbers $h^{p,q}(Y, w)$ motivated by mirror symmetry considerations, and the numbers $i^{p,q}(Y, w)$ defined using ordinary mixed Hodge theory.

To define these new invariants, the authors require that the LG model be tamely compactifiable in the following sense.

DEFINITION 5.1 ([**KKP**]). *A tame compactified Landau–Ginzburg model* is the data $((Z, f), D_Z)$, where

(1) Z is a smooth projective variety and $f\colon Z \to \mathbb{P}^1$ is a flat morphism.
(2) $D_Z = (\cup_i D_i^h) \cup (\cup_j D_j^v)$ is a reduced normal crossings divisor such that
  (i) $D^v = \cup_j D_j^v$ is a scheme theoretical pole divisor of f, i.e. $f^{-1}(\infty) = D^v$. In particular $ord_{D_j^v}(f) = -1$ for all j;
  (ii) each component D_i^h of $D^h = \cup_i D_i^h$ is smooth and horizontal for f, i.e. $f|_{D_i^h}$ is a flat morphism;
  (iii) The critical locus $crit(f) \subset Z$ does not intersect D^h.
(3) D_Z is an anticanonical divisor on Z.
  One says that $((Z, f), D_Z)$ is *a compactification of the Landau–Ginzburg model* (Y, w) if in addition the following holds:
(4) $Y = Z \setminus D_Z$, $f|_Y = w$.

REMARK 5.2. A caveat about algebraic compactifications should be mentioned here. [**BCG**] showed that the choice of compactification may have strong effects on the Hodge diamonds. In fact, they give examples when the topology of the compactified regular fibre for f_H changes drastically according to the choice of homogenisation of the ideal cutting out the orbit as an affine variety. For the case of the maximal adjoint orbit of $\mathfrak{sl}(3, \mathbb{C})$, namely the one diffeomorphic to the cotangent bundle of the full flag $F(1, 2)$, [**BCG**] give examples of two algebraic compactifications of such an orbit which produce in one case $h^{1,4} = h^{4,1} = 16$ and in the other case $h^{1,4} = h^{4,1} = 1$. Such radical difference being produced simply by the choice of homogenisation of the ideal defining the adjoint orbit. Thus, one must be very careful when using compactifications to study the Hodge theory of noncompact varieties. This remark also highlights the importance of the careful definition of a tame compactification.

Assume that we are given a Landau–Ginzburg model (Y, w) with a tame compactification $((Z, f), D_Z)$ as above. Denote by $n = \dim Y = \dim Z$ the (complex) dimension of Y and Z. Choose a point $b \in \mathbb{C}$ which is near ∞ and such that the fibre $Y_b = w^{-1}(b) \subset Y$ is smooth. Let us briefly recall the definitions, for more details see [**LP**].

5.2.1. $f^{p,q}(Y, w)$. Let $\Omega^{\bullet}_Z(log\, D_Z)$ denote the logarithmic de Rham complex. For each $a \geq 0$ define the *a sheaf* $\Omega^a_Z(log\, D_Z, f)$ *of f-adapted logarithmic forms* as:

$$\Omega^a_Z(log\, D_Z, f) = \{\alpha \in \Omega^a_Z(log\, D_Z) \mid df \wedge \alpha \in \Omega^{a+1}_Z(log\, D_Z)\}.$$

DEFINITION 5.3. *The Landau–Ginzburg Hodge numbers* $f^{p,q}(Y, w)$ are defined as follows:

$$f^{p,q}(Y, w) = \dim \mathrm{H}^p(Z, \Omega^q_Z(log\, D_Z, f)).$$

5.2.2. $h^{p,q}(Y, w)$.

DEFINITION 5.4. [**LP**, Def. 8] Assume that (Y, w) is a Landau–Ginzburg model of Fano type. Consider the relative cohomology $\mathrm{H}^{\bullet}(Y, Y_b)$ with the nilpotent operator N and the induced canonical filtration W. *The Landau–Ginzburg numbers* $h^{p,q}(Y, w)$ are defined as follows:

$$h^{p,n-q}(Y, w) = \dim gr^{W,n-a}_{2(n-p)}\, \mathrm{H}^{n+p-q}(Y, Y_b) \quad \text{if } a = p - q \geq 0,$$

$$h^{p,n-q}(Y, w) = \dim gr^{W,n+a}_{2(n-q)}\, \mathrm{H}^{n+p-q}(Y, Y_b) \quad \text{if } a = p - q < 0$$

where gr is the standard grading of the mixed Hodge structure by the weight filtration.

5.2.3. $i^{p,q}(Y, w)$. For each $\lambda \in \mathbb{C}$ one has the corresponding sheaf $\phi_{w-\lambda}\mathbb{C}_Y$ of vanishing cycles for the fibre Y_λ. The sheaf $\phi_{w-\lambda}\mathbb{C}_Y$ is supported on the fibre Y_λ and is equal to zero if λ is not a critical value of w.

DEFINITION 5.5. Assume that the horizontal divisor $D^h \subset Z$ is empty, i.e. assume that the map $w\colon Y \to \mathbb{C}$ is proper. Then

(1) *the Landau–Ginzburg Hodge numbers* $i^{p,q}(Y, w)$ are defined as follows:

$$i^{p,q}(Y, w) = \sum_{\lambda \in \mathbb{C}} \sum_k i^{p,q+k} \mathbb{H}^{p+q-1}(Y_\lambda, \phi_{w-\lambda}\mathbb{C}_Y).$$

(2) In the general case denote by $j\colon Y \hookrightarrow Z$ the open embedding and define similarly

$$i^{p,q}(Y,w) = \sum_{\lambda\in\mathbb{C}}\sum_k i^{p,q+k}\mathbb{H}^{p+q-1}(Y_\lambda, \phi_{w-\lambda}\mathbf{R}j_*\mathbb{C}_Y).$$

Here $\mathbb{H}$ de notes the hypercohomology of the constructible complexes of the sheaves of vanishing cycles, having higher derived images $\phi_{w-\lambda}\mathbf{R}j_*\mathbb{C}_Y$.

5.3. The KKP conjecture. The *KKP conjecture* states that the three invariants coincide, that is,

$$f^{p,q}(Y,w) = h^{p,q}(Y,w) = i^{p,q}(Y,w).$$

For Y a specific rational surface with a map $w\colon Y \to \mathbb{C}$ such that the generic fibre is an elliptic curve [**LP**] Lunts and Przyjalkowski proved the equality $f^{p,q}(Y,w) = h^{p,q}(Y,w)$ and gave an example where $i^{p,q}(Y,w) \neq h^{p,q}(Y,w)$. Thus, in full generality the conjecture is false. Nevertheless, Cheltsov and Przyjalkowski proved KKP the conjecture for Fano threefolds [**CP**] and Ballico, Gasparim, Rubilar, and San Martin proved the KKP conjecture for minimal semisimple adjoint orbits [**BGRSM**].

DEFINITION 5.6. In the cases when the KKP conjecture holds true, the invariants then define a new diamond, which we call the *KKP diamond.*

We now give some examples of Hodge diamonds and KKP diamonds computed in [**BGRSM**].

5.3.1. *An example in 2 dimensions.* Consider the semisimple adjoint orbit $\mathcal{O}_2$ of $\mathfrak{sl}(2,\mathbb{C})$. Hence, $\mathcal{O}_2$ can be viewed as the affine hypersurface of $\mathbb{C}^3$ cut out by the equation $x^2 + yz = 1$, see [**BBGGSM**, Sec. 2].

The Hodge diamond of $\mathcal{O}_2$ is:

$$\begin{array}{ccccc} & & 0 & & \\ & 0 & & 0 & \\ \infty & & 0 & & 0 \\ & \infty & & 0 & \\ & & \infty & & \end{array}$$

$\mathcal{O}_2$

We know by Cor. 3.9 that $\mathcal{O}_2$ is a deformation of $\mathrm{T}^*\mathbb{P}^1$. The Hodge diamond of $\mathrm{T}^*\mathbb{P}^1$ is:

$$\begin{array}{ccccc} & & 0 & & \\ & 0 & & 0 & \\ \infty & & 1 & & 0 \\ & \infty & & 0 & \\ & & \infty & & \end{array}$$

$\mathrm{T}^*\mathbb{P}^1$

The KKP Hodge Diamond of the Landau–Ginzburg model on $\mathrm{LG}(\mathcal{O}_2)$ obtained from Thm. 4.3 was calculated in [**BGRSM**, Sec. 7] and is:

$$
\begin{array}{ccccc}
 & & 0 & & \\
 & 0 & & 0 & \\
0 & & 2 & & 0 \\
 & 0 & & 0 & \\
 & & 0 & &
\end{array}
$$

LG(2)

5.3.2. *An example in 4 dimensions.* Our LG model is $(\mathcal{O}_3, f_H) =: \mathrm{LG}(3)$ obtained from Thm. 4.3 using the adjoint orbit $\mathcal{O}_3$ of $\mathrm{Diag}(2,-1,-1)$ with potential f_H corresponding to the choice of $H = \mathrm{Diag}(1,0,-1)$.

The adjoint orbit $\mathcal{O}_3$ is a noncompact affine variety of dimension 4 that has the following Hodge diamond:

$$
\begin{array}{ccccccccc}
 & & & & 0 & & & & \\
 & & & 0 & & 0 & & & \\
 & & 0 & & 0 & & 0 & & \\
 & 0 & & 0 & & 0 & & 0 & \\
\infty & & 0 & & 0 & & 0 & & 0 \\
 & \infty & & 0 & & 0 & & 0 & \\
 & & \infty & & 0 & & 0 & & \\
 & & & \infty & & 0 & & & \\
 & & & & \infty & & & &
\end{array}
$$

$\mathcal{O}_3$

The adjoint orbit $\mathcal{O}_3$ is a deformation of the cotangent bundle of $\mathbb{P}^2$, and $T^*\mathbb{P}^2$ has the following Hodge diamond:

$$
\begin{array}{ccccccccc}
 & & & & 0 & & & & \\
 & & & 0 & & 0 & & & \\
 & & 0 & & 0 & & 0 & & \\
 & 0 & & 0 & & 0 & & 0 & \\
\infty & & 0 & & 1 & & 0 & & 0 \\
 & \infty & & 0 & & 0 & & 0 & \\
 & & \infty & & 1 & & 0 & & \\
 & & & \infty & & 0 & & & \\
 & & & & \infty & & & &
\end{array}
$$

$\mathrm{T}^*\mathbb{P}^2$

LG(3) admits a tame compactification, and the corresponding KKP diamond calculated in [**BGRSM**, Sec. 7] is:

$$\begin{array}{ccccccccc} & & & & 0 & & & & \\ & & & 0 & & 0 & & & \\ & & 0 & & 0 & & 0 & & \\ & 0 & & 0 & & 0 & & 0 & \\ 0 & & 0 & & 3 & & 0 & & 0 \\ & 0 & & 0 & & 0 & & 0 & \\ & & 0 & & 0 & & 0 & & \\ & & & 0 & & 0 & & & \\ & & & & 0 & & & & \end{array}$$

LG(3)

5.3.3. *The general case.* [**BGRSM**] proved the KKP conjecture for minimal adjoint orbits of $\mathfrak{sl}(n, \mathbb{C})$. They calculated the KKP diamond of LG(n) to be:

$$\begin{array}{ccccccc} & & & 0 & & & \\ & & 0 & & 0 & & \\ & & & \vdots & & & \\ 0 & \cdots & & n & & \cdots & 0 \\ & & & \vdots & & & \\ & & 0 & & 0 & & \\ & & & 0 & & & \end{array}$$

In conclusion, we have shown that our new concept of deformations of complex structures can be applied to many interesting examples. Furthermore, we have described classical Hodge theoretical invariants of cotangent bundles of projective spaces, and have compared them to the classical Hodge theoretical invariants of the nontrivial affine deformations of them, namely the minimal semisimple adjoint orbits of $\mathfrak{sl}(n, \mathbb{C})$. Finally we have described Landau–Ginzburg models on these adjoint orbits, and presented their KKP diamonds, that is, diamonds containing the three new invariants defined by [**KKP**], which as proved by [**BGRSM**], coincide for such orbits. How KKP diamonds vary under deformations is an interesting question that remains open. In fact, there are many delicate and intricate open questions about the deformation theory of Landau–Ginzburg models both in complex and in symplectic geometry.

6. Acknowledgments

We thank Severin Barmeier and the referee for suggesting several improvements to the text.

References

[BV] V. Balaji and P. A. Vishwanath, *On the deformation of certain moduli spaces of vector bundles*, Amer. J. Math. **115** (1993), no. 2, 279–303, DOI 10.2307/2374860. MR1216433

[BBGGSM] E. Ballico, S. Barmeier, E. Gasparim, L. Grama, and L. A. B. San Martin, *A Lie theoretical construction of a Landau-Ginzburg model without projective mirrors*, Manuscripta Math. **158** (2019), no. 1-2, 85–101, DOI 10.1007/s00229-018-1024-1. MR3895750

[BCG] E. Ballico, B. Callander, and E. Gasparim, *Compactifications of adjoint orbits and their Hodge diamonds*, J. Algebra Appl. **17** (2018), no. 6, 1850099, 16, DOI 10.1142/S0219498818500998. MR3805710

[BGK] E. Ballico, E. Gasparim, and T. Köppe, *Vector bundles near negative curves: moduli and local Euler characteristic*, Comm. Algebra **37** (2009), no. 8, 2688–2713, DOI 10.1080/00927870802562351. MR2543513

[BG1] Severin Barmeier and Elizabeth Gasparim, *Classical deformations of noncompact surfaces and their moduli of instantons*, J. Pure Appl. Algebra **223** (2019), no. 6, 2543–2561, DOI 10.1016/j.jpaa.2018.09.006. MR3906562

[BG2] S. Barmeier, E. Gasparim, *Quantizations of local surfaces and rebel instantons*, arXiv:1904.09455.

[BGRSM] E. Ballico, E. Gasparim, F. Rubilar, L. A. B. San Martin, *KKP conjecture for minimal adjoint orbits*, arXiv:1901.07939.

[Bq] Olivier Biquard, *Sur les équations de Nahm et la structure de Poisson des algèbres de Lie semi-simples complexes* (French), Math. Ann. **304** (1996), no. 2, 253–276, DOI 10.1007/BF01446293. MR1371766

[Bo] F. A. Bogomolov, *The decomposition of Kähler manifolds with a trivial canonical class* (Russian), Mat. Sb. (N.S.) **93(135)** (1974), 573–575, 630. MR0345969

[CP] I. Cheltsov, V. Przyjalkowski, *Katzarkov–Kontsevich–Pantev Conjecture for Fano threefolds*, arxiv:1809.09218.

[GGSM1] Elizabeth Gasparim, Lino Grama, and Luiz A. B. San Martin, *Symplectic Lefschetz fibrations on adjoint orbits*, Forum Math. **28** (2016), no. 5, 967–979, DOI 10.1515/forum-2015-0039. MR3543704

[GGSM2] Elizabeth Gasparim, Lino Grama, and Luiz A. B. San Martin, *Adjoint orbits of semi-simple Lie groups and Lagrangian submanifolds*, Proc. Edinb. Math. Soc. (2) **60** (2017), no. 2, 361–385, DOI 10.1017/S0013091516000286. MR3653862

[GKRS] Elizabeth Gasparim, Thomas Köppe, Francisco Rubilar, and Bruno Suzuki, *Deformations of noncompact Calabi-Yau threefolds* (English, with English and Spanish summaries), Rev. Colombiana Mat. **52** (2018), no. 1, 41–57, DOI 10.15446/recolma.v1n52.74554. MR3863276

[GS] E. Gasparim, B. Suzuki, *Infinite dimensionality of deformation families for noncompact threefolds*, arXiv:1909.01842v2.

[H] James E. Humphreys, *Introduction to Lie algebras and representation theory*, Graduate Texts in Mathematics, vol. 9, Springer-Verlag, New York-Berlin, 1978. Second printing, revised. MR499562

[KKP] Ludmil Katzarkov, Maxim Kontsevich, and Tony Pantev, *Bogomolov-Tian-Todorov theorems for Landau-Ginzburg models*, J. Differential Geom. **105** (2017), no. 1, 55–117. MR3592695

[Ko] Kunihiko Kodaira, *Complex manifolds and deformation of complex structures*, Grundlehren der Mathematischen Wissenschaften [Fundamental Principles of Mathematical Sciences], vol. 283, Springer-Verlag, New York, 1986. Translated from the Japanese by Kazuo Akao; With an appendix by Daisuke Fujiwara. MR815922

[Kv] A. G. Kovalev, *Nahm's equations and complex adjoint orbits*, Quart. J. Math. Oxford Ser. (2) **47** (1996), no. 185, 41–58, DOI 10.1093/qmath/47.1.41. MR1380949

[Kr1] P. B. Kronheimer, *A hyper-Kählerian structure on coadjoint orbits of a semisimple complex group*, J. London Math. Soc. (2) **42** (1990), no. 2, 193–208, DOI 10.1112/jlms/s2-42.2.193. MR1083440

[Kr2] P. B. Kronheimer, *Instantons and the geometry of the nilpotent variety*, J. Differential Geom. **32** (1990), no. 2, 473–490. MR1072915

[LP] Valery Lunts and Victor Przyjalkowski, *Landau-Ginzburg Hodge numbers for mirrors of del Pezzo surfaces*, Adv. Math. **329** (2018), 189–216, DOI 10.1016/j.aim.2018.02.024. MR3783413

[M] Marco Manetti, *Lectures on deformations of complex manifolds (deformations from differential graded viewpoint)*, Rend. Mat. Appl. (7) **24** (2004), no. 1, 1–183. MR2130146

[NR] M. S. Narasimhan and S. Ramanan, *Deformations of the moduli space of vector bundles over an algebraic curve*, Ann. of Math. (2) **101** (1975), 391–417, DOI 10.2307/1970933. MR384797

[R] F. Rubilar, *Deformaciones de estructuras complejas de 3-variedades Calabi–Yau*, Master Thesis, Universidad Católica del Norte, Chile (2017).

[SM] L. A. B. San Martin, *Álgebras de Lie*, Segunda edição, Editora Unicamp, ISBN 978-85-268-0876-8 (2010), 454 pp.

Depto. Matemáticas, Univ. Católica del Norte, Antofagasta, Chile.
Email address: etgasparim@gmail.com

Depto. Matemáticas, Univ. Católica del Norte, Antofagasta, Chile.
Email address: francisco.rubilar.arriagada@gmail.com

Contemporary Mathematics
Volume **766**, 2021
https://doi.org/10.1090/conm/766/15379

A brute force computer aided proof of an existence result about extremal hyperbolic surfaces

Ernesto Girondo and Cristian Reyes

Abstract. Extremal compact hyperbolic surfaces contain a packing of discs of the largest possible radius permitted by the topology of the surface. It is well known that arithmetic conditions on the uniformizing group are necessary for the existence of a second extremal packing in the same surface, but constructing explicit examples of this phenomenon is a complicated task. We present a brute force computational procedure that can be successfully used to produce such examples in all cases.

1. Introduction

It is well known that the radius of a metric disc embedded in a compact hyperbolic surface cannot be larger than certain bound depending only in the topology of the surface (cf. [**1**], [**6**], [**8**]). Recently, this result has been extended to embeddings of k-packings (i.e. collections of a number k of pairwise disjoint metric discs of given radius) in hyperbolic surfaces, first in the orientable case (see [**4**]) and then in the non-orientable case ([**9**]). By the term *extremal surface* we mean a (orientable or non-orientable) surface S with an embedded k-packing reaching the upper bound for the radius, given by

$$\cosh R = \frac{1}{2\sin\left(\frac{k\pi}{6(k-\chi)}\right)}.$$

Extremal surfaces may be studied from the point of view of discrete groups of Möbius transformations. We can regard such a surface S as a quotient $S \simeq \mathbb{D}/K$ of the universal covering space $\mathbb{D}$, the unit disc, by a *uniformizing group* K, which is a Fuchsian group or a proper NEC group depending on whether S is orientable or not.

It has been observed that extremality translates into algebraic, and even arithmetic properties of the group K. First, a necessary and sufficient condition for the existence of an extremal k-packing in S is that the group K must be contained in a (Fuchsian or NEC) triangle group $\Delta = \Delta(2,3,N)$ where $N = 6(k-\chi)/k$, the index being $6(k-\chi)$ or $12(k-\chi)$ for the orientable and the non-orientable case, respectively ([**4**], [**9**]). It follows that the topology does not impose any restriction to extremality for $k = 1, 2, 3$ and 6. On the other hand, when k does not divide

2020 *Mathematics Subject Classification.* Primary 30F50; Secondary 20H10.
Partially supported by grant MTM2016-79497-P.
This work was funded by the CONICYT PFCHA/Becas Chile 72180175.

6χ, embedded k-packings cannot be as dense as they are in hyperbolic space, and therefore extremal k-packings do not exist for such topological classes of hyperbolic surfaces.

Actual existence of extremal surfaces for the admissible pairs (k, χ) (or, equivalently (k, g), where g stands for the orientable or non-orientable genus) has been shown either as a consequence of known results about inclusions of Fuchsian groups ([**4**], where a well-known theorem by Edmonds et al. [**3**] is used) or as a result of a clever combinatorial procedure called *edge-grafting* ([**9**]).

There is a natural question where, surprisingly, arithmetic seems to play a role related to extremality. Asking if a extremal surface of genus g may contain several distinct extremal k-packings translates into asking if the uniformizing group K can be contained in two different (conjugate) copies of the corresponding triangle group Δ. A combination of known results by Margulis, Singerman and Takeuchi shows that extremal k-packings are often necessarily unique within their surfaces, the possible exceptions corresponding to values (k, g) such that the triangle group Δ is arithmetic (see [**14**] for the definition of arithmeticity). This way, first in [**4**] for the orientable case and later in [**9**] for the non-orientable case, it was proved the following

THEOREM 1.1. *If X is a compact hyperbolic k-extremal surface of genus g and*

$$N := \frac{6(k-\chi)}{k} \not\in \{7, 8, 9, 10, 11, 12, 14, 16, 18, 24, 30\}$$

then the extremal k-packing is unique in X.

Summarizing, a combination of topology/combinatorics and arithmetics restricts the possible pairs (k, g) for which there could be an extremal k-surface of genus g with several extremal k-packings to values such that

- Orientable case: k divides $12g + 6k - 12$ and
  $N = \frac{12g+6k-12}{k} \in \{7, 8, 9, 10, 11, 12, 14, 16, 18, 24, 30\}$.
- Non orientable case: k divides $6g + 6k - 12$ and
  $N = \frac{6g+6k-12}{k} \in \{7, 8, 9, 10, 11, 12, 14, 16, 18, 24, 30\}$.

Notice that these are necessary conditions. Showing that they are also sufficient is far from being obvious. In the first cases considered in the literature, namely those with $k = 1$ (see [**7**], [**8**]), the strategy employed was as follows. First, determine all possible extremal surfaces, something that can be done by finding a representative of all conjugacy classes of Fuchsian/proper NEC groups of the appropriate index inside the triangle group Δ with the help of some algebra package such as GAP. Then, perform an exhaustive metric study (similar to what was done for instance in [**7**]; see Section 3 below for some hints about what we mean with this) in order to determine if a second extremal k-packing exists in any of the surfaces.

The serious problem for extending this strategy to $k > 1$ is that sometimes it is virtually impossible to list the huge number of subgroups $K < \Delta$ involved.

2. The main result

Let us denote $\mathcal{N} = \{7, 8, 9, 10, 11, 12, 14, 16, 18, 24, 30\}$. We shall prove the following theorem:

THEOREM 2.1. *Let $g \geq 3$ and $k \geq 1$ be such that k divides $6g + 6k - 12$. Then there exists a compact non-orientable k-extremal surface of genus g with more than* *one extremal k-packing if and only if* $\frac{6g+6k-12}{k} \in \mathcal{N}$. *In all other cases extremal k-packings are unique within compact non-orientable k-extremal surfaces of genus g.*

A simple argument involving the double cover of non-orientable, k-extremal surfaces (see the last section of [**9**]), together with explicit constructions for two particular cases, very similar to what we will do in the next section, show that

THEOREM 2.2. *Let* $g \geq 2$ *and* $k \geq 1$ *be such that* k *divides* $12g + 6k - 12$. *Then there exists a compact Riemann surface of genus* g *that is* k*-extremal and has more than one extremal* k*-packing if and only if* $N := \frac{12g+6k-12}{k} \in \mathcal{N}$. *In all other cases extremal* k*-packings are unique within compact Riemann* k*-extremal surfaces of genus* g.

Note that by the characterization of compact non-orientable k-extremal surfaces in terms of triangle groups (cf. [**9**]), we know that if such a surface $S \simeq \mathbb{H}/K$ admits two extremal k-packings, then we must have two inclusions

$$K \leq \Delta_1^{\pm} = \Delta_1^{\pm}\left(2, 3, \frac{6g+6k-12}{k}\right), \quad K \leq \Delta_2^{\pm} = \Delta_2^{\pm}\left(2, 3, \frac{6g+6k-12}{k}\right)$$

both with index $12g + 12k - 24$.

As a consequence, a compact non-orientable k-extremal surface whose NEC group K is contained in a non-arithmetic extended triangle group $\Delta^{\pm}$, has a unique extremal k-packing (see [**9**]). This happens whenever $\frac{6g+6k-12}{k} \notin \mathcal{N}$.

The difficult statement is the converse, since there is apparently no theoretical reason ensuring the existence of the uniformizing group K with the required properties for all the pairs (k, g) involved. We are forced to find explicitly one example for every such pair (k, g).

With the same terminology as in [**9**], for $N \in \mathcal{N}$ we call a pair (k, g) such that $\frac{6g+6k-12}{k} = N$ *primitive* if both k and g are minimal among all the pairs (k, g) verifying the same relation with N. We denote (k_N, g_N) the primitive pair for a given N. The existence of extremal non-orientable k_N-extremal surfaces of genus g_N for the eleven primitive pairs (one for each value $N \in \mathcal{N}$) is enough to ensure the existence of k-extremal surfaces of genus g for all the pairs (k, g) in Theorem 2.1, see Lemma 1 in [**9**].

Notice that four of the eleven required surfaces can be found already in the literature, since they correspond to the cases for which $k_N = 1$ (see [**8**], [**10**], [**11**], [**12**] and [**13**]). These surfaces correspond to $N = 12, 18, 24$ or 30. For the cases $N = 7, 8, 9, 10$ it could still be possible to do an exhaustive analysis of all compact non-orientable k_N-extremal of genus g_N (similar to that one done in [**8**]), since we can compute all proper NEC uniformizing subgroups of the extended triangle group $\Delta^{\pm}(2, 3, N)$ with the required index. But the computation of all the corresponding subgroups of $\Delta^{\pm}(2, 3, N)$ with $N = 11, 14, 16$ is a inadequate approach, since there are sometimes hundreds of thousands of such surfaces. We propose a different strategy, that is the brute-force procedure we refer to in the title of this paper, and we devote the rest of the text to explain the details.

3. The brute force construction

For a given N, any compact non-orientable k_N-extremal surface of genus g_N is uniformized by a proper NEC subgroup K of index $2k_N N$ inside a triangle group

$\Delta^{\pm}(2,3,N)$. Such a group admits a fundamental domain F consisting of a union of k_N regular N-gons of angle $2\pi/3$, the extremal k_N-packing being simply induced by the discs inscribed to these polygons.

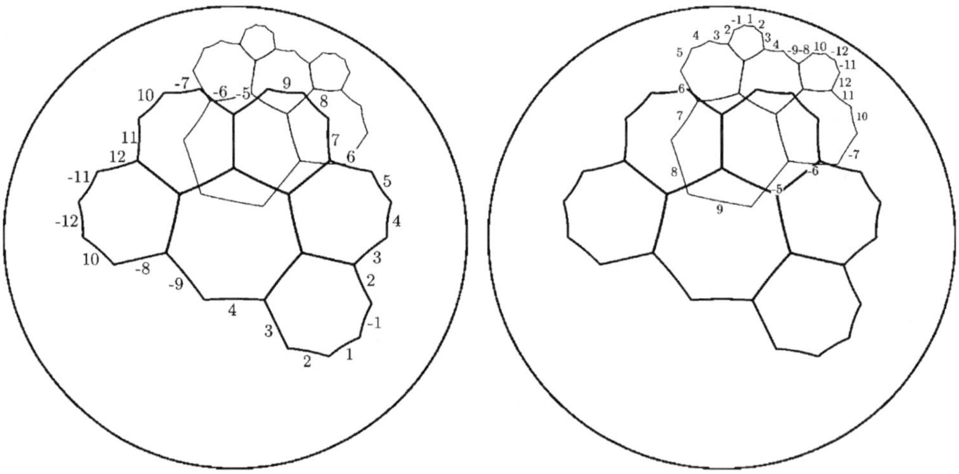

FIGURE 1. A primitive example with two different extremal k-packings for the case $N = 7$. The figure shows two different fundamental domains F, F' for the same proper NEC uniformizing group K, both decomposed as the union of six regular heptagons of angle $2\pi/3$. The numbers indicate side-pairing transformations generating K, and a minus sign means that the corresponding transformation is orientation-reversing. Surprisingly, both sets of side pairings generate the same group K (see Section 5).

If the resulting extremal surface $S \simeq \mathbb{D}/K$ admits a second extremal k_N-packing, the group K would have a second fundamental domain F' isometric to F but not K-related to it (Figure 1 shows an example of this phenomenon in the case $N = 7$, where $k_7 = 6$ and $g_7 = 3$; see Section 5 for more information about this surface). The main remark here is that the k_N centers of the N-gons forming F' must be very special points, since they must respect very strong metric conditions with respect to K. The following lemma, taken from [**8**], describes such conditions.

LEMMA 3.1. *Let S be a compact non-orientable k-extremal surface of genus g uniformized by K, and let $\pi : \mathbb{D} \longrightarrow \mathbb{D}/K \simeq S$ be the natural projection. Let $p = [z]_K$ be the center of one of the k discs forming an extremal k-packing inside S, and denote $N = \frac{6g+6k-12}{k}$ and $\mathcal{T}$ a tessellation of $\mathbb{D}$ by regular N-gons of angle $2\pi/3$. Then, for every $\gamma \in K$ the distance $d(z, \gamma(z))$ must agree with the distance between some pair of polygon centers of $\mathcal{T}$. We call these* admissible distances.

We need to compute as many admissible distances as we can, in order to have a precise idea about the possible displacement of centers of discs belonging to *hidden* extremal k_N-packings. The following observation is useful for this purpose:

LEMMA 3.2. *Consider a tessellation $\mathcal{T}$ of $\mathbb{D}$ given by copies of a regular N-gon P_0 of angle $\frac{2\pi}{3}$ centered at the origin 0. Consider a triangle $\triangle ABC$ of angles*

$\pi/3, \pi/N$ and $\pi/2$ with vertices at a vertex A of P_0, the origin $B = 0$ and an edge midpoint C, as in Figure 2. Denote by a, b, c the reflections with respect to the lines $\overline{BC}, \overline{CA}$ and $\overline{AB}$ respectively, and let $R_m = (ca)^m ab$.

Given any polygon P in $\mathcal{T}$, there is a transformation of the form $R_{i_1} R_{i_2} \cdots R_{i_k}$ that sends P_0 to P, where $0 \leq i_j < N$ for $j = 1, \dots, k$.

PROOF. If we select an arbitrary polygon P in the tessellation, we can always connect the origin to its center by a hyperbolic polygonal curve Γ joining centers of a sequence of pairwise adjacent polygons $P_0, P_1, \dots, P_k = P$. Figure 2 illustrates the construction for an example in the case $N = 7$.

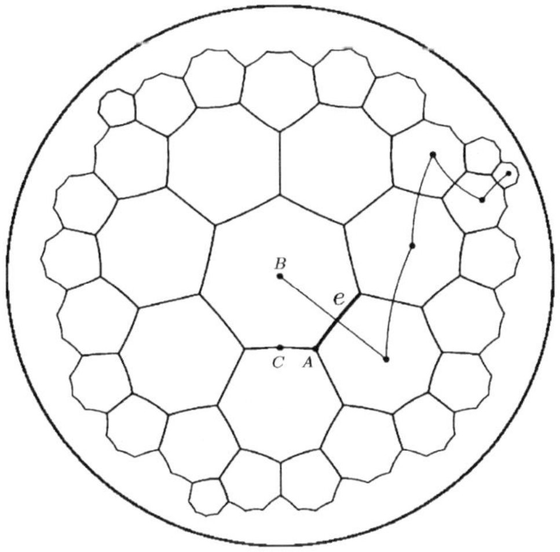

FIGURE 2. The tesselation $\mathcal{T}$ and a sketch of the construction for the case $N = 7$. In this case $k = 5$ and $i_1 = 1, i_2 = 6, i_3 = 4, i_4 = 1, i_5 = 5$.

Now, label the edges of the central polygon counterclockwise, with the label 0 at the edge containing the point C, and let i_1 be the label of the edge e where Γ cuts the central polygon P_0. We assign then the label 0 to e considered as an edge of the polygon P_1 meeting P_0 along e, and complete the labelling $1, 2, \dots$ of the remaining edges of P_1 counterclockwise. Next, define i_2 as the label of the edge of P_1 where Γ leaves P_1, and so on. This way we construct a sequence of numbers $i_1, i_2, \dots, i_k$. We claim that the transformation $R_{i_1} R_{i_2} \cdots R_{i_k}$ sends P_0 to $P_k = P$.

The proof can be done by induction in the index k. Taking $k = 1$ corresponds to the case when P is a direct neighbor of the central polygon P_0, therefore this case is absolutely obvious.

Now, assume that the statement is true for polygons whose center can be joined to the origin by polygonal lines with k segments. Assume Γ is a polygonal line connecting the origin to the center of certain polygon P, crossing the sequence $P_0, P_1, \dots, P_k, P_{k+1} = P$, and define $R_{i_1} R_{i_2} \cdots R_{i_k} R_{i_{k+1}}$ as above. By the induction hypothesis, the transformation $R = R_{i_1} R_{i_2} \cdots R_{i_k}$ sends P_0 to P_k, therefore $RR_{i_{k+1}}$ sends P_0 to one of the neighbors of P_k that, by construction, is precisely P_{k+1}. $\square$

Note that the set of N^k transformations $\mathcal{R}_k = \{R_{m_1} R_{m_2} \cdots R_{m_k},\ 0 \leq m_i < N\}$ contains elements that move the central polygon P_0 to any polygon that can be connected to P_0 with a polygonal Γ of k steps. A collection of admissible distances of the tessellation $\mathcal{T}$ can be constructed by computing the distance $d(0, R(0))$ for every R in $\mathcal{R}_k$. Larger values of k yield larger subsets of the full set of admissible distances.

The locus of points which are moved certain prescribed distance by a hyperbolic transformation g can be expressed in terms of the translation length T and the distance to the axis A of g (invariant geodesic) by the following well-known formula for the displacement function (cf. [**2**]):

$$\sinh \frac{d(z, g(z))}{2} = \cosh d(z, A) \sinh \frac{T}{2}.$$

One can easily prove a similar result for glide-reflections, for which we still call translation length the displacement of points belonging to the axis:

PROPOSITION 3.3. *If g is a glide-reflection with translation length T and axis A, then*

$$\cosh \frac{d(z, g(z))}{2} = \cosh d(z, A) \cosh \frac{T}{2}.$$

PROOF. Since any glide reflection is conjugate to a map of the form $h : z \mapsto -k\overline{z}$, $k > 1$, all we have to do is to prove the result for such transformations h. In order to check that the formula holds, we compute the three magnitudes involved using well-known expressions for the hyperbolic sine or cosine of the hyperbolic distance between points of the upper half-plane $\mathbb{H}$ (see [**2**], Theorem 7.2.1).

First, for the term on the left of the equation we have

$$\cosh \frac{d(z, h(z))}{2} = \frac{|z - \overline{h(z)}|}{2\sqrt{\mathrm{Im}(z)\mathrm{Im}(h(z))}} = \frac{|z||1+k|}{|y|2\sqrt{k}},$$

where $z = x + iy$. Now, since the axis A' of h is the imaginary axis, we can compute

$$\cosh d(z, A') = \cosh d(z, i|z|) = 1 + \frac{|z - i|z||^2}{2\mathrm{Im}(z)\mathrm{Im}(i|z|)} = \frac{|z|}{y}$$

and, finally, we obtain T using that $z = i$ belongs to the axis, so that

$$\cosh \frac{T}{2} = \cosh \frac{d(i, ki)}{2} = \frac{|i + ki|}{2\sqrt{\mathrm{Im}(i)}\sqrt{\mathrm{Im}(ki)}} = \frac{(1+k)}{2\sqrt{k}}.$$

$\square$

From the last two propositions we see that the locus of points with a prescribed displacement under a hyperbolic isometry or a glide reflection coincides with the locus of points at certain distance from the axis of the transformation. This set has two components, one at each side of the axis, which are arcs of generalized circles (i.e. arcs of circumference or of straight lines), see [**2**], Section 7.20. We call such sets *bananas* for obvious reasons (see Figure 5).

We have now all the ingredients needed for our brute force project. Assume we want to decide if there exists a compact hyperbolic k-extremal surface S of genus g with more than one extremal k-packing. The plan is as follows:

(1) We know that $S \simeq \mathbb{D}/K$, and that we can obtain a fundamental domain F for K joining k regular N-gons of angle $2\pi/3$, where $N := \frac{6(k-\chi)}{k}$. There are different possible combinatorial configurations for the relative position of the k polygons, so we choose one to start with. Note that we may need to change to a different configuration if the process is not successful at some of the next steps.
(2) For the chosen fundamental domain, we compute the set $\mathcal{P}$ of all the possible side-pairing transformations between all the possible pairs of edges of the polygons, admitting orientation reversing transformations if we want to construct a non-orientable surface. The set $\mathcal{P}$ is usually very large.
(3) We compute a set as large as possible of admissible distances associated to a tessellation by regular N-gons of angle $2\pi/3$.
(4) We choose a starting pair of transformations $p_1, p_2 \in \mathcal{P}$, and compute the set $\mathcal{C}$ of points of F that show certain admissible displacements under p_1 and p_2, which will be our *candidates*. By what has been said above, there are finitely many candidate points for every choice of F, p_1 and p_2, and there are finitely many choices of such a triple. Again, if our construction does not produce the surface we look for with this choice of p_1, p_2, we change to a different starting pair of transformations.
(5) We compute the displacement of points $x \in \mathcal{C}$ under all the transformations in $\mathcal{P}$, in order to determine the subset $\mathcal{P}_x \subset \mathcal{P}$ consisting of those transformations moving x an admissible distance.
(6) We check if there are enough transformations in $\mathcal{P}_x$ in order to construct a full set of side-pairing transformations between pairs of edges of F such that every vertex cycle is surrounded by a full 2π angle. If this is the case, the group K generated by these transformations uniformizes a k-extremal surface of genus g, with an obvious extremal k-packing formed by the discs inscribed to the k-polygons that form F. Additionally, the point x may be the center of one of the discs of a second *hidden* k-extremal disc.
(7) We determine if there exists another configuration F' of k regular N-polygons which is also a fundamental domain for K, one of the polygon centers being the candidate x. If this is the case, a second extremal k-packing exists in $S = \mathbb{D}/K$. In practice this is a difficult step unless we can show the existence of an automorphism of S moving the obvious k-packing to the hidden one.

4. Full details of the construction for the case $N = 14$.

We illustrate now our strategy for the case $N = 14$, and recall that our study can be easily adapted with minor modifications to the seven relevant values of N, namely $N = 7, 8, 9, 10, 11, 14$ and 16.

The primitive $N = 14$ example we are looking for is a 3-extremal compact non-orientable surface S of genus 6, which has as fundamental region the connected union F of three regular 14-gons with angle $\frac{2\pi}{3}$.

(1) We begin choosing the region F in Figure 3, consisting of three regular 14-gons with angle $\frac{2\pi}{3}$ with a common vertex. We label the edges of each polygon counterclockwise from 0 to 13. There is no guarantee about being able to construct our surface starting with the domain F: if some of the

next steps of the process fails, we would have to come back to this point and start again with a different shaped F.

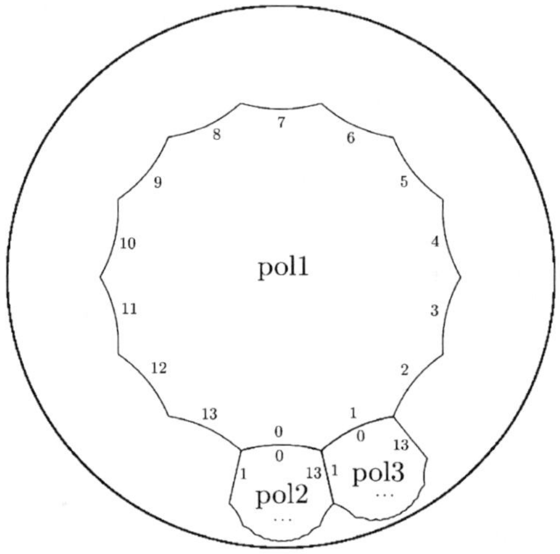

FIGURE 3. The starting fundamental region F.

(2) We compute all the conformal and anticonformal transformations that can be used to identify edges of F, collecting all this information in a list that we call $\mathcal{P}$ (see [**5**] for the computer code and the results). We can identify an element p of $\mathcal{P}$ by a triple $(\pm, m_i, n_j)$ meaning that γ maps the edge labeled i in polygon m to the edge labeled j in the polygon n, the first entry indicating if γ is conformal $(+)$ or anticonformal $(-)$; see an example in Figure 4.
(3) We compute a set D of admissible distances provided by the transformations in $\mathcal{R}_5$ (recall the notation for $\mathcal{R}_5$ after the proof of Lemma 3.2). See [**5**] again to see the code employed and the data obtained.
(4) We choose the following starting pair of side-pairing transformations:

$$p_1 = (-, 1_{13}, 1_{10}), \quad p_2 = (+, 1_2, 2_7).$$

Now we compute the set of points showing an admissible displacement under p_1 and p_2; in Figure 5 we show some of the corresponding bananas (the code is once more available at [**5**]). Figure 6 shows some of the points in the set $\mathcal{C}$ of candidate points obtained as intersection of the bananas of p_1 and p_2.
(5) For a point $x \in \mathcal{C}$, we collect in a list $\mathcal{P}_x$ all the side pairings p in $\mathcal{P}$ which have a banana that passes through x. For example, for the point x labeled ■ in Figure 6, which has coordinates $x \simeq 0.324 - 0.478i$, we find $\mathcal{P}_x = \{(-, 1_{13}, 1_{10}), (+, 1_2, 2_7), (-, 1_{11}, 2_1), (+, 2_8, 3_{13})\}$.
(6) We see, for instance, that in the case $x \simeq 0.324 - 0.478i$ the list $\mathcal{P}_x$ does not contain a side pairing involving the edge 1_3, so this point x fails to be a good candidate.

Only two of the points of $\mathcal{C}$ marked in Figure 6 satisfy the needed condition that $\mathcal{P}_x$ must contain enough transformations so that we can

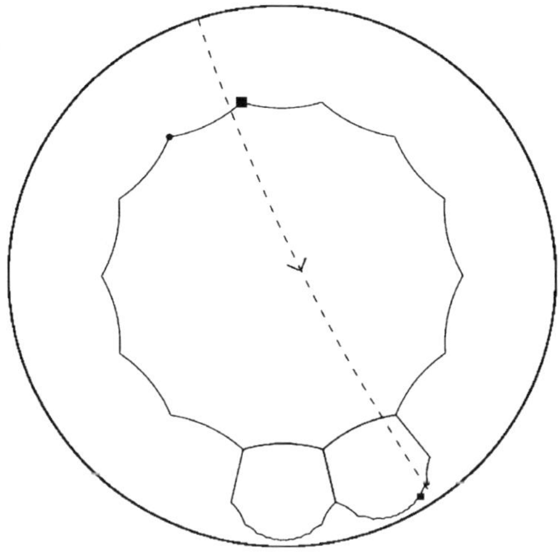

FIGURE 4. The transformation $(-, 1_8, 3_{11})$ is anticonformal and is determined by its action on the segment joining the points • and ■. The dashed line is its (oriented) axis.

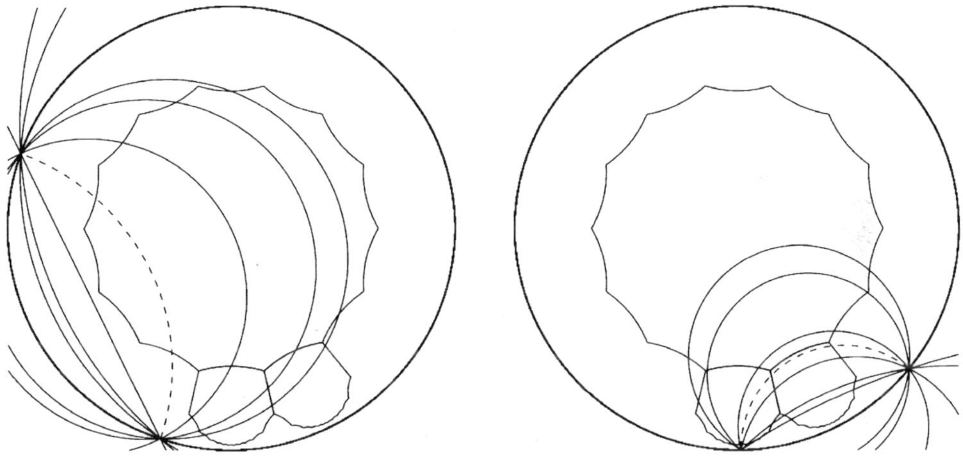

FIGURE 5. The starting set of bananas corresponding to p_1 (left) and p_2 (right). The dashed lines indicate the axes of these transformations.

choose a side-pairing for every edge of F. One is obviously the origin and the other one is $x \simeq 0.516 - 0.248i$, marked $\star$ in Figure 6. The list $\mathcal{P}_x$ is huge, and contains several side pairings for every edge. For example, if we focus on the edge 1_3, we find in $\mathcal{P}_x$ the following side pairings:

$$(+, 1_3, 1_{11}), (-, 1_3, 1_8), (+, 1_3, 2_{10}), (-, 1_3, 2_7), (+, 1_3, 3_{11}), (-, 1_3, 3_8)$$

Considering all the elements of $\mathcal{P}_x$ that correspond to each edge, all we have to do is to choose, by direct inspection, a set of transformations that would identifiy in pairs all the edges of F and generate a group K

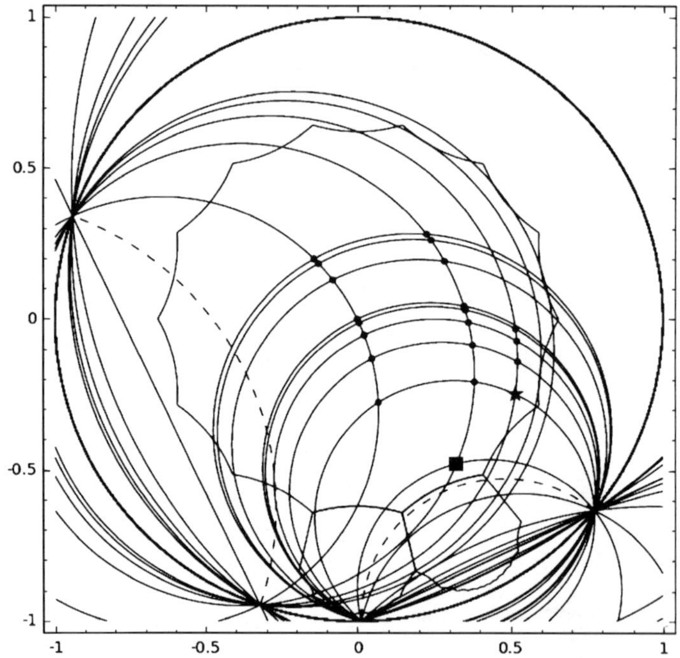

FIGURE 6. The construction of part of the set $\mathcal{C}$ of candidate points via intersection of bananas of p_1 and p_2.

uniformizing a compact surface. The precise set we find is shown in Table 1, where $G_i = (ca)^i(ab)(ca)(ab)(ca)^{-i}$. Note that G_0, G_1 are rotations of order 14 fixing the center of pol2 and pol3 respectively.

(7) The surface $S = \mathbb{D}/K$ has a point $[x]_K = [0.516 - 0.248i]_K$ that may play a special role, as it satisfies the metric requirements that a center of one of the three discs of an extremal 3-packing should satisfy.

Figure 7, extracted from the SAGE files available in [**5**], shows how miraculously one banana corresponding to every side pairing transformation passes through this point.

Showing that a second 3-packing actually exists could be quite difficult, but we may check if, as it happens in the classical $k = 1$ case (cf. [**7**]), an automorphism of S maps the original 3-packing outside of the original tessellation. This is exactly what happens in our case: the elliptic element τ of order 2 fixing the midpoint M between $x \simeq 0.516 - 0.248i$ and $(ca)^2b(0)$ normalizes the group K, thus it induces an automorphism of S. In order to prove this claim we show that the conjugate by τ of every side pairing s_j in Table 1 belongs indeed to K. The second column of the table in Figure 8 shows, for a given generator s of K, the element $k \in K$ such that $\tau s \tau k = 1$ (see the corresponding SAGE file in [**5**] where these identities are computed).

TABLE 1. A set of side pairings generating a surface NEC group K uniformizing a 3-extremal compact non-orientable surface of genus 6 (N=14).

	Side pairing	Word in the generators of $\Delta^{\pm}(2,3,14)$
s_1	$(-,1_{13},1_{10})$	$(ca)^{-4}b(ca)$
s_2	$(+,1_2,2_7)$	$G_0^7(ca)^{-2}$
s_3	$(-,2_1,1_{11})$	$(ca)^{-3}aG_0^{-1}$
s_4	$(+,3_{13},2_8)$	$G_0^9G_1^2$
s_5	$(+,1_{12},1_7)$	$(ca)^7ab(ca)^2$
s_6	$(-,1_8,2_2)$	$G_0^2a(ca)^6$
s_7	$(+,1_5,2_3)$	$G_0^3(ca)^9$
s_8	$(-,1_9,1_6)$	$(ca)^6b(ca)^5$
s_9	$(+,1_4,3_5)$	$G_1^5(ca)^{-3}$
s_{10}	$(+,1_3,2_{10})$	$G_0^{-4}(ca)^{-3}$
s_{11}	$(+,2_9,3_6)$	$G_1^5G_0^4$
s_{12}	$(+,3_{12},3_7)$	$G_1^7(ca)(ab)(ca)^{-1}G_1^2$
s_{13}	$(+,2_6,2_{11})$	$G_0^{-3}abG_0^{-6}$
s_{14}	$(+,2_4,3_4)$	$G_1^3G_0^{-5}$
s_{15}	$(+,3_3,3_{11})$	$G_1^{-3}(ca)(ab)(ca)^{-1}G_1^{-3}$
s_{16}	$(+,2_5,3_{10})$	$G_1^{-5}G_0^{-6}$
s_{17}	$(+,2_{12},3_9)$	$G_1^{-6}G_0$
s_{18}	$(+,3_2,3_8)$	$G_1^{-6}(ca)(ab)(ca)^{-1}G_1^{-2}$

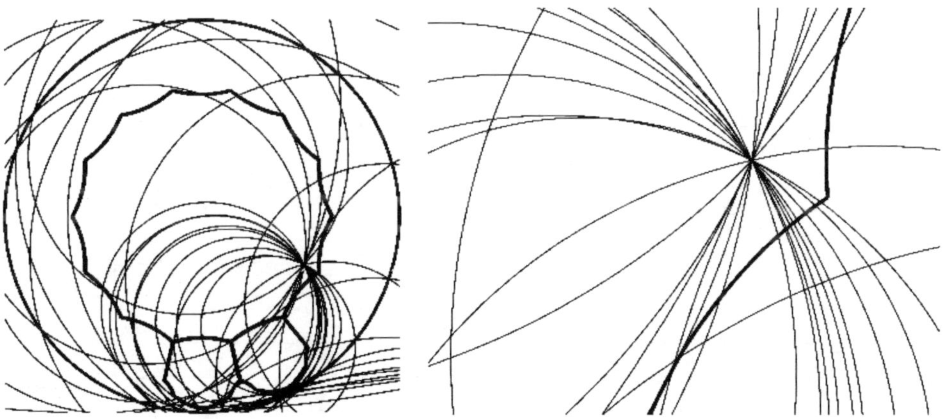

FIGURE 7. The point $0.516-0.248i$ belongs to one banana for each of the side-pairing generators of the group K constructed. In the picture on the right we have zoomed in to show how miraculously all the arcs meet there.

5. End of the proof of Theorem 2.1

The proof of Theorem 2.1 would be complete if we were able to successfully apply the procedure described at the end of Section 3 for the primitive pairs (k_N, g_N)

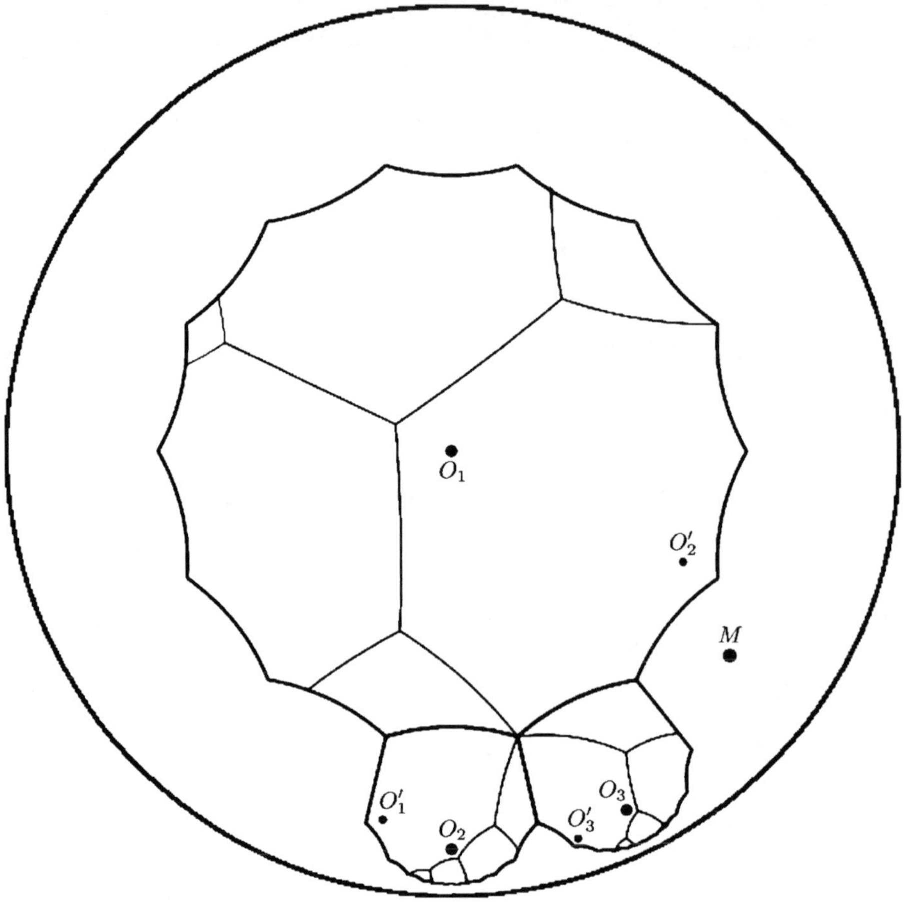

Side pairing s	$k = \tau s^{-1} \tau$	Side pairing s	$k = \tau s^{-1} \tau$
s_1	$s_2^{-1} s_7 s_6^{-1} s_2$	s_{10}	$s_2^{-1} s_{11}^{-1} s_{14} s_2$
s_2	$s_7^{-1} s_2$	s_{11}	$s_2^{-1} s_{14}^{-1} s_{11} s_{14} s_2$
s_3	$s_2^{-1} s_7 s_6^{-1} s_2$	s_{12}	$s_2^{-1} s_{16}^{-1} s_{12} s_{16} s_2$
s_4	$s_7^{-1} s_2$	s_{13}	$s_2^{-1} s_7 s_2^{-1} s_{11}^{-1} s_{14} s_2$
s_5	$s_2^{-1} s_6 s_1 s_2$	s_{14}	$s_2^{-1} s_7 s_{14} s_2$
s_6	$s_2^{-1} s_3^{-1} s_7^{-1} s_2$	s_{15}	$s_2^{-1} s_{14}^{-1} s_{16} s_2$
s_7	$s_2^{-1} s_7^{-1} s_2$	s_{16}	$s_2^{-1} s_7 s_{16} s_2$
s_8	$s_2^{-1} s_3^{-1} s_2$	s_{17}	$s_2^{-1} s_{14}^{-1} s_{11} s_4 s_{16} s_2$
s_9	$s_2^{-1} s_{14} s_2$	s_{18}	$s_2^{-1} s_{14}^{-1} s_{11} s_4 s_{16} s_2$

FIGURE 8. Our example for $N = 14$ is a 3-extremal non-orientable surface $S = \mathbb{D}/K$ of genus 6 with two different extremal 3-packings: the centers O_j' of the second packing are the image of the centers O_j of the first packing by τ, an order two automorphism of S induced by an elliptic transformation fixing $M \simeq 0.620 - 0.459i$. The thin lines are the part of the tessellation associated to the second extremal packing that lies inside the fundamental domain F. The data in the table justifies that τ normalizes the group K.

corresponding to the values $N = 7, 8, 9, 10, 11, 16$ (see the last paragraph of Section 2), thus constructing a 6-extremal surface of genus 3, a 3-extremal surface of genus 3, a 2-extremal surface of genus 3, a 3-extremal surface of genus 4, a 6-extremal surface of genus 7 and a 3-extremal surface of genus 7 (these are the cases $N = 7, 8, 9, 10, 11, 16$, respectively), all them with multiple extremal disc packings.

Figures 9 to 14 below contain all the data that describes such surfaces: a table showing the side pairing transformations generating the corresponding subgroup $K < \Delta(2, 3, N)$, the result of conjugating these generators by an order two elliptic element τ fixing certain specific point M (showing that τ induces an automorphism of the quotient surface $\mathbb{D}/K$), and a picture showing the centers O_j of the starting extremal disc packing, the location of the fixpoint M and the location of $O'_j = \tau(O_j)$, the centers of the discs forming the second extremal disc packing.

In the tables in Figures 9 to 14 we have employed a notation similar to the one used for the case $N = 14$ in the previous section: a, b, c are the reflections at the vertices of a triangle with angle $\pi/2$, $\pi/3$ and π/N respectively (the generators of the extended triangle group $\Delta^{\pm}(2, 3, N)$), and $G_i = (ca)^i(ab)(ca)(ab)(ca)^{-i}$. The thin lines in the figures show how the Dirichlet cells corresponding to the second extremal packing look like inside the fundamental domain of K adapted to the original extremal packing. All these examples can be constructed by the brute force procedure described in Section 3 that we have applied in detail to the case $N = 14$ in Section 4 (see [**5**] for the SAGE code and the computations).

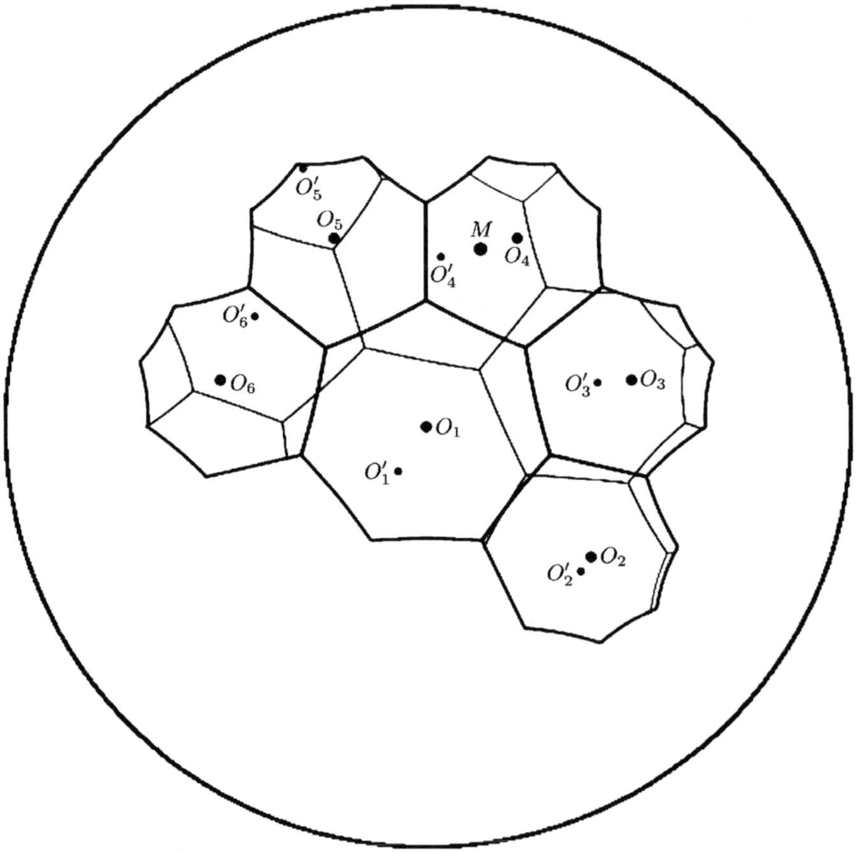

	Side pairing	Word in the generators of $\Delta^{\pm}(2,3,7)$	$k = \tau s^{-1}\tau$
s_1	$(-, 3_3, 3_4)$	$G_1^4(ca)b(ca)^{-1}G_1^4$	$s_9^{-1}s_1s_9$
s_2	$(+, 3_2, 3_5)$	$G_1^5(ca)ab(ca)^{-1}G_1^5$	$(s_6s_9)^{-1}$
s_3	$(+, 3_1, 4_2)$	$G_2^2(ca)^2ab(ca)^{-1}G_1^{-1}$	$(s_6s_9)^{-1}$
s_4	$(+, 1_0, 4_3)$	$G_2^3(ca)^2$	$(s_6s_9)^{-1}$
s_5	$(-, 4_4, 5_5)$	$G_3^{-2}(ca)^3b(ca)^{-2}G_2^3$	s_6
s_6	$(-, 4_5, 6_2)$	$G_4^2(ca)^4b(ca)^{-2}G_2^2$	s_6
s_7	$(-, 5_2, 6_3)$	$G_4^3(ca)^4b(ca)^{-3}G_3^{-2}$	s_6
s_8	$(-, 5_3, 7_6)$	$G_5^{-1}(ca)^5b(ca)^{-3}G_3^{-3}$	s_9
s_9	$(-, 5_4, 1_6)$	$(ca)^6a(ca)^{-3}G_3^3$	s_9
s_{10}	$(+, 6_4, 7_5)$	$G_5^4G_4^2$	$s_7^{-1}s_9$
s_{11}	$(-, 6_5, 7_3)$	$G_5^2G_4^{-1}(ca)^4a(ca)^{-4}G_4^2$	$s_7^{-1}s_{11}^{-1}s_9$
s_{12}	$(-, 7_2, 7_4)$	$G_5^4(ca)^5b(ca)^{-5}G_5^{-2}$	$s_7^{-1}s_{11}^{-1}s_9$

FIGURE 9. Data for a surface solving the case $k = 6$, $g = 3$, $N = 7$. Here $M \simeq 0.129 + 0.422i$.

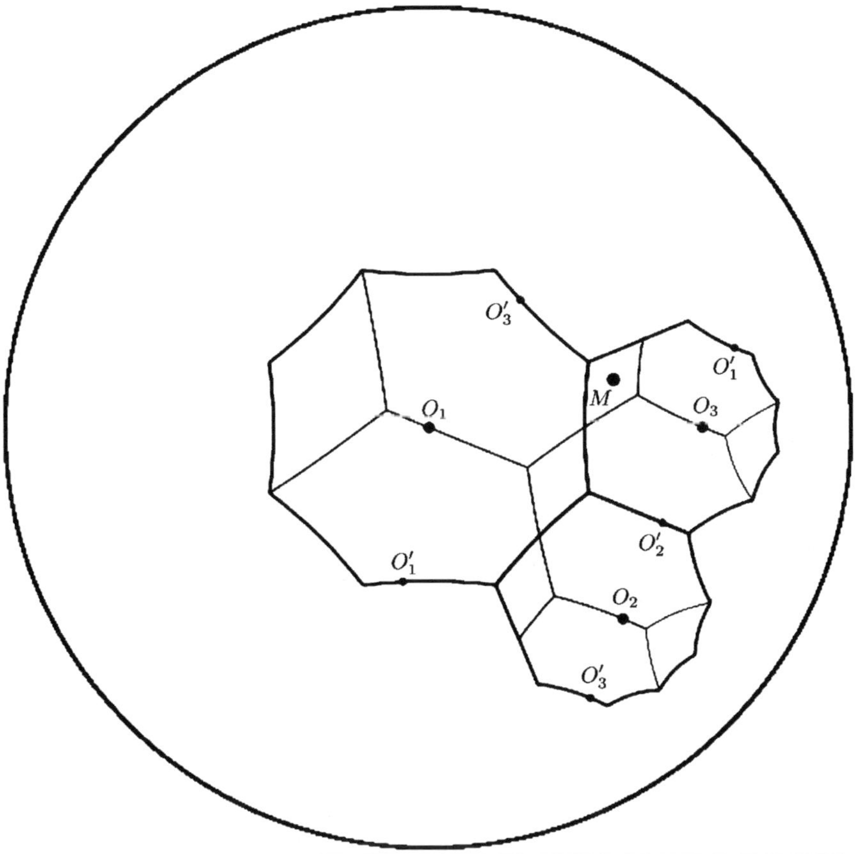

	Side pairing	Word in the generators of $\Delta^{\pm}(2,3,8)$	$k = \tau s^{-1}\tau$
s_1	$(-, 3_4, 3_6)$	$G_1^{-2}(ca)b(ca)^{-1}G_1^4$	$s_8^{-1}s_4s_8$
s_2	$(+, 3_3, 4_3)$	$G_2^2G_1^4$	s_3s_8
s_3	$(-, 3_2, 1_3)$	$(ca)^2(ca)a(ca)^{-1}G_1^{-2}$	s_3
s_4	$(-, 3_5, 4_2)$	$G_2G_1^{-1}(ca)a(ca)^{-1}G_1^3$	$s_8^{-1}s_4s_8$
s_5	$(-, 1_0, 4_6)$	$G_2^{-2}(ca)^2a$	s_5
s_6	$(-, 1_7, 1_5)$	$(ca)^{-3}b(ca)$	$s_9^{-1}s_7^{-1}$
s_7	$(+, 1_6, 4_5)$	$G_2^{-3}(ca)^4$	s_7
s_8	$(-, 1_4, 4_2)$	$G_2^{-4}(ca)^2a(ca)^4$	s_8
s_9	$(-, 4_7, 3_1)$	$G_1(ca)^{-1}(ca)^2b(ca)^{-2}G_2$	s_5^{-1}

FIGURE 10. Data for a surface solving the case $k = 3$, $g = 3$, $N = 8$. Here $M \simeq 0.433 + 0.113i$.

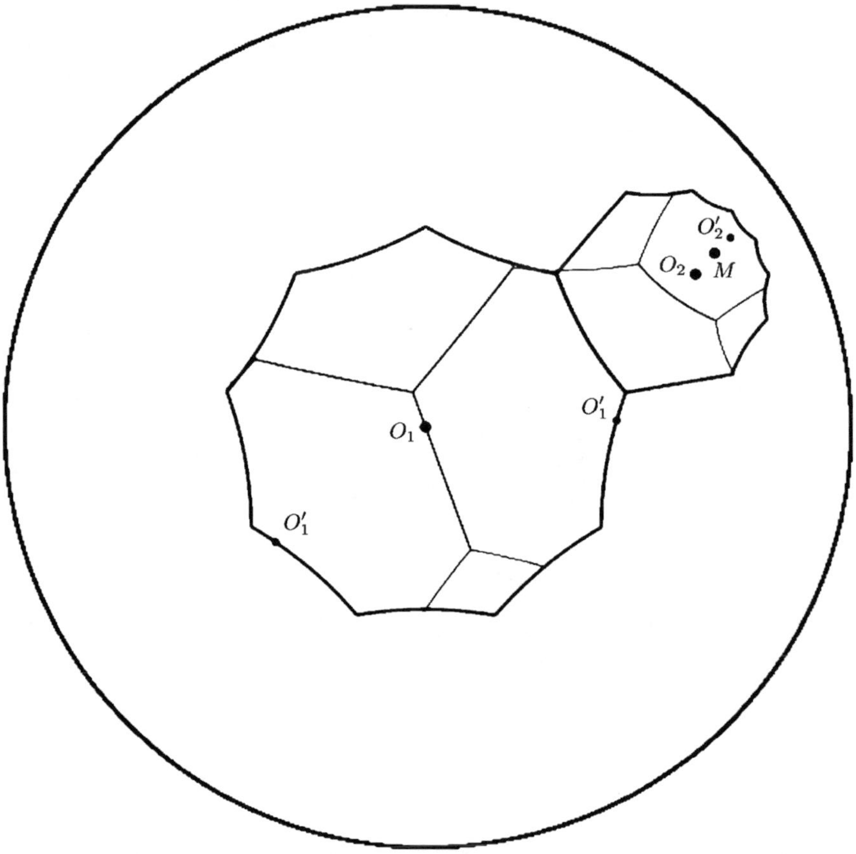

	Side pairing	Word in the generators of $\Delta^{\pm}(2,3,9)$	$k = \tau s^{-1} \tau$
s_1	$(-, 1_0, 1_1)$	$(ca)b$	$(s_5 s_1^{-1} s_8^{-2})^{-1}$
s_2	$(+, 1_8, 1_2)$	$(ca)^2 ab(ca)$	$(s_5 s_8^{-2})^{-1}$
s_3	$(+, 1_7, 5_1)$	$G_3(ca)^5$	$(s_7^2 s_8^{-2})^{-1}$
s_4	$(-, 1_6, 5_3)$	$G_3^3(ca)^3 a(ca)^3$	$(s_7 s_8^{-2})^{-1}$
s_5	$(+, 1_5, 5_5)$	$G_3^5(ca)^7$	$(s_3 s_5^{-1})^{-1}$
s_6	$(-, 1_4, 5_7)$	$G_3^{-2}(ca)^3 a(ca)^5$	s_8
s_7	$(-, 5_8, 5_6)$	$G_3^{-3}(ca)^3 b(ca)^{-3} G_3$	s_8
s_8	$(-, 5_2, 5_4)$	$G_3^4(ca)^3 b(ca)^{-3} G_3^{-2}$	s_6

FIGURE 11. Data for a surface solving the case $k = 2$, $g = 3$, $N = 9$. Here $M \simeq 0.677 + 0.415i$.

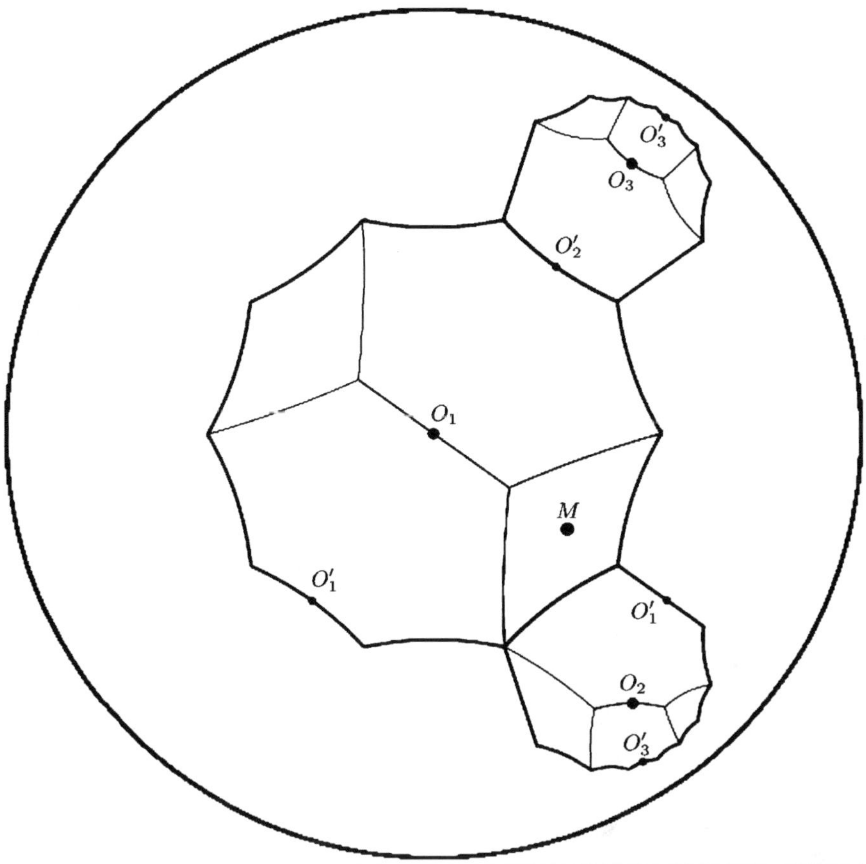

	Side pairing	Word in the generators of $\Delta^{\pm}(2,3,10)$	$k=\tau s^{-1}\tau$
s_1	$(-,1_0,1_2)$	$(ca)^2b$	s_1
s_2	$(-,1_3,3_1)$	$(cb)^2cacac$	s_1^{-1}
s_3	$(+,1_5,3_6)$	$cb(ca)^6ab(ac)^5$	s_3
s_4	$(+,1_6,6_6)$	$(ca)^4ab(ca)^6ab(ac)^6$	$s_{12}^{-1}s_4s_5^{-1}$
s_5	$(+,1_7,3_3)$	$cb(ca)^3ab(ac)^7$	s_{12}^{-1}
s_6	$(-,1_8,6_2)$	$(ca)^2((ca)^2ab)^2(ca)^7c$	$s_1s_5^{-1}$
s_7	$(-,1_9,3_9)$	$cb(ca)^9ab(ca)^8c$	s_1
s_8	$(-,3_2,6_1)$	$(ca)^4abcab(ac)^2abac$	s_1
s_9	$(+,3_4,6_5)$	$(ca)^4ab(ca)^5(ab)(ac)^4abac$	$s_4s_5^{-1}$
s_{10}	$(+,3_5,6_7)$	$(ca)^4ab(ca)^7(ab)(ac)^5abac$	$s_4s_3^{-1}$
s_{11}	$(+,3_7,6_9)$	$(ca)^4ab(ca)^9(ab)(ac)^7abac$	s_3^{-1}
s_{12}	$(+,3_8,6_3)$	$ca((ca)^3ab)^2(ac)^8abac$	s_5^{-1}
s_{13}	$(+,6_4,6_8)$	$(ca)^4ab(ca)^8(ab(ac)^4)^2$	$s_1^{-1}s_6s_3^{-1}$

FIGURE 12. Data for a surface solving the case $k=3$, $g=4$, $N=10$. Here $M\simeq 0.310-0.225i$.

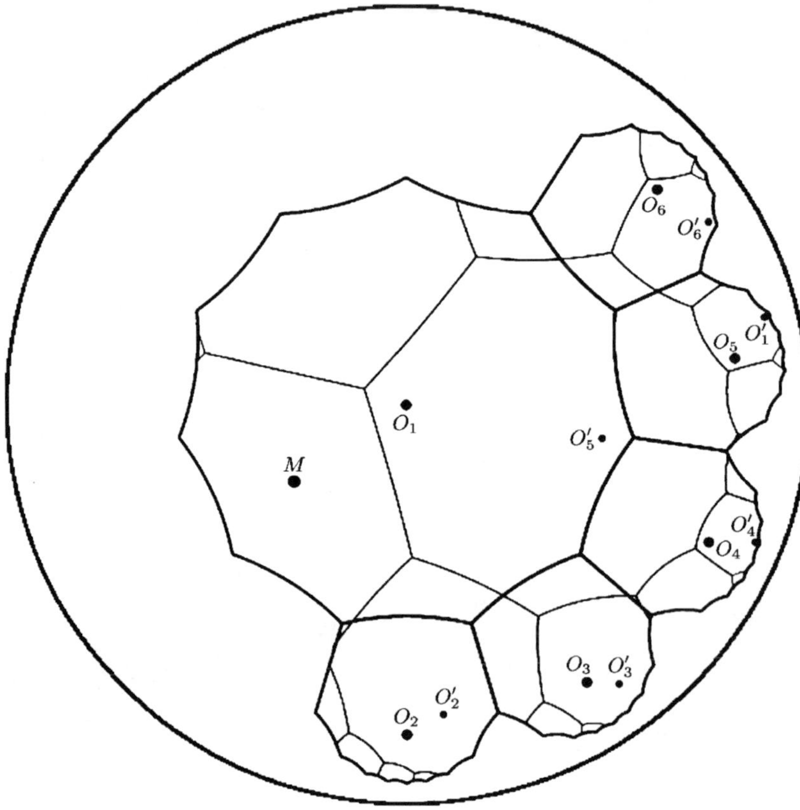

	Side pairing	Word in the generators of $\Delta^{\pm}(2,3,11)$	$k=\tau s^{-1}\tau$
s_1	$(-,1_{10},1_7)$	$(ca)^{-4}b(ca)$	s_1
s_2	$(+,1_5,4_5)$	$G_2^5(ca)^{-3}$	$s_1^{-1}s_{20}s_6$
s_3	$(-,1_8,2_1)$	$G_0a(ca)^3$	s_1^{-1}
s_4	$(+,6_{10},4_6)$	$G_2^6(ca)^{-2}(ca)^4ab(ca)^{-4}G_4$	$s_1^{-1}s_{20}s_6$
s_5	$(+,1_6,5_8)$	$G_3^{-3}(ca)^{-3}$	$s_1^{-1}s_6$
s_6	$(-,1_9,5_7)$	$G_3^{-4}(ca)^3a(ca)^2$	s_6
s_7	$(+,4_4,5_9)$	$G_3^{-3}G_2^6$	$s_6^{-1}s_{20}^{-1}s_6$
s_8	$(+,4_3,6_2)$	$G_4^2(ca)^2(ca)^2ab(ca)^{-2}G_2^{-3}$	$s_6^{-1}s_{20}^{-1}s_6$
s_9	$(+,2_2,5_6)$	$G_3^{-5}(ca)^3abG_0^{-2}$	s_1s_6
s_{10}	$(+,2_5,2_9)$	$G_0^{-2}abG_0^6$	$s_1s_{10}s_1^{-1}$
s_{11}	$(+,2_4,3_2)$	$G_1^2(ca)abG_0^{-4}$	$s_1s_{10}s_1^{-1}$
s_{12}	$(-,2_3,3_7)$	$G_1^{-4}(ca)bG_0^{-3}$	$s_1s_{13}^{-1}s_6$
s_{13}	$(-,3_6,5_5)$	$G_3^5(ca)^2(ca)b(ca)^{-1}G_1^{-6}$	$s_1s_{12}^{-1}s_6$
s_{14}	$(-,3_3,3_8)$	$G_1^8(ca)b(ca)^{-1}G_1^{-3}$	$s_1s_{10}^{-1}s_{12}s_1^{-1}$
s_{15}	$(-,4_2,6_6)$	$G_4^6(ca)^2(ca)^2b(ca)^{-2}G_2^{-2}$	$s_6^{-1}s_{13}s_{18}^{-1}s_6$
s_{16}	$(-,3_9,6_5)$	$G_4^5(ca)^3(ca)b(ca)^{-1}G_1^{-9}$	$s_6^{-1}s_{13}s_{18}^{-1}s_6$
s_{17}	$(-,6_7,6_3)$	$G_4^3(ca)^4b(ca)^{-4}G_4^{-7}$	$s_6^{-1}s_{17}s_6$
s_{18}	$(+,3_4,6_4)$	$G_4^4(ca)^3(ca)ab(ca)^{-1}G_1^{-4}$	$s_1s_{10}^{-1}s_{18}^{-1}s_6$
s_{19}	$(-,3_5,6_8)$	$G_4^8(ca)^3(ca)b(ca)^{-1}G_1^{-5}$	$s_1s_{12}^{-1}s_7^{-1}s_6$
s_{20}	$(+,5_4,6_9)$	$G_4^9(ca)(ca)^3ab(ca)^{-3}G_3^{-4}$	$s_6^{-1}s_7^{-1}s_6$
s_{21}	$(+,4_7,5_3)$	$G_3^3(ca)(ca)^2ab(ca)^{-2}G_2^{-7}$	$s_6^{-1}s_{20}^{-1}s_2^{-1}s_6$
s_{22}	$(+,2_7,5_2)$	$G_3^2(ca)^3abG_0^{-7}$	$s_1s_{22}^{-1}s_6$
s_{23}	$(+,2_8,4_9)$	$G_2^9(ca)^2abG_0^{-8}$	$s_1s_{22}^{-1}s_6$
s_{24}	$(+,2_6,4_8)$	$G_2^8(ca)^2abG_0^{-6}$	$s_1s_{10}s_{22}^{-1}s_6$

FIGURE 13. Data for a surface solving the case $k=6$, $g=7$, $N=11$. Here $M\simeq -0.277-0.191i$.

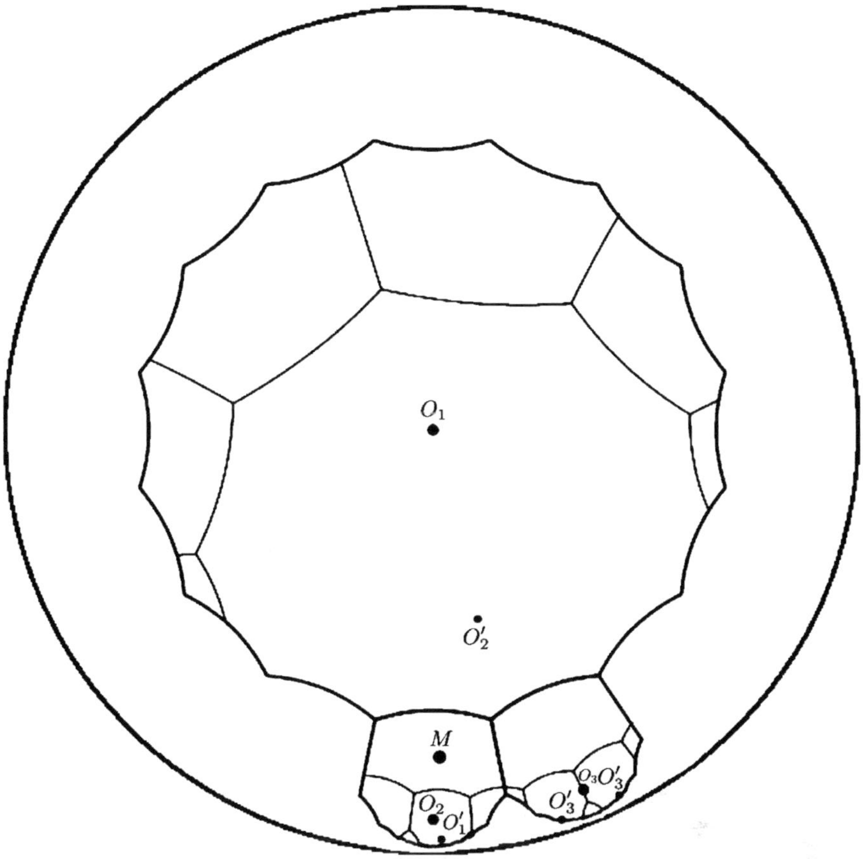

	Side pairing	Word in the generators of $\Delta^{\pm}(2,3,16)$	$k = \tau s^{-1} \tau$
s_1	$(-, 1_{15}, 1_{12})$	$(ca)^{-4}b(ca)$	s_{11}^{-1}
s_2	$(-, 2_1, 1_{13})$	$(ca)^{-3}aG_0^{-1}$	s_{11}^{-1}
s_3	$(+, 1_2, 1_5)$	$(ca)^5ab(ca)^{-2}$	s_9^{-1}
s_4	$(+, 3_{15}, 1_6)$	$(ca)^5G_1$	s_9^{-1}
s_5	$(-, 1_{14}, 3_6)$	$G_1^6(ca)a(ca)^2$	$s_{18}s_{13}^{-1}s_{20}$
s_6	$(+, 2_2, 3_5)$	$G_1^4G_0^{-3}$	$s_{11}^{-1}s_{18}s_{13}^{-1}s_{20}$
s_7	$(+, 1_{11}, 3_7)$	$G_1^7(ca)^6$	$s_{17}s_{13}^{-1}s_{20}$
s_8	$(-, 2_6, 2_7)$	$G_0^7bG_0^{-6}$	s_{14}^{-1}
s_9	$(+, 2_5, 2_8)$	$G_0^8abG_0^{-5}$	s_3^{-1}
s_{10}	$(-, 2_{12}, 3_2)$	$G_1G_0^{-1}aG_0^4$	s_1
s_{11}	$(-, 2_{14}, 2_{11})$	$G_0^{-5}bG_0^2$	s_1^{-1}
s_{12}	$(-, 3_{10}, 3_{14})$	$G_1^{-2}(ca)b(ca)^{-1}G_1^6$	$s_{20}^{-1}s_{19}$
s_{13}	$(-, 1_7, 3_{11})$	$G_1^{-5}(ca)a(ca)^9$	$s_{15}^{-1}s_{20}$
s_{14}	$(-, 1_3, 1_4)$	$(ca)^4b(ca)^{-3}$	s_8^{-1}
s_{15}	$(+, 2_9, 3_8)$	$G_1^7G_0^6$	$s_{13}^{-1}s_{20}$
s_{16}	$(+, 1_8, 3_3)$	$G_1^3(ca)^9$	$s_{15}^{-1}s_6$
s_{17}	$(+, 1_{10}, 2_{10})$	$G_0^{10}(ca)^6$	s_{17}
s_{18}	$(-, 1_9, 2_{13})$	$G_0^{-3}a(ca)^7$	$s_{15}^{-1}s_5$
s_{19}	$(+, 2_4, 3_9)$	$G_1^{-8}G_0^{-5}$	$s_{12}s_{20}$
s_{20}	$(-, 2_3, 3_{13})$	$G_1^{-4}G_0^{-1}aG_0^{-3}$	s_{20}
s_{21}	$(-, 3_4, 3_{12})$	$G_1^{-4}(ca)b(ca)^{-1}G_1^{-4}$	$s_6^{-1}s_{20}$

FIGURE 14. Data for a surface solving the case $k = 3$, $g = 7$, $N = 16$. Here $M \simeq 0.015 - 0.773i$.

Acknowledgment

We are indebted to the referee for many valuable suggestions that helped us improve an earlier version of the paper.

References

[1] Christophe Bavard, *Disques extrémaux et surfaces modulaires* (French, with English and French summaries), Ann. Fac. Sci. Toulouse Math. (6) **5** (1996), no. 2, 191–202. MR1413853

[2] Alan F. Beardon, *The geometry of discrete groups*, Graduate Texts in Mathematics, vol. 91, Springer-Verlag, New York, 1983. MR698777

[3] Allan L. Edmonds, John H. Ewing, and Ravi S. Kulkarni, *Torsion free subgroups of Fuchsian groups and tessellations of surfaces*, Invent. Math. **69** (1982), no. 3, 331–346, DOI 10.1007/BF01389358. MR679761

[4] Ernesto Girondo, *Extremal disc packings in compact hyperbolic surfaces*, Rev. Mat. Complut. **31** (2018), no. 2, 467–478, DOI 10.1007/s13163-017-0252-3. MR3785805

[5] ———, *Personal webpage at Universidad Autónoma de Madrid*, `http://matematicas.uam.es/~ernesto.girondo/`, May 2019.

[6] Ernesto Girondo and Gabino González-Diez, *On extremal discs inside compact hyperbolic surfaces* (English, with English and French summaries), C. R. Acad. Sci. Paris Sér. I Math. **329** (1999), no. 1, 57–60, DOI 10.1016/S0764-4442(99)80461-3. MR1703263

[7] Ernesto Girondo and Gabino González-Diez, *Genus two extremal surfaces: extremal discs, isometries and Weierstrass points*, Israel J. Math. **132** (2002), 221–238, DOI 10.1007/BF02784513. MR1952622

[8] Ernesto Girondo and Gou Nakamura, *Compact non-orientable hyperbolic surfaces with an extremal metric disc*, Conform. Geom. Dyn. **11** (2007), 29–43, DOI 10.1090/S1088-4173-07-00157-9. MR2295996

[9] Ernesto Girondo and Cristian Reyes, *Extremal k-packings in compact non-orientable surfaces*, arXiv e-prints (2019).

[10] Gou Nakamura, *Compact non-orientable surfaces of genus 4 with extremal metric discs*, Conform. Geom. Dyn. **13** (2009), 124–135, DOI 10.1090/S1088-4173-09-00194-5. MR2497316

[11] Gou Nakamura, *Compact non-orientable surfaces of genus 5 with extremal metric discs*, Glasg. Math. J. **54** (2012), no. 2, 273–281, DOI 10.1017/S0017089511000589. MR2911368

[12] Gou Nakamura, *Compact Klein surfaces of genus 5 with a unique extremal disc*, Conform. Geom. Dyn. **17** (2013), 39–46, DOI 10.1090/S1088-4173-2013-00251-8. MR3027523

[13] Gou Nakamura, *Compact non-orientable surfaces of genus 6 with extremal metric discs*, Conform. Geom. Dyn. **20** (2016), 218–234, DOI 10.1090/ecgd/298. MR3513567

[14] Kisao Takeuchi, *Arithmetic triangle groups*, J. Math. Soc. Japan **29** (1977), no. 1, 91–106, DOI 10.2969/jmsj/02910091. MR429744

Departamento de Matemáticas, Universidad Autónoma de Madrid. 28049, Madrid (Spain)

Email address: `ernesto.girondo@uam.es`

Departamento de Matemáticas, Universidad Autónoma de Madrid. 28049, Madrid (Spain)

Email address: `cristianr.reyes@estudiante.uam.es`

Contemporary Mathematics
Volume **766**, 2021
https://doi.org/10.1090/conm/766/15380

Complex surfaces with mutually non-biholomorphic universal covers

Gabino González-Diez and Sebastián Reyes-Carocca

ABSTRACT. It is known that the universal cover of compact Riemann surface is either the projective line, the complex plane or the unit disk. In this article we construct a very explicit family of complex surfaces that gives rise to uncountably many mutually non-biholomorphic universal covers. The slope of these surfaces, which are going to be total spaces of Kodaira fibrations, is also determined.

1. Introduction and statement of results

Let X be a complex projective variety and let k be a subfield of the field of the complex numbers. We shall say that X is defined over k or that k is a *field of definition* for X if there exists a collection of homogenous polynomials with coefficients in k so that the variety they define is isomorphic to X.

Let S be a non-singular minimal projective surface of general type. Based on results due to Bers and Griffiths [**3, 14**] on uniformization of complex projective varieties and on results of Shabat [**18, 19**] on automorphism groups of universal covers of families of Riemann surfaces, in [**13**] the authors succeeded in proving that whether or not S is defined over a given algebraically closed field depends only on the holomorphic universal cover of its Zariski open subsets.

A special kind of complex surfaces as before are the total space of the so-called *Kodaira fibrations*. A Kodaira fibration consists of a compact complex surface S, a compact Riemann surface B and a surjective holomorphic map $S \to B$ everywhere of maximal rank such that the fibers are connected and not mutually isomorphic Riemann surfaces. The genus g of the fibre is called the genus of the fibration and it is known that necessarily $g \geq 3$.

Kodaira fibrations were introduced by Kodaira himself in [**17**] to show that the signature of a differentiable fiber bundle need not be multiplicative. Further properties of Kodaira fibrations were obtained by Atiyah, Catanese, Hirzebruch and Kas, among others; see, for example, [**1, 5, 15, 16**]. For explicit constructions of Kodaira fibrations we refer to the articles [**4, 6, 7, 10, 20**].

2020 *Mathematics Subject Classification.* Primary 32J25, 14J20, 14J25.

Key words and phrases. Riemann surface, moduli and Teichmüller space, complex surfaces and their universal covers, field of definition.

Partially supported by Spanish Government Research Project MTM2016-79497.

Partially supported by Fondecyt Grant 11180024, 1190991 and Redes Grant 2017-170071.

Let $S \to B$ be a Kodaira fibration. According to [**13**, Theorem 8] (together with [**13**, Remark 9]) the relation between fields of definition and universal covers takes a neater form in this case. For these surfaces whether or not S is defined over a given algebraically closed field depends only on the holomorphic universal cover of S instead of on the whole collection of Zariski open subsets. In a more concrete way, for $i = 1, 2$, let $S_i \to B_i$ be a Kodaira fibration. In [**12**, Theorem 1.1] it was proved that if the holomorphic universal covers of S_1 and S_2 are biholomorphically equivalent, then S_1 and S_2 are defined over the same algebraically closed fields.

In this paper, for any integer $g \geq 7$, we construct a collection of Kodaira fibrations of genus g and then we apply [**12**, Theorem 1.1] to show that it contains an uncountable subfamily whose corresponding holomorphic universal covers are mutually non-biholomorphic.

THEOREM 1.1. *For any $\lambda \in \mathbb{C} - \{0, -\frac{27}{4}\}$ and $r \geq 8$ even, there is a Kodaira fibration $S_{\lambda,r} \to B$ of genus $g = 3 + \frac{r}{2}$ such that if λ and μ are algebraically independent transcendental complex numbers, then the universal covers of $S_{\lambda,r}$ and $S_{\mu,r}$ are non-biholomorphically equivalent bounded domains of $\mathbb{C}^2$.*

In particular, for each $r \geq 8$ even, there are in this family uncountably many Kodaira fibrations whose corresponding holomorphic universal covers are pairwise non-biholomorphically equivalent.

As pointed out in [**12**] and [**13**], we remark that, in this respect, complex surfaces are very much in contrast with compact Riemann surfaces, for which the universal cover only depends on the genus.

As mentioned above, Kodaira fibrations were the first known examples of fiber bundles whose signature is not multiplicative. We recall that the signature $\tau(S)$ of an algebraic surface S is related to the ratio of its Chern numbers

$$\upsilon(S) := c_1^2(S)/c_2(S)$$

by the formula

$$\tau(S) = e(S)(\upsilon(S) - 2)/3,$$

where $e(S)$ stands for the Euler characteristic of S. Thus, the vanishing of the signature is equivalent to the *slope* $\upsilon(S)$ being equal to 2. In the case of our surfaces $S_{\lambda,r}$ we have:

PROPOSITION 1.2.

$$\upsilon(S_{\lambda,r}) = 2 + \frac{3}{2(4+r)}.$$

It is known that the slope of a Kodaira surface lies in the interval $(2, 3)$ and that the limit values 2 and 3 correspond to surfaces whose universal cover is the bidisk and the 2-ball respectively (see, for example, [**5**, Section 2.1.2]). From this point of view, the fact that the limit of $\upsilon(S_{\lambda,r})$ tends to 2 when r tends to infinity could be interpreted as an indication that the universal covers of the surfaces $S_{\lambda,r}$, which are known to be bounded domains of $\mathbb{C}^2$, approach the bidisk when the integer r is allowed to take large values. As far as we know, the examples with largest slope have been constructed by Catanese and Rollenske [**6**] and reach the value $2 + 2/3$.

2. Preliminaries

2.1. Uniformization of Kodaira surfaces. Let $f : S \to B$ be a Kodaira fibration, let $\pi : \mathbb{H} \to B$ be the universal covering map of B and let Γ be the covering group so that $B \cong \mathbb{H}/\Gamma$. By considering the pull-back

$$h : \pi^* S \to \mathbb{H}$$

of f by π, we obtain a new family of Riemann surfaces in which, for each $t \in \mathbb{H}$, the fiber $h^{-1}(t)$ agrees with the Riemann surface $f^{-1}(\pi(t))$.

Teichmüller theory together with results due to Bers and Griffiths [**3**, **14**] enable us to choose uniformizations $h^{-1}(t) = D_t/K_t$ possessing the following properties:

(a) K_t is a Kleinian group acting on a bounded domain D_t of $\mathbb{C}$ which is biholomorphically equivalent to a disk.

(b) The union of all these disks

$$\mathscr{B} := \cup_{t\in\mathbb{H}} D_t$$

is a bounded domain of $\mathbb{C}^2$ which is biholomorphic to the universal cover of S, that is, $S \cong \mathscr{B}/\mathbb{G}$, where $\mathbb{G} < \mathrm{Aut}(\mathscr{B})$ is the covering group.

(c) The group $\mathbb{G}$ is endowed with a surjective homomorphism of groups $\Theta : \mathbb{G} \to \Gamma$ which induces an exact sequence of groups

$$1 \longrightarrow \mathbb{K} \longrightarrow \mathbb{G} \xrightarrow{\Theta} \Gamma \longrightarrow 1,$$

where the subgroup $\mathbb{K}$ preserves each *quasi-disk* D_t and acts on it as the Kleinian group K_t for each $t \in \mathbb{H}$.

We note that $\mathscr{B}$ carries a fibration structure itself $\mathscr{B} \to \mathbb{H}$ whose fiber over $t \in \mathbb{H}$ is D_t (the domain $\mathscr{B}$ is called a *Bergman domain* in Bers' terminology; see [**3**, p. 284]). The situation is summarized in the following commutative diagram

$$\begin{array}{ccc} \pi^*S & \longrightarrow & S \cong \mathscr{B}/\mathbb{G} \\ h\big\downarrow & & \big\downarrow f \\ \mathbb{H} & \xrightarrow{\pi} & B \cong \mathbb{H}/\Gamma \end{array}$$

2.2. Basic Teichmüller theory. Let C_0 be a compact Riemann surface of genus g admitting an automorphism τ_0 of order two with r fixed points such that $C_0/\langle\tau_0\rangle$ has genus two. By the Riemann-Hurwitz formula, r is necessarily even and

$$g = 3 + \tfrac{r}{2}.$$

We denote by R_0^* the Riemann surface obtained from $R_0 := C_0/\langle\tau_0\rangle$ by removing the r branch values $\{q_1, \dots, q_r\}$ of the associated covering map $C_0 \to R_0$. Let us denote by $\mathrm{Mod}_g(\tau_0)$ the normalizer of $\langle\tau_0\rangle$ viewed as a subgroup of the mapping class group

$$\mathrm{Mod}(C_0) = \mathrm{Mod}_g\,.$$

The Teichmüller space $T_{2,r}$ of the Riemann surface R_0^* is naturally embedded in the Teichmüller space T_g of C_0 as the fixed locus of $\langle\tau_0\rangle$. Therefore, the group $\mathrm{Mod}_g(\tau_0)$ acts on $T_{2,r}$ in a obvious way, and this action induces a faithful action of $\mathrm{Mod}_g(\tau_0)/\langle\tau_0\rangle$ on $T_{2,r}$. This permits to realize $\mathrm{Mod}_g(\tau_0)/\langle\tau_0\rangle$ as a finite index subgroup of the full group of holomorphic automorphisms of $T_{2,r}$ which is the mapping class group

$$\mathrm{Mod}(R_0^*) = \mathrm{Mod}_{2,r}\,.$$

In a similar way, we denote by $\mathscr{P}_{2,r}$ the *pure mapping class group*, which is the subgroup of $\mathrm{Mod}_{2,r}$ consisting of those mapping classes that preserve each distinguished point q_i for $1 \le i \le r$. Analogously, we denote by $\mathscr{P}_g(\tau_0)$ the *pure relative modular group*, which is the subgroup of $\mathrm{Mod}_g(\tau_0)$ consisting of those mapping classes that preserve each fixed point of τ_0.

We recall that just as the quotient

$$\mathscr{M}_{g,n} = T_{g,n}/\,\mathrm{Mod}_{g,n}$$

yields the moduli space of Riemann surfaces of genus g with a set of n distinguished points, the quotient

$$\mathscr{M}_{g,n}^{pure} = T_{g,n}/\mathscr{P}_{g,n}$$

yields the moduli space of Riemann surfaces of genus g with an ordered set of n distinguished points. In other words, two $(n+1)$-tuples $(C; p_1, \ldots, p_n)$ and $(D; q_1, \ldots, q_n)$ represent the same point in $\mathscr{M}_{g,n}$ if and only if there is an isomorphism $\alpha : C \to D$ which sends the set $\{p_1, \ldots, p_n\}$ to the set $\{q_1, \ldots, q_n\}$ whereas they represent the same point in $\mathscr{M}_{g,n}^{pure}$ only if α sends p_i exactly to q_i for $i = 1, \ldots, n$.

Next, we introduce the *level three* normal subgroup $\mathrm{Mod}_g[3]$ of Mod_g consisting of those (homotopy classes of) homeomorphisms of C_0 which induce the identity map on the homology group $\mathrm{H}_1(C_0, \mathbb{Z}/3\mathbb{Z})$. Since by Serre's lemma the group $\mathrm{Mod}_g[3]$ does not contain any non-trivial element of finite order and since it has finite index in Mod_g, the homomorphism $\mathscr{P}_g(\tau_0) \to \mathscr{P}_{2,r}$ maps injectively the group

$$\mathscr{P}_g(\tau_0) \cap \mathrm{Mod}_g[3]$$

into a finite index subgroup of $\mathscr{P}_{2,r}$, henceforth denoted by $\mathscr{P}_g(3, \tau_0)$. See [**11**, Proposition 1.1].

We refer to the article [**9**] for more details.

3. Proof of the Theorem

3.1. Construction of the surfaces $S_{\lambda,r}$. Let λ be an arbitrary complex number different from 0 and $-\frac{27}{4}$. Consider the (affine) elliptic curve

$$E_\lambda \, : \, y^2 = x^3 + \lambda x + \lambda$$

endowed with the standard group law $\oplus$, with the point ∞ at infinity playing the role of neutral element. The map $\pi : (x, y) \mapsto (x^2, y)$ induces a two-fold covering map between the genus two Riemann surface

$$X_\lambda \, : \, y^2 = x^6 + \lambda x^2 + \lambda$$

and E_λ with branch points $s_+ = (0, \lambda^{1/2})$ and $s_- = (0, -\lambda^{1/2})$.

Clearly, we can inductively construct a sequence of $\overline{\mathbb{Q}(\lambda)}$−rational points

$$e_2, \ldots, e_r, \ldots \in E_\lambda$$

such that all the points e_i in this sequence as well as their differences $e_i \ominus e_j$ are different from the points $\pi(s_+) \ominus \pi(s_-)$, its inverse $\pi(s_-) \ominus \pi(s_+)$ and the identity element $\infty \in E_\lambda$. Now, for any natural number $r \geq 1$ we consider the complex one-dimensional analytic space

$$C_{\lambda,r} := \{(p_1, \ldots, p_r) \in X_\lambda^r : \pi(p_i) = \pi(p_1) \oplus e_i \text{ for each } 2 \leq i \leq r\}.$$

Note that

$$C_{\lambda,r} \subset X_\lambda^r - \Delta_r$$

where Δ_r stands for the diagonal subset of X_λ^r consisting of the r-tuples with two coincident coordinates.

CLAIM. $C_{\lambda,r}$ is a compact Riemann surface of genus $r2^{r-1}+1$.

Let $\Psi : X_\lambda^r \to E^{r-1}$ be the holomorphic map defined by

$$(p_1, \dots, p_r) \mapsto (\pi(p_2) \ominus \pi(p_1) \ominus e_2, \dots, \pi(p_r) \ominus \pi(p_1) \ominus e_r).$$

It is clear that $C_{\lambda,r} = \Psi^{-1}(\infty, \dots, \infty)$. An easy computation shows that the Jacobian matrix of Ψ at a point $(p_1, \dots, p_r)$ is of the form

$$J = \frac{\partial(\Psi_1, \dots, \Psi_{r-1})}{\partial(z_1, \dots, z_r)} = \begin{pmatrix} -\pi'(p_1) & \pi'(p_2) & 0 & \cdots & 0 & 0 \\ -\pi'(p_1) & 0 & \pi'(p_3) & \cdots & 0 & 0 \\ \vdots & \vdots & \vdots & \ddots & \vdots & \vdots \\ -\pi'(p_1) & 0 & 0 & \cdots & \pi'(p_{r-1}) & 0 \\ -\pi'(p_1) & 0 & 0 & \cdots & 0 & \pi'(p_r) \end{pmatrix}$$

Now, $\pi'(p_i) = 0$ if and only if p_i is one the branch points s_+, s_-. But the choice of the points e_i ensures that at most one of the entries of $(p_1, \dots, p_r)$ can be a branch point. Therefore J has maximal rank $r-1$ at each point of $C_{\lambda,r}$, showing that $C_{\lambda,r}$ is a (possibly non-connected) non-singular complex curve. Now, the map

$$\pi_{\lambda,r} : (p_1, \dots, p_r) \mapsto (p_1, \dots, p_{r-1})$$

is clearly a holomorphic two-fold ramified cover between $C_{\lambda,r}$ and $C_{\lambda,r-1}$. This implies that $C_{\lambda,r}$ is a connected Riemann surface. Moreover, it is easy to see that this double cover has exactly 2^r branch points, namely all points $(p_1, \dots, p_r)$ with $p_r = s_+$ or s_-. From here the value of the genus follows from the Riemann-Hurwitz formula and induction on r. This completes the proof of our claim.

Now the strategy is to associate to each point $(p_1, \dots, p_r) \in C_{\lambda,r}$ a two-fold branched cover of X_λ ramified over $p_1, \dots, p_r$. But, in order to show that this idea leads to a well-defined Kodaira fibration, we must proceed carefully.

Let $\mathbb{H} \to C_{\lambda,r}$ be the universal cover of $C_{\lambda,r}$ and Γ its covering group so that $C_{\lambda,r} \cong \mathbb{H}/\Gamma$. The map

$$X_\lambda^r - \Delta_r \to \mathscr{M}_{2,r}^{pure} \quad \text{defined by} \quad (p_1, \dots, p_r) \mapsto [X_\lambda, \{p_1, \dots, p_r\}]$$

is a holomorphic map [**10**, Lemma 1]; let us denote by Φ its restriction to $C_{\lambda,r}$. Following [**11**, Proposition 3.1], if $r \geq 7$ then $\mathscr{P}_{2,r}$ acts freely on $T_{2,r}$ and therefore, by the theory of covering spaces, there exists a holomorphic lift

$$\hat{\Phi} : \mathbb{H} \to T_{2,r} \quad \text{of} \quad \Phi : C_{\lambda,r} \to \mathscr{M}_{2,r}^{pure}$$

together with a group homomorphism $\Theta : \Gamma \to \mathscr{P}_{2,r}$ defined by

$$\Theta(\alpha) \circ \hat{\Phi} = \hat{\Phi} \circ \alpha \quad \text{with} \quad \alpha \in \Gamma.$$

Let us now set

$$\Gamma[3] := \Theta^{-1}(\mathscr{P}_g(3, \tau_0))$$

and $C_{\lambda,r}[3] := \mathbb{H}/\Gamma[3]$. Then, the natural projection

$$\zeta_{\lambda,r} : C_{\lambda,r}[3] \to C_{\lambda,r}$$

is a finite unbranched covering map of compact Riemann surfaces, and there is an obvious commutative diagram as follows:

$$\begin{array}{ccccc} C_{\lambda,r}[3] = \mathbb{H}/\Gamma[3] & \longrightarrow & T_{2,r}/\mathscr{P}_g(3,\tau_0) & \longrightarrow & T_g/\mathscr{M}od_g[3] \\ \zeta_{\lambda,r}\downarrow & & \downarrow & & \\ C_{\lambda,r} = \mathbb{H}/\Gamma & \xrightarrow{\ \Phi\ } & T_{2,r}/\mathscr{P}_{2,r} = \mathscr{M}_{2,r}^{pure} & & \end{array}$$

Let us denote by $\mathscr{M}_g[3]$ the quotient space $T_g/\operatorname{Mod}_g[3]$. This is called the *moduli space with level three* structure. Clearly, $\mathscr{M}_g[3]$ is a finite Galois cover of $\mathscr{M}_g$ with covering group

$$\operatorname{Mod}_g/\operatorname{Mod}_g[3] \cong \operatorname{Sp}(2g, \mathbb{Z}/3\mathbb{Z}).$$

Recall that the moduli space $\mathscr{M}_g$ comes equipped with a fibration $\pi_g : \mathscr{C}_g \to \mathscr{M}_g$ (called the *universal curve*), whose fiber above a point $[F] \in \mathscr{M}_g$ is a Riemann surface which is not isomorphic to F but to $F/\operatorname{Aut}(F)$. However, the fact that $\operatorname{Mod}_g[3]$ is torsion free, implies that the corresponding level three universal curve $\pi_{g,3} : \mathscr{C}_g[3] \to \mathscr{M}_g[3]$ has the property that the fiber over a point of $\mathscr{M}_g[3]$ parametrising a Riemann surface F is isomorphic to the Riemann surface F itself. This means that for any non-constant holomorphic map $\varphi : B \to \mathscr{M}_g[3]$ the pull-back $\varphi^*(\mathscr{C}_g[3])$ is a genus g Kodaira fibration with base B. Thus, if

$$h_{\lambda,r} : C_{\lambda,r}[3] \to \mathscr{M}_g[3]$$

denotes the holomorphic map defined by the first row of the above diagram then we get a Kodaira fibration a genus $g = 3 + \frac{r}{2}$

$$f_{\lambda,r} : S_{\lambda,r} := h_{\lambda,r}^* \mathscr{C}_g[3] \to B := C_{\lambda,r}[3].$$

simply by considering the pull-back of $\pi_{g,3}$ by $h_{\lambda,r}$.

REMARK 3.1. For later use, we observe that each element $h \in \operatorname{Mod}_g$ induces a fiber-preserving automorphism of the universal curve $\mathscr{C}_g[3] \to \mathscr{M}_g[3]$. When $h = \tau_0$ this automorphism, let us call it τ, preserves each of the fibres in the image of $T_{2,r}/\mathscr{P}_g(3,\tau_0)$. Therefore τ induces an automorphism of the fibration $f_{\lambda,r}$ which restricts to an automorphism τ_b on each fibre

$$F_b := f_{\lambda,r}^{-1}(b) \quad \text{in such a way that} \quad F_b/\langle\tau_b\rangle \cong X_\lambda.$$

3.2. Universal covers pairwise non-biholomorphic. As the curve $C_{\lambda,r}$ is defined over $\overline{\mathbb{Q}(\lambda)}$, we can proceed analogously as in the proof of Theorem 4.1 in [**8**] to ensure that curve B is also defined over $\overline{\mathbb{Q}(\lambda)}$. Consequently, we can apply Theorem 1.2 in [**12**] to infer that $S_{\lambda,r}$ is also defined over $\overline{\mathbb{Q}(\lambda)}$.

We now assume that λ is a trascendental complex number. We claim that if μ is any transcendental complex number which is algebraically independent of λ, then $S_{\lambda,r}$ cannot be defined over the field $\overline{\mathbb{Q}(\mu)}$. Indeed, if that were the case, we could apply Theorem 2.12 in [**8**] to deduce that $S_{\lambda,r}$ is actually defined over a number field and this, in turn, would imply that E_λ is defined over a number field too (see [**8**, Theorem 4.4]). But this is impossible since the minimum field of definition of an elliptic curve is known to be the field generated by its $j-$invariant, which in this case is

$$j(E_\lambda) = \lambda^3/(4\lambda^3 + 27\lambda^2).$$

We now apply Theorem 1.1 in [**12**] to prove the first statement of the theorem. The last part of the theorem follows directly from the existence of uncountably many pairwise algebraically independent transcendental complex numbers.

4. Proof of the Proposition

We recall that if S is an algebraic surface, then $c_2(S) = e(S)$, the Euler class of S, and

$$c_1^2(S) = K^2$$

where K^2 stands for the self-intersection of the canonical divisor K of S. Moreover, if $S \to B$ is a Kodaira fibration of genus g over a curve of genus γ, it is known that

$$e(S) = (2g-2)(2\gamma-2).$$

Thus, for our surface $S_{\lambda,r}$

$$\upsilon(S_{\lambda,r}) = K^2/(2\gamma-2)(4+r)$$

and therefore to prove the equality

$$\upsilon(S_{\lambda,r}) = 2 + \frac{3}{2(4+r)},$$

we only need to check that

$$K^2 = (8+2r)(2\gamma-2) + 3(\gamma-1). \tag{4.1}$$

We begin by noting that Remark 3.1 implies that, after passing to a smooth cover of the base if necessary, we can assume that

$$S_{\lambda,r}/\langle\tau\rangle \cong B \times X_\lambda.$$

Let us consider the holomorphic map

$$g_{\lambda,r} : S_{\lambda,r} \to S_{\lambda,r}/\langle\tau\rangle \to X_\lambda$$

induced by the projection $B \times X_\lambda \to X_\lambda$. Its ramification divisor R is clearly the fixed set of τ. By a further change of base (which does not alter the slope of the fibration) we can get a new Kodaira fibration, which we still denote $f_{\lambda,r} : S_{\lambda,r} \to B$, with the property that R decomposes as a union of r disjoint sections $R_1 \dots, R_r$ of $f_{\lambda,r}$ such that at each fibre F_b, the section R_j picks out the j-th branch point of the restriction of $g_{\lambda,r}$ to F_b, that is to say the j-th fixed point of the automorphism τ_b (see Remark 3.1).

Let us denote by $\zeta_{\lambda,r} : B \to C_{\lambda,r}$ the smooth cover obtained after the successive changes of base performed on our initial Kodaira fibration and by $\pi_j : C_{\lambda,r} \to X_\lambda$ the projection onto the j-th coordinate. Then we see that

(i) the restriction of $g_{\lambda,r}$ to R_j is given by

$$g_{\lambda,r}|_{R_j} = \pi_j \circ \zeta_{\lambda,r} \circ f_{\lambda,r}|_{R_j} \tag{4.2}$$

(ii) the restriction of $g_{\lambda,r}$ to a fibre F_b is a degree 2 covering $F_b \to X_\lambda$ whose branching value set is $\{\pi_1 \circ \zeta_{\lambda,r}(b), \cdots, \pi_r \circ \zeta_{\lambda,r}(b)\}$.

Let us now consider the holomorphic forms on X_λ given by

$$w_1 = x dx/y \quad \text{and} \quad w_2 = dx/y$$

and let us denote by $D_1 = s_1 + t_1$ and $D_2 = s_2 + t_2$ their corresponding divisors. It is easily checked that $\{s_1, t_1\} = \{(0, \pm\lambda^{1/2})\}$ and that s_2, t_2 are the two points at infinity of X_λ.

Similarly we choose two holomorphic forms v_1 and v_2 on B with simple zeroes and denote their divisors as

$$\mathrm{div}(v_k) = \Sigma_{i=1}^{2\gamma-2} b_i^k.$$

Without loss of generality, we can suppose that:

(a) $\{b_i^1\}$ and $\{b_i^2\}$ are disjoint sets.

(b) For no of triple (k, i, j) with $k = 1, 2$; $i = 1, \ldots, 2\gamma - 2$ and $j = 1, \ldots, = r$, the point $\pi_j \circ \zeta_{\lambda,r}(b_i^k)$ equals s_1, t_1, s_2 or t_2.

In view of the observation (ii) above this is equivalent to saying that the branching values of the restriction of $g_{\lambda,r}$ to the fibres $F_{b_i^k}$ are different from s_1, t_1, s_2 or t_2.

Next we define the following two sections of the canonical bundle K

$$W_k = f_{\lambda,r}^* v_k \wedge g_{\lambda,r}^* w_k; \quad k = 1, 2.$$

Clearly their corresponding divisors of $S_{\lambda,r}$ can be expressed as

$$\mathrm{div}(W_k) = \Sigma_{i=1}^{2\gamma-2} F_{b_i^k} + g_{\lambda,r}^{-1}(D_k) + R$$

We can now compute K^2 as the intersection $\mathrm{div}(W_1) \cdot \mathrm{div}(W_2)$ (see, for example, [**2**]). Setting $R = \Sigma_{j=1}^r R_j$ and taking into account that $\Sigma_{i=1}^{2\gamma-2} b_i^1$ and $\Sigma_{i=1}^{2\gamma-2} b_i^2$ (resp. D_1 and D_2) are disjoint divisors of B (resp. of X_λ) we first get

$$K^2 = \begin{cases} \cancel{\Sigma F_{b_i^1} \cdot \Sigma F_{b_i^2}} & + \ \Sigma F_{b_i^1} \cdot g_{\lambda,r}^{-1}(D_2) & + \ \Sigma F_{b_i^1} \cdot \Sigma R_j & + \\ g_{\lambda,r}^{-1}(D_1) \cdot \Sigma F_{b_i^2} & + \ \cancel{g_{\lambda,r}^{-1}(D_1) \cdot g_{\lambda,r}^{-1}(D_2)} & + \ g_{\lambda,r}^{-1}(D_1) \cdot \Sigma R_j & + \\ \Sigma R_j \cdot \Sigma F_{b_i^2} & + \ \Sigma R_j \cdot g_{\lambda,r}^{-1}(D_2) & + \ \Sigma R_j \cdot \Sigma R_j & \end{cases}$$

Next we make two observations. The first one is that $F_{b_i^k} \cdot R_j = 1$. The second one is that, by the property (b) above, the branching values of the restriction of $g_{\lambda,r}$ to the fibres $F_{b_i^1}$ do not lie in D_2 which implies that the divisors $F_{b_i^1}$ and $g_{\lambda,r}^{-1}(D_2)$ meet transversally at two points. It follows that

$$\Sigma F_{b_i^1} \cdot g_{\lambda,r}^{-1}(D_2) = \Sigma F_{b_i^1} \cdot g_{\lambda,r}^{-1}(s_2) + \Sigma F_{b_i^1} \cdot g_{\lambda,r}^{-1}(t_2) = 2(2\gamma - 2) + 2(2\gamma - 2),$$

and the analogous statement for $g_{\lambda,r}^{-1}(D_1) \cdot \Sigma F_{b_i^2}$. This leaves us with

$$K^2 = (2r + 8)(2\gamma - 2) + \Sigma R_j^2 + \Sigma g_{\lambda,r}^{-1}(D_1) \cdot R_j + \Sigma R_j \cdot g_{\lambda,r}^{-1}(D_2) \tag{4.3}$$

Now, Noether's formula for the genus of a curve embedded in a surface (see, for example, [**2**, p.12]) applied to $B \cong R_j \subset S_{\lambda,r}$ yields the equality

$$R_j \cdot K = (2\gamma - 2) - R_j^2$$

But the intersection numbers $R_j \cdot K$ can also be expressed using again the representation of the canonical divisor $K = \mathrm{div}(W_k)$, namely

$$\begin{aligned} R_j \cdot K = R_j \cdot \mathrm{div}(W_k) &= R_j \cdot \Sigma F_{b_i^k} + R_j \cdot g_{\lambda,r}^{-1}(D_k) + R_j \cdot R \\ &= (2\gamma - 2) + R_j \cdot g_{\lambda,r}^{-1}(D_k) + R_j^2 \end{aligned}$$

From these two expresions for $R_j \cdot K$ we infer that

$$R_j^2 = -\frac{1}{2} R_j \cdot g_{\lambda,r}^{-1}(D_1) = -\frac{1}{2} R_j \cdot g_{\lambda,r}^{-1}(D_2);$$

hence our expression (4.3) for K^2 turns into

$$K^2 = (2r+8)(2\gamma-2) + \frac{3}{2}\Sigma R_j \cdot g_{\lambda,r}^{-1}(D_1).$$

It follows that in order to prove the identity (4.1) it is enough to show that $\Sigma R_j \cdot g_{\lambda,r}^{-1}(D_1) = 2(\gamma-1)$. Since $D_1 = s_1 + t_1$ all we have to do amounts to check that $R_j \cdot g_{\lambda,r}^{-1}(s_1) = g_{\lambda,r}^{-1}(t_1) \cdot R_j = (\gamma-1)/r$. Thus, the result will follow from the following lemma.

LEMMA 4.1.

(1) $|g_{\lambda,r}^{-1}(s_1) \cap R_j| = |g_{\lambda,r}^{-1}(t_1) \cap R_j| = \dfrac{\gamma-1}{r}$.

(2) *The divisor* R_j *meets transversally each of the divisors* $g_{\lambda,r}^{-1}(s_1)$ *and* $g_{\lambda r}^{-1}(t_1)$.

PROOF. (1) By (4.2) $f_{\lambda,r}$ induces a bijection between the finite sets $g_{\lambda,r}^{-1}(s_1) \cap R_j$ and $(\pi_j \circ \zeta_{\lambda,r})^{-1}(s_1)$, its inverse being the map that sends each $b \in (\pi_j \circ \zeta_{\lambda,r})^{-1}(s_1)$ to the (unique) intersection point of F_b and R_j. Hence $|g_{\lambda,r}^{-1}(s_1^1) \cap R_j| = \deg(\zeta_{\lambda,r})|\pi_j^{-1}(s_1)|$. Now, since only one of the r coordinates of a point of $C_{\lambda,r}$ can be equal to either s_1 or t_1, we infer that these two points are not branch values of π_j and that $|\pi_j^{-1}(s_1)| = |\pi_j^{-1}(t_1)| = 2^{r-1} = \deg(\pi_j)$. Therefore

$$|g_{\lambda,r}^{-1}(s_1) \cap R_j| = |g_{\lambda,r}^{-1}(t_1) \cap R_j| = 2^{r-1}\deg(\zeta_{\lambda,r}).$$

Now, we apply the Riemann-Hurwitz formula to the smooth covering $\zeta_{\lambda,r} : B \to C_{\lambda,r}$ to see that $2\gamma - 2 = 2^r \deg(\zeta_{\lambda,r})$, and the result follows.

(2) To show that R_j meets transversally the divisor given by the equation $g_{\lambda,r} = s_1$ (resp. $g_{\lambda,r} = t_1$) at each point P in the intersection, we must check that the restriction of $g_{\lambda,r}$ to R_j is non-singular at P. This follows at once from the expression $g_{\lambda,r}|_{R_j} = \pi_j \circ \zeta_{\lambda,r} \circ f_{\lambda,r}|_{R_j}$ given in (4.2) since $f_{\lambda,r}|_{R_j} : R_j \to B$ is an isomorphism, $\zeta_{\lambda,r} : B \to C_{\lambda,r}$ is a smooth cover and, as noted above, any point of $C_{\lambda,r}$ in the fibre $\pi_j^{-1}(s_1)$ (resp. $\pi_j^{-1}(t_1)$), such as $\zeta_{\lambda,r} \circ f_{\lambda,r}|_{R_j}(P)$, is a regular point of π_j. □

REMARK 4.2. The referee has kindly pointed out to us that a shorter computation of K^2 can be achieved by using tools more specific of the theory of algebraic surfaces such as the ones employed by Catanese and Rollenske in [**6**].

Acknowledgments

The construction of these families goes back to old work by William J. Harvey and the first author. William J. Harvey is the supervisor of the first author and the supervisor's supervisor of the second. We would like to take this opportunity to express our hearty gratitude to him.

References

[1] M. F. Atiyah, *The signature of fibre-bundles*, Global Analysis (Papers in Honor of K. Kodaira), Univ. Tokyo Press, Tokyo, 1969, pp. 73–84. MR0254864

[2] Arnaud Beauville, *Surfaces algébriques complexes* (French), Société Mathématique de France, Paris, 1978. Avec une sommaire en anglais; Astérisque, No. 54. MR0485887

[3] Lipman Bers, *Uniformization, moduli, and Kleinian groups*, Bull. London Math. Soc. **4** (1972), 257–300, DOI 10.1112/blms/4.3.257. MR348097

[4] Jim Bryan and Ron Donagi, *Surface bundles over surfaces of small genus*, Geom. Topol. **6** (2002), 59–67, DOI 10.2140/gt.2002.6.59. MR1885589

[5] Fabrizio Catanese, *Kodaira fibrations and beyond: methods for moduli theory*, Jpn. J. Math. **12** (2017), no. 2, 91–174, DOI 10.1007/s11537-017-1569-x. MR3694930

[6] Fabrizio Catanese and Sönke Rollenske, *Double Kodaira fibrations*, J. Reine Angew. Math. **628** (2009), 205–233, DOI 10.1515/CRELLE.2009.024. MR2503241

[7] Gabino González-Diez, *Galois action on universal covers of Kodaira fibrations*, Duke Math. J. **169** (2020), no. 7, 1281–1303, DOI 10.1215/00127094-2019-0078. MR4094736

[8] Gabino González-Diez, *Variations on Belyi's theorem*, Q. J. Math. **57** (2006), no. 3, 339–354, DOI 10.1093/qmath/hai021. MR2253591

[9] G. González Díez and W. J. Harvey, *Moduli of Riemann surfaces with symmetry*, Discrete groups and geometry (Birmingham, 1991), London Math. Soc. Lecture Note Ser., vol. 173, Cambridge Univ. Press, Cambridge, 1992, pp. 75–93, DOI 10.1017/CBO9780511565793.010. MR1196918

[10] Gabino González Díez and William J. Harvey, *On complete curves in moduli space. I, II*, Math. Proc. Cambridge Philos. Soc. **110** (1991), no. 3, 461–466, 467–472, DOI 10.1017/S0305004100070535. MR1120481

[11] G. González-Díez and W. J. Harvey, *Surface groups inside mapping class groups*, Topology **38** (1999), no. 1, 57–69, DOI 10.1016/S0040-9383(97)00104-3. MR1644059

[12] Gabino González-Diez and Sebastián Reyes-Carocca, *The arithmeticity of a Kodaira fibration is determined by its universal cover*, Comment. Math. Helv. **90** (2015), no. 2, 429–434, DOI 10.4171/CMH/359. MR3351751

[13] Gabino González-Diez and Sebastián Reyes-Carocca, *Families of Riemann surfaces, uniformization and arithmeticity*, Trans. Amer. Math. Soc. **370** (2018), no. 3, 1529–1549, DOI 10.1090/tran/6988. MR3739184

[14] Phillip A. Griffiths, *Complex-analytic properties of certain Zariski open sets on algebraic varieties*, Ann. of Math. (2) **94** (1971), 21–51, DOI 10.2307/1970733. MR310284

[15] F. Hirzebruch, *The signature of ramified coverings*, Global Analysis (Papers in Honor of K. Kodaira), Univ. Tokyo Press, Tokyo, 1969, pp. 253–265. MR0258060

[16] Arnold Kas, *On deformations of a certain type of irregular algebraic surface*, Amer. J. Math. **90** (1968), 789–804, DOI 10.2307/2373484. MR242199

[17] Arnold Kas, *On deformations of a certain type of irregular algebraic surface*, Amer. J. Math. **90** (1968), 789–804, DOI 10.2307/2373484. MR242199

[18] G. B. Šabat, *The complex structure of domains that cover algebraic surfaces* (Russian), Funkcional. Anal. i Priložen. **11** (1977), no. 2, 67–75, 96. MR0454053

[19] G. B. Shabat, *Local reconstruction of complex algebraic surfaces from universal coverings* (Russian), Funktsional. Anal. i Prilozhen. **17** (1983), no. 2, 90–91. MR705058

[20] Chris Zaal, *Explicit complete curves in the moduli space of curves of genus three*, Geom. Dedicata **56** (1995), no. 2, 185–196, DOI 10.1007/BF01267642. MR1338958

Departamento de Matemáticas, Universidad Autónoma de Madrid, Spain.
Email address: gabino.gonzalez@uam.es

Departamento de Matemática y Estadística, Universidad de La Frontera, Chile.
Email address: sebastian.reyes@ufrontera.cl

Contemporary Mathematics
Volume **766**, 2021
https://doi.org/10.1090/conm/766/15381

Hodge theory and moduli

Phillip Griffiths

Outline

I.A. Introductory comments. Algebraic geometry is frequently seen as a very interesting and beautiful subject, but one that is also very difficult to get into. This is partly due to its breadth, as traditionally algebra (commutative and homological), topology, analysis, differential geometry, Lie theory — and more recently combinatorics, logic, categorical and derived algebraic techniques, ... — are used to study it. I have tried to make these notes accessible to a general audience by reviewing some of the most basic concepts, illustrating the material with elementary examples and informal geometric and heuristic arguments, and with occasional side comments for experts in the subject.

Although algebra, both commutative and homological, are the central tools in algebraic geometry, their use in many current areas of research (birational geometry and the minimal model program) is frequently quite technical and will not be extensively discussed in these notes. On the other hand, partly because Hodge theory involves analysis, Lie theory and differential geometry as well as a wide variety of homological methods, it is perhaps less prevalent in the more algebraic works in the field. One objective of these talks is to illustrate how, in partnership with the more algebraic and homological methods, Hodge theory may be used to study interesting and important geometric questions. In the appendix we have sketched how this may be carried out in a particular example.

Notes based on the lecture presented at the conference "Geometry at the frontier" held at Pucon, Chile during November 2018.
Parts of this paper represents joint work in progress with Mark Green, Radu Laza and Colleen Robles.

In summary the theme of these notes is to discuss and illustrate how complex analysis, differential geometry and Lie theory may be combined to study a basic problem in algebraic geometry. Although the main techniques in much of contemporary algebraic geometry are algebraic, some of the most interesting questions in the subject require non-algebraic methods for their study and this is what we hope to illustrate.[1]

As a warm up and to establish some notations we recall that an affine algebraic variety is given by the solution space over the complex numbers to polynomial equations

$$(*) \qquad f_i(x_1, \dots, x_n) = 0 \qquad i = 1, \dots, m.$$

The "elementary" examples are

linear spaces $ax + by + c = 0$

conics $ax^2 + 2bxy + cy^2 + ex + fy + g = 0$

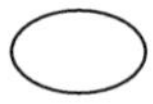

quadrics $Q(x) = \sum_{i,j=1}^{n} a_{ij}x_ix_j + \sum_{i=1}^{n} b_ix_i + c = 0, \quad a_{ij}, a_{ji}$

The first non-elementary examples are cubics

$$y^2 = 4x^3 + ax + b$$

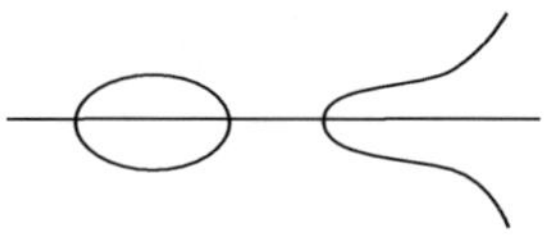

The first two elementary examples and the non-elementary example are algebraic curves. One of course considers higher dimensional varieties; surfaces, threefolds,... We will be primarily concerned with curves and surfaces.

The above algebraic curves are all affine algebraic varieties in $\mathbb{C}^2$. In general one adds to an affine variety the asymptotes or "points at infinity" to obtain the projective space $\mathbb{P}^n = \mathbb{C}^n \cup \mathbb{P}^{n-1}_\infty$. Equivalently, $\mathbb{P}^n$ is the quotient of $\mathbb{C}^{n+1}\backslash\{0\}$ by the scaling action $z_i \to \lambda z_i$ where $\lambda \in \mathbb{C}^*$. Geometrically $\mathbb{P}^n$ is the set of lines through the origin in $\mathbb{C}^{n+1}$. The picture of the projective plane $\mathbb{P}^2$ is

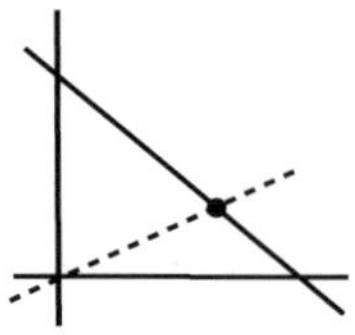

line at infinity

[1]This paper is written in an informal style, usually without precise definitions and statements of results. At the end there is a fairly extensive set of references where the more standard presentation of the material to algebraic geometers may be found. In particular we suggest [**CMSP17**] as a general reference relating Hodge theory and algebraic geometry and where several examples relevant to these notes are discussed in detail.

and the picture in $\mathbb{P}^2$ of the hyperbola in $\mathbb{C}^2$ given by

$$x^2 - y^2 = 1$$

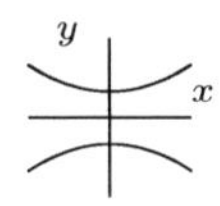

is something like

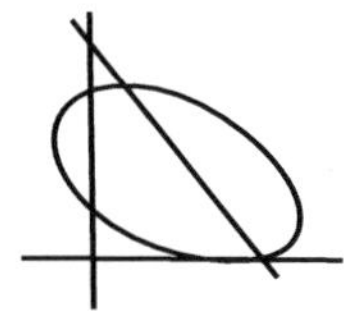

The equations of the completion in $\mathbb{P}^n$ of an affine variety given in $\mathbb{C}^n$ by $(*)$ are obtained by homogenizing: set $x_i = z_i/z_0$ and clear denominators to obtain

$$f_i(z_0, z_1, \ldots, z_n) = 0$$

where $f_i(\lambda z_0, \lambda z_1, \ldots, \lambda z_n) = \lambda^{d_i} f_i(z_0, z_i, \ldots, z_n)$, $d_i = \deg f_i$. Then the hyperbola above becomes

$$z_1^2 - z_2^2 = z_0^2.$$

Later on we will consider varieties in weighted projective spaces $\mathbb{P}(a_0, a_1, \ldots, a_n)$ where $\lambda \in \mathbb{C}^*$ acts on z_i by $\lambda(z_i) = \lambda^{a_i} z_i$, and in the quotient $\mathbb{C}^{n+1}\backslash\{0\}/\mathbb{C}^*$ the algebraic varieties are defined by weighted homogeneous polynomials. Here the weights a_i are positive integers and $\gcd(a_1, \ldots, a_N) = 1$. The following will be used in our discussion of the I-surfaces:

Example: $\mathbb{P}(1,1,2)$ is embedded in $\mathbb{P}^3$ by

$$[x_0, x_1, y] \to [x_0^2, x_0 x_1, x_1^2, y].$$

The image is

$$z_0 z_2 = z_1^2, \text{ which is}$$

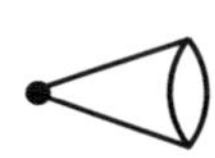

Two algebraic varieties are considered to be *equivalent* if there is a "change of variables" that transforms one into the other. Thus if the discriminant $b^2 - ac \neq 0$ all conics are equivalent to the circle

$$z_1^2 + z_2^2 = z_0^2.$$

Initially changes of variables were linear transformations (including projections); later on rational changes of variables became allowed.

Historically algebraic varieties arose from two sources: projective geometry (lines and linear spaces, conics and higher dimensional quadrics), and from complex analysis. In complex analysis the issue was to understand the integrals

$$\int r(x, y(x))dx, \qquad f(x, y(x)) = 0$$

of algebraic functions, and the "functions" defined by inverting the integrals

$$w = \int^{x(w)} r(x, y(x))dx.$$

Here $f(x,y)$ is a polynomial and $r(x,y)$ is a rational function; $y(x)$ is a multi-valued "algebraic function" defined by $f(x, y(x)) = 0$. The integral depends on choosing a path γ of integration in the complex plane and a branch $y(x)$ of $f(x, y(x)) = 0$ along γ.[2] Thus

$$w = \int^{\sin w} \frac{dx}{\sqrt{1-x^2}} = \int^{\sin w} \frac{dx}{y(x)}$$

where $x^2 + y^2 = 1$ and $\sqrt{dx^2 + dy^2} = dx/y$ gives the parametrization $w \to (\sin w, \cos w)$ of the circle by arclength in terms of "elementary" functions (trigonometric functions and logarithms). However parametrizing the ellipse by arclength led to integrals such as

$$w = \int^{p(w)} \frac{dx}{\sqrt{4x^3 + ax + b}}$$

which gave non-elementary functions and led Euler, Legendre, Abel, Gauss, Jacobi, Riemann,... to the beginnings of the rich and deep interplay between analysis and algebraic geometry. This evolved into modern *Hodge theory*, and it is this interface between analysis and algebraic geometry that is a main theme of these talks.

Part of the richness of the subject of algebraic geometry are the multiple perspectives that may be used in its study:

- *geometric*;
- *algebraic* — e.g., as we will briefly discuss, in birational geometry the algebraic classification of certain classes of singularities

  of algebraic varieties plays a central role;[3]
- *analytic*; we have mentioned complex analysis and the integrals of algebraic functions;
- *topological*; one always has in mind the first picture

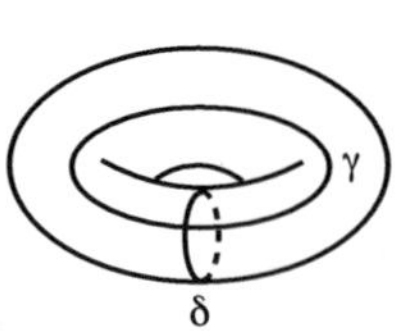

  that is encountered.

Example (running example for illustrative purposes): The compact 1-dimensional complex manifold associated to the algebraic curve

$$X = \{y^2 = (x - a_1)(x - a_2) \cdots (x - a_{2g+1})\}$$

where the a_i are distinct is a compact Riemann surface of genus g which can be studied from all of these perspectives.

---

[2]A more correct notation would be $w = \int^{(x(w),y(w))} r(x, y(x))dx$ where $(x(w), y(w))$ is the point on the curve $f(x, y) = 0$ in $\mathbb{C}^2$ that is given by the integral along the path γ in the curve. To simplify the notation the second coordinate $y(w)$ will be implicit in what follows.

[3]Cf. [**Kol13b**] in which the extent and complexity of this story are on full display.

For $g = 1$ as in the cubic above, in addition to the topological one the pictures are

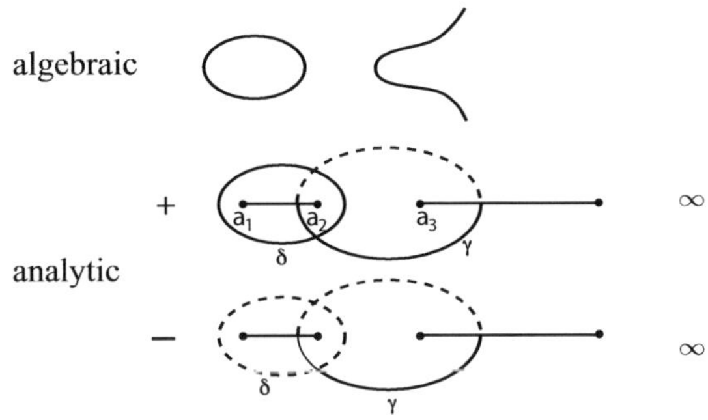

where a_1, a_2, a_3 are the roots of the cubic equation in the RHS of the above equation. The $+$ and $-$ represent the two possible values of $y(x)$ with the understanding that analytic construction across a slit changes the sign of $y(x) = \sqrt{(x-a_1)(x-a_2)\cdots(x-a_{2g+1})}$.

The integral

$$\int^w \frac{dx}{y}$$

is determined up to the *periods*

$$\pi_1 = \int_\delta \frac{dx}{y}, \quad \pi_2 = \int_\gamma \frac{dx}{y};$$

one may show that $\pi_1 \neq 0$ and $\mathrm{Im}(\pi_2/\pi_1) > 0$.

Incidentally for the hyperbola $y^2 = x^2 - 1$ the picture is

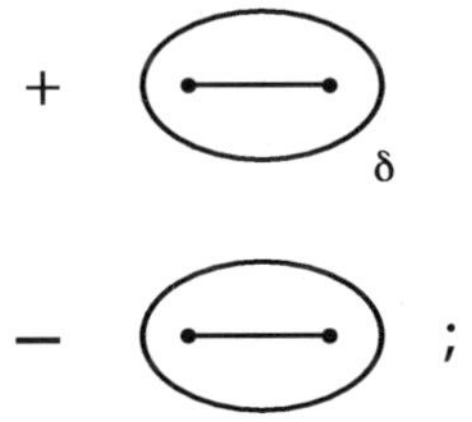

;

there is only one cycle δ. In this case there is only one period $\int_\delta \frac{dx}{y}$.

For another analytic perspective, as will be further discussed below the function $p(w)$ given by inverting the elliptic integral is a doubly periodic entire function (doubly due to the periods $\int_\delta dx/y$ and $\int_\gamma dx/y$)[4] that leads to the parametrization of the cubic curve

$$\begin{array}{ccc} \mathbb{C} & \longrightarrow & X \\ \cup & & \cup \\ w & \longrightarrow & (p(w), p'(w)) \end{array}$$

[4]For the hyperbola above there is only one period and inverting the integral gives singly periodic trigonometric functions.

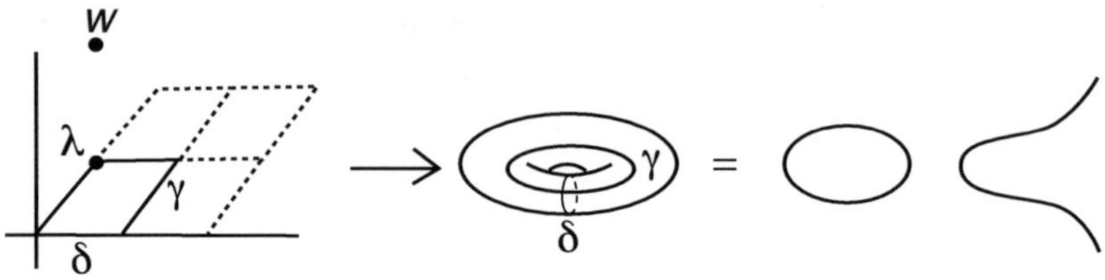

We note that $y(w) = p'(w)$ arises from

$$dw = d\int^{p(w)} \frac{dx}{y} = \frac{p'(w)dw}{y(w)}.$$

Here the ratio $\lambda = \pi_2/\pi_1$ of the periods is determined up to

$$\lambda \to \frac{a\lambda + b}{c\lambda + d}$$

with $\left(\begin{smallmatrix} a & b \\ c & d \end{smallmatrix}\right) \in \mathrm{SL}_2(\mathbb{Z})$ reflecting the choices of the basis $\delta, \gamma \in H_1(X, \mathbb{Z})$ with $(\delta, \gamma) = 1$.

I.B. Classification problem; first case of relation between moduli and Hodge theory. A central problem in algebraic geometry is to classify the equivalence classes of algebraic varieties. For this there are two types of parameters:

- discrete (e.g., the genus $g = (1/2)b_1(X)$ for smooth algebraic curves)
- continuous (*moduli*; the smooth curves of genus g form a $(3g-3+\rho)$-dimensional family $\mathcal{M}_g$, where $\rho = \dim \mathrm{Aut}(X)$).

Example (curves):

$g = 0$ there is only one equivalence class; for example using the birational map given by stereographic projection all non-singular conics are projectively equivalent and are birationally equivalent to a line $\mathbb{P}^1$

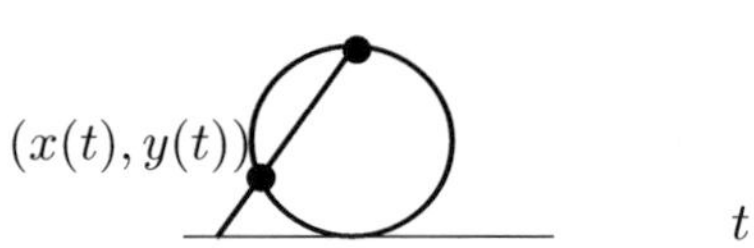

where $x(t), y(t)$ are rational functions of t.

$g = 1$ $\dim \mathcal{M}_1 = 1$, from the theory of elliptic curves on has that

$$j = \frac{1728a^3}{\Delta}, \quad \Delta = a^3 - 27b^2$$

gives a set-theoretically 1-1 map $\mathcal{M}_1 \to \mathbb{C}$; thus the curves

$$\begin{cases} x_0^3 + x_1^3 + x_2^3 = x_0x_1x_1 & (\text{in } \mathbb{P}^2) \\ Q_1(x_0, x_1, x_2, x_3) = Q_2(x_0, x_1, x_2, x_3) = 0 & (\text{in } \mathbb{P}^3) \end{cases}$$

are both equivalent to $y^2 = 4x^3 + ax + b$ for a unique value of j. Here $Q_1(x) = 0$ and $Q_2(x) = 0$ are smooth quadrics that intersect transitively.[5]

In general one hopes that

[5]We are finessing the subtle issues that arise when the curve has non-trivial automorphisms. We refer to [**Kol13a**] and the references given there for a discussion of the structure of moduli spaces at points where the corresponding algebraic variety has automorphisms.

(i) a moduli space $\mathcal{M}$ will be an algebraic variety, generally *not* complete (compact);

(ii) there will be a canonical completion $\overline{\mathcal{M}}$ corresponding to adding *certain* singular varieties.

In these notes we will assume (i) and will be primarily concerned with (ii). *What does* (ii) *mean?*

We imagine a family of plane curves

$$X_t = \{f(x,y,t) = 0\}, \quad t \in \Delta$$

that are smooth for $t \in \Delta^*$ but may be singular for $t = 0$. A picture like

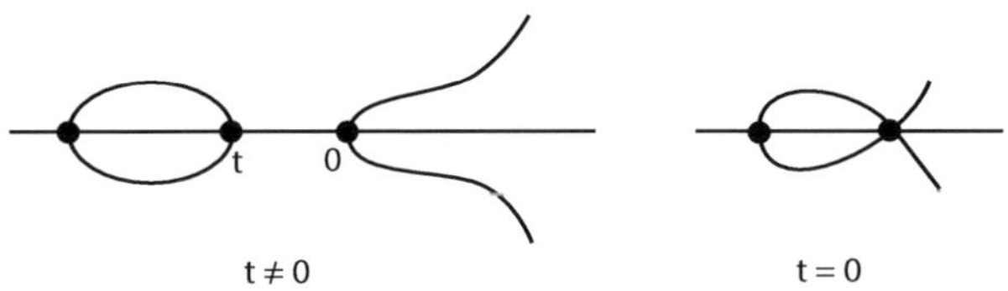

will give such a family.[6]

Applying coordinate changes depending on t can give a family $\widetilde{X}_t$ such that $\widetilde{X}_t$ is equivalent to X_t for $t \neq 0$ but $\widetilde{X}_0$ is quite different from X_0. For example we can have

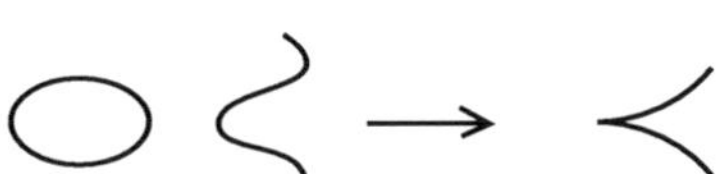

and even

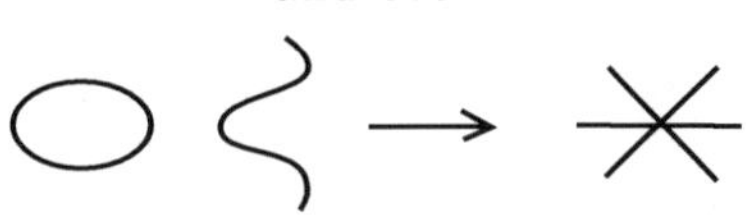

How can we say what a **canonical** *choice for* X_0 *should be?*

Historically one suggested answer to this question was provided by Hodge theory; i.e., considering the *period matrix* associated to the curve. For the example $y^2 = x(x-t)(x-1)$

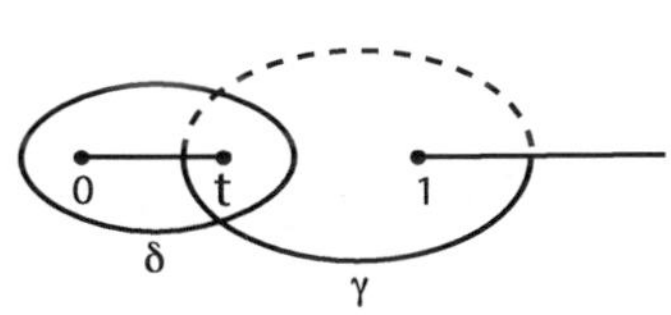

as t turns around 0 by following what γ does we get a picture like the figure on the left

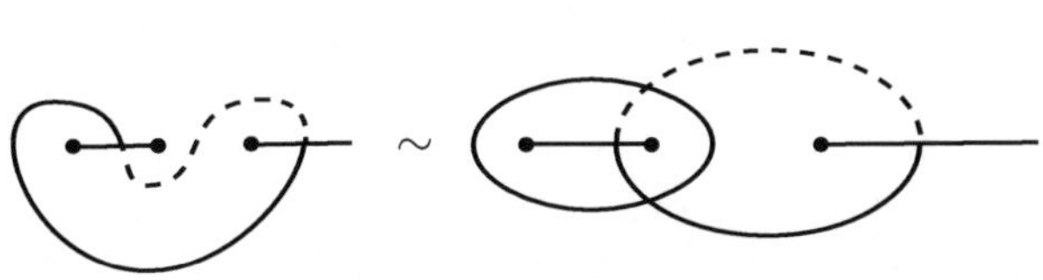

and denoting homology by $\sim$ the picture on the right leads to the result that as t turns around the origin in homology we have the *Picard-Lefschetz transformation*

$$\begin{cases} \delta \to \delta \\ \gamma \to \gamma + \delta. \end{cases}$$

[6]Below we will give equations for such a picture.

For the periods by $\pi_1 = \int_\delta dx/y$, $\pi_2 = \int_\gamma dx/y$, using elementary complex analysis one may show that

- $\pi_1(t)$ is non-zero and holomorphic for $t \in \Delta$;
- $\pi_2(t) = \pi_1(t)\frac{\log t}{2\pi i} +$ (holomorphic function of $t \in \Delta$).

In general for *any* family X_t, $t \in \Delta^*$, of smooth genus 1 curves we will have periods π_1, π_2 as above where

$$\lambda = \pi_2/\pi_1, \qquad \operatorname{Im}\lambda > 0;$$

here we are thinking of λ as a point in $\mathrm{SL}_2(\mathbb{Z})\backslash\mathbb{H}$ where $\mathbb{H} = \{w \in \mathbb{C} : \operatorname{Im} w > 0\}$ is the upper half plane. The periods are locally holomorphic functions of $t \in \Delta^*$, and as t turns around the origin the cycles δ, γ will undergo a *monodromy transformation*

$$\begin{cases} \left(\begin{smallmatrix} \pi_2 \\ \pi_1 \end{smallmatrix}\right) \to \left(\begin{smallmatrix} a & b \\ c & d \end{smallmatrix}\right)\left(\begin{smallmatrix} \pi_2 \\ \pi_1 \end{smallmatrix}\right), \text{ where} \\ T = \left(\begin{smallmatrix} a & b \\ c & d \end{smallmatrix}\right) \in \mathrm{SL}_2(\mathbb{Z}) \text{ is the monodromy matrix.} \end{cases}$$

For w a lifting of λ to $\mathbb{H}$ this gives a diagram

$$\begin{array}{ccccc} z \in & \mathbb{H} & \xrightarrow{\ w\ } & \mathbb{H} \\ \downarrow & \downarrow & & \downarrow \\ e^{2\pi i z} = & t \in \Delta^* & \xrightarrow{\ \pi\ } & \mathrm{SL}_2(\mathbb{Z})\backslash\mathbb{H} \end{array}$$

where $\pi(t) = \lambda$ and

$$w(z+1) = Tw(z).$$

LEMMA 1. *The eigenvalues μ of T satisfy $|\mu| = 1$.*

Since the characteristic polynomial of T has integral coefficients, by the Gelfand-Schneider theorem from analytic number theory

$$\mu = e^{2\pi i p/q}$$

is root of unity. Replacing t by t^q gives that T is unipotent, and for a suitable choice of generators of $H_1(X_t, \mathbb{Z})$ we may assume that

$$T = \begin{pmatrix} 1 & m \\ 0 & 1 \end{pmatrix}, \qquad m \in \mathbb{Z}^+.$$

Thus *any* monodromy matrix is equivalent to a power of a Picard-Lefschetz transformation.

LEMMA 2. *Given a holomorphic mapping $w : \mathbb{H} \to \mathbb{H}$ satisfying*

$$w(z+1) = w(z) + m,$$

it follows that

$$w(z) = mz + u(e^{2\pi i z})$$

where u is bounded as $\operatorname{Im} z \to \infty$.

Taking $m = 1$ for simplicity this gives

$$\pi(t) = \frac{\log t}{2\pi i} + u(t),$$

where $u(t)$ is holomorphic in all of Δ.

Both of these lemmas are proved by complex analysis arguments using the Schwarz lemma in the form

w is distance decreasing in the $\mathrm{SL}_2(\mathbb{R})$*-invariant hyperbolic (Poincaré) metric* $\frac{dw\, d\overline{w}}{|w|^2}$

and the observation that

the length of the circles $|t| = \epsilon$ *tends to zero as* $\epsilon \to 0$.

Sketch of the proof of Lemma 1: Let $z_n \in \mathbb{H}$ be a sequence with $\operatorname{Re} z_n = 0$, $\operatorname{Im} z_n \to \infty$ and set

$$w_n = w(z_n).$$

Then

$$w(z_n + 1) = Tw(z_n) = Tw_n.$$

The hyperbolic distance $d(z_n, z_n + 1) \to 0$, and by the distance decreasing property of w we have

$$d(w_n, Tw_n) \to 0.^7$$

Now $w_n = A_n \cdot w$ for some $A_n \in \mathrm{SL}_2(\mathbb{R})$ and fixed $w \in \mathbb{H}$, and using the invariance of the metric from

$$d(w, A_n^{-1} T A_n w) \to 0$$

a little argument shows that by passing to a subsequence we will have

$$A_n^{-1} T A_n \to H = \{\text{isotropy group of } w\}.$$

Since H is compact, all its eigenvalues have absolute value 1 and this implies the same for T.[8] Since all that we have used is that the ratio of the periods $\lambda(t)$ of *any* holomorphic family of genus 1 curve parametrized by Δ^* has the same analytic behavior as for the above family has a pair of roots of the cubic coming together at $t = 0$ we may draw the

Conclusion: *The periods of an arbitrary family of genus* 1 *algebraic curves over* Δ^* *have the same asymptotic behavior as a family acquiring nodal singularities given locally analytically by*

$$x^2 = y^2 + tf(x, y).$$

The local pictures are

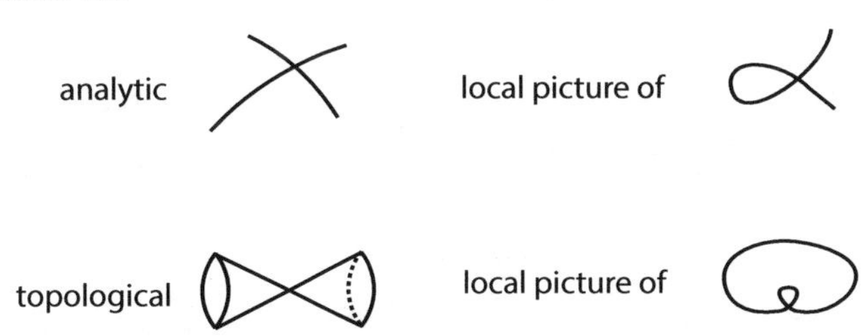

This analysis of the asymptotics of the period matrix (Hodge structure) extends to that of algebraic curves of any genus $g \geqq 2$ and provided an early suggestion as to what $\overline{\mathcal{M}}_g$ should be.

[7]The horizontal segments from z_n to $z_n + 1$ map down to circles in the punctured disc.

[8]Since $\mathrm{SL}(\mathbb{R})$ acts transitively on $\mathbb{H}$, the isotropy group of any point is conjugate in $\mathrm{SL}_2(\mathbb{R})$ to the isotropy group $\{\left(\begin{smallmatrix}\cos\theta & \sin\theta \\ -\sin\theta & \cos\theta\end{smallmatrix}\right)\}$ of $i \in \mathbb{H}$.

The schematic of $\overline{\mathcal{M}}_2$ is

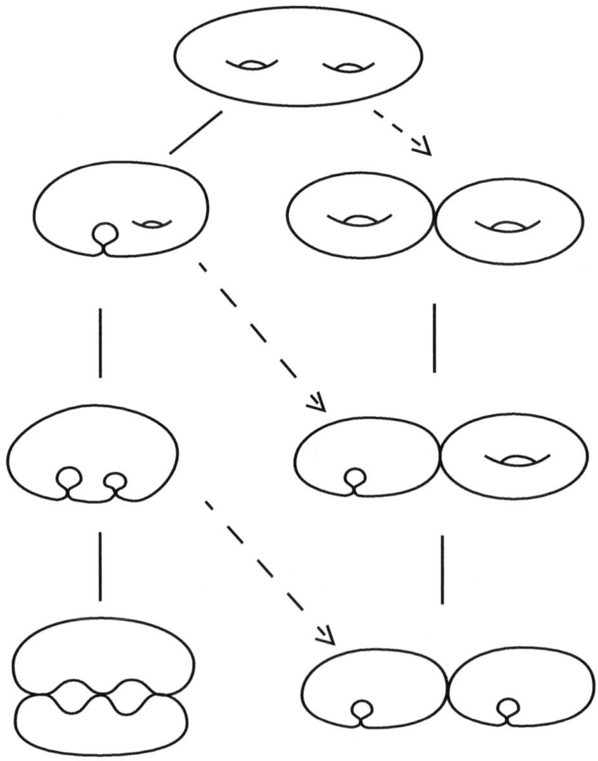

This gives the stratification of $\overline{\mathcal{M}}_2$ together with the incidence (degeneration) relations among the strata. (The solid and dotted arrows will be explained later.) We will see that this stratification is captured by the Hodge structures and their limits.

The objective of these notes is to discuss how this picture might be extended to some completed moduli spaces of varieties of general type (analogues of curves of genus $g \geqq 2$) and to illustrate how this works for the first non-classical algebraic surfaces (called I-surfaces and which have the invariants $p_g(X) = 2$, $q(X) = 0$, $K_X^2 = 1$).

To jump ahead and anticipate some of the main points to be made; with notations and terminology to be explained, there are first the general results (some of which are work in progress).

- For a given class of surfaces of general type the moduli space $\mathcal{M}$ exists and has a canonical completion $\overline{\mathcal{M}}$.
- There is a period mapping

$$\mathcal{M} \xrightarrow{\Phi} \mathcal{P} \subset \Gamma\backslash D$$

  that associates to each surface X the Hodge structure on $H^2(X, \mathbb{Z})$.
- There is a canonical minimal completion $\overline{\mathcal{P}}$ of $\mathcal{P}$ to projective variety, and the period mapping extends to

$$\overline{\mathcal{M}} \xrightarrow{\Phi_e} \overline{\mathcal{P}}.$$

There are then the specific results for the I-surface X. We will foreshadow these here with the explanation of the notations and terminology to also be given later. The objective is to give something of the flavor of what is to come. We begin with the

Picture/equations:

- 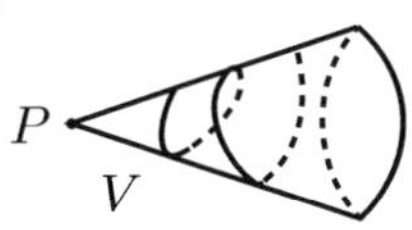

$\mathbb{P}(1,1,2) \hookrightarrow \mathbb{P}^3$ given by $(t_0, t_1) \hookrightarrow [t_0^2, t_0t_1, t_1^2, y]$
$X \to \mathbb{P}(1,1,2)$is a 2:1 map branched over the vertex P and the intersection of $\mathbb{P}(1,1,2)$ with a quintic surface $V \in |\mathcal{O}_{\mathbb{P}^3}(5)|$;
- X is realized as the weighted hypersurface in $\mathbb{P}(1,1,2,5)$ given by an equation $z^2 = F_{10}(t_0, t_1, y)$.[9]

For the moduli space, $\mathcal{M}_I$ is smooth and

- $\dim \mathcal{M}_I = h^1(T_X) = 28$;[10]
- the period domain D_I is a homogeneous contact manifold with $\dim D_I = 57 = 2\dim \mathcal{M}_I + 1$;
- $\Phi = \mathcal{M}_I \to \Gamma_I \backslash D_I$ and Φ_* is injective (local Torelli)

$$\Downarrow$$

$$\Phi(\mathcal{M}_I) = \text{contact submanifold } \mathcal{P} \hookrightarrow \Gamma_I \backslash D_I.$$

The contact structure on $\mathcal{P}$ will be explained below. In the appendix by analyzing the Hodge structure at the boundary of moduli we will sketch a proof of generic local Torelli.

On the Hodge theoretic side we have the

Picture of the stratification of $\overline{\mathcal{P}}$ ($N =$ logarithm of monodromy):

[9]This is the bicanonical model of a smooth I-surface.

[10]In fact, it can be shown that for a smooth X, $h^0(T_X) = h^2(T_X) = 0$, and from the Hirzebruch Riemann-Roch theorem the Euler characteristic $\chi(T_X) = -28$. One may also "count parameters" in the above equation for X and arrive at the same conclusion.

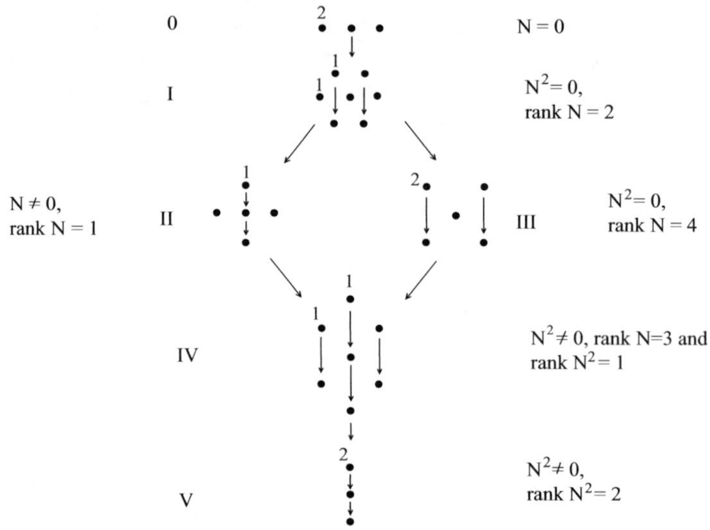

The main result, here stated informally, is

The Hodge theoretic stratification of $\overline{\mathcal{P}}$ *uniquely determines the stratification of* $\overline{\mathcal{M}}_I^{\mathrm{Gor}}$.[11]

Rather than display the whole picture, the following table is just the part for simple elliptic singularities (types I and III). They have $N^2 = 0$ since for the semi-stable-reduction (SSR) of such a degeneration only double curves (and no triple points) occur; all of the other types occur if we include cusp singularities.

stratum	dimension	minimal resolution $\widetilde{X}$	$\sum_{i=1}^{k}(9-d_i)$	k	codim in $\overline{\mathcal{M}}_I$
I_0	28	canonical singularities	0	0	0
I_2	20	blow up of a K3-surface	7	1	8
I_1	19	minimal elliptic surface with $\chi(\widetilde{X}) = 2$	8	1	9
$\mathrm{III}_{2,2}$	12	rational surface	14	2	16
$\mathrm{III}_{1,2}$	11	rational surface	15	2	17
$\mathrm{III}_{1,1,R}$	10	rational surface	16	2	18
$\mathrm{III}_{1,1,E}$	10	blow up of an Enriques surface	16	2	18
$\mathrm{III}_{1,1,2}$	2	ruled surface with $\chi(\widetilde{X}) = 0$	23	3	26
$\mathrm{III}_{1,1,1}$	1	ruled surface with $\chi(\widetilde{X}) = 0$	24	3	27

The particular K3, elliptic surfaces, rational surfaces and ruled surfaces can be specified. Note that the last column is the sum of the two columns preceding it.

[11] Here $\overline{\mathcal{M}}_I^{\mathrm{Gor}}$ is the completed moduli space for Gorenstein I-surfaces. Work in progress suggests that the Hodge theoretic structure of $\overline{\mathcal{P}}$ may go a long way towards determining the full stratification of $\overline{\mathcal{M}}_I$ by singularity type.

(This may be explained using Hodge theory; in the appendix we will sketch how this may be done for I_2.)

II.C. Moduli and Hodge theory. We have seen that Hodge theory, in the classical form of periods of integrals of algebraic functions together with some complex analysis and differential geometry, suggests what singular curves should be included to compactify the moduli space $\mathcal{M}_g$, leading to an essentially smooth $\overline{\mathcal{M}}_g$.

For surfaces (and higher dimensional) varieties of general type the story thus far is both similar and different,[12] especially in the non-classical (term to be explained) case.

Birational geometry tells us that it is possible to define a moduli space $\mathcal{M}$ with a canonical completion $\overline{\mathcal{M}}$. It does not

(i) tell us what the singular surfaces X corresponding to the boundary $\partial\mathcal{M} = \overline{\mathcal{M}}\backslash\mathcal{M}$ are;[13]
(ii) tell us the stratification of $\overline{\mathcal{M}}$; and
(iii) in contrast to the curve case, $\overline{\mathcal{M}}$ may be highly singular along $\overline{\mathcal{M}}\backslash\mathcal{M}$, and it does not suggest how to desingularize it.

We will explain and illustrate how Hodge theory, in partnership with birational geometry, helps us understand points (i)–(iii).

A. Moduli Invariants of a smooth projective variety X are basically

- discrete
  - Kodaira dimension $\kappa(X)$
  - topological (Chern numbers); and
- continuous (moduli).

X is a compact, complex manifold and a basic invariant is the space $H^0(K_X^m)$ of global holomorphic forms expressed locally in holomorphic coordinates $x_1, \dots, x_n$ as

$$\varphi = f(x)(dx_1 \wedge \cdots \wedge dx_n)^m$$

where $f(x)$ is holomorphic and transforms by the mth power of the Jacobian determinant when we change coordinates.

The *Kodaira dimension* $\kappa(X)$ is defined by

$$\dim H^0(mK_X) = h^0(mK_X) = Cm^{\kappa(X)} + \cdots, C > 0.$$

By convention we set $\kappa(X) = -\infty$ if all $h^0(mK_X) = 0$.

The purpose of this part of these notes is to give an informal introduction to moduli, to describe two simple classes of algebraic curves and surfaces, and to illustrate the semi-log-canonical (slc) singularities that arise for surfaces and to begin the discussion of how Hodge theory relates to them.

Examples: When $n = 1$ and X is the smooth algebraic curve (compact Riemann surface) with affine equation

$$y^2 = \prod_{i=1}^{2g+1}(x - a_i), \quad a_i \text{ distinct},$$

---

[12]Cf. the papers [**Kol13a**], [**Kol13b**] and the recent Séminaire Bourbaki by Benoist in the references.

[13]In fact, even though the non-Gorenstein isolated singularities are all rational, there does not yet seem to be a practical bound on their index.

we may picture X as a 2-sheeted branched covering of $\mathbb{P}^1$

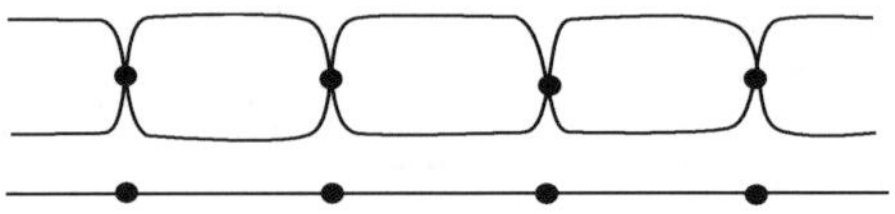

and the holomorphic 1-forms on X are

$$\varphi = \frac{p(x)dx}{y}, \qquad \deg p(x) \leqq g-1.$$

For $m \geqq 2$ there are similar expressions $\frac{q(x)dx^m}{y}$ where $\deg q(x) \leqq (2g-2)m - g$ for the holomorphic m-forms.

The genus $g(X) = g$ and the number of parameters of X's given as above is $2g+1-2 = 2g-1$; a general curve of genus of genus g is represented this way for $g = 1, 2$. The φ's above give the space $H^0(K_X) = H^0(\Omega^1_X)$ of holomorphic differentials on X, and the first de Rham cohomology group is a direct sum

$$H^1_{\mathrm{DR}}(X) \cong H^0(\Omega^1_X) \oplus \overline{H^0(\Omega^1_X)}$$

which using de Rham's theorem gives the *Hodge structure* on $H^1(X, \mathbb{C}) \cong H^1_{\mathrm{DR}}(X, \mathbb{C})$. Thus for $h^0(K_X) = \dim H^0(K_X)$ we have

$$h^0(K_X) = \left(\frac{1}{2}\right) b_1(X) = g,$$

the first result in Hodge theory relating the algebro-geometric invariant $h^0(K_X)$ to the topological invariant $b_1(X)$.

Now take X to be the smooth algebraic surface with affine equation

$$(*) \qquad z^2 = f(x,y), \quad \deg f(x,y) = 2k$$

where $f(x,y) = 0$ defines a smooth algebraic curve $C \subset \mathbb{P}^2$.[14] The holomorphic 2-forms on X are

$$\varphi = \frac{p(x,y)dx \wedge dy}{z}, \qquad \deg p(x,y) \leqq k-3.^{15}$$

There are formulas similar to the above in the curve case for the $H^0(mK_X)$'s.

For $H^0(\Omega^2_X) = H^0(K_X)$ the direct sum decomposition

$$(**) \qquad H^2_{\mathrm{DR}}(X) \cong H^0(\Omega^2_X) \oplus H^1(\Omega^1_X) \oplus \overline{H^0(\Omega^2_X)}, \;\; H^1(\Omega_X) = \overline{H^1(\Omega^1_X)}$$

gives the Hodge structure on $H^2(X, \mathbb{C}) \cong H^2_{\mathrm{DR}}(X)$.[16]

For an initial explanation of the $H^1(\Omega^1_X)$ term, if Q is the bilinear form on $H^2(X, \mathbb{C})$ given by the cup-product, then

$$\begin{array}{c} F^2H^2(X,\mathbb{C}) := H^0(\Omega^2_X) \\ \cap \\ F^1H^2(X,\mathbb{C}) := H^0(\Omega^2_X) \oplus H^1(\Omega^1_X) \end{array}$$

where under the cup product Q in cohomology

$$F^1 = F^{2\perp}$$

[14]This surface is similar to but both simpler and more complicated than the I-surface.

[15]Here $p(x,y)$ is arbitrary. Later in these notes we will encounter the case $k=3$, i.e., X is a 2-sheeted covering branched over a sextic curve in $\mathbb{P}^2$: this is a K3 surface with $h^0(\Omega^2_X) = 1$.

[16]In contrast to the curve case, for any $k \geqq 4$ a general surface in the moduli space $\mathcal{M}$ of surfaces of the above numerical type is equivalent to one given by $(*)$.

and

$$F^1 \cap \overline{F}^1 = H^1(\Omega^1_X)$$

so that the Hodge decomposition (**) is determined by $H^0(\Omega^2_X)$ and Q. Thus in both these cases, $H^0(\Omega^n_X)$ together with Q determines the Hodge decomposition and resulting Hodge structure on the cohomology $H^n(X, \mathbb{C})$, $n = \dim X$.

The *Kodaira number* $\kappa(X)$ for curves is

$$\kappa(X) = \begin{cases} -\infty \text{ for} & g(X) = 0 \\ 0 \text{ for} & g(X) = 1 \\ 1 \text{ for} & g(X) \geqq 2. \end{cases}$$

For the above algebraic surfaces

$$\kappa(X) = \begin{cases} -\infty & k \leqq 2 \qquad \text{(rational)} \\ 0 & k = 3 \qquad \text{(K3)} \\ 2 & k \geqq 4 \qquad \text{(general type)} \end{cases}$$

to get $\kappa(X) = 1$ you have to allow C to be quite singular.

General type surfaces are those with $\kappa(X) = 2$; for these the important numerical invariants are

- $p_g = h^0(K_X) = h^0(\Omega^2_X) =$ geometric genus;
- $q = h^0(\Omega^1_X) =$ irregularity;
- $K^2_X = c_1(X)^2$.

They are related by

- $p_g - q + 1 = \frac{1}{12}(K^2_X + \chi_{\text{top}}(X))$ (Noether's formula);
- $p_g \leqq \frac{K^2_X}{2} + 2$ (Noether's inequality).[17]

THEOREM ([**KSB88**], [**Ale94**]). *For general type surfaces with given numerical invariants there exists a moduli space $\mathcal{M}$ with a canonical completion $\overline{\mathcal{M}}$.*

As noted above the proof is via birational geometry. It describes in principle what the singularities of a surface X corresponding to a boundary point in $\overline{\mathcal{M}} \backslash \mathcal{M}$ can be.[18] For surfaces there is no description, nor examples that I know other than the work of [**FPR15a, FPR15b, FPR17**], of the global structure of what the X's can be and how they fit together.

Some guiding questions are

- *How does Hodge theory limit what the singular X's can be?*
- *Which Hodge-theoretically possible degenerations are realized algebro-geometrically?*
- *Does the Hodge theoretic stratification capture the algebro-geometric one?*

Given a family $\mathcal{X}^* \xrightarrow{\pi} \Delta^*$ of smooth surfaces $X_t = \pi^{-1}(t)$ for $t \neq 0$, by the theorem there is defined a *unique* limit surface $X_0 = X$ that fills in the family $\mathcal{X} \to \Delta$ where the conditions

[17] For the above surface we have $K^2_X = 2$ and $p_g = 3$, so that it is extremal for Noether's inequality. For the I-surface we have $K^2_X = 1$ and $p_g = 2$ so that it is also extremal.

[18] In the paper [**Kol13a**] by Kollár in the references there is a fairly short list of the singularity types that can occur. However within each non-Gorenstein type there is an invariant, the *index*. The recent paper [**RU19**] seems to be a promising approach to bounding the index in terms of K^2_X.

- $\mathcal{X}$ has *canonical singularities* over X_{sing};[19] and
- $\mathcal{X}$ is of relative general type and minimal (more precisely, the relative dualizing sheaf $\omega_{\mathcal{X}/\Delta}$ is $\mathbb{Q}$-Cartier and relatively ample).

The first condition is local along X; the second is global.

As is the case for any analytic variety, $\overline{\mathcal{M}}$ has a *stratification*

- $\overline{\mathcal{M}}$ is a union of irreducible subvarieties $\mathcal{Z}_i$;
- the incidence relation $\mathcal{Z}_j \subset \overline{\mathcal{Z}}_i$ means that singular varieties parametrized by $\mathcal{Z}_i$ can degenerate further into those parametrized by $\mathcal{Z}_j$.

The proof of the general existence theorem does not suggest what the stratification should be; again aside from [**FPR15a, FPR15b, FPR17**] I know of no other examples where it has been analyzed.

To give some flavor of how Hodge theory helps to organize the singularities, we note that $\mathcal{X}^* \to \Delta^*$ is topologically a fibre bundle over the circle, and thus there is a monodromy operator (here $t_0 \in \Delta^*$ is a base point)

$$T : H^2(X_{t_0}, \mathbb{Z}) \to H^2(X_{t_0}, \mathbb{Z}).$$ [20]

Denoting by

$$T = T_{\text{ss}} T_{\text{u}}$$

the Jordan decomposition of T where T_{ss} is semi-simple and T_{u} is unipotent with logarithm N, using analytic arguments arising from Hodge theory that extend the one given above in the case of elliptic curves leads to a proof the *monodromy theorem* [**PS08**], [**Sch73**]

$$T_{\text{ss}}^m = 1 \text{ (i.e., the eigenvalues of } T \text{ are roots of unity)}$$
$$N^3 = 0 \text{ (i.e., the Jordan blocks of } T \text{ have length} \leqq 2).$$

A crude Hodge theoretic classification of the singularities of X is given by

- normal (a): $N = 0$
- normal (b): $N \neq 0$, $N^2 = 0$
- normal (c): $N^2 \neq 0$
- non-normal (a): $N \neq 0$ but $N^2 = 0$[21]
- non-normal (b): $N^2 \neq 0$.

This may be refined by putting in the ranks of N and of N^2.

A much finer invariant is given by also including the conjugacy class of T_{ss}, usually expressed in terms of the spectrum. And if we include the extension data in the limiting mixed Hodge structure (LMHS), we obtain even more Hodge-theoretic information.[22]

The following is an informal discussion of typical singularities of each of the above types.

---

[19]For normal $\mathcal{X}$ this means that for U any open set in $\mathcal{X}$ any holomorphic $\omega \in H^0(\mathcal{U} \cap \mathcal{X}_{\text{reg}}, K_{\mathcal{X}})$ satisfies

$$\int_{\mathcal{X}_{\text{reg}}} \omega \wedge \overline{\omega} < \infty$$

.

[20]We will consider integral cohomology modulo torsion.

[21]In the examples I know of, non-normal $\implies N \neq 0$.

[22]The above crude Hodge theoretic classification is extracted from its associated graded. The LMHS will be defined below.

Normal (a): Then the monodromy is finite and the Hodge structures on the $H^2(X_t, \mathbb{C})$ extend across $t = 0$.[23] These include a number of finite quotient singularities; typically among them are those denoted

$$\frac{1}{d}(1, a), \qquad \text{gcd } (d, a) = 1$$

and which is given by the quotient of $\mathbb{C}^2$ acted on by the cyclic group generated by

$$(x, y) \to (\zeta x, \zeta^a y)$$

where $\zeta = e^{2\pi i/d}$. Among these are the Wahl singularities $\frac{1}{n^2}(1, na - 1)$. For $n = 2$ this is a cone over a rational normal curve C in $\mathbb{P}^4$. It is noteworthy in that for this singularity $T = \mathrm{Id}$.

Normal (b): simple elliptic singularities. Here (X, p) is a normal surface X having an isolated singular point p, the minimal resolution $(\widetilde{X}, C) \to (X, p)$ is then given by contracting an elliptic curve $C \subset \widetilde{X}$ with $C^2 = -d$ where $d > 0$ is the *degree* of the elliptic singularity. To say that $(\widetilde{X}, \widetilde{C})$ is minimal means that there are no (-1)-curves not meeting $\widetilde{C}$.[24] The assumption that (X, p) is smoothable implies that $1 \leqq d \leqq 9$.

For $d \geqq 3$, one may think of the cone over an elliptic normal curve in $\mathbb{P}^{d-1}$. One typically pictures such a singularity as

There are two types of restriction here:

(i) the cone is over a smooth elliptic curve as opposed to a cone over a curve of genus $g \geqq 2$;
(ii) the restriction $d \leqq 9$ for the elliptic curve.

An analytic explanation may be given for (i); as noted above, (ii) is the condition that the isolated singularity be smoothable.

Note: We are finessing the subtlety that in order to fit the desingularization $\widetilde{X}$ of X into a family $\widetilde{\mathfrak{X}} \to \widetilde{\Delta}$ we have to do semi-stable-reduction (SSR), which involves a base change $t = \tilde{t}^m$ where $T_{\mathrm{ss}}^m = \mathrm{Id}$. The fibre over the origin in $\widetilde{\mathfrak{X}} \to \widetilde{\Delta}$ has $\widetilde{X}$ as one component. The other component is a rational surface Y meeting $\widetilde{X}$ along C, and $p_g(Y) = 0$ so that $\lim_{t\to 0} H^0(\Omega^2_{X_t})$ lives on $\widetilde{X}$.[25]

For the normal (b) degeneration a similar argument applies except now for $\omega_t \in H^0(\Omega^2_{X_t})$ and $\lim_{t\to 0} \omega_t := \omega \in H^0(\Omega^2_{\widetilde{X}})$

$$\mathrm{Res}_C(\omega) \in H^0(\Omega^1_C) \cong \mathbb{C}.$$

---

[23]There is a Riemann extension theorem for a family of Hodge structures over Δ^* having finite monodromy. Thus even though X_0 may be singular, there is associated to it a polarized Hodge structure.

[24]These are curves $E \cong \mathbb{P}^1$ with $E^2 = -1$.

[25]An additional subtlety is that in order to have some degree of uniqueness one may want to allow $\widetilde{\mathfrak{X}}$ to have cyclic quotient singularities of a "simpler" type than the ones that were started with. The meaning of $\lim_{t\to 0} H^0(\Omega^2_{X_t})$ will be discussed below.

If there are e elliptic singularities, this type of argument leads to the bound

$$e \geqq \operatorname{rank} N.$$

Another type of Hodge theoretic argument gives

$$e \leqq \operatorname{rank} N + 1.$$

We will see that both bounds are sharp for I-surfaces.

We will also see that for I-surfaces the degrees of the elliptic singularities are determined by the eigenvalues of T_{ss}. An explicit such singular surface will be discussed in the appendix.

Normal (c): cusp singularity. (X, p) where the minimal resolution $(X, D) \to (X, p)$ has for D a cycle of $\mathbb{P}^1$'s E_i with all $E_i^2 \leqq -2$

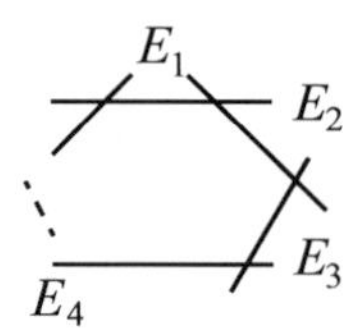

and the least one $E_i^2 \leqq -3$. It seems plausible, and may in fact be known, that the $-E_i^2$ are determined by the spectrum of T_{ss}.

For the cusp, $\operatorname{Res}_{E_i}(\omega)$ is a 1-form on $\mathbb{P}^1$ with log poles at the two interssection points. At a point of $E_i \cap E_{i+1}$ the residues are opposite. Hence for each cusp the p_g can drop by at most 1 in the limit.

Non-normal (a): X has a smooth double curve C with pinch points (Whitney swallowtail given locally by $x^2y = z^2$).

Non-normal (b): Informally stated, X has a nodal double curve with pinch points whose minimal resolution has a cycle of $\mathbb{P}^1$'s. These surfaces are frequently constructed by a gluing construction that will be illustrated below.

Example of non-normal (a): Let $C \subset \mathbb{P}^2$ be a smooth plane quartic having an involution $\tau : C \to C$ with quotient $D = C/\tau$ an elliptic curve. Then $X = \mathbb{P}^2/\tau$ is a surface having a smooth double curve D with pinch points at the 4 branch points of $C \to D$. The desingularization $\widetilde{X}$ of X is $\mathbb{P}^2$ and by pulling back 2-forms one has that

$$H^0(K_X) \cong H^0(\Omega^2(\log C))^- \cong H^0(\mathcal{O}_{\mathbb{P}^2}(1))^-$$

are the τ anti-invariant 2-forms on $\mathbb{P}^2$ having a log pole on C. Thus

$$\begin{cases} h^0(K_X) = 2 \\ K_X^2 = 1 \end{cases}$$

so that X "looks like" an I-surface. In fact X can be smoothed to such ([**FPR15a, FPR15b, FPR17**]).

non-normal (b) ([LR16]): This is a degeneration of the preceeding example. Before explaining it we will give a general contextual comment.

Conjecturally an I-surface analogue of the "most degenerate," meaning no equisingular deformations, genus 2 curve

or \$ (dollar bill curve)

obtained by identifying three distinct points on each of two $\mathbb{P}^1$'s is the surface obtained by identifying pairs of lines in a quadrilateral

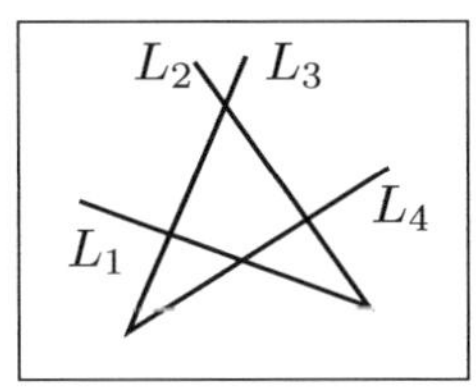

Here to obtain a well-defined involution τ of the quartic curve given by the quadrilateral we identify L_1 and L_2 by $\left\{\begin{array}{l} 12 \longleftrightarrow 21 \\ 13 \longleftrightarrow 24 \\ 14 \longleftrightarrow 23 \end{array}\right\}$ and similarly for L_3 and L_4. In more detail, to identify two $\mathbb{P}^1$'s we need to say how three points on each are identified. Setting $ij = L_i \cap L_j$ the identification of L_1 with L_2 is described by the procedure in the brackets. The construction is illustrated by the picture

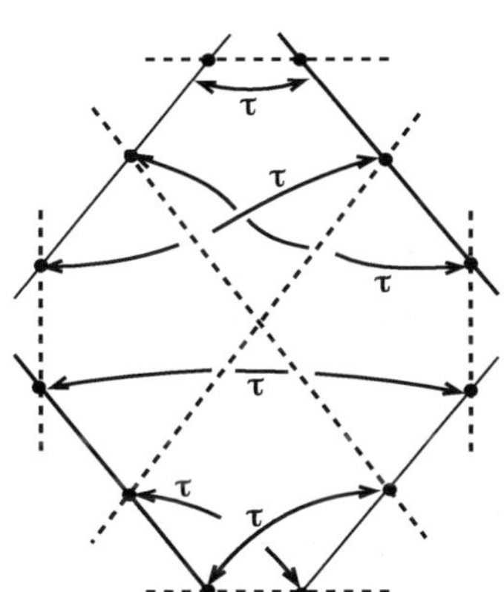

Here the dotted lines are the blowups of the intersection points of the original four lines. On this blowup $\mathbb{P}^2$ the involution τ is well defined, and the cycles of $\mathbb{P}^1$'s are obtained from those in the picture. We will return to this example later.

B. Hodge theory: Traditionally there have been two principal ways in which Hodge theory interacts with algebraic geometry:

- topology; as previously noted many of the deeper aspects of the topology of an algebraic variety X are proved via Hodge theory;[26]

[26]This was true initially when X is smooth. Using mixed Hodge theory it is now the case when X is arbitrary (singular, non-complete or both), and as will be discussed below it is also the case when we have a degeneration $X_t \to X$ leading for example to a proof of the above monodromy theorem and definition of the definition of $\lim_{t\to 0} H^n(X_t)$. There is also a very rich and beautiful Hodge theory associated to isolated hypersurface singularities.

- geometry; the Hodge structure on cohomology and its 1st order variations have been used to study the geometry of an algebraic variety X, especially the algebraic cycles that lie in X and in varieties constructed from X.

A central point of these notes is to possibly add a third point to this list:

- it is now well understood how Hodge structures can degenerate to a *limiting mixed Hodge structure*; this can then be used to guide and complement the study of algebraic varieties acquiring singularities as occur in moduli.

> *What is meant by a Hodge structure (HS), a mixed Hodge structure (MHS) and a limiting mixed Hodge structure (LMHS)?*

Traditionally a HS or a MHS was given by a period matrix

$$\left\| \int_{\Gamma_\alpha} \omega_i \right\|$$

where the ω_i are rational (meromorphic) differential forms on an algebraic variety X and the Γ_α are cycles (including relative ones).[27] When X is smooth and the ω_i are regular (holomorphic) n-forms this gives a holomorphic part

$$H^0(\Omega^n_X) = H^{n,0}(X) \subset H^n_{\mathrm{DR}}(X) \cong H^n(X, \mathbb{C})$$

of the cohomology of X. As noted above, when $n = \dim X = 1, 2$ using conjugation and the cup product in cohomology the holomorphic part $H^0(\Omega^n_X)$ determines the *Hodge decomposition*

$$\begin{cases} H^n(X, \mathbb{C}) = \underset{p+q=n}{\oplus} H^{p,q}(X) \\ \overline{H^{p,q}(X)} = H^{q,p}(X) \end{cases}$$

on cohomology, where using the isomorphism given by de Rham's theorem

$$H^{p,q}(X) = \left\{ \begin{array}{l} \text{cohomology classes represented by } C^\infty \\ \text{differential forms of type } (p,q) \end{array} \right\}$$

One defines a *Hodge structure of weight n* $(V, F^\bullet)$ to be given by a $\mathbb{Q}$-vector space V and a decreasing *Hodge filtration*

$$F^n \subset F^{n-1} \subset \cdots \subset F^0 = V_{\mathbb{C}}$$

that satisfies

$$F^p \oplus \overline{F}^{n-p+1} \xrightarrow{\sim} V_{\mathbb{C}}$$

We note that much of the use of Hodge theory in birational geometry centers around the implications (primarily to vanishing theorems) of the surjectivity of the natural map $H^1(X, \mathbb{C}) \twoheadrightarrow H^1(\mathcal{O}_X)$. There is a natural splitting of this map. Also, of course the mere existence of functorial mixed Hodge structures and the strictness property of morphisms between them is of frequent use (cf. [**KK10**]). The actual geometry arising from Hodge structures together with their extensions and 1st order variations has thus far not played a particularly significant role.

[27] Classically (Riemann, Picard, Lefschetez,...) there were differentials of 1st 2nd and 3rd kinds. It is now understood that the first kind deals with the holomorphic part of the Hodge theory of smooth varieties, the second kind with the full cohomology of smooth varieties and the third with the mixed Hodge theory of singular varieties.

for $0 \leqq p \leqq n$.[28] The relations

$$\begin{cases} F^p = \underset{p' \geqq p}{\oplus} V^{p',q} \\ V^{p,q} = F^p \cap \overline{F}^q \end{cases}$$

give a 1-1 correspondence between Hodge filtrations and Hodge decompositions

$$\begin{cases} V_{\mathbb{C}} = \underset{p+q=n}{\oplus} V^{p,q} \\ \overline{V}^{p,q} = V^{q,p}. \end{cases}$$

The reason for using Hodge filtrations is that the $F^p(X)$ vary holomorphically with X. In practice there will also be a lattice $V_{\mathbb{Z}} \subset V$ that represents integral cohomology.

When X is of dimension n the cup products on cohomology and relations extending

$$\begin{cases} \int_X \omega \wedge \omega' = 0 & \text{(because } \omega \wedge \omega' = 0) \\ c_n \int_X \omega \wedge \overline{\omega} > 0 & \text{(because } c_n \omega \wedge \overline{\omega} > 0)^{29} \end{cases}$$

for holomorphic n-forms lead to the definition of a *polarized Hodge structure* $(V, Q, F^\bullet)$ where

$$Q : V \otimes V \to \mathbb{Q}, \qquad Q(u,v) = (-1)^n Q(v,u)$$

and the following two *Hodge-Riemann bilinear relations* are satisfied:

(I) $Q(F^p, F^{n-p+1}) = 0$;
(II) $i^{p-q} Q(V^{p,q}, \overline{V}^{p,q}) > 0$.

A *mixed Hodge structure* is given by $(V, W_\bullet, F^\bullet)$ where the increasing weight filtration

$$W_0 \subset W_1 \subset \cdots \subset W_m$$

is defined over $\mathbb{Q}$, and where the Hodge filtration $F^\bullet$ induces on the graded quotients

$$\operatorname{Gr}_n^W V = W_n(V)/W_{n-1}(V)$$

a Hodge structure of weight n. Here $F^p \operatorname{Gr}_n^W V = F^p \cap W_n(V)/W_{n-1}(V)$.

Mixed Hodge structures have wonderful linear algebra properties. They form an abelian category,[30] and any morphism $\varphi : (V, W, F) \to (V', W', V')$ is *strict* in the sense that $\varphi(V) \cap W'_k = \varphi(W_k)$ and $\varphi(V) \cap F'^p = \varphi(F^p)$.

The basic results connecting Hodge theory to the cohomology of algebraic varieties are

- for X smooth and complete, $H^n(X, \mathbb{Q})$ has a Hodge structure of weight n (Hodge).

As noted above, for $m = n = 1$ the HS is determined by the period matrix

$$\Omega = \left\| \int_{\gamma_i} \omega_\alpha \right\| \begin{array}{l} \diagup \quad \omega_\alpha \in H^0(\Omega_X^1) \quad (\dim = g) \\ \\ \diagdown \quad \gamma_i \in H_1(X, \mathbb{Z}) \quad (\cong \mathbb{Z}^{2g}) \end{array}$$

---

[28]One may think of F^p as represented by differential forms of degree n having in holomorphic local coordinates $z_1, \dots, z_n$ at least p dz_i's.

[30]Polarized Hodge structures constitute a semi-simple category. Although many Hodge structures occurring in geometry have no natural polarization, the proofs of the deeper results in Hodge theory and its applications to topology require the Hodge structures to be polarizable.

For $m = n = 2$ the HS on $H^2(X)$ is determined by $H^0(\Omega_X^2) = F^1$ by $F^2 = F^{1\perp}$; as in the curve case $H^0(\Omega_X^2)$ is given by the period matrix for the holomorphic 2-forms. Thus for both curves and surfaces the PHS is determined by the classical period matrix.

For a general complete algebraic variety X, $H^m(X, \mathbb{Q})$ has a mixed Hodge structure where the weight filtration is $W_0 \subset \cdots \subset W_m$ (Deligne).

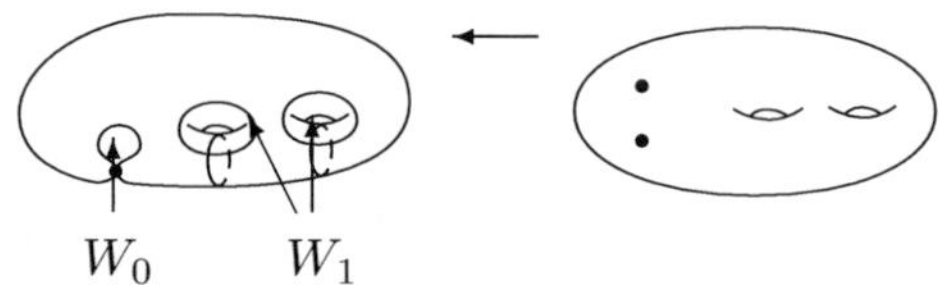

The picture is meant to suggest that MHS on $H^m(X, \mathbb{Q})$ is constructed from the pure HS's on the strata of a desingularization of X; this is (very non-trivially using homological algebra constructions — cf. [**CMSP17**] and [**PS08**]) indeed the case.

The use of Hodge theory to study a degenerating family $X_t \to X_0 = X$ of algebraic varieties leads to the notion of a very special type of mixed Hodge structures, namely that of a *limiting mixed Hodge structure* $(V, W(N), F^\bullet)$. Here $W(N)$ is the *monodromy weight filtration* constructed from the logarithm N of the unipotent part of monodromy. Assuming $N^{n+1} = 0$, $N^n \neq 0$, it is the unique filtration

$$W_0(N) \subset W_1(N) \subset \cdots \subset W_{2n}(N)$$

satisfying

$$\begin{cases} N : W_k(N) \to W_{k-2}(N) \\ N^k : W_{n+k}(N) \xrightarrow{\sim} W_{n-k}(N).^{31} \end{cases}$$

Then a LMHS is given by a MHS $(V, W(N), F^\bullet_{\lim})$ where

$$N : F^p_{\lim} \to F^{p-1}_{\lim}.$$

The associated graded Gr(LMHS)$\cong \overset{2m}{\underset{\ell=0}{\oplus}} H^\ell$ where H^ℓ is a HS of weight ℓ; the picture is a Hodge diamond. Here $m = 2$ and N is the vertical arrows; the dots are the $H^{p,q}$'s:

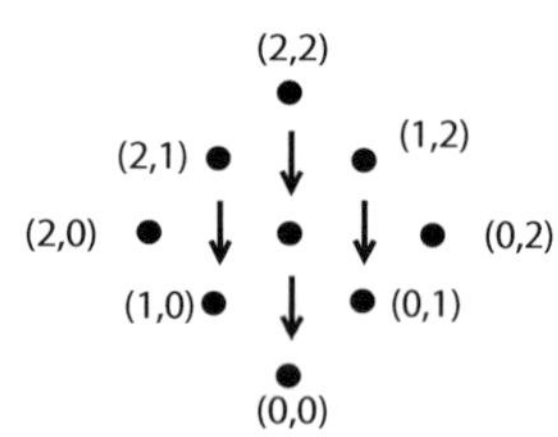

We will set $h^{p,q}$ = dimension of the (p, q) dot.

THEOREM (Schmid).[32] *Given $\mathfrak{X} \to \Delta$ as above*

$$\lim_{t \to 0} H^m(X_t) = \text{LMHS}.$$

The proof is a combination of

- Lie theory

[32]Cf. [**Sch73**] and [**CKS86**] in the references. An algebraic approach may be found in in [**PS08**].

- complex analysis
- differential geometry

In simplest terms, the period matrices of a degenerating family of varieties have entries that are polynomials in $\log t$ and with holomorphic functions as coefficients. Monodromy is given by analytic continuation of $\log t$ around $t = 0$. The above ingredients, especially Lie theory, are then used to give a precise description of the asymptotic behavior of the period matrix.

The following is a typical picture one has in mind of a family of varieties X_t parametrized by the disc Δ and which are smooth for $t \in \Delta^* = \{t \in \Delta : t \neq 0\}$ while X_0 has acquired singularities.

$$\begin{array}{ccc} \mathcal{X}^* & \subset & \mathcal{X} \\ \downarrow & & \downarrow \\ \Delta^* & \subset & \Delta \end{array}$$

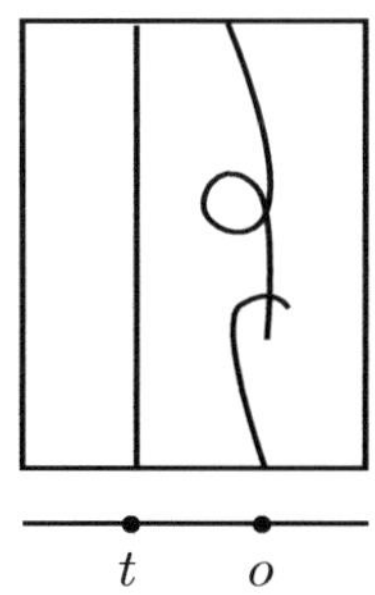

The following picture represents a family of genus 2 curves acquiring a node:

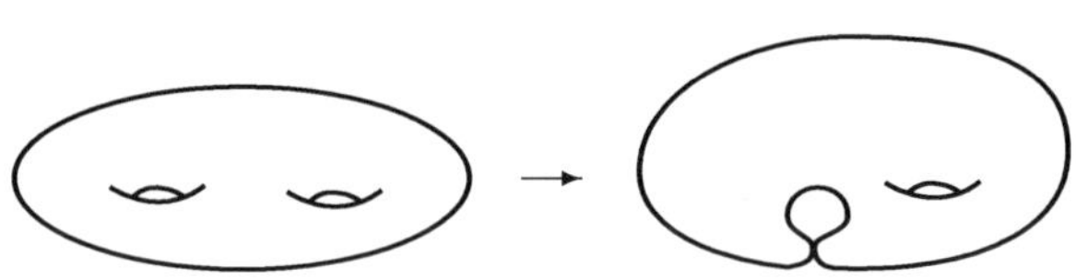

X_t X_0

Associated to a general family of degenerating smooth varieties is a monodromy operator $T : H^n(X_t) \to H^n(X_t)$

$$\begin{cases} T = T_{ss}T_u \quad \text{(Jordan decomposition)} \\ T_{ss}^k = I, T_u = e^N \ \text{ with } \ N^{n+1} = 0. \end{cases}$$

It is the basic topological invariant associated to the family $\mathcal{X}^* \to \Delta^*$.

For the above degeneration of a genus 2 curve the LMHS is

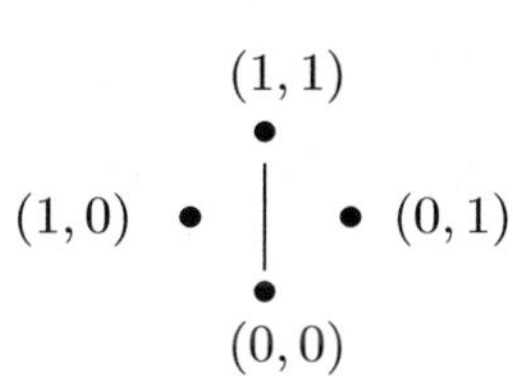

The solid vertical line represents the action of N on the associated graded to the LMHS (cf. Chapter 4 in [**CMSP17**]).

Referring to the picture of $\overline{\mathcal{M}}_2$ in the introduction, the solid lines in the diagram represent degenerations with $N \neq 0$.

The following illustrates the type of pictures one has in mind for a LMHS:

- 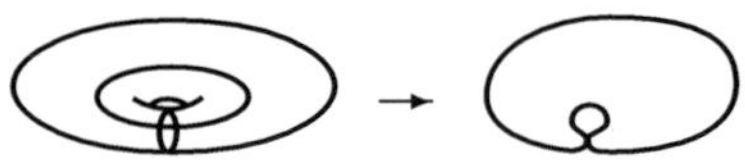 topological picture
- 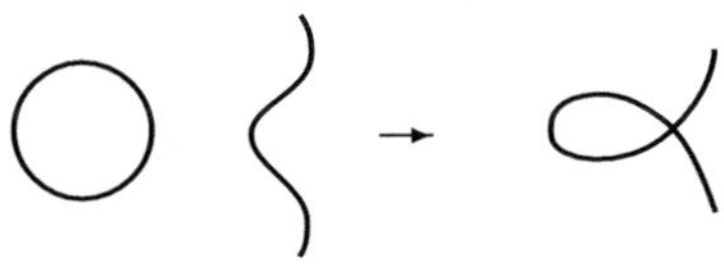 $y^2 = x(x-1)(x-t)$ algebraic picture
- $X = \mathbb{C}/\Lambda$, $\Lambda = \{1, \lambda\}$

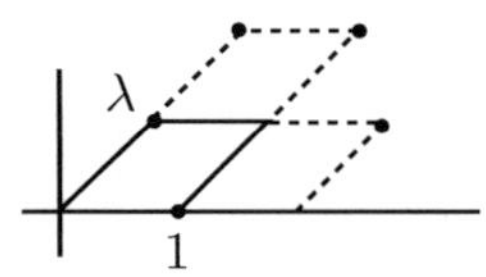

analytic or Hodge-theoretic picture

Here λ is determined up to $\lambda \to \frac{a\lambda+b}{c\lambda+d}$ where $\left(\begin{smallmatrix} a & b \\ c & d \end{smallmatrix}\right) \in \mathrm{SL}_2(\mathbb{Z})$ and $\mathcal{M}_1 \cong \mathrm{SL}(2,\mathbb{Z})\backslash\mathbb{H}$, $\mathbb{H} = \{\lambda : \operatorname{Im}\lambda > 0\}$

In this case $V = \left(\begin{smallmatrix} * \\ * \end{smallmatrix}\right)$, $Q = \left(\begin{smallmatrix} 0 & 1 \\ -1 & 0 \end{smallmatrix}\right)$, $F^1 = \left[\begin{smallmatrix} \lambda \\ 1 \end{smallmatrix}\right] \in \mathbb{P}^1$, HR II corresponds to $\operatorname{Im}\lambda > 0$ and the space of PHS's is $\mathbb{H} \subset \mathbb{P}^1$. The monodromy $T = \left(\begin{smallmatrix} 1 & 1 \\ 0 & 1 \end{smallmatrix}\right)$ is translation in the upper half plane, and as $\lambda \to i\infty$ we have $F^1 \to \left[\begin{smallmatrix} 1 \\ 0 \end{smallmatrix}\right] = F^1_{\lim}$.[33]

$$\lambda_t = \frac{\log t}{2\pi i}$$

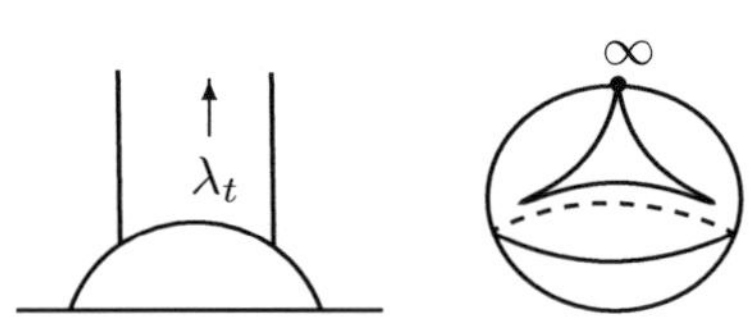

How does Lie theory enter? The actors in the story are

- *Period domain* $D = \{F^\bullet = \text{flag}\left\{F^n \subset \cdots \subset F^0 = V_{\mathbb{C}}\right\} \text{ in } V_{\mathbb{C}} : (V, Q, F^\bullet) = \text{PHS}\right\}$;
- *compact dual* $\check{D} = \left\{F^\bullet \text{ is a flag with } Q(F^p, F^{n-p+1}) = 0\right\}$;
- $G = \operatorname{Aut}(V, Q)$ is a $\mathbb{Q}$-algebraic group;

[33]This picture is not indicative of what happens in the non-classical case. The $F'_{\lim}$ will not in general lie in the boundary of the period domain in its compact dual (see below).

- $G_{\mathbb{R}}$ acts transitively on D and $G_{\mathbb{C}}$ acts transitively on $\check{D}$ so that we have

$$\begin{array}{l} D = G_{\mathbb{R}}/H \text{ with } H \text{ compact} \\ \cap \\ \check{D} = G_{\mathbb{C}}/P \text{ with } P \text{ parabolic } \begin{pmatrix} * & * & * & * \\ \circ & * & * & * \\ \circ & \circ & * & * \\ \circ & \circ & \circ & * \end{pmatrix} \end{array}$$

where D is an open $G_{\mathbb{R}}$-orbit in $\check{D}$.

Examples:

- $m = 1$: $D = \mathrm{Sp}(2g, \mathbb{R})/\mathcal{U}(g) = \mathcal{H}_g$ where $g = h^{1,0}$;
- $m = 2$: $D = \mathrm{SO}(2k, \ell)/\mathcal{U}(k) \times \mathrm{SO}(\ell)$ where $k = h^{2,0}$, $\ell = h^{1,1}$.

The classical case is when we have

$$\begin{array}{l} D = \text{ Hermitian symmetric domain (HSD)} \\ \parallel \\ G_R/K, \ K = \text{ maximal compact.} \end{array}$$

In algebraic geometry these arise as

$m = 1$ (curves, abelian varieties);
$m = 2$ is HSD $\iff$ $k = 1$ (K3's).[34]

Thus $h^{2,0} \geqq 2$ is non-classical. For $n \geqq 3$ and X Calabi-Yau, the D corresponding to $H^n(X)$ is also non-classical.

Period domains have sub-domains corresponding to PHS's with additional structure; e.g.,

$$\begin{array}{ccc} D' & \subset & D \\ \parallel & & \\ \left\{ \begin{array}{c} \text{reducible PHS's} \\ \text{that are } \oplus\text{'s} \end{array} \right\} & & \end{array}$$

This is what the dotted lines represent in the diagram in the Section I.A for $\overline{\mathcal{M}}_2$. In general one has Mumford-Tate sub-domains of D, defined to be those PHS's with a given algebra of Hodge tensors.

Period mappings arise from holomorphic mappings

$$\Phi : B \to \left\{ \begin{array}{c} \text{equivalence} \\ \text{classes of} \\ \text{PHS's} \end{array} \right\} = \Gamma \backslash D$$

where B is a complex manifold and $\Gamma \subset G_{\mathbb{Z}}$ contains the monodromy group. One may think of B as the parameter space for a family of smooth algebraic varieties X_b, $b \in B$, whose cohomology groups can be identified with $H^n(X_{b_0})$ for a base point $b_0 \in B$ up to the action by monodromy of $\pi_1(B, b_0)$ on $H^n(X_{b_0})$.

Example: As noted above the first non-classical case is weight $n = 2$ when $h^{2,0} = 2$. In this case D has an invariant contact structure and the image of any period mapping Φ is an integral variety of that structure. This means that if the contact structure is given by a 1-form θ, which up to scaling is invariant by $G_{\mathbb{R}}$, then

$$\Phi^*(\theta) = \Phi^*(d\theta) = 0.$$

[34]In this case D is a type IV HSD; it may be equivariantly embedded on $\mathcal{H}_g$. Also the unit ball may be equivariantly in $\mathcal{H}_g$. Thus the classical case should probably be defined as referring to algebraic varieties whose PHS's lie in a Mumford-Tate sub-domain of $\mathcal{H}_g$.

In general the differential constraint satisfied by period mappings in the non-classical case is the basic new phenomenon that occurs. Thus $\Phi(\mathcal{M})$ cannot contain an open set in D, Γ need not be arithmetic, etc.

Note: For families of algebraic surfaces with $p_g = 2$ we think of the period mapping Φ as given by a holomorphically varying $2 \times k$ matrix Ω satisfying HRI expressed by $\Omega Q {}^t\Omega = 0$. Differentiating this relation gives that the 2×2 matrix $d\Omega Q {}^t\Omega = -{}^t(d\Omega Q {}^t\Omega)$ is skew symmetric; writing

$$d\Omega Q {}^t\Omega = \begin{pmatrix} 0 & \theta \\ -\theta & 0 \end{pmatrix}$$

the 1-form θ gives the contact structure.

Using Lie theory the set of equivalence classes of LMHS's has been classified [**Ker15**]; they form a stratified object. As noted above one may informally say that we know how Hodge structures degenerate; the strategy is then to use this information to help understand how algebraic varieties degenerate.

Examples: For $n = 1$ the stratification may be pictured as

$$I_0 - I_1 - \cdots - I_g \; .$$

reflecting (for $g = 2$)

For $n = 2$ with $h^{2,0} = 1$ the picture is

$$\mathrm{I} \;\text{———}\; \mathrm{II} \;\text{———}\; \mathrm{III}$$

with corresponding Hodge diamonds

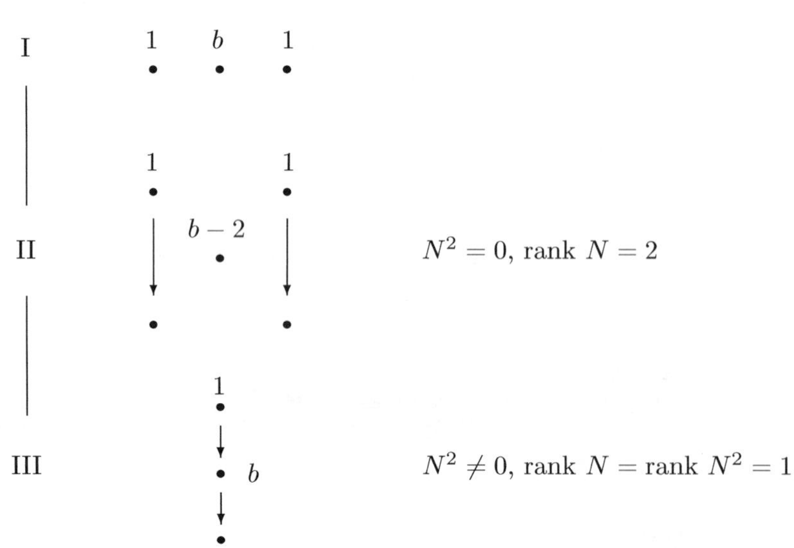

In the case of moduli of K3's, case I corresponds to smooth polarized K3 surfaces.[35] In case II, Hodge theory suggests is that the surface acquires an elliptic double curve. Case III is the case when the LMHS is of Hodge-Tate type. Dating from the work of Kulikov in the 70's, the moduli of K3's is a much studied and very beautiful story (cf. [**Fri84**] and [**Laz16**]).

For $n = 2$ the picture is

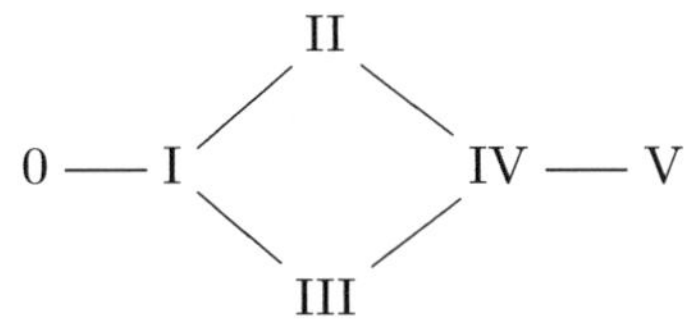

The detailed picture with the ranks of N and N^2 filled in when $h^{2,0} = 2$ will be given below. In all of these examples the Roman numerals reflect the associated graded to the equivalence classes of LMHS's. The stratification is linear and transitive in the classical case, transitive but not linear in the non-classical $n = 2$ case, and neither transitive nor linear in the general $n \geqq 3$ case.

Within each of the above strata there is a refined stratification given by PHS's with "additional" Hodge tensors (Mumford-Tate sub-domains).

We emphasize that the Hodge-theoretic stratification has the following ingredients:

(a) the equivalence classes of LMHS's over $\mathbb{Q}$;
(b) the spectra (basically the eigenvalues) of the semi-simple parts of monodromy;
(c) for each equivalence class of LMHS's over $\mathbb{Q}$, within the PHS's given by the associated graded there is a stratification by Mumford-Tate sub-domains.

For the I-surface example we will see that the part of the stratification of $\overline{\mathcal{M}}_I$ that has been determined is faithfully captured by using all of the Hodge-theoretic ingredients (i), (ii), (iii) above.

Example: Curves with $N = 0$. Using the proper Mumford-Tate sub-domain given by PHS's that are non-trivial direct sums over $\mathbb{Z}$ one may Hodge-theoretically detect the degeneration

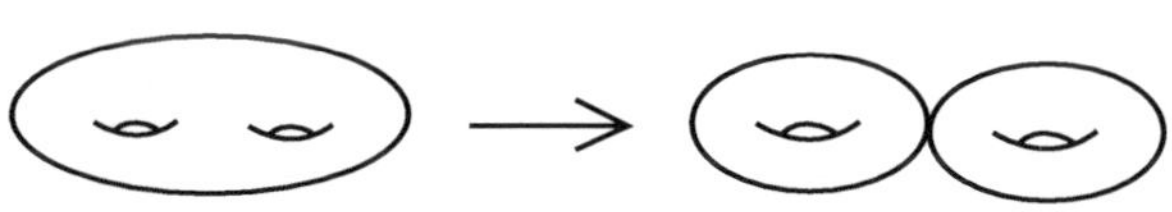

which has trivial monodromy. In general, in the stratification of $\overline{\mathcal{M}}_2$ pictured in the first section of these notes the solid lines refer to degenerations where $N \neq 0$ and the dotted lines to degenerations where the Jacobian of the normalization splits further into a direct sum of principally polarized abelian varieties.

Example: $n = 2$. At least in some examples one may Hodge theoretically detect a degeneration to a $\frac{1}{d}(1, a)$ singularity where $N = 0$. The first case is the Wahl singularity $\frac{1}{4}(1, 1)$ where $T = \text{Id}$; then for I-surfaces X with one such singularity there is an outline of an argument that the image in the period domain picks up an extra Hodge class. It is known that having this singularity defines a non-empty boundary divisor in moduli, and then the tentative result is that where the image

[35] Suitably interpreted it also includes the case where X is nodal.

of the period mapping meets the Mumford-Tate domain corresponding to the extra Hodge class will define that boundary divisor. This will be further discussed below.

III.A. Generalities on Hodge theory and moduli. The first point is that there is a moduli space

$$\mathcal{H} = \Gamma\backslash D$$

for Γ-equivalence classes of PHS's (think of $D = \mathbb{H}$ and $\Gamma \subset \mathrm{SL}_2(\mathbb{Z})$).

The second point is that there is a *period mapping*

$$(*) \qquad \Phi : \mathcal{M} \to \mathcal{P} \subset \Gamma\backslash D$$

where Γ contains the *global monodromy group* given by the image of the monodromy representation

$$\rho : \pi_1(\mathcal{M}) \to G_{\mathbb{Z}}.$$

There are several important technicalities here, some dealing with the singularities of $\mathcal{M}$ and some with the presence of X's with extra automorphisms. And of course the general X may not be smooth but rather will have canonical singularities. In the example of the I-surface to be discussed next these issues can be addressed directly.

The following are statements that have been proved at the set-theoretic level and full results established under various assumptions; complete proofs are a work in progress (cf. [[**Gri69**]] for a discussion of this).

THEOREM A. *The image $\mathcal{P} = \Phi(\mathcal{M}) \subset \Gamma\backslash D$ is a quasi-projective variety that has a canonical projective completion $\overline{\mathcal{P}}$.*[36] *Set-theoretically, $\overline{\mathcal{P}}$ is obtained from $\mathcal{P}$ by attaching the associated graded to the limiting mixed Hodge structures arising from Φ in $(*)$.*

We shall call $\overline{\mathcal{P}}$ the *Satake-Baily-Borel* (SBB) *completion of* $\mathcal{P}$. In the classical case using the Borel extension theorem the $\overline{\mathcal{P}}$ is induced from the classical Satake-Baily-Borel compactification of arithmetic quotients of HSD's.

In general $\overline{\mathcal{P}}$ is the minimal natural Hodge theoretic completion of $\mathcal{P}$. One may think of it as throwing out the extension data in the LMHS. For Γ arithmetic and with the assumption of the existence of a fan, Kato-Usui in [**KU09**] have constructed a universal maximal completion of $\Gamma\backslash D$, one in which the extension data is included. It may be thought of as a Hodge-theoretic toroidal compactification of $\Gamma\backslash D$.

Note: The proof of the above theorem (if completed) will have the following algebro-geometric implication: Let

$$Y \xrightarrow{f} Z$$

be a morphism of smooth, projective varieties and assume that the relative dualizing sheaf $\omega_{Y/Z}$ is a line bundle. Then

$$(**) \qquad \Lambda =: \det f_*(\omega_{Z/Y}) \text{ is } \textit{semi-ample.}$$

It is known that Λ is nef, and if local Torelli holds for a general point of Z, then Λ is big [[**Gri69**]]. The freeness seems more subtle (witness the abundance conjecture).

[36]Cf. [BBT] and [BK] for an interesting "model-theoretic" proof of the result that when Γ is arithmetic $\mathcal{P}$ is quasi-projective.

One may ask: Once you know that Λ is big and nef, why don't the standard methods of birational geometry (the minimal-model-program, including the base-point-free theorem) apply to give a proof? The interesting answer is that the signs needed in the base-point-free theorem are opposite to those that occur in the above situation.

The connection of this statement with the above theorem is that if

$$\overline{\mathcal{X}} \xrightarrow{f} \overline{\mathcal{M}}$$

is a versal family of general type varieties, then

$$\overline{\mathcal{P}} = \mathrm{Proj}(\Lambda).$$

Thus assuming $(**)$ one may define the SBB completion of the image of the period mapping without using any Hodge theory.[37]

The second work-in-progress result is

THEOREM B. *The period mapping* Φ *extends to*

$$(**) \qquad \Phi_e : \overline{\mathcal{M}} \to \overline{\mathcal{P}}.$$

The above two structural statements provide a conceptual framework for the use of Hodge theory to partner with and help guide the standard algebro-geometric methods used to study the boundary structure for the KSBA moduli spaces for surfaces of general type.[38] How this works will now be illustrated.

III.B. *I*-surfaces and their period mappings. Murphy's law (Vakil): Whatever nasty property a scheme can have already occurs for the moduli spaces of general type surfaces. Thus unlike curves one should select "special" surfaces to study. In geometry extremal cases are frequently interesting; Noether's inequality

$$p_g(X) \leqq \frac{K_X^2}{2} + 2$$

suggests studying surfaces close to extremal. The 1st non-classical case is given by the

Definition: An *I-surface* X is a regular $(q(X) = 0)$ general type surface that satisfies

$$p_g(X) = 2, K_X^2 = 1.$$

Here we are assuming that X is either smooth or has canonical (du Val) singularities. The KSBA moduli space for these surfaces will be denoted by $\mathcal{M}_I$.

One studies general type surfaces via their pluri-canonical maps

$$(\natural) \qquad \varphi_{mK_X} : X \dashrightarrow \mathbb{P}H^0(mK_X)^* \cong \mathbb{P}^{P_m - 1}$$

and pluricanonical rings $R(X) = \oplus H^0(mK_X)$.

Instead of $(\natural)$ it is frequently better to use weighted projective spaces corresponding to when we add new generators to $R(X)$. From

$$P_m(X) = m(m-1)/2 + 3, \qquad m \geqq 2$$

[37]Of course even if $(**)$ is proved algebro-geometrically, just taking the Proj of $\det(f_* \omega_{X/Y})$ does not seem to even roughly suggest what singularities are added on the boundary.

[38]It is not necessary to have proofs in order to use the statements of these results to help to guide how one may understand moduli.

and Kodaira-Kawamata-Viehweg vanishing (cf. [**Dem12**]) one has for the I-surface

$$\varphi_{K_X} : X \dashrightarrow \mathbb{P}^1, \quad |K_X| = \text{ pencil of hyperelliptic curves}$$
$$\varphi_{2K_X} : X \to \mathbb{P}(1,1,2) \hookrightarrow \mathbb{P}^3 \text{ of degree 2;}$$
$$\varphi_{5K_X} : X \hookrightarrow \mathbb{P}(1,1,2,5) \hookrightarrow \mathbb{P}^{12} \text{ an embedding.}$$

Here the homogenous coordinates of the mappings φ_{mK_X} are given by the generators of $R(X)$ in the indicated degrees. If $C \in |K_X|$ is a general smooth fibre, then by adjunction

$$2K_X\big|_C = K_C.$$

Thus the images $\varphi_{2K_X}(C) = \varphi_{K_C}(C)$ are canonical curves. The I-surface was important classically since φ_{4K_X} is not birational, while for any general type surface φ_{5K_X} always is birational.

For the I-surface we have the following equations and picture:

- 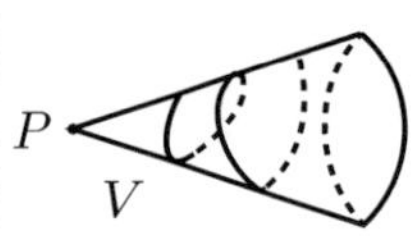

$\mathbb{P}(1,1,2) \hookrightarrow \mathbb{P}^3$given by $(t_0, t_1) \hookrightarrow [t_0^2, t_0t_1, t_1^2, y]$
$X =$ 2:1 map branched over P and $V \in |\mathcal{O}_{\mathbb{P}^3}(5)|$
- The canonical pencil $|K_X|$ has base point P and is given by the two sheeted coverings of the lines L through the vertex in the quadric and branched over P and $L \cap V$.
- The equation of X is $z^2 = F_5(t_0, t_1, y)z + F_{10}(t_0, t_1, y)$ (weighted hypersurface in $\mathbb{P}(1,1,2,5)$).

One may extend the definition to include surfaces X whose canonical Weil divisor class K_X is $\mathbb{Q}$-Cartier and which have semi-log-canonical singularities and which have the numerical properties of smooth I-surfaces (cf. [**Kol13a**]).

Example: At the other extreme to the smooth I-surfaces is the surface

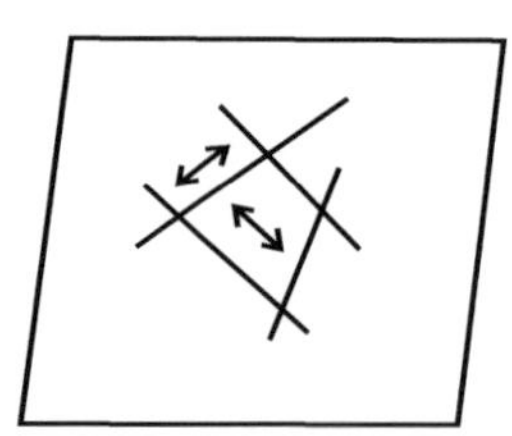

discussed above and which is obtained from $(\mathbb{P}^2, 4 \text{ lines})$ by identifying opposite pairs of lines. For this surface the equation is (cf. [**LR16**])

$$z^2 = y(t_0^2 - y)^2(t_1^2 - y)^2.$$

Geometrically it is a double cover of a quadric cone in $\mathbb{P}^3$ branched over the vertex, a plane section, and two double plane sections. A general curve in $|K_X|$ is

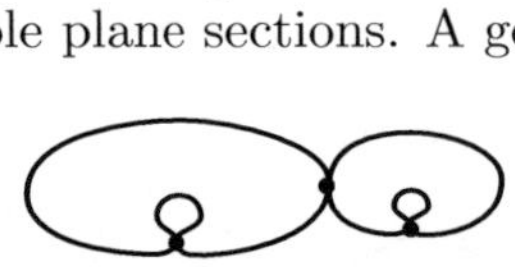

Concerning the moduli space $\mathcal{M}_I$ for smooth I-surfaces we have

- $\mathcal{M}_I$ is smooth and
  - $\dim \mathcal{M}_I = h^1(T_X) = 28$,
  - $\dim D_I = 57 = 2\dim \mathcal{M}_X + 1$,
- $\Phi : \mathcal{M}_I \to \Gamma_I \backslash D_I$ has injective differential Φ_* (local Torelli)

and this implies that

- $\Phi(\mathcal{M}_I)$ is a contact submanifold $\mathcal{P} \hookrightarrow \Gamma_I \backslash D_I$,
- Γ_I is arithmetic; it is not known is whether $\Gamma = G_{\mathbb{Z}}$ or not.

Elaborating a bit on a previous statement, in general, in the non-classical case there is a non-trivial homogeneous sub-bundle $E \subset TD$ such that any period mapping satisfies

$$\Phi_* : T\mathcal{M} \to E \subset T(\Gamma \backslash D);^{39}$$

thus the image can never be an open set in $\Gamma \backslash D$. Moreover, although it is always the case that

$$\operatorname{vol} \Phi(\mathcal{M}) < \infty,$$

It can happen that $\Gamma \subset G_{\mathbb{Z}}$ is a thin subgroup, i.e., a subgroup with $[\Gamma : G_{\mathbb{Z}}] = \infty$.

Stratification of the space of Gr(LMHS)'s: For curves with $\Gamma = \mathrm{Sp}(2g, \mathbb{Z})$ we have for LMHS's

$$\begin{array}{ccccccc} I_0 & \text{---} & I_1 & - \; - & I_2 & \text{---} \cdots \text{---} & I_g \\ \updownarrow & & \updownarrow & & \updownarrow & & \updownarrow \\ \mathcal{H}_g & & \mathcal{H}_{g-1} & & \mathcal{H}_{g-2} & & \mathcal{H}_0 . \end{array}$$

Note that I_{g-m} corresponds to $N : \mathrm{Gr}_2 \to \mathrm{Gr}_0$ with $N^2 = 0$, rank $N = m$.

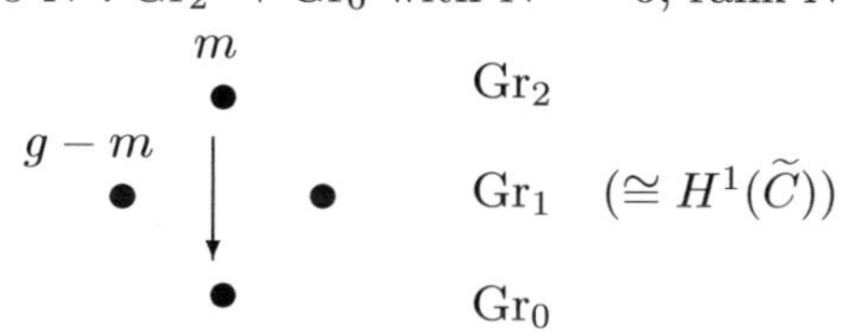

For each boundary component we have the stratification

$$H^1 = \oplus H^1_i .$$

The composite of these induces a stratification of $\overline{\mathcal{M}}_g$ by

$$\{\# \text{ nodes}, \; \# \text{ components}\}.$$

Of course this is just the beginning of the story of $\overline{\mathcal{M}}_g$.

For K3 surfaces there is an extensive literature dealing with the stratification of the image of the period mapping extended to include singular K3's for which $N = 0$ (Heegard divisor, etc.); cf. [**Laz16**] for an analysis of degree 2 K3's and further references. The stratification diagram for the possible LMHS's was given above.

For surfaces with $p_g = 2$ a refinement of the earlier picture by $N \neq 0$, $N^2 \neq 0$ of the classification of Gr(LMHS)'s/$\mathbb{Q}$ is

---

39 E is defined by the differential constraint

$$\overset{\bullet}{F}{}^p \subseteq F^{p-1}.$$

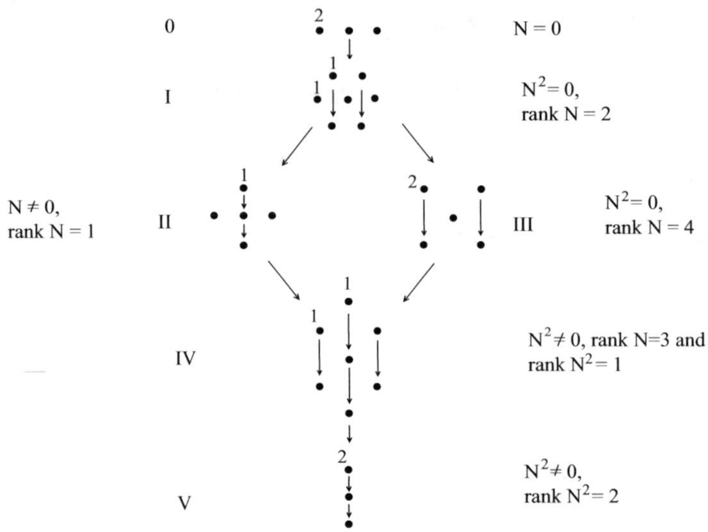

For the refined Hodge-theoretic stratification of Gr(LHHS/$\mathbb{Z}$)'s we use $T_{\mathrm{ss}} \to \{$conjugacy class $[T_{\mathrm{ss}}]$ of T_s in $\Gamma\}$. Within each of these strata we use Mumford-Tate sub-domains appearing in Gr(LMHS)'s in $\overline{\mathcal{M}}_I$.

We begin by considering the Gorenstein part $\overline{\mathcal{M}}_I^{\mathrm{Gor}} \subset \overline{\mathcal{M}}_I$. One reason for this is the general result

if $X_t \to X$ is a KSBA degeneration of a surface where all the singularities of X are isolated and non-Gorenstein, then $N = 0$.

Hence only Gorenstein singularities can non-trivially contribute to the LMHS/$\mathbb{Q}$.[40]

Heuristically the reason for this is the following.

- For the resolution of the singularity of a non-Gorenstein slc singularity one has a divisor $D = \sum E_i$ where the E_i are $\mathbb{P}^1$'s and the dual graph is a chain or perhaps a Dynkin-like diagram with forks; there are no cycles.
- For a KSBA degeneration $X_t \to X$ with $\widetilde{X} \to X$ a desingularization, and $\omega_t \in H^0(\Omega^2_{X_t})$, the limit $\lim_{t\to 0} \omega_t = \omega \in H^0(\Omega^2_{\widetilde{X}}(\log D))$ and then $\mathrm{Res}_D\, \omega$ gives a meromorphic 1-form on the E_i's with log poles on $E_i \cap E_j$ and thus $\mathrm{Res}_D\, \omega = 0$. It follows that $p_g(\widetilde{X}) = p_g(X)$, which then implies that $N = 0$.[41]

The following results from coupling the classification in [**FPR15a**, **FPR15b**, **FPR17**] with the analysis of the LMHS's in the various cases.

THEOREM B. *The Hodge theoretic stratification of $\overline{\mathcal{P}}$ given by the above diagram via the extended period mapping uniquely determines the stratification of $\overline{\mathcal{M}}_I^{\mathrm{Gor}}$. Moreover, any Hodge-theoretic degenerations that are possible as pictured in the table below are actually realized by Gorenstein degenerations of I-surfaces.*

---

[40] We recall that for a normal surface X, Gorenstein means that the canonical Weil divisor class K_X is a line bundle. In general the *index* is the least integer m such that mK_X is a line bundle. For the $\frac{1}{4}(1,1)$ singularity the index is 2. A central general question in moduli is to determine a *useful* bound on the index.

[41] This is a consequence of the Clemens-Schmid exact sequence; cf. [**CMSP17**] and [**PS08**].

Rather than display the whole table the following is just the part for simple elliptic singularities (types I_k and III_k). They have $N^2 = 0$ since for the semi-stable-reduction (SSR) of a degeneration only double curves (and no triple points) occur; all of the other types occur if we include cusp singularities.

In the following,

- X is irreducible (since K_X is a line bundle with $K_X^2 = 1$ and any component of X will have positive K_X^2)
- d_i = degree of elliptic singularity
- $k = \#$ elliptic singularities — in general, as previously noted using Hodge theory one may show that $k \leqq p_g + 1$
- $\widetilde{X}$ = minimal desingularization of X — in a SSR given by $\widetilde{\mathcal{X}} \to \widetilde{\Delta}$ the surface $\widetilde{X}$ will appear as one component of the fibre over the origin.

In the following table, in the 1st column subscripts denote the degrees of the elliptic singularities, which one can show are uniquely determined by the $[T_{\mathrm{ss}}]$'s.[42] We will explain the $\sum(9 - d_i)$ column below in the appendix.

stratum	dimension	minimal resolution $\widetilde{X}$	$\sum_{i=1}^{k}(9 - d_i)$	k	codim in $\overline{\mathcal{M}}_I$
I_0	28	canonical singularities	0	0	0
I_2	20	blow up of a K3-surface	7	1	8
I_1	19	minimal elliptic surface with $\chi(\widetilde{X})$=2	8	1	9
$\mathrm{III}_{2,2}$	12	rational surface	14	2	16
$\mathrm{III}_{1,2}$	11	rational surface	15	2	17
$\mathrm{III}_{1,1,R}$	10	rational surface	16	2	18
$\mathrm{III}_{1,1,E}$	10	blow up of an Enriques surface	16	2	18
$\mathrm{III}_{1,1,2}$	2	ruled surface with $\chi(\widetilde{X})$=0	23	3	26
$\mathrm{III}_{1,1,1}$	1	ruled surface with $\chi(\widetilde{X})$=0	24	3	27

Note that the last column is the sum of the two columns preceding it.

Appendix: We will will use Hodge theory as a guide to study the boundary component

$$\overline{\mathcal{M}}_2^{\mathrm{Gor}} \subset \overline{\mathcal{M}}_I$$

whose general point is a normal I-surface X that arises from a KSBA degeneration

$$(*) \qquad \mathcal{X} \to \Delta$$

whose monodromy logarithm N satisfies

$$\begin{cases} N^2 = 0 \\ \operatorname{rank} N = 2 \end{cases}$$

[42]The T_{ss} contains the Coxeter element of the Dynkin diagram. By inspecting the table of Coxeter elements for simple elliptic singularities with indices 1 and 2 one may check this statement.

and where the degree of the specialization $(*)$ is 2 (we will explain below how to define the degree of the specialization). As above, $\overline{\mathcal{M}}^{\mathrm{Gor}}$ denotes the part of $\overline{\mathcal{M}}$ parametrizing surfaces with only Gorenstein singularities.

With these assumptions we will see that X is irreducible with a single normal singular point p. Hodge theory then gives that p is a simple elliptic singularity whose degree is in this case defined to be degree of the specialization. In this situation one has the following diagram:

$$(**) \qquad \begin{array}{ccc} & (\widetilde{X}, \widetilde{C}) & \\ {}^{g}\swarrow & & \searrow^{f} \\ (X_{\min}, C) & & (X, p) \end{array}$$

where

- $(\widetilde{X}, \widetilde{C})$ is the minimal resolution of (X, p) where $\widetilde{C}$ is a smooth curve that contracts to p under the mapping f and that has self-intersection -2. To say that $\widetilde{X}$ is minimal means that there are no (-1)-curves in $\widetilde{X}$ that do not meet $\widetilde{C}$;
- $X_{\min}$ is the minimal model of $\widetilde{X}$ (all (-1)-curves in $\widetilde{X}$ have been contracted) and $C = g(\widetilde{C})$.

We will use Hodge theory as a guide to show that

(a) $X_{\min}$ is a K3 surface with a degree 2 polarization. Thus $X_{\min} \to \mathbb{P}^2$ is a 2:1 covering branched over a smooth sextic curve $B \in |\mathcal{O}_{\mathbb{P}^2}(6)|$, and $C \subset X_{\min}$ is the inverse image of a tangent line to B. Thus C is irreducible with a single node; the arithmetic genus $p_a(C) = 2$ and the normalization is $\widetilde{C} \to C$ where $\widetilde{C}$ is a smooth elliptic curve.

(b) Any such configuration $(X_{\min}, C)$ gives rise to a diagram $(**)$. From this we will infer that
  - $\dim \overline{\mathcal{M}}_2 = 20$;
  - a suitably interpreted version of local Torelli holds for the extended period mapping $\Phi_e : \overline{\mathcal{M}}_2 \to \overline{\mathcal{P}}$.

(c) A semi-stable resolution of the KSBA degeneration may be obtained by smoothing a normal crossing surface $\widetilde{X} \cup_{\widetilde{C}} Y$ where Y is a degree 2 del Pezzo surface. For a fixed X the number of parameters in such surfaces $\widetilde{X} \cup_{\widetilde{C}} Y$ is 7.[43] These parameters correspond to the extension data in the limiting mixed Hodge structure associated to the KSBA degeneration. Inserting these amounts to blowing up $\overline{\mathcal{M}}_2$ in $\overline{\mathcal{M}}_I$ and leads to a desingularization of $\overline{\mathcal{M}}_I$ along $\overline{\mathcal{M}}_2$.

We will also see that the arguments used in the above lead to a proof of generic local Torelli for the period mapping

$$\Phi : \mathcal{M}_I \to \mathcal{P} \subset \Gamma \backslash D.$$

As mentioned above, using other methods one may also prove that local Torelli holds everywhere for this period mapping; a result that was independently obtained by Carlson-Toledo and by Pearlstein-Zhang. It seems quite possible that the methods

[43] This is the "7" that appears in the $\sum_{i=1}^{k}(9 - d_i)$ column in the above table.

of Friedman (cf. [**Fri84**]) may be used to prove generic global Torelli, meaning that the period mapping $\Phi : \mathcal{M}_I \to \mathcal{P}$ has degree one.

Conclusion: *Using Hodge theory as a guide one may determine the structure of the boundary component* $\overline{\mathcal{M}}_2$ *of* $\overline{\mathcal{M}}_I$ *as well as a desingularization of* $\overline{\mathcal{M}}_I$ *along* $\overline{\mathcal{M}}_2$.

Using the results of [**FPR15a, FPR15b, FPR17**], together with them we are working to extend the above conclusion to the other boundary components of $\overline{\mathcal{M}}_I^{\mathrm{Gor}}$ as listed in the above table. We will also discuss the non-Gorenstein components below.

Step one: The Hodge diamond for the LMHS associated to a semi-stable reduction $\widetilde{X} \to \Delta$ arising from a KSBA family with X as central fibre in

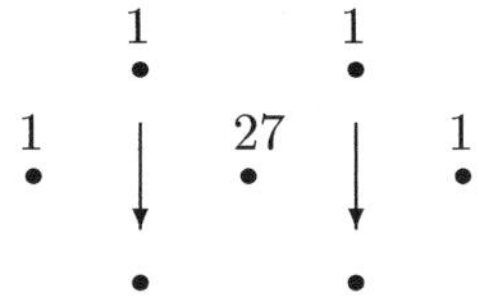

where the dimensions are written in above the dots. This suggests (does not prove) that the central fibre $\widetilde{X}_0$ of $\widetilde{\mathcal{X}} \to \Delta$ is a normal crossing surface of the form

$$\widetilde{X}_0 = \widetilde{X} \cup_{\widetilde{C}} Y$$

where

- $\widetilde{X}$ is a desingularization of X;
- $\widetilde{C}$ is a smooth double curve.

In this situation the LMHS is computed from the groups $H^p(\widetilde{C})$, $H^q(\widetilde{X})$ using Gysin and restriction mappings (cf. [**PS08**]). Assuming that $H^1(\widetilde{X}) = 0$ and $H^1(Y) = 0$ this gives

- $H^1 := \mathrm{Gr}^1(\mathrm{LMHS}) = H^1(\widetilde{C})$;
- $H^2 := \mathrm{Gr}^2(\mathrm{LMHS})$ is the middle cohomology of the complex

$$(\natural) \qquad H^0(\widetilde{C})(-1) \xrightarrow{\mathrm{Gy}} H^2(\widetilde{X}) \oplus H^2(Y) \xrightarrow{\mathrm{Re}} H^2(\widetilde{C})$$

where Gy is the direct sum of Gysin maps and Re is the signed restriction map.

Note: The condition [**Fri83**] that $\widetilde{X} \cup_{\widetilde{C}} Y$ be smoothable is

$$(\natural\natural) \qquad N_{\widetilde{C}/\widetilde{X}} \cong N^*_{\widetilde{C}/Y}.$$

Denoting by $\widetilde{C}_{\widetilde{X}}$ the curve $\widetilde{C}$ considered in $\widetilde{X}$ and similarly for $\widetilde{C}_Y$, $(\natural\natural)$ gives

$$\widetilde{C}^2_{\widetilde{X}} + \widetilde{C}^2_Y = 0,$$

which is exactly the condition that

$$\mathrm{Re} \circ \mathrm{Gy} = 0$$

in $(\natural)$ above.

Step two: $\widetilde{C}$ is a smooth elliptic curve and therefore can be realized as a smooth cubic in $\mathbb{P}^2$. In order to have $-\deg N_{\widetilde{C}/\widetilde{X}} = \deg N_{\widetilde{C}/Y} = 2$ in $(\natural\natural)$ to obtain Y we must blow up $\mathbb{P}^2$ in 7 points p_i and take the proper transform of the cubic curve to obtain

$$\widetilde{C} \subset Y = \mathrm{Bl}_{\{p_i\}}\mathbb{P}^2.$$[44]

There is still one parameter to adjust to have (♮♮) as an equality of line bundles.

From dimensions in the Hodge diamond and $\dim H^2(Y) = 8$ we obtain

$$\begin{cases} \dim H^{2,0}(\widetilde{X}) = 1 \\ \dim H^{1,1}(\widetilde{X}) = 21. \end{cases}$$

Now we use intersection numbers. From the assumption that $(\widetilde{X}, \widetilde{C}) \to (X, p)$ is a minimal resolution of an I-surface and adjunction we obtain

$$\begin{cases} (K_{\widetilde{X}} + \widetilde{C})^2 = 1 \\ K_{\widetilde{X}} \cdot \widetilde{C} + \widetilde{C}^2 = 0 \implies K_{\widetilde{X}} \cdot \widetilde{C} = 2. \end{cases}$$

From this we infer that $K^2_{\widetilde{X}} = -1$, and since $h^{2,0}(\widetilde{X}) = 1$ the line bundle $K_{\widetilde{X}}$ is $[E]$ for E a (-1)-curve in $\widetilde{X}$. This gives the diagram $(**)$ which may be pictured as

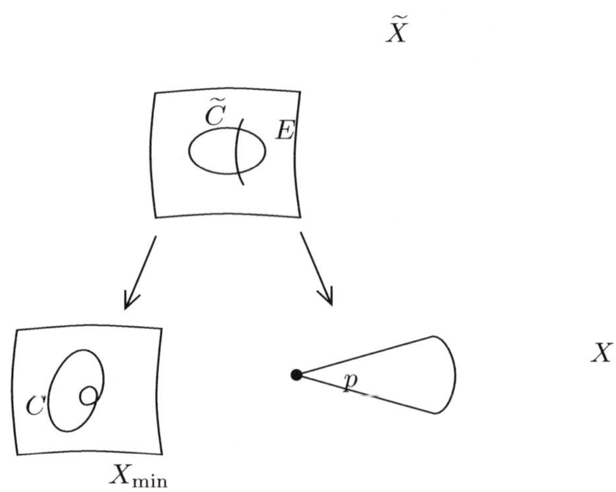

We then have

- $K_{X_{\min}} \cong \mathcal{O}_{X_{\min}}$ (since $K_{\widetilde{X}} = [E]$);
- $h^{1,1}(X_{\min}) = 20$;
- $C^2 = 2 \implies p_a(C) = 2,\ g(\widetilde{C}) = 1$.

Step three: We now observe that this construction is reversible. Given a K3 surface $X_{\min}$ with a degree 2 polarization and a curve C with $p_a(C) = 2$, $g(\widetilde{C}) = 1$ we have the situation described in (iii) above. We then construct $\widetilde{X}$ by blowing up the node on C to obtain $\widetilde{C}$ with $\widetilde{C}^2 = -2$. Contracting it then gives (X, p).

We next count parameters. First we have

$$\begin{cases} \dim X_{\min}\text{'s} = 19 \\ \dim C\text{'s} = 1 \end{cases}$$

[44]In [**Fri84**] type II degenerations are constructed by first gluing two $\mathbb{P}^2$'s together along a common cubic curve C. To obtain the condition (♮♮) one has to blow up 18 points on C. As shown in [**Fri84**] there are exactly four ways to group these points into two groups that specify in which $\mathbb{P}^2$ the blowups occur. It may be that in the situation being discussed here there is a unique such choice.

so that the boundary component $\overline{\mathcal{M}}_2$ has dimension 20. This gives the dimension count in (ii).

For the dimension count for the semi-stable surfaces $\widetilde{X} \cup_{\widetilde{C}} Y$ with a fixed (X, p) we have

$$+\begin{cases} \text{dimension of the spaces of } p_i\text{'s in } \mathbb{P}^2 = 14 \\ \text{dimension of cubics through the } p_i \quad = \underline{\ \ 2\ } \\ \hphantom{\text{dimension of cubics through the } p_i \quad = } 16 \end{cases}$$

and

$$-\begin{cases} \text{parameter to have } (\natural\natural) & 1 \\ \text{dimension of } \mathrm{Aut}(\mathbb{P}^2) & \underline{8} \\ & 9 \end{cases}$$

This gives

$$\left\{\begin{matrix} \text{dimension of the surfaces } \widetilde{X} \cup_{\widetilde{C}} Y \\ \text{with a fixed } (X_{\min}, C) \end{matrix}\right\} = 16 - 9 = 7 = 28 - 1.$$

Letting $(X_{\min}, C)$ now vary one may show that the space of $\widetilde{X} \cup_{\widetilde{C}} Y$'s forms a smooth variety of dimension 27 biregularly equivalent to the blowup of $\overline{\mathcal{M}}_I$ along the 20 dimensional $\overline{\mathcal{M}}_2$.

Step four: Finally how does Hodge theory enter via Torelli to relate to the above parameter counts? With the details to appear in a later work the rough idea is

- H^1 determines the elliptic curve $\widetilde{C}$;
- $H^2 = H^{2'} \oplus H^{2''}$ where $H^{2'}$ is the Hodge structure of a K3 with a degree 2 polarization, and $H^{2''}$ is a lattice of rank in $\mathrm{Hg}(H^2)$ of rank 7;
- the extension data in $\mathrm{Ext}^1_{\mathrm{MHS}}(H^{2''}, H^1)$ gives the set of 7 points p_i on $\widetilde{C}$.

There are a number of subtleties involving the quadratic forms on weight 2 PHS's which are a work in progress. We refer especially to [**Fri84**] where similar constructions are carried out in the case of K3's, and where a first instance of the use of extension data to provide parameters for boundary components in moduli of surfaces appeared. In summary, Hodge theory suggests where to look — the seven parameters arise from the possible extension data for GR(LMHS) — and following [**FPR15a, FPR15b, FPR17**] in the references), and on discusssions that the four of us have had with them related to a joint project that is in progress.

References

[Ale94] Valery Alexeev, *Boundedness and K^2 for log surfaces*, Internat. J. Math. **5** (1994), no. 6, 779–810, DOI 10.1142/S0129167X94000395. MR1298994

[CKS86] Eduardo Cattani, Aroldo Kaplan, and Wilfried Schmid, *Degeneration of Hodge structures*, Ann. of Math. (2) **123** (1986), no. 3, 457–535, DOI 10.2307/1971333. MR840721

[CMSP17] James Carlson, Stefan Müller-Stach, and Chris Peters, *Period mappings and period domains*, Cambridge Studies in Advanced Mathematics, vol. 168, Cambridge University Press, Cambridge, 2017. Second edition of [MR2012297]. MR3727160

[Dem12] Jean-Pierre Demailly, *Analytic methods in algebraic geometry*, Surveys of Modern Mathematics, vol. 1, International Press, Somerville, MA; Higher Education Press, Beijing, 2012. MR2978333

[Fri83] Robert Friedman, *Global smoothings of varieties with normal crossings*, Ann. of Math. (2) **118** (1983), no. 1, 75–114, DOI 10.2307/2006955. MR707162

[Fri84] Robert Friedman, *A new proof of the global Torelli theorem for K3 surfaces*, Ann. of Math. (2) **120** (1984), no. 2, 237–269, DOI 10.2307/2006942. MR763907

[Gri69] Phillip A. Griffiths, *Hermitian differential geometry, Chern classes, and positive vector bundles*, Global Analysis (Papers in Honor of K. Kodaira), Univ. Tokyo Press, Tokyo, 1969, pp. 185–251. MR0258070

[Ker15] C. Kerr, M. andRobles, *Partial orders and polarized relations on limited mixed Hodge structures*, 2015, in preparation.

[KK10] János Kollár and Sándor J. Kovács, *Log canonical singularities are Du Bois*, J. Amer. Math. Soc. **23** (2010), no. 3, 791–813, DOI 10.1090/S0894-0347-10-00663-6. MR2629988

[Kol13a] János Kollár, *Moduli of varieties of general type*, Handbook of moduli. Vol. II, Adv. Lect. Math. (ALM), vol. 25, Int. Press, Somerville, MA, 2013, pp. 131–157. MR3184176

[Kol13b] János Kollár, *Singularities of the minimal model program*, Cambridge Tracts in Mathematics, vol. 200, Cambridge University Press, Cambridge, 2013. With a collaboration of Sándor Kovács. MR3057950

[KSB88] J. Kollár and N. I. Shepherd-Barron, *Threefolds and deformations of surface singularities*, Invent. Math. **91** (1988), no. 2, 299–338, DOI 10.1007/BF01389370. MR922803

[KU09] Kazuya Kato and Sampei Usui, *Classifying spaces of degenerating polarized Hodge structures*, Annals of Mathematics Studies, vol. 169, Princeton University Press, Princeton, NJ, 2009. MR2465224

[Laz16] Radu Laza, *The KSBA compactification for the moduli space of degree two $K3$ pairs*, J. Eur. Math. Soc. (JEMS) **18** (2016), no. 2, 225–279, DOI 10.4171/JEMS/589. MR3459951

[LR16] Wenfei Liu and Sönke Rollenske, *Geography of Gorenstein stable log surfaces*, Trans. Amer. Math. Soc. **368** (2016), no. 4, 2563–2588, DOI 10.1090/tran/6404. MR3449249

[PS08] Chris A. M. Peters and Joseph H. M. Steenbrink, *Mixed Hodge structures*, Ergebnisse der Mathematik und ihrer Grenzgebiete. 3. Folge. A Series of Modern Surveys in Mathematics [Results in Mathematics and Related Areas. 3rd Series. A Series of Modern Surveys in Mathematics], vol. 52, Springer-Verlag, Berlin, 2008. MR2393625

[RU19] Julie Rana and Giancarlo Urzúa, *Optimal bounds for T-singularities in stable surfaces*, Adv. Math. **345** (2019), 814–844, DOI 10.1016/j.aim.2019.01.029. MR3901675

[Sch73] Wilfried Schmid, *Variation of Hodge structure: the singularities of the period mapping*, Invent. Math. **22** (1973), 211–319, DOI 10.1007/BF01389674. MR382272

[FPR15a] M. Franciosi, R. Pardini, and S. Rollenske, *Computing invariants of semi-log-canonical surfaces*, Math. Z. 280 no. 3-4 (2015), 1107–1123.

[FPR15b] M. Franciosi, R. Pardini and S. Rollenske, *Log-canonical pairs and Gorenstein stable surfaces with $K_X^2 = 1$*, Compos. Math. 151 (2015), no. 8, 1529–1542.

[FPR17] M. Franciosi, R. Pardini and S. Rollenske, *Gorenstein stable surfaces with $K_X^2 = 1$ and $p_g > 0$*, Math. Nachr. 290 (2017), no. 5-6, 794–814.

IMSA (Institute of the Mathematical Sciences of the Americas) at the University of Miami, and the Institute for Advanced Study

Email address: pg@ias.edu

Contemporary Mathematics
Volume **766**, 2021
https://doi.org/10.1090/conm/766/15382

Constructing Jacobian varieties with many elliptic curves

Rubén A. Hidalgo

ABSTRACT. Given two elliptic curves, E_1 and E_2, Earle provided an explicit genus two Riemann surface R_2 such that $JR_2 \cong_{isog.} E_1 \times E_2$. In this paper, given $s \geq 3$ elliptic curves $E_1, \ldots, E_s$, we construct an explicit closed Riemann surface R_s, of genus $g = 1 + 2^{s-2}(s-2)$, such that $JR_s \cong_{isog.} E_1 \times \cdots \times E_s \times A$, where A is also a product of at least $s(s-3)/2$ elliptic curves and jacobian varieties of some hyperelliptic Riemann surfaces, all of these curves explicitly given in terms of the given elliptic curves. In particular, for every triple E_1, E_2, E_3 of elliptic curves this provides an explicit Riemann surface R_3 of genus three with $JR_3 \cong_{isog.} E_1 \times E_2 \times E_3$.

1. Introduction

Let R be a closed Riemann surface of genus $g \geq 2$ and JR be its jacobian variety. As JR is a particular type of principally polarized abelian variety (p.p.a.v.), it follows from the Poincaré reducibility theorem, $JR \cong_{isog.} A_1^{n_1} \times \cdots \times A_s^{n_s}$ (the symbol $\cong_{isog.}$ stands for *isogenous*), where each A_j is a simple abelian variety and each n_j is a positive integer (the factors A_j and the integers n_j are uniquely determined up to isogeny and permutation). If $g \leq 4$, then it is possible to assume all these factors A_j to be (isogenous to) jacobian varieties. For $g \geq 5$, this is in general not possible (in fact, it might be that JR is a non-simple p.p.a.v., but not isogenous to a product of at least two jacobian varieties).

In recent years there has been an interest in constructing examples of closed Riemann surfaces whose jacobian varieties are isogenous to a product with many elliptic factors (see, for instance, [**7, 8, 12, 15, 18–24, 26**]). Such an interest come from different contexts, for instance, in cryptography [**1, 6**] and in arithmetics and number theory [**2, 10, 24**]. For many values of $g \leq 1297$, in [**9**], Ekedahl and Serre constructed examples of closed Riemann surfaces of genus g whose jacobian varieties are isogenous to a product of g elliptic curves (completely reduced). In [**22**] there has been constructed more such kind of examples for some of the missing genera. It is still not known if such kind of examples exist in every genus or even if they exist for infinitely many values of g. In this paper we are interested in the following (which is somehow related to the above) questions.

Q: *(1) Let $E_1, \ldots, E_r$, where $r \geq 2$, be elliptic curves. Find explicit equations (in terms of these elliptic curves) of a closed Riemann surface R of minimal genus*

2020 *Mathematics Subject Classification.* Primary 30F10, 30F20, 14H40, 14H37.

Key words and phrases. Riemann surface, Jacobian variety, isogeny.

Partially supported by Project Fondecyt 1190001.

$e(E_1,\dots,E_r)$ with $JR \cong_{isog.} E_1\times\cdots\times E_r\times A$, where A is also a product of jacobian varieties. (2) Determine the maximum $e(r)$ of all these minimal genera obtained by varying all the r elliptic curves.

Observe that $e(E_1,\dots,E_r)\geq r$ and equality works for Ekedahl-Serre's examples. We may think of $e(E_1,\dots,E_r)-r$ as the defect for $Z=E_1\times\cdots\times E_r\in\mathcal{H}_r$ (where $\mathcal{H}_r$ is the upper-half Siegel space) to be isogenous to a jacobian variety. The $\mathrm{Sp}(2r,\mathbb{Q})$-orbit of Z is dense and any point in this orbit is isogenous to Z. As the dimension of $\mathcal{H}_r$ is $r(r+1)/2$ and that of its jacobian sublocus is $3(r-1)$, it follows that, for $r\in\{2,3\}$, there is a dense set of jacobians being isogenous to Z. This, in particular, asserts that $e(E_1,E_2)=2$ and $e(E_1,E_2,E_3)=3$. (For every r this also asserts that the jacobian variety of every surface of genus r is a limit point of the $\mathrm{Sp}(2r,\mathbb{Q})$-orbit of Z, in particular, that $\mathrm{Sp}(2r,\mathbb{Q})$-translates of $\mathcal{H}_1^r$ provides a dense subset.) Given any two elliptic curves E_1, E_2, Earle [**7, 8**] (see also [**11–13, 18**]) described an explicit equation of a genus two Riemann surface R_2, in terms of the two given elliptic curves, such that $JR_2\cong_{isog.} E_1\times E_2$ (we recall this in Theorem 2 for completeness). In this paper, given three elliptic curves E_1,E_2,E_3, we describe explicitly a genus three closed Riemann surface R_3, in terms of the given elliptic curves, such that $JR_3\cong_{isog.} E_1\times E_2\times E_3$ (Theorem 4).

In Section 2 we recall a method due to Kani and Rosen [**16**] for finding isogenous decompositions of the jacobian varieties of closed Riemann surfaces admitting non-trivial groups of automorphisms. In Section 3 we apply such a method to two explicit constructions. In the first construction, given $s\geq 3$ elliptic curves $E_1,\dots E_s$, we obtain an explicit closed Riemann surface R_s, of genus $g=1+2^{s-2}(s-2)$, such that $JR_s\cong_{isog.} E_1\times\cdots\times E_s\times A$, where A is a product of at least $s(s-3)/2$ elliptic curves and some jacobian varieties of hyperelliptic curves, each of them explicitly in terms of the given s elliptic curves (Theorem 3). As a consequence, we obtain that: (i) $e(r)\leq 1+2^{(r-2)/2}r$, for $r\geq 4$ even, and (ii) $e(r)\leq 1+2^{(r-3)/2}(r-1)$, for $r\geq 5$ odd (Theorem 5). We conjecture these inequalities to be sharp, but unfortunately we do not have a reasonable reason for it. Our constructions also allows us to produce (i) an explicit 2-dimensional family of Riemann surfaces of genus nine whose jacobian varieties are isogenous to the product of nine elliptic curves (Example 5.2), (ii) a 2-dimensional family of Riemann surfaces of genus five and (iii) a 1-dimensional family of Riemann surfaces of genus thirteen, whose jacobian varieties are isogenous to the product of only elliptic curves (Examples 5.1 and 5.3).

2. Some preliminaries

2.1. Some notation and remarks. Let $\Delta_1=\mathbb{C}-\{0,1\}$ and, for $s\geq 2$, we set $\Delta_s=\{(\lambda_1,\dots,\lambda_s)\in\mathbb{C}^s:\lambda_j\in\Delta_1;\ \lambda_i\neq\lambda_j,\ i\neq j\}$. If $\lambda\in\Delta_1$, then we set the elliptic curve $E_\lambda: y^2=x(x-1)(x-\lambda)$. It is known that E_λ and E_μ are isomorphic if and only if there is some $T\in\mathbb{G}=\langle u(\lambda)=1/\lambda, V(\lambda)=1-\lambda\rangle\cong\mathfrak{S}_3$ so that $\mu=T(\lambda)$. It can be seen that given two elliptic curves E_1 and E_2, there is some $(\lambda_1,\lambda_2)\in\Delta_2$ such that E_j and E_{λ_j} are isomorphic for $j=1,2$. As for every $\lambda\in\Delta_1$ there exist infinitely many values $\mu\in\Delta_1$ (in fact a dense subset) so that E_μ and E_λ are isogenous, it follows that given $s\geq 3$ elliptic curves $E_1,\dots,E_s$, there are infinitely many tuples $(\lambda_1,\dots,\lambda_s)\in\Delta_s$ so that E_{λ_j} and E_j are isogenous for each $j=1,\dots,s$.

2.2. The jacobian variety of a closed Riemann surface. A *polarized abelian variety* of dimension $g\geq 1$ is a pair $A=(T,Q)$, where $T=\mathbb{C}^g/L$ is a complex torus of dimension g and Q (called a *polarization* of A) is a positive-definite Hermitian product in $\mathbb{C}^g$ such that $\operatorname{Im}(Q)$ has integral values over elements of the lattice L. There is basis of L for which $\operatorname{Im}(Q)$ can be represented by the matrix

$$\begin{pmatrix} 0 & D \\ -D & 0 \end{pmatrix}$$

where D is a diagonal matrix with diagonal entries given by non-negative integers $d_1, \ldots, d_g$, where d_j divides d_{j+1}. The tuple $(d_1, \ldots, d_g)$ is called the *polarization type.* When $d_1 = \cdots = d_g = 1$, the polarization is called *principal* and that the abelian variety is *principally polarized.* (Not every torus T can be given the structure of a polarized abelian variety; a necessary and sufficient condition is the existence of a holomorphic embedding of T into a complex projective space.) Two tori $T_1 = \mathbb{C}^g/L_1$ and $T_2 = \mathbb{C}^g/L_2$ are called *isogenous* ($T_1 \cong_{isog.} T_2$) if there exits a non-constant surjective morphism (i.e., holomorphic and also a group homomorphism) $h : T_1 \to T_2$ (so it has a finite kernel); h is called an *isogeny.* In particular, we may talk of *isogenous abelian varieties* (the polarization plays no role in this definition). Let R be a closed Riemann surface of genus $g \geq 1$. Its first homology group $H_1(R, \mathbb{Z})$ is isomorphic to $\mathbb{Z}^{2g}$ and its complex vector space $H^{1,0}(R)$ of its holomorphic 1-forms is isomorphic to $\mathbb{C}^g$. There is a natural injective map $\iota : H_1(R, \mathbb{Z}) \hookrightarrow \left(H^{1,0}(R)\right)^* : \alpha \mapsto \int_\alpha \cdot$, where $\left(H^{1,0}(R)\right)^*$ denotes the dual space of $H^{1,0}(R)$. The image $\iota(H_1(R, \mathbb{Z}))$ is a lattice in $\left(H^{1,0}(R)\right)^*$ and the quotient g-dimensional torus $JR = \left(H^{1,0}(R)\right)^* / \iota(H_1(R, \mathbb{Z}))$ is called the *jacobian variety* of R. The intersection product in $H_1(R, \mathbb{Z})$ induces a principal polarization on JR; that is, JR is a principally polarized abelian variety.

2.3. Kani-Rosen's decomposition result. An abelian variety is called *simple* if it is not isogenous to a product of at least two abelian varieties of smaller dimensions. Poincaré irreducibility theorem asserts that for every abelian variety A there exist simple polarized abelian varieties $A_1, \ldots, A_s$ and positive integers $n_1, \ldots, n_s$ such that $A \cong_{isog.} A_1^{n_1} \times \cdots A_s^{n_s}$. Moreover, the factors A_j and the integers n_j are unique up to isogeny and permutation of the factors. In general, to describe these simple factors seems to be a very difficult problem. When the abelian variety A admits a non-trivial group G of automorphisms, then there is a method to compute factors (non-necessarilly simple ones) by using the rational representations of G (the group algebra decomposition) [**4, 17, 25**]. In the particular case that $A = JR$, for R a closed Riemann surface, the following decomposition, in terms of automorphisms of R, is due to Kani and Rosen.

THEOREM 1 (Kani-Rosen's decomposition [**16**]). *Let R be a closed Riemann surface of genus $g \geq 1$ and let $H_1, \ldots, H_s < \operatorname{Aut}(R)$ such that: (1) $H_iH_j = H_jH_i$, for all $i, j = 1, \ldots, s$; (2) $g_{H_iH_j} = 0$, for $1 \leq i < j \leq s$; and (3) $g = \sum_{j=1}^{s} g_{H_j}$. Then $JR \cong_{isog.} JR_{H_1} \times \cdots \times JR_{H_s}$.*

2.4. Generalized Humbert curves. The constructions of the corresponding Riemann surfaces of Theorems 3 and 6 will be suitable quotients of certain types of Riemann surfaces called generalized Humbert curves. We proceed to recall them and some of their properties (details can be found in [**3**]).

Let S be a closed Riemann surface. We say that S is a *generalized Humbert curve of type $n \geq 2$* if it admits a group $H \cong \mathbb{Z}_2^n$ of conformal automorphisms such that S/H has genus zero (in particular, it has exactly $n+1$ cone points, each one of order two). The group H is called a *generalized Humbert group of type* n. By the Riemann-Hurwitz formula, the genus of S is $g_n = 1 + 2^{n-2}(n-3)$. If $n \geq 4$, then: (i) $g_n \geq 5$, (ii) S is non-hyperelliptic [**3**] and (iii) it has a unique generalized Humbert group of type n [**14**]. Let $\pi : S \to \widehat{\mathbb{C}}$ be a branched regular covering whose deck group is H. By post-composing it by a suitable Möbius transformation, we may assume its branch values to be given by $\infty, 0, 1, \lambda_1, \ldots, \lambda_{n-2}$, where $\lambda_i \in \mathbb{C} - \{0, 1\}$ are pairwise different. It can be observed that the following is a non-singular complex projective algebraic curve

$$C(\lambda_1, \ldots, \lambda_{n-2}) := \left\{ \begin{array}{rcl} x_1^2 + x_2^2 + x_3^2 & = & 0 \\ \lambda_1 x_1^2 + x_2^2 + x_4^2 & = & 0 \\ \vdots & \vdots & \vdots \\ \lambda_{n-2} x_1^2 + x_2^2 + x_{n+1}^2 & = & 0 \end{array} \right\} \subset \mathbb{P}^n.$$

REMARK 1. Set $\lambda_0 = 1$ and, for $j \in \{0, 1, \ldots, n-2\}$, let $C_j := \{\lambda_j x_1^2 + x_2^2 + x_{3+j}^2 = 0\} \subset \mathbb{P}^2$ be the classical degree two Fermat curve. The rational map $\pi_j : C_j \to \widehat{\mathbb{C}}$, defined by $\pi_j([x_1 : x_2 : x_{3+j}]) = -(x_2/x_1)^2$, is a branched regular covering with deck group $H_j = \langle a([x_1 : x_2 : x_{3+j}]) = [-x_1 : x_2 : x_{3+j}], b([x_1 : x_2 : x_{3+j}]) = [x_1 : -x_2 : x_{3+j}] \rangle \cong \mathbb{Z}_2^2$, whose branch values are ∞, 0 and λ_j. If we consider the fiber product of all these curves, with the given maps, we obtain a reducible projective algebraic curve with 2^{n-2} irreducible components (all of them isomorphic to $C(\lambda_1, \ldots, \lambda_{n-2})$).

If we consider the linear automorphisms of $\mathbb{P}^n$ given by

$$b_j([x_1 : \cdots : x_{n+1}]) := [x_1 : \cdots : x_{j-1} : -x_j : x_{j+1} : \cdots : x_{n+1}],\ j = 1, \ldots, n+1,$$

then $b_1 b_2 \cdots b_{n+1} = 1$ and $H_0 = \langle b_1, \ldots, b_n \rangle \cong \mathbb{Z}_2^n$ is a subgroup of Aut $(C(\lambda_1, \ldots, \lambda_{n-2}))$. The set of fixed points of b_j in $C(\lambda_1, \ldots, \lambda_{n-2})$ is given by the intersection

$$\mathrm{Fix}(b_j) := \{x_j = 0\} \cap C(\lambda_1, \ldots, \lambda_{n-2}),$$

which is of cardinality 2^{n-1}. The only non-trivial elements of H_0 acting with fixed points on $C(\lambda_1, \ldots, \lambda_{n-1})$ are the non-trivial powers of the elements $b_1, \ldots, b_{n+1}$.

The map $\pi_0 : C(\lambda_1, \ldots, \lambda_{n-2}) \to \widehat{\mathbb{C}} : [x_1 : \cdots : x_{n+1}] \mapsto -(x_2/x_1)^2$ is a regular branched cover with deck group H_0 and whose branch values are $\infty, 0, 1, \lambda_1, \ldots, \lambda_{n-2}$, each one of order two. In other words, $C(\lambda_1, \ldots, \lambda_{n-2})$ is a generalized Humbert curve of type n and H_0 is its generalized Fermat group of type n. It was noticed in [**3**] that S and $C(\lambda_1, \ldots, \lambda_{n-2})$ are isomorphic Riemann surfaces.

REMARK 2. Let $\mathcal{F}$ be the collection of all subsets of $\{\infty, 0, 1, \lambda_1, \ldots, \lambda_{n-2}\}$ of even cardinality at least four. For each $I \in \mathcal{F}$ we set R_I the Riemann surface defined by the algebraic curve $y^2 = \prod_{\alpha \in I}(x - \alpha)$ (if $\alpha = \infty$, then we delete the factor $(x - \alpha)$). If $|I| = 4$, then R_I is an elliptic curve and if $|I| = 2m \geq 6$, then R_I is a hyperelliptic Riemann surface of genus $g = m - 1$. In [**5**] (as an application of Kani-Rosen's result) it was observed that $JS \cong_{isog.} \prod_{I \in \mathcal{F}} JR_I$.

2.5. The known genus two situation. As already said in the introduction, given two elliptic curves E_1 and E_2, there is a closed Riemann surface R_2 of genus two with JR_2 isogenous to $E_1 \times E_2$. This can be tracked back to [**7, 8, 11–13**].

THEOREM 2. *Let E_1 and E_2 two elliptic curves and let $(\lambda_1, \lambda_2) \in \Delta_2$ be so that E_j is isomorphic to E_{λ_j}, for $j = 1, 2$. If $\eta_1 = (\lambda_1 - 1)/(\lambda_2 - 1)$, $\eta_2 =$*

$\lambda_2(\lambda_1-1)/\lambda_1(\lambda_2-1)$, and R_2 is the genus two Riemann surface defined by the hyperelliptic curve $y^2=(x^2-1)(x^2-\eta_1)(x^2-\eta_2)$, then $JR_2 \cong_{isog.} E_1\times E_2$.

PROOF. Consider the genus two curve $C:\ y^2=(x^2-1)(x^2-\eta_1)(x^2-\eta_2)$ and its order two automorphisms $a_1(x,y)=(-x,y)$ and $a_2(x,y)=(-x,-y)$. Then $H=\langle a_1,a_2\rangle\cong\mathbb{Z}_2^2$ and a_2a_1 is the hyperelliptic involution. Set $H_1=\langle a_1\rangle$ and $H_2=\langle a_2\rangle$. If we consider the elliptic curves $F_1:\ y^2=(x-1)(x-\eta_1)(x-\eta_2)$ and $F_2:\ y^2=x(x-1)(x-\eta_1)(x-\eta_2)$, then $P_j:C\to F_j$, where $P_1(x,y)=x^2$ and $P_2(x,y)=(x^2,xy)$, are two-fold branched coverings with respective deck groups H_1 and H_2. As $H_1H_2=H=H_2H_1$ and C/H is the Riemann sphere (with exactly 5 cone points, these being ∞, 0, 1, η_1 and η_2), we may apply Kani-Rosen's result (Theorem 1) to obtain that $JC\cong_{isog.} F_1\times F_2$. As, for $L(x)=(1-\eta_1)(x-\eta_2)/(1-\eta_2)(x-\eta_1)$, one has that $L(1)=1$, $L(\eta_1)=\infty$, $L(\eta_2)=0$, $L(\infty)=(1-\eta_1)/(1-\eta_2)=\lambda_1$ and $L(0)=\lambda_2$, we observe that $E_1\cong F_1$ and $E_2\cong F_2$. □

REMARK 3 (The curve R_2 in terms of fiber products). The fiber product of (E_{λ_1},π_1) and (E_{λ_2},π_2), where $\pi_j(x,y)=x$, (in projective coordinates) is

$$X_2:=\big\{([x_1:y_1:z_1],[x_2:y_2:z_2])\in\mathbb{P}^2\times\mathbb{P}^2:\ x_1/z_1=x_2/z_2,$$
$$y_1^2z_1=x_1(x_1-z_1)(x_1-\lambda_1z_1),\quad y_2^2z_2=x_2(x_2-z_2)(x_2-\lambda_2z_2)\big\}.$$

This is an irreducible algebraic curve with exactly 3 nodes, these being at the points $\infty:=([0:1:0],[0:1:0])$, $p:=([0:0:1],[0:0:1])$ and $q:=([1:0:1],[1:0:1])$. After desingularization, this produces a genus two Riemann surface R, and each of these three special points induces two different points (∞_1, ∞_2, p_1, p_2, q_1 and q_2) on R. On X_2 we have the order two automorphisms

$$a_1([x_1:y_1:z_1],[x_2:y_2:z_2])=([x_1:-y_1:z_1],[x_2:y_2:z_2]),$$
$$a_2([x_1:y_1:z_1],[x_2:y_2:z_2])=([x:y_1:z_1]),[x_2:-y_2:z_2]),$$

with $H=\langle a_1,a_2\rangle\cong\mathbb{Z}_2^2$. The projection $\pi:X_2\to\widehat{\mathbb{C}}:([x_1:y_1:z_1],[x_2:y_2,z_2])\mapsto x_1/z_1$, is a Galois branched cover with H as its deck group. The branch values of π are ∞, 0, 1, λ_1 and λ_2. The Riemann surface R has genus two and admits the group H as group of conformal automorphisms. If $H_j=\langle a_j\rangle$, it can be seen that the orbifold R/H_j has underlying Riemann surface $E_{\lambda_{3-j}}$. The surface R is another model for the genus two Riemann surface R_2 described in the above theorem.

3. The constructions

3.1. First construction. Let us assume $s\geq 3$ and consider the set of cardinality $2^{s-1}-1$

$$V_s=\{\alpha=(\alpha_1,\ldots,\alpha_s)\in\{0,1\}^s-\{(0,\ldots,0)\}:\alpha_1+\cdots+\alpha_s \text{ is even}\}.$$

If $P:=(\lambda,\mu_{1,1},\mu_{1,2},\mu_{2,1},\mu_{2,2},\ldots,\mu_{s-2,1},\mu_{s-2,2})\in\Delta_{2s-3}$ and $\alpha=(\alpha_1,\ldots,\alpha_s)\in V_s$, then we set the following elliptic curves

$$E_1(P):\ y^2=x(x-1)(x-\lambda),\ E_2(P):\ y^2=(x-1)(x-\lambda)(x-\mu_{1,1})(x-\mu_{1,2}),$$
$$E_3(P):\ y^2=(x-\mu_{1,1})(x-\mu_{1,2})(x-\mu_{2,1})(x-\mu_{2,2}),$$
$$E_4(P):\ y^2=(x-\mu_{2,1})(x-\mu_{2,2})(x-\mu_{3,1})(x-\mu_{3,2}),$$
$$\vdots$$
$$E_{s-1}(P):\ y^2=(x-\mu_{s-3,1})(x-\mu_{s-3,2})(x-\mu_{s-2,1})(x-\mu_{s-2,2}),$$

$$E_s(P):\ y^2 = x(x-\mu_{s-2,1})(x-\mu_{s-2,2}),$$

the complex numbers $\eta_0 = -(\mu_{s-2,2})^{-1}$, $\eta_1 = (1-\mu_{s-2,2})^{-1}$, $\eta_2 = (\lambda-\mu_{s-2,2})^{-1}$, $\eta_3 = (\mu_{s-2,1}-\mu_{s-2,2})^{-1}$ and, for each $t=1,2$ and $j=1,\dots,s-3$, $\eta_{j,t} = (\mu_{j,t}-\mu_{s-2,2})^{-1}$, and the following non-zero complex number

$$K_\alpha = (-\mu_{s-2,2})^{\alpha_1}(\mu_{s-2,2}-1)^{\alpha_2}(\mu_{s-2,2}-\lambda)^{\alpha_2}(\mu_{s-2,2}-\mu_{s-2,1})^{\alpha_s} \times \prod_{k=1}^{s-3}(\mu_{s-2,2}-\mu_{k,1})^{\alpha_{k+2}}(\mu_{s-2,\mu_{k,2}}-\lambda)^{\alpha_{k+2}}.$$

THEOREM 3. *Let $s\geq 3$ and $P=(\lambda,\mu_{1,1},\mu_{1,2},\mu_{2,1},\mu_{2,2},\dots,\mu_{s-2,1},\mu_{s-2,2})\in\Delta_{2s-3}$. Let $X\subset\mathbb{C}^{2^{s-1}}$ be the affine curve defined by the following $(2^{s-1}-1)$ elliptic/hyperelliptic equations*

$$\left\{\begin{array}{c} w_\alpha^2 = K_\alpha z^{\alpha_1}(z-\eta_0)^{\alpha_1}(z-\eta_1)^{\alpha_2}(z-\eta_2)^{\alpha_2}(z-\eta_3)^{\alpha_s}\prod_{k=1}^{s-3}(z-\eta_{k,1})^{\alpha_{k+2}}(z-\eta_{k,2})^{\alpha_{k+2}},\\ \alpha=(\alpha_1,\dots,\alpha_s)\in V_s. \end{array}\right\}.$$

Then X defines a closed Riemann surface R_s of genus $g=1+2^{s-2}(s-2)$ whose jacobian variety is isogenous to the product of the jacobian varieties of the following $\sum_{j=1}^{[s/2]}\binom{s}{2j}$ elliptic/hyperelliptic curves

$$C_{i_1,\dots,i_k}: \nu^2 = (\upsilon-\rho_{i_1,1})(\upsilon-\rho_{i_1,2})\cdots(\upsilon-\rho_{i_k,1})(\upsilon-\rho_{i_k,2}),$$

where $2\leq k\leq s$ is even, the tuples $(i_1,\dots,i_k)$ satisfy $1\leq i_1<i_2<\dots<i_k\leq s$, and

$$\rho_{i_j,1} = \left\{\begin{array}{cc} \infty, & i_j=1\\ 1, & i_j=2\\ \mu_{r-2,1}, & i_j=r\geq 3\end{array}\right. \qquad \rho_{i_j,2} = \left\{\begin{array}{cc} 0, & i_j=1\\ \lambda, & i_j=2\\ \mu_{r-2,2}, & i_j=r\geq 3\end{array}\right.$$

where, if $\rho_{i_j,1}=\infty$, then the factor $(u-\rho_{i_j,1})$ is deleted from the above expression. In particular, JR_s contains, in its isogenous factors above, the elliptic curves $E_1(P),\dots,E_s(P)$.

REMARK 4. (1) JR_s contains at least $s(s-1)/2$ elliptic curves in its isogenous decomposition. (2) Given s elliptic curves, $\widehat{E}_1^*,\dots,\widehat{E}_s^*$, it is possible to find a tuple $P\in\Delta_{2s-3}$ such that $\widehat{E}_j^*$ is isogenous to $E_j(P)$, for every $j=1,\dots,s$. So, if all these elliptic curves $E_j(P)$ are isogenous to a given one E, then $JR_s\cong_{isog.} E^s\times A$, where A is isogenous to the product of other elliptic curves and the jacobian varieties of hyperelliptic Riemann surfaces.

3.2. Two applications. In Theorem 3, for $s=3$, the surface R_3 has genus three and $JR_3\cong_{isog.} E_1(P)\times E_2(P)\times E_3(P)$. Next, we observe that we may chose the tuple $P\in\Delta_3$ such that the three elliptic curves $E_1(P), E_2(P)$ and $E_3(P)$ may be chosen to be any three given elliptic curves.

THEOREM 4. *Let E_1, E_2 and E_3 be three elliptic curves. Choose $(\lambda_1,\lambda_2,\lambda_3)\in\Delta_3$ so that E_j is isogenous to E_{λ_j}, for $j=1,2,3$. If μ is a root of*

$$\lambda_2\lambda_3\mu^2-(\lambda_1\lambda_2+\lambda_2\lambda_3+\lambda_1\lambda_3-\lambda_1-\lambda_3+1)\mu+\lambda_1\lambda_2=0,$$

then

$$R_3 := \left\{\begin{array}{ccl} w_1^2 &=& \mu(\lambda_3\mu-1)(\lambda_3\mu-\lambda_1)(\lambda_3-1)z\left(z-\dfrac{1}{\lambda_1-\lambda_3\mu}\right)\left(z-\dfrac{1}{\mu(1-\lambda_3)}\right)\\ w_2^2 &=& -\lambda_3\mu^2(\lambda_3-1)z\left(z+\dfrac{1}{\lambda_3\mu}\right)\left(z-\dfrac{1}{1-\lambda_3\mu}\right)\\ w_3^2 &=& -\lambda_3\mu^2(\lambda_3\mu-1)(\lambda_3-1)z^2\left(z+\dfrac{1}{\lambda_3\mu}\right)\left(z-\dfrac{1}{\mu(1-\lambda_3)}\right)\end{array}\right\},$$

defines a closed Riemann surface of genus three with $JR_3 \cong_{isog.} E_1 \times E_2 \times E_3$.

PROOF. If $\lambda = \lambda_1$, $\mu_{1,1} = \mu$ and $\mu_{1,2} = \lambda_3\mu$, then $(\lambda, \mu_{1,1}, \mu_{1,2}) \in \Delta_3$ and

$$E_{\lambda_1} = F_1 : y^2 = x(x-1)(x-\lambda_1),\ E_{\lambda_2} \cong F_2 : y^2 = (x-1)(x-\lambda_1)(x-\mu)(x-\lambda_3\mu),$$
$$E_{\lambda_3} \cong F_3 : y^2 = x(x-\mu)(x-\lambda_3\mu).$$

Theorem 3, applied to the triple $(\lambda, \mu_{1,1}, \mu_{1,2})$, asserts that $JR_3 \cong_{isog.} F_1 \times F_2 \times F_3$, so isogenous to the product $E_1 \times E_1 \times E_3$. In this case the Riemann surface R_3 is described by the curve

$$\left\{ \begin{array}{lll} w_1^2 & = & \mu(\lambda_3\mu - 1)(\lambda_3\mu - \lambda_1)(\lambda_3 - 1)z\left(z - \dfrac{1}{\lambda_1 - \lambda_3\mu}\right)\left(z - \dfrac{1}{\mu(1-\lambda_3)}\right) \\ w_2^2 & = & -\lambda_3\mu^2(\lambda_3 - 1)z\left(z + \dfrac{1}{\lambda_3\mu}\right)\left(z - \dfrac{1}{1-\lambda_3\mu}\right) \\ w_3^2 & = & -\lambda_3\mu^2(\lambda_3\mu - 1)(\lambda_3 - 1)z^2\left(z + \dfrac{1}{\lambda_3\mu}\right)\left(z - \dfrac{1}{\mu(1-\lambda_3)}\right) \end{array} \right\},$$

the group $H = \langle a_1, a_2 \rangle \cong \mathbb{Z}_2^2$, where

$$a_1(z, w_1, w_2, w_3) = (z, w_1, -w_2, -w_3),\ a_2(z, w_1, w_2, w_3) = (z, -w_1, w_2, -w_3),$$

each of the three automorphisms a_1, a_2 and $a_3 = a_1a_2$ acting with exactly four fixed points, and the corresponding regular branched cover with H as deck group is

$$\pi : R_3 \to \widehat{\mathbb{C}} : (z, w_1, w_2, w_3) \mapsto (\lambda_3\mu z + 1)/z.$$

The non-trivial proper subgroups of H are $H_{1,2} = \langle a_1a_2 \rangle$, $H_{1,3} = \langle a_1a_3 \rangle = \langle a_2 \rangle$ and $H_{2,3} = \langle a_2a_3 \rangle = \langle a_1 \rangle$. The quotients $R_3/\langle a_1 \rangle$, $R_3/\langle a_2 \rangle$ and $R_3/\langle a_1a_2 \rangle$ are of genus one and they correspond, respectively, to the elliptic curves E_{λ_2}, E_{λ_1} and E_{λ_3}. □

Another consequence of the construction provided by Theorem 3 is the following upper bound for $e(r)$ (which we conjecture to be sharps).

THEOREM 5.

$$e(r) \leq \left\{ \begin{array}{ll} 1 + 2^{(r-2)/2}r, & r \geq 4 \quad \textit{even} \\ 1 + 2^{(r-3)/2}(r-1), & r \geq 5 \quad \textit{odd} \end{array} \right.$$

PROOF. Assume $r \geq 3$ is odd and write $r = 2s - 3$, where $s \geq 3$. Let us fix $(\lambda_1, \ldots, \lambda_r) \in \Delta_r$. Set $\lambda = \lambda_1$ and, for $j = 1, \ldots, s-2$, we set $\mu_{j,2} = \lambda_{j+1}\mu_{j,1}$ and $\mu_{j,1}$ a root of the polynomial

$$\lambda_{j+1}(1-\lambda_{s-2+j})\mu_{j,1}^2 + (\lambda_{s-2+j} - \lambda_{j+1} - \lambda_1 + \lambda_1\lambda_{j+1}\lambda_{s-2+j})\mu_{j,1} + (1 - \lambda_1\lambda_{s-2+j}) = 0.$$

As a value λ_j may be changed to some other value λ'_j (inside an infinite set of values) so that E_{λ_j} and $E_{\lambda'_j}$ are isogenous, we may assume $(\lambda, \mu_{1,1}, \mu_{1,2}, \mu_{2,1}, \mu_{2,2}, \ldots, \mu_{s-2,1}, \mu_{s-2,2}) \in \Delta_{2s-3}$. If R_s is the Riemann surface constructed in Theorem 3, then JR_s is isogenous to a product of certain explicit jacobian varieties. It can be seen, from the proof of that theorem, that some of these factors are (isogenous to) the elliptic curves

$$\begin{array}{lll} y^2 & = & x(x-1)(x-\lambda) \\ y^2 & = & x(x-\mu_{j,1})(x-\mu_{j,2}), \quad j = 1, \ldots, s-2, \\ y^2 & = & (x-1)(x-\lambda)(x-\mu_{j,2}), \quad j = 1, \ldots, s-2. \end{array}$$

The choice we have made for the tuple $(\lambda, \mu_{1,1}, \mu_{1,2}, \mu_{2,1}, \mu_{2,2}, \dots, \mu_{s-2,1}, \mu_{s-2,2}) \in \Delta_{2s-3}$ ensures that they are all isomorphic to the elliptic curves $y^2 = x(x-1)(x-\rho)$, where $\rho \in \{\lambda_1, \dots, \lambda_r\}$. The case $r \geq 4$ even can be worked similarly, but in this case we add an extra elliptic curve to the r given ones in order to obtain the result as a consequence of the odd situation. □

3.3. The second construction. The Riemann surface obtained in the construction given in Theorem 3 is one of two (isomorphic) factors of a fiber product. The next theorem describes a similar construction, but in this case, this surface is an irreducible fiber product.

THEOREM 6. *If $(\lambda_1, \dots, \lambda_r) \in \Delta_r$, then the affine algebraic curve*

$$Y = \left\{ \begin{array}{c} w_1^2 = (\lambda_r - \lambda_1)u + 1,\ w_2^2 = (\lambda_r - \lambda_2)u + 1, \cdots, w_{r-1}^2 = (\lambda_r - \lambda_{r-1})u + 1, \\ w_r^2 = -u(\lambda_r u + 1)((\lambda_r - 1)u + 1) \end{array} \right\}$$

defines a closed Riemann surface R of genus $g = 1 + 2^{r-2}(r-1)$ such that $JR \cong_{isog.} E_{\lambda_1} \times \cdots \times E_{\lambda_r} \times A_{g-r}$, where A_{g-r} is the product of certain explicit elliptic/hyperelliptic Riemann surfaces.

4. Proof of Theorems 3 and 6

4.1. Proof of Theorem 3. Set $E_j = E_j(P)$, where $j = 1, \dots s$. Let us consider the generalized Humbert curve of type $2s-1$

$$D: \left\{ \begin{array}{rclrcl} z_1^2 + z_2^2 + z_3^2 & = & 0, & \lambda z_1^2 + z_2^2 + z_4^2 & = & 0, \\ \mu_{1,1} z_1^2 + z_2^2 + z_5^2 & = & 0, & \mu_{1,2} z_1^2 + z_2^2 + z_6^2 & = & 0, \\ \vdots & \vdots & \vdots & \vdots & \vdots & \vdots \\ \mu_{k,1} z_1^2 + z_2^2 + z_{2k+3}^2 & = & 0, & \mu_{k,2} z_1^2 + z_2^2 + z_{2k+4}^2 & = & 0, \\ \vdots & \vdots & \vdots & \vdots & \vdots & \vdots \\ \mu_{s-2,1} z_1^2 + z_2^2 + z_{2s-1}^2 & = & 0, & \mu_{s-2,2} z_1^2 + z_2^2 + z_{2s}^2 & = & 0. \end{array} \right\}.$$

As observed in Section 2.4, D is a closed Riemann surface of genus $g_D = 1 + 2^{2s-2}(s-2)$, with generalized Humbert group $\langle b_1, \dots, b_{2s-1} \rangle = H_0 \cong \mathbb{Z}_2^{2s-1}$ of type $2s-1$.

Let us consider the surjective homomorphism

$$\begin{array}{c} \theta : H_0 \to H = \langle a_1, \dots, a_{s-1} \rangle \cong \mathbb{Z}_2^{s-1} \\ b_{2k-1}, b_{2k} \mapsto a_k,\ k = 1, \dots, s-1, \\ b_{2s-1}, b_s \mapsto a_s = a_1 a_2 \cdots a_{s-1}. \end{array}$$

The kernel of θ is given by

$$K = \langle b_1 b_2, b_3 b_4, \dots, b_{2k-1} b_{2k}, \dots, b_{2s-3} b_{2s-2}, b_1 b_3 b_5 \cdots b_{2s-1} \rangle \cong \mathbb{Z}_2^s$$

and, as the only non-trivial elements acting with fixed points are $b_1, \dots, b_{2s-1}$ and $b_{2s} = b_1 b_2 \cdots b_{2s-1}$, it follows that K acts freely on D. It follows that $R = D/K$ is a closed Riemann surface of genus (by the Riemann-Hurwitz formula) $g_R = 1 + 2^{s-2}(s-2)$. In order to write equations for R, we consider the affine model D^0 of D by taking $z_{2s} = 1$. Now, we need to compute a set of generators of $\mathbb{C}[z_1, \dots, z_{2s-1}]^K$, the algebra of K-invariant polynomials. Since the linear action of K is given by diagonal matrices, a set of generators can be found to be $t_1 = z_1^2, t_2 = z_2^2, \dots, t_{2s-1} = z_{2s-1}^2$, together with the monomials of the form $t_\alpha = (z_1 z_2)^{\alpha_1} (z_3 z_4)^{\alpha_2} \cdots (z_{2s-3} z_{2s-2})^{\alpha_{s-1}} z_{2s-1}^{\alpha_s}$, where $\alpha = (\alpha_1, \dots, \alpha_s) \in V_s$. As V_s has cardinality $2^{s-1} - 1$, the number of the above set of generators is $N = 2^{s-1} +$

$2s-2$. Using the map $\Phi : D^0 \to \mathbb{C}^N$, whose coordinates are $t_1, \ldots, t_{2s-1}$ and the monomials t_α, $\alpha \in V_s$, one obtains that the Riemann surface induced by $\Phi(D^0)$ is isomorphic to an affine model of R and that its equations are given by

$$\Phi(D^0) = \left\{ \begin{array}{rr} t_1 + t_2 + t_3 = 0, & \lambda t_1 + t_2 + t_4 = 0, \\ \mu_{1,1} t_1 + t_2 + t_5 = 0, & \mu_{1,2} t_1 + t_2 + t_6 = 0, \\ \vdots & \vdots \\ \mu_{k,1} t_1 + t_2 + t_{2k+3} = 0, & \mu_{k,2} t_1 + t_2 + t_{2k+4} = 0, \\ \vdots & \vdots \\ \mu_{s-2,1} t_1 + t_2 + t_{2s-1} = 0, & \mu_{s-2,2} t_1 + t_2 + 1 = 0, \\ t_\alpha^2 = (t_3 t_4)^{\alpha_2} \cdots (t_{2s-3} t_{2s-2})^{\alpha_{s-1}} t_{2s-1}^{\alpha_s}; & \alpha = (\alpha_1, \ldots, \alpha_s) \in V_s \end{array} \right\}.$$

The first linear equations permit us to write $t_2, \ldots, t_{2s-1}$ in terms of t_1 as follows:

$$t_2 = -1 - \mu_{s-2,2} t_1, \; t_3 = 1 + (\mu_{s-2,2} - 1) t_1, \; t_4 = 1 + (\mu_{s-2,2} - \lambda) t_1$$
$$t_{2k+3} = 1 + (\mu_{s-2,2} - \mu_{k,1}) t_1, \quad t_{2k+4} = 1 + (\mu_{s-2,2} - \mu_{s,2}) t_1, \quad k = 1, \ldots, s-3,$$
$$t_{2s-1} = 1 + (\mu_{s-2,2} - \mu_{s-2,1}) t_1.$$

We may then eliminate the variables $t_2, \ldots, t_{2s-1}$ and just keep the variables t_1 and $t_{\alpha_1, \ldots, \alpha_s}$. Let us set $t_1 = z$ and $t_\alpha = w_\alpha$, for $\alpha \in V_s$. In these new 2^{s-1} coordinates, the above curve is isomorphic to the one given by

$$C = \left\{ \begin{array}{c} w_\alpha^2 = z^{\alpha_1} (-1 - \mu_{s-2,2} z)^{\alpha_1} (1 + (\mu_{s-2,2} - 1) z)^{\alpha_2} (1 + (\mu_{s-2,2} - \lambda) z)^{\alpha_2} \cdots \\ \cdots (1 + (\mu_{s-2,2} - \mu_{k,1}) z)^{\alpha_{k+2}} (1 + (\mu_{s-2,2} - \mu_{k,2}) z)^{\alpha_{k+2}} \cdots \\ \cdots (1 + (\mu_{s-2,2} - \mu_{s-3,1}) z)^{\alpha_{s-1}} (1 + (\mu_{s-2,2} - \mu_{s-3,2}) z)^{\alpha_{s-1}} (1 + (\mu_{s-2,2} - \mu_{s-2,1}) z)^{\alpha_s}, \\ \alpha = (\alpha_1, \ldots, \alpha_s) \in V_s. \end{array} \right\}$$

By making the choices as described in the hypothesis of the theorem for K_α and the values of η_0, η_1, η_2, η_3 and $\eta_{k,j}$, then the above curve can be written in the desired algebraic form. If $\Phi_1 : D^0 \to \mathbb{C}^{2^{s-1}}$ is the map whose coordinates are z and w_α, where $\alpha \in V_s$, then $\Phi_1(D^0) = C$. If $\alpha = (\alpha_1, \ldots, \alpha_s)$ and $j = 1, \ldots, s-1$, then the induced automorphisms a_j acts by multiplication by -1 at coordinates w_α if $\alpha_j = 1$ and acts by the identity on the rest of coordinates. The map $\pi : C \to \widehat{\mathbb{C}} : (z, \{w_{\alpha \in V_s}\}) \mapsto (1 + \mu_{s-2,2} z)/z$ is a regular branched cover with H as its deck group and its satisfies that $P = \pi \circ \Phi_1$. The branch locus of π is the set $\{\infty, 0, 1, \lambda, \mu_{1,1}, \mu_{1,2}, \ldots, \mu_{s-2,1}, \mu_{s-2,2}\}$. All the above permit to observe the following.

LEMMA 1. *The only non-trivial elements of H acting with fixed points are $a_1, \ldots, a_{s-1}$ and $a_s = a_1 a_2 \cdots a_{s-1}$. Moreover, $\pi(\mathrm{Fix}(a_1)) = \{\infty, 0\}$, $\pi(\mathrm{Fix}(a_2)) = \{1, \lambda\}$, $\pi(\mathrm{Fix}(a_k)) = \{\mu_{k-2,1}, \mu_{k-2,2}\}$, for $k = 3, \ldots, s-1$, and $\pi(\mathrm{Fix}(a_s)) = \{\mu_{s-2,1}, \mu_{s-2,2}\}$. It can be seen that, for $j = 1, \ldots, s$, $\mathrm{Fix}(a_j)$ has cardinality 2^{s-1}.*

If $2 \leq k \leq s$ is even and $1 \leq i_1 < i_2 < \cdots < i_k \leq s$, then we consider the subgroup

$$H_{i_1, i_2, \ldots, i_k} = \langle a_{i_1} a_{i_2}, a_{i_1} a_{i_3}, \ldots, a_{i_1} a_{i_k}, a_j; j \in \{1, \ldots, s\} - \{i_1, \ldots, i_k\} \rangle \cong \mathbb{Z}_2^{s-2}.$$

If $k = 2$, then we have $s(s-1)/2$ such subgroups. Between them are $H_{1,2} = \langle a_1 a_2, a_3, \ldots, a_s \rangle$, $H_{2,3} = \langle a_2 a_3, a_4, \ldots, a_1 \rangle$, $H_{3,4} = \langle a_3 a_4, a_5, \ldots, a_2 \rangle, \ldots$, $H_{s-1,s} = \langle a_{s-1} a_s, a_1, \ldots, a_{s-2} \rangle$ and $H_{s,1} = \langle a_s a_1, a_2, \ldots, a_{s-1} \rangle$. The quotient orbifold $C/H_{j,j+1}$ has underlying Riemann surface structure $E_j \cong E_{\lambda_j}$, for $j = 1, \ldots, s-1$, and $C/H_{s,1}$ has underlying Riemann surface structure $E_s \cong E_{\lambda_s}$.

LEMMA 2. *With the above notations, the following hold for the above defined subgroups.*

(1) *Any two such subgroups* $H_{i_1,i_2,\ldots,i_k}$ *and* $H_{j_1,j_2,\ldots,j_l}$ *commute.*
(2) *The quotient* $C/H_{i_1,i_2,\ldots,i_k}$ *has genus* $k-1$ *and its underlying Riemann surface is given by the (elliptic) hyperelliptic curve*

$$\nu^2 = (\upsilon - \rho_{i_1,1})(\upsilon - \rho_{i_1,2}) \cdots (\upsilon - \rho_{i_k,1})(\upsilon - \rho_{i_k,2}),$$

where

$$\rho_{i_j,1} = \begin{cases} \infty, & i_j = 1 \\ 1, & i_j = 2 \\ \mu_{r-2,1}, & i_j = r \geq 3 \end{cases} \qquad \rho_{i_j,2} = \begin{cases} 0, & i_j = 1 \\ \lambda, & i_j = 2 \\ \mu_{r-2,2}, & i_j = r \geq 3 \end{cases}$$

In the case that $\rho_{i_j,1} = \infty$, *then the factor* $(u - \rho_{i_j,1})$ *is deleted from the above expression.*
(3) *The group generated by any two different such subgroups is* H.

PROOF. Property (1) holds trivially as H is an abelian group. Property (2) follows from Riemann-Hurwitz formula and Lemma 1. Property (3) is clear as in the product we obtain all the generators. $\square$

The next result states that the sum of the genera appearing in all quotients of the form $C/H_{i_1,i_2,\ldots,i_k}$ is equal to the genus of C (that is, the genus of R). Recall that we are considering k even and $2 \leq k \leq s$.

LEMMA 3. $\sum_{k=2}^{s} \binom{r}{k}(k-1)\left(\frac{1+(-1)^k}{2}\right) = 1 + 2^{s-2}(s-2)$.

PROOF. If $f(x) = \frac{(1+x)^s}{2x} = \frac{1}{2}\sum_{k=0}^{s}\binom{s}{k}x^{k-1}$, then $f'(x) = \frac{(1+x)^{s-1}((s-1)x-1)}{2x^2} = \frac{1}{2}\sum_{k=0}^{s}\binom{s}{k}(k-1)x^{k-2}$, and $2^{s-2}(s-2) = f'(1) + f'(-1) = \frac{1}{2}\sum_{k=0}^{s}\binom{s}{k}(k-1)(1+(-1)^k) = -1 + \sum_{k=2}^{s}\binom{s}{k}\frac{(1+(-1)^k)}{2}(k-1)$. $\square$

We may apply Theorem 1 for C using all the subgroups $H_{i_1,\ldots,i_k}$ in order to obtain that JC (so JR) is isogenous to the product of the jacobian varieties of all Riemann surfaces $C/H_{i_1,\ldots,i_k}$ as desired. The equations of these curves are provided in Lemma 2.

REMARK 5 (A fiber product description). Let us consider the s elliptic curves $E_j = E_j(P)$, where $j = 1, \ldots s$. If we consider the degree two maps $\pi_j : E_j \to \widehat{\mathbb{C}}$ defined as $\pi_j(x,y) = x$, then we may perform the fiber product of the s pairs $(E_1, \pi_1), \ldots, (E_s, \pi_s)$. Such a fiber product is given by an affine curve $Z \subset \mathbb{C}^{s+1}$, formed of the tuples $(x, y_1, \ldots, y_s)$ so that $(x, y_j) \in E_j$, for $j = 1, \ldots, s$. The curve Z is reducible and contains two irreducible components, both of them being isomorphic, and it admits the group of automorphisms $N = \langle f_1, \ldots, f_s \rangle \cong \mathbb{Z}_2^s$, where $f_j(x, y_1, \ldots, y_s) = (x, y_1, \ldots, y_{j-1}, -y_j, y_{j+1}, \ldots, y_s)$. The two irreducible factors are permuted by some elements of N and each one is invariant under a subgroup isomorphic to $\mathbb{Z}_2^{s-1}$. It can be shown that the Riemann surface defined by any of these two irreducible components is isomorphic to X as defined in Theorem 3.

4.2. Proof of Theorem 6. Let $(\lambda_1,\ldots,\lambda_r)\in\Delta_r$, $E_{\lambda_j}: y_j^2=x_j(x_j-1)(x_j-\lambda_j)$ and $\pi_j: E_{\lambda_j}\to\widehat{\mathbb{C}}$ defined by $\pi_j(x_j,y_j)=x_j$. The locus of branch values of π_j is the set $\{\infty,0,1,\lambda_j\}$. Let us consider the generalized Humbert curve $F=C(\lambda_1,\ldots,\lambda_r)\subset\mathbb{P}^{r+2}$ of type $r+2$, which is of genus $g=1+2^r(r-1)$, with corresponding generalized Humbert group $\langle b_1,\ldots,b_{r+2}\rangle=H_0\cong\mathbb{Z}_2^{r+2}$ of type $r+2$. As the only non-trivial elements of H_0 acting with fixed points are $b_1,\ldots,b_{r+2}$ and $b_{r+3}=b_1b_2\cdots b_{r+2}$, the subgroup $K^*=\langle b_1b_2,b_2b_3\rangle\cong\mathbb{Z}_2^2$ acts freely on F. Then $Y=F/K^*$ is a closed Riemann surface of genus $g_Y=1+2^{r-2}(r-1)$. The quotient group $L=H_0/\langle b_1b_2,b_2b_3\rangle$ induces the group of automorphisms of Y. It can be seen that $L=\langle c_1,\ldots,c_r\rangle\cong\mathbb{Z}_2^r$, where

$$c_j(x,y_1,\ldots,y_r)=(x,y_1,\ldots,y_{j-1},-y_j,y_{j+1},\ldots,y_r),\ j=1,...,r.$$

The map $\pi: Y\to\widehat{\mathbb{C}}:(x,y_1,\ldots,y_r)\mapsto x$, is a regular branched cover, with H as its deck group, and its branch locus is the set $\{\infty,0,1,\lambda_1,\ldots,\lambda_r\}$.

REMARK 6. With the above description, we obtain a set of equations for Y as

$$\left\{\begin{array}{c} w_1^2=(\lambda_r-\lambda_1)u+1,\ w_2^2=(\lambda_r-\lambda_2)u+1,\cdots,w_{r-1}^2=(\lambda_r-\lambda_{r-1})u+1,\\ w_r^2=-u(\lambda_r u+1)((\lambda_r-1)u+1)\end{array}\right\}.$$

The equations for E_{λ_j} can also be written as (for $j=1,\ldots,r-1$)

$$\left\{\begin{array}{c} w_1^2=(\lambda_r-\lambda_1)u+1,\cdots,w_{j-1}^2=(\lambda_r-\lambda_{j-1})u+1\\ w_j^2=(\lambda_r-\lambda_j)u+1,\cdots,w_{r-1}^2=(\lambda_r-\lambda_{r-1})u+1\\ w_r^2=-u(\lambda_r u+1)((\lambda_r-1)u+1)((\lambda_r-\lambda_j)u+1)\end{array}\right\}$$

and for E_{λ_r} as

$$\left\{\begin{array}{c} w_1^2=(\lambda_r-\lambda_1)u+1,\cdots,w_{r-1}^2=(\lambda_r-\lambda_{r-1})u+1\\ w_r^2=-u(\lambda_r u+1)((\lambda_r-1)u+1)\end{array}\right\}.$$

LEMMA 4. *The only non-trivial elements of L acting with fixed points are $c_1,\ldots,c_r$ and $c_{r+1}=c_1c_2\cdots c_r$. Moreover, $\pi(\mathrm{Fix}(c_j))=\lambda_j$, for $j=1,\ldots,r$, and $\pi(\mathrm{Fix}(c_{r+1}))=\{\infty,0,1\}$.*

PROOF. A non-trivial element of L has the form $c(x,y_1,\ldots,y_r)=(x,(-1)^{\alpha_1}y_1,\ldots,(-1)^{\alpha_r}y_r)$, where $\alpha_1,\ldots,\alpha_r\in\{0,1\}$ and $\alpha_1+\cdots+\alpha_r>0$. A point $(x,y_1,\ldots,y_r)\in C$ is a fixed point of c if and only if $y_j=0$ for $\alpha_j=1$. The equality $y_j=0$ is equivalent to havig $x\in\{\infty,0,1,\lambda_j\}$. The values $x\in\{\infty,0,1\}$ produce fixed points for c_{r+1}. Also, as we are assume the values λ_j to be different, it follows that the only possibility is to have only one j with $\alpha_j=1$. □

It can be seen that, for $j=1,\ldots,r$, $\mathrm{Fix}(c_j)$ has cardinality 2^{r-1} and that $\mathrm{Fix}(c_1\cdots c_r)$ has cardinality $3\times 2^{r-1}$. In particular, for $r=3$, the surface Y is hyperelliptic with $c_1c_2c_3$ as its hyperelliptic involution.

If either $1\leq k\leq r$ is odd or $4\leq k\leq r$ is even, and $\{i_1,\ldots,i_k\}\subset\{1,\ldots,r\}$ with $1\leq i_1<i_2<\cdots<i_k\leq r$, then we consider the subgroup

$$L_{i_1,i_2,\ldots,i_k}=\langle c_{i_1}c_{i_2},c_{i_1}c_{i_3},\ldots,c_{i_1}c_{i_k},c_j;j\in\{1,\ldots,r\}-\{i_1,\ldots,i_k\}\rangle\cong\mathbb{Z}_2^{r-1}.$$

Note that for $k=1$ we have the subgroups $L_j=\langle c_1,\ldots,c_{j-1},c_{j+1},\ldots,c_r\rangle\cong\mathbb{Z}_2^{r-1}$. If

$$Q_j: Y\to E_{\lambda_j}:(x,y_1,\ldots,y_r)\mapsto(x,y_j),$$

then Q_j is a regular branched cover with deck group being L_j. The branch locus of Q_j is the set

$$\{(\lambda_j, y_i) : y_i^2 = \lambda_i(\lambda_i - 1)(\lambda_i - \lambda_j),\ i = 1, \dots, r, i \neq j\}.$$

LEMMA 5. *With the above notations, the following hold for the above defined subgroups.*

(1) *Any two such subgroups commute.*
(2) *The quotient $Y/L_{i_1,i_2,\dots,i_k}$ is an orbifold of genus $(k+1)/2$ if k is odd and genus $(k-2)/2$ if k is even. Moreover, the underlying Riemann surface is given by the hyperelliptic curve*

$$w^2 = \begin{cases} z(z-1)(z-\lambda_{i_1})(z-\lambda_{i_2})\cdots(z-\lambda_{i_k}), & \text{if } k \text{ is odd,} \\ (z-\lambda_{i_1})(z-\lambda_{i_2})\cdots(z-\lambda_{i_k}), & \text{if } k \text{ is even.} \end{cases}$$

(3) *The group generated by any two different such subgroups is L.*

PROOF. Property (1) holds trivially as L is an abelian group. Property (2) follows from Riemann-Hurwitz formula and Lemma 4. Property (3) is clear. □

The next result states that the sum of the genera appearing in all quotients of the form $Y/L_{i_1,i_2,\dots,i_k}$ is equal to the genus of Y.

LEMMA 6. $\sum_{k=1}^{r}\binom{r}{k}\frac{(1-(-1)^k)}{2}\frac{(k+1)}{2}+\sum_{k=4}^{r}\binom{r}{k}\frac{(1+(-1)^k)}{2}\frac{(k-2)}{2} = 1+2^{r-2}(r-1).$

PROOF. If $f_1(x) = x\frac{(1+x)^r}{4} = \frac{1}{4}\sum_{k=0}^{r}\binom{r}{k}x^{k+1}$ and $f_2(x) = \frac{(1+x)^r}{4x^2} = \frac{1}{4}\sum_{k=0}^{r}\binom{r}{k}x^{k-2}$, then $f_1'(x) = \frac{(1+x)^r(1+(1+r)x)}{4} = \frac{1}{4}\sum_{k=0}^{r}\binom{r}{k}(k+1)x^k$ and $f_2'(x) = \frac{(1+x)^{r-1}(r-2x(1+x))}{4x^2} = \frac{1}{4}\sum_{k=0}^{r}\binom{r}{k}(k-2)x^{k-3}$. Then $2^{r-3}(r+2) = f_1'(1) - f_1'(-1) = \frac{1}{4}\sum_{k=0}^{r}\binom{r}{k}(k+1)(1-(-1)^k) = \sum_{k=1}^{r}\binom{r}{k}\frac{(1-(-1)^k)}{2}\frac{(k+1)}{2}$ and $2^{r-3}(r-4) = f_2'(1) + f_2'(-1) = \frac{1}{4}\sum_{k=0}^{r}\binom{r}{k}(k-2)(1+(-1)^k) = -1 + \sum_{k=4}^{r}\binom{r}{k}\frac{(1+(-1)^k)}{2}\frac{(k-2)}{2}$. By adding these two equalities we obtain the desired result. □

We may now apply Theorem 1 for Y using the subgroups $L_{i_1,\dots,i_k}$ in order to obtain that JY is isogenous to a product of the form $E_{\lambda_1} \times \cdots \times E_{\lambda_r} \times A_{g-r}$, where A_{g-r} is the product of the jacobian varieties of all elliptic/hyperelliptic Riemann surfaces $Y/L_{i_1,\dots,i_k}$, for $k \geq 2$.

REMARK 7 (A fiber product description of Y). We may generalize the fiber product description done for $r = 2$ in Remark 3, for every $r \geq 2$, as follows. Let $\pi_j : E_{\lambda_j} \to \widehat{\mathbb{C}}$ be defined by $\pi_j(x_j, y_j) = x_j$. The locus of branch values of π_j is the set $\{\infty, 0, 1, \lambda_j\}$. An affine model of the fiber product of the r pairs $(E_{\lambda_1}, \pi_1), \dots, (E_{\lambda_r}, \pi_r)$ is

$$\widehat{Y} = \left\{(x, y_1, \dots, y_r) : y_j^2 = x(x-1)(x-\lambda_j),\ j = 1, \dots, r\right\},$$

which is irreducible and has singular points at those points with first coordinate $x \in \{\infty, 0, 1\}$. The Riemann surface T defined by $\widehat{Y}$ (after desingularization) admits a group of conformal automorphisms $H \cong \mathbb{Z}_2^r$ so that T/H is the Riemann sphere with conical points (each one of order two) at ∞, 0, 1, $\lambda_1, \dots, \lambda_r$. In particular, it has genus $g = 1 + 2^{r-2}(r-1)$. It can be checked that T is isomorphic to the surface Y of Theorem 6.

5. Explicit examples

5.1. Example of genus five. In Theorem 6, assume $r = 3$. In this case, Y has genus five and JY is isogenous to $E_{\lambda_1} \times E_{\lambda_2} \times E_{\lambda_3} \times JT$, where $E_{\lambda_1} : y^2 = x(x-1)(x-\lambda_1)$, $E_{\lambda_2} : y^2 = x(x-1)(x-\lambda_2)$, $E_{\lambda_3} : y^2 = x(x-1)(x-\lambda_3)$ and $T : y^2 = x(x-1)(x-\lambda_1)(x-\lambda_2)(x-\lambda_3)$. The corresponding subgroups of H are in this case $L_1 = \langle c_2, c_3 \rangle$, $L_2 = \langle c_1, c_3 \rangle$, $L_3 = \langle c_1, c_2 \rangle$ and $L_{1,2,3} = \langle c_1c_2, c_1c_3 \rangle$.

COROLLARY 1. *If $\lambda_3 = \lambda_1/\lambda_2$, then JY isogenous to the product of 5 elliptic curves. So, it provides a 2-dimensional family of curves Y of genus five with JY isogenous to the product of five elliptic curves.*

PROOF. If $\lambda_3 = \lambda_1/\lambda_2$, then T admits the involution $(x,y) \mapsto (\lambda_1/x, \lambda_1^{3/2}y/x^3)$, with exactly two fixed points. It follows that JT is isogenous to the product of two elliptic curves. ∎

5.2. Example of genus nine. Next, we provide a two-dimensional family of genus nine Riemann surfaces whose jacobian varieties are isogenous to the product of nine elliptic curves. For $(\lambda, \mu) \in \Delta_2$ set $\mu_{1,1} = \mu$, $\mu_{1,2} = \lambda/\mu$, $\mu_{2,1} = \lambda(\mu - 1)/(\mu - \lambda)$, $\mu_{2,2} = (\mu-\lambda)/(\mu-1)$,

$$K_1 = (\mu_{2,2}-\mu_{1,1})(\mu_{2,2}-\mu_{1,2})(\mu_{2,2}-\mu_{2,1}), \qquad K_2 = (\mu_{2,2}-1)(\mu_{2,2}-\lambda)(\mu_{2,2}-\mu_{2,1}),$$
$$K_3 = (\mu_{2,2}-1)(\mu_{2,2}-\lambda)(\mu_{2,2}-\mu_{1,1})(\mu_{2,2}-\mu_{1,2}), \qquad K_4 = -\mu_{2,2}(\mu_{2,2}-\mu_{2,1}),$$
$$K_5 = -\mu_{2,2}(\mu_{2,2}-\mu_{1,1})(\mu_{2,2}-\mu_{1,2}), \qquad K_6 = -\mu_{2,2}(\mu_{2,2}-1)(\mu_{2,2}-\lambda).$$

Let R be the genus nine Riemann surface defined by the curve

$$\left\{\begin{array}{lcl}
w_1^2 & = & K_1\left(z-\dfrac{1}{\mu_{1,1}-\mu_{2,2}}\right)\left(z-\dfrac{1}{\mu_{1,2}-\mu_{2,2}}\right)\left(z-\dfrac{1}{\mu_{2,1}-\mu_{2,2}}\right)\\
w_2^2 & = & K_2\left(z-\dfrac{1}{1-\mu_{2,2}}\right)\left(z-\dfrac{1}{\lambda-\mu_{2,2}}\right)\left(z-\dfrac{1}{\mu_{2,1}-\mu_{2,2}}\right)\\
w_3^2 & = & K_3\left(z-\dfrac{1}{1-\mu_{2,2}}\right)\left(z-\dfrac{1}{\lambda-\mu_{2,2}}\right)\left(z-\dfrac{1}{\mu_{1,1}-\mu_{2,2}}\right)\left(z-\dfrac{1}{\mu_{1,2}-\mu_{2,2}}\right)\\
w_4^2 & = & K_4z\left(z+\dfrac{1}{\mu_{2,2}}\right)\left(z-\dfrac{1}{\mu_{2,1}-\mu_{2,2}}\right)\\
w_5^2 & = & K_5z\left(z+\dfrac{1}{\mu_{2,2}}\right)\left(z-\dfrac{1}{\mu_{1,1}-\mu_{2,2}}\right)\left(z-\dfrac{1}{\mu_{1,2}-\mu_{2,2}}\right)\\
w_6^2 & = & K_6z\left(z+\dfrac{1}{\mu_{2,2}}\right)\left(z-\dfrac{1}{1-(\mu_{2,2}}\right)\left(z-\dfrac{1}{\lambda-\mu_{2,2}}\right)\\
w_7^2 & = & w_3^2w_4^2
\end{array}\right\}$$

By Theorem 3 (and Lemma 2), $JR \cong_{isog.} E_1 \times \cdots \times E_6 \times JT$, where $E_1 : y^2 = x(x-1)(x-\lambda)$, $E_2 : y^2 = (x-1)(x-\lambda)(x-\mu_{1,1})(x-\mu_{1,2})$, $E_3 : y^2 = (x-\mu_{1,1})(x-\mu_{1,2})(x-\mu_{2,1})(x-\mu_{2,2})$, $E_4 : y^2 = x(x-\mu_{2,1})(x-\mu_{2,2})$, $E_5 : y^2 = x(x-\mu_{1,1})(x-\mu_{1,2})$, $E_6 : y^2 = (x-1)(x-\lambda)(x-\mu_{2,1})(x-\mu_{2,2})$ and $T : y^2 = x(x-1)(x-\lambda)(x-\mu_{1,1})(x-\mu_{1,2})(x-\mu_{2,1})(x-\mu_{2,2})$. The group $J = \langle f_1(x) = \lambda/x, f_2(x) = \lambda(x-1)/(x-\lambda)\rangle \cong \mathbb{Z}_2^2$ keeps invariant the set $\{\infty, 0, 1, \lambda, \mu_{1,1}, \mu_{1,2}, \mu_{2,1}, \mu_{2,2}\}$. In this way, T admits the following automorphisms

$$A_1(x,y) = \left(\lambda/x, \lambda^2y/x^4\right), \; A_2(x,y) = \left(\lambda(x-1)/(x-\lambda), \lambda^2(\lambda-1)^2y/(x-\lambda)^4\right).$$

We may see that $\langle A_1, A_2\rangle \cong \mathbb{Z}_2^2$ and that each of the involutions A_1, A_2 and $A_1 \circ A_2$ acts with exactly 4 fixed points on T. The quotients $T/\langle A_1\rangle$, $T/\langle A_2\rangle$ and $T/\langle A_1 \circ A_2\rangle$ have genus one. We may apply Theorem 1 to T using the three cyclic groups of order two in order to see that JT is isogenous to the product of three elliptic curves. In this way, JR is isogenous to the product of nine elliptic curves.

5.3. Example of genus thirteen. Set, in Theorem 6, the case $r=4$. In this case, Y has genus 13 and JY is isogenous to $E_{\lambda_1}\times E_{\lambda_2}\times E_{\lambda_3}\times E_{\lambda_4}\times JT_1\times JT_2\times JT_3\times JT_4\times E_5$, where $E_{\lambda_1}: y^2=x(x-1)(x-\lambda_1)$, $E_{\lambda_2}: y^2=x(x-1)(x-\lambda_2)$, $E_{\lambda_3}: y^2=x(x-1)(x-\lambda_3)$, $E_{\lambda_4}: y^2=x(x-1)(x-\lambda_4)$, $T_1: y^2=x(x-1)(x-\lambda_1)(x-\lambda_2)(x-\lambda_3)$, $T_2: y^2=x(x-1)(x-\lambda_1)(x-\lambda_2)(x-\lambda_4)$, $T_3: y^2=x(x-1)(x-\lambda_1)(x-\lambda_3)(x-\lambda_4)$, $T_4: y^2=x(x-1)(x-\lambda_2)(x-\lambda_3)(x-\lambda_4)$ and $E_5: y^2=(x-\lambda_1)(x-\lambda_2)(x-\lambda_3)(x-\lambda_4)$. The corresponding subgroups of H are in this case $L_1=\langle c_2,c_3,c_4\rangle$, $L_2=\langle c_1,c_3,c_4\rangle$, $L_3=\langle c_1,c_2,c_4\rangle$, $L_4=\langle c_1,c_2,c_3\rangle$, $L_{1,2,3}=\langle c_1c_2,c_1c_3,c_4\rangle$, $L_{1,2,4}=\langle c_1c_2,c_1c_4,c_3\rangle$, $L_{1,3,4}=\langle c_1c_3,c_1c_4,c_2\rangle$, $L_{2,3,4}=\langle c_2c_3,c_2c_4,c_1\rangle$ and $L_{1,2,3,4}=\langle c_1c_2,c_1c_3,c_1c_4\rangle$.

COROLLARY 2. *If $\lambda_3=\lambda_1/\lambda_2$, $\lambda_4=\lambda_1(\lambda_2-1)/(\lambda_2-\lambda_1)$ and $\lambda_2^2(1+\lambda_1)-4\lambda_1\lambda_2+\lambda_1(1+\lambda_1)=0$, then JY isogenous to the product of 13 elliptic curves.*

PROOF. If $a_1(z)=\lambda_1/z$ and $a_2(z)=\lambda_1(z-1)/(z-\lambda_1)$, then the group generated by them is isomorphic to $\mathbb{Z}_2^2$. Since a_1 permutes in pairs the elements in $\{\infty,0,1,\lambda_1,\lambda_2,\lambda_3\}$, it follows that JT_1 is isogenous to the product of two elliptic curves. Similarly, as a_2 permutes in pairs the elements in $\{\infty,0,1,\lambda_1,\lambda_2,\lambda_4\}$, it follows that JT_2 is isogenous to the product of two elliptic curves and as a_2a_1 permutes in pairs the elements in $\{\infty,0,1,\lambda_1,\lambda_3,\lambda_4\}$, it follows that JT_3 is isogenous to the product of two elliptic curves. In this way, under the above assumptions, JY is isogenous to the product of 11 elliptic curves and JT_4. If we also assume that $\lambda_2^2(1+\lambda_1)-4\lambda_1\lambda_2+\lambda_1(1+\lambda_1)=0$, then $a_3(z)=\lambda_2(z-\lambda_3)/(z-\lambda_2)$ permutes in pairs the elements of the set $\{\infty,0,1,\lambda_2,\lambda_3,\lambda_4\}$. In this case, JT_4 is also isogenous to the product of two elliptic curves □

REMARK 8. Examples of values as above are $\lambda_1=2$ and $\lambda_2=(4+i\sqrt{2})/3$; so $\lambda_3=(4-i\sqrt{2})/3$ and $\lambda_4=-i\sqrt{2}$.

Acknowledgments

The author would like to express his deep gratitude to the referee for supplying very useful comments, suggestions and corrections.

References

[1] Daniel J. Bernstein and Tanja Lange, *Hyper-and-elliptic-curve cryptography*, LMS J. Comput. Math. **17** (2014), no. suppl. A, 181–202, DOI 10.1112/S1461157014000394. MR3240803

[2] Gabriel Cardona, *$\mathbb{Q}$-curves and abelian varieties of* GL_2*-type from dihedral genus 2 curves*, Modular curves and abelian varieties, Progr. Math., vol. 224, Birkhäuser, Basel, 2004, pp. 45–52. MR2058641

[3] Angel Carocca, Víctor González-Aguilera, Rubén A. Hidalgo, and Rubí E. Rodríguez, *Generalized Humbert curves*, Israel J. Math. **164** (2008), 165–192, DOI 10.1007/s11856-008-0025-2. MR2391145

[4] Angel Carocca and Rubí E. Rodríguez, *Jacobians with group actions and rational idempotents*, J. Algebra **306** (2006), no. 2, 322–343, DOI 10.1016/j.jalgebra.2006.07.027. MR2271338

[5] Mariela Carvacho, Rubén A. Hidalgo, and Saúl Quispe, *Jacobian variety of generalized Fermat curves*, Q. J. Math. **67** (2016), no. 2, 261–284, DOI 10.1093/qmath/haw009. MR3509993

[6] Iwan Duursma and Negar Kiyavash, *The vector decomposition problem for elliptic and hyperelliptic curves*, J. Ramanujan Math. Soc. **20** (2005), no. 1, 59–76. MR2140763

[7] Clifford J. Earle, *Some Jacobian varieties which split*, Complex analysis Joensuu 1978 (Proc. Colloq., Univ. Joensuu, Joensuu, 1978), Lecture Notes in Math., vol. 747, Springer, Berlin, 1979, pp. 101–107. MR553033

[8] Clifford J. Earle, *The genus two Jacobians that are isomorphic to a product of elliptic curves*, The geometry of Riemann surfaces and abelian varieties, Contemp. Math., vol. 397, Amer. Math. Soc., Providence, RI, 2006, pp. 27–36, DOI 10.1090/conm/397/07459. MR2217995

[9] Torsten Ekedahl and Jean-Pierre Serre, *Exemples de courbes algébriques à jacobienne complètement décomposable* (French, with English and French summaries), C. R. Acad. Sci. Paris Sér. I Math. **317** (1993), no. 5, 509–513. MR1239039

[10] Gerhard Frey and Ernst Kani, *Curves of genus* 2 *covering elliptic curves and an arithmetical application*, Arithmetic algebraic geometry (Texel, 1989), Progr. Math., vol. 89, Birkhäuser Boston, Boston, MA, 1991, pp. 153–176. MR1085258

[11] P. Gaudry and É. Schost, *On the invariants of the quotients of the Jacobian of a curve of genus 2*, Applied algebra, algebraic algorithms and error-correcting codes (Melbourne, 2001), Lecture Notes in Comput. Sci., vol. 2227, Springer, Berlin, 2001, pp. 373–386, DOI 10.1007/3-540-45624-4_39. MR1913484

[12] Tsuyoshi Hayashida and Mieo Nishi, *Existence of curves of genus two on a product of two elliptic curves*, J. Math. Soc. Japan **17** (1965), 1–16, DOI 10.2969/jmsj/01710001. MR201434

[13] C. Hermite. Sur un example de réduction d'intégrales abéliennes aux fonctions elliptiques. *Ann. Soc. Sci. Bruxelles* **1**: (1876), 1–16.

[14] Rubén A. Hidalgo, Aristides Kontogeorgis, Maximiliano Leyton-Álvarez, and Panagiotis Paramantzoglou, *Automorphisms of generalized Fermat curves*, J. Pure Appl. Algebra **221** (2017), no. 9, 2312–2337, DOI 10.1016/j.jpaa.2016.12.011. MR3631719

[15] Ernst Kani, *Elliptic curves on abelian surfaces*, Manuscripta Math. **84** (1994), no. 2, 199–223, DOI 10.1007/BF02567454. MR1285957

[16] E. Kani and M. Rosen, *Idempotent relations and factors of Jacobians*, Math. Ann. **284** (1989), no. 2, 307–327, DOI 10.1007/BF01442878. MR1000113

[17] H. Lange and S. Recillas, *Abelian varieties with group action*, J. Reine Angew. Math. **575** (2004), 135–155, DOI 10.1515/crll.2004.076. MR2097550

[18] Kay Magaard, Tanush Shaska, and Helmut Völklein, *Genus 2 curves that admit a degree 5 map to an elliptic curve*, Forum Math. **21** (2009), no. 3, 547–566, DOI 10.1515/FORUM.2009.027. MR2526800

[19] Ryo Nakajima, *On splitting of certain Jacobian varieties*, J. Math. Kyoto Univ. **47** (2007), no. 2, 391–415, DOI 10.1215/kjm/1250281052. MR2376963

[20] Jennifer Paulhus, *Decomposing Jacobians of curves with extra automorphisms*, Acta Arith. **132** (2008), no. 3, 231–244, DOI 10.4064/aa132-3-3. MR2403651

[21] Jennifer Paulhus, *Elliptic factors in Jacobians of hyperelliptic curves with certain automorphism groups*, ANTS X—Proceedings of the Tenth Algorithmic Number Theory Symposium, Open Book Ser., vol. 1, Math. Sci. Publ., Berkeley, CA, 2013, pp. 487–505, DOI 10.2140/obs.2013.1.487. MR3207428

[22] Jennifer Paulhus and Anita M. Rojas, *Completely decomposable Jacobian varieties in new genera*, Exp. Math. **26** (2017), no. 4, 430–445, DOI 10.1080/10586458.2016.1199981. MR3684576

[23] Gonzalo Riera and Rubí E. Rodríguez, *The period matrix of Bring's curve*, Pacific J. Math. **154** (1992), no. 1, 179–200. MR1154738

[24] Christophe Ritzenthaler and Matthieu Romagny, *On the Prym variety of genus 3 covers of genus 1 curves* (English, with English and French summaries), Épijournal Géom. Algébrique **2** (2018), Art. 2, 8. MR3781951

[25] Anita M. Rojas, *Group actions on Jacobian varieties*, Rev. Mat. Iberoam. **23** (2007), no. 2, 397–420, DOI 10.4171/RMI/500. MR2371432

[26] Takuya Yamauchi, *On* $\mathbb{Q}$*-simple factors of Jacobian varieties of modular curves*, Yokohama Math. J. **53** (2007), no. 2, 149–160. MR2302608

Departamento de Matemática y Estadística, Universidad de La Frontera. Temuco, Chile

Email address: ruben.hidalgo@ufrontera.cl

Contemporary Mathematics
Volume **766**, 2021
https://doi.org/10.1090/conm/766/15383

Weil's Galois descent theorem from a computational point of view

Rubén A. Hidalgo and Sebastián Reyes-Carocca

Abstract. Let $\mathcal{L}/\mathcal{K}$ be a finite Galois extension and let X be an affine algebraic variety defined over $\mathcal{L}$. Weil's Galois descent theorem provides necessary and sufficient conditions for X to be definable over $\mathcal{K}$, that is, for the existence of an algebraic variety Y defined over $\mathcal{K}$ together with a birational isomorphism $R : X \to Y$ defined over $\mathcal{L}$. Weil's proof does not provide a method to construct the birational isomorphism R. The aim of this paper is to give an explicit construction of R.

1. Introduction

Let $\mathcal{K}$ be a perfect field and let $\mathcal{C}$ be an algebraic closure of it. An affine algebraic variety $X \subset \mathcal{C}^n$ is said to be *defined* over a subfield $\mathcal{R}$ of $\mathcal{C}$ if its corresponding ideal of polynomials $\mathbb{I}(X)$ can be generated by a finite collection of polynomials with coefficients in $\mathcal{R}$. Let us assume that X is defined over a subfield $\mathcal{L}$ of $\mathcal{C}$ which is a finite Galois extension of $\mathcal{K}$. Under this assumption, we will say that X is *definable* over $\mathcal{K}$, with respect to the Galois extension $\mathcal{L}/\mathcal{K}$, if there is an algebraic variety Y defined over $\mathcal{K}$ and a birational isomorphism $R : X \to Y$ defined over $\mathcal{L}$.

To decide whether or not X is definable over $\mathcal{K}$ is, in general, a difficult task. For instance, if $\mathcal{L}$ is the field of complex numbers and $\mathcal{K}$ is the field of real numbers, then there are known explicit examples of complex algebraic curves which are not definable over the reals. These examples were provided by Shimura [**21**] and Earle [**5, 6**] and later by Huggins [**12**] (in the hyperelliptic case) and by the first author [**9**] and Kontogeorgis [**15**] (in the non-hyperelliptic situation).

The natural action of $\mathrm{Gal}(\mathcal{L}/\mathcal{K})$ on the ring of polynomials with coefficients in $\mathcal{L}$ induces a well-defined action $(\sigma, X) \to X^\sigma$ on the set of birational isomorphism classes of algebraic varieties. A collection of birational isomorphisms

$$\{f_\sigma : X \to X^\sigma : \sigma \in \mathrm{Gal}(\mathcal{L}/\mathcal{K})\}$$

defined over $\mathcal{L}$ satisfying the so-called *Weil's co-cycle condition*

$$f_{\tau\sigma} = f_\sigma^\tau \circ f_\tau \quad \text{for each} \quad \sigma, \tau \in \mathrm{Gal}(\mathcal{L}/\mathcal{K})$$

is called a *Galois descent datum* for X with respect to $\mathcal{L}/\mathcal{K}$.

2020 *Mathematics Subject Classification.* Primary 14E99, 14A10, 12F10.
Key words and phrases. Algebraic varieties, Galois extensions, Fields of definition.
Partially supported by Fondecyt Grant 1190001.
Partially supported by Fondecyt Grant 11180024, 1190991 and Redes Grant 2017-170071.

Assume that X is definable over $\mathcal{K}$. Namely, suppose the existence of a birational isomorphism $R : X \to Y$ defined over $\mathcal{L}$ where Y is an affine algebraic variety defined over $\mathcal{K}$. Then $Y = Y^\sigma$ for each $\sigma \in \text{Gal}(\mathcal{L}/\mathcal{K})$, and the collection

$$\{(R^\sigma)^{-1} \circ R : X \to X^\sigma : \sigma \in \text{Gal}(\mathcal{L}/\mathcal{K})\}$$

is a Galois descent datum for X with respect to $\mathcal{L}/\mathcal{K}$. In other words, the existence of such a Galois descent datum for X is a necessary condition for X to be definable over $\mathcal{K}$. Conversely, Weil in [**22**] proved that the existence of such a Galois descent datum is also sufficient condition. More precisely:

THEOREM 1.1 (Weil's Galois descent theorem). *Let $\mathcal{K}$ be a perfect field, let $\mathcal{C}$ be an algebraic closure of $\mathcal{K}$ and let $\mathcal{L}$ be a subfield of $\mathcal{C}$ that is a finite Galois extension of $\mathcal{K}$. Set $\Gamma = \text{Gal}(\mathcal{L}/\mathcal{K})$ and assume that X is an affine algebraic variety defined over $\mathcal{L}$.*

(a) If X admits a Galois descent datum $\{f_\sigma\}_{\sigma\in\Gamma}$ with respect to $\mathcal{L}/\mathcal{K}$, then there exists an algebraic variety Y, defined over $\mathcal{K}$, and there exists a birational isomorphism $R : X \to Y$, defined over $\mathcal{L}$, such that $R = R^\sigma \circ f_\sigma$ for every $\sigma \in \Gamma$. Moreover, if all the isomorphisms f_σ are biregular then R can be chosen to be biregular.

(b) If there is another birational isomorphism $\hat{R} : X \to \hat{Y}$, defined over $\mathcal{L}$, where $\hat{Y}$ is defined over $\mathcal{K}$, such that $\hat{R} = \hat{R}^\sigma \circ f_\sigma$ for every $\sigma \in \Gamma$, then there exists a birational isomorphism $J : Y \to \hat{Y}$, defined over $\mathcal{K}$, such that $\hat{R} = J \circ R$.

Weil's proof does not provide an algorithm to construct the isomorphism R explicitly. However, in the proof of [**22**, Proposition 1], it was observed that if $\{f_\sigma\}$ is a Galois descent datum, if each f_σ is biregular and if an explicitly birational map R as before is known, then there is an explicit method to obtain a new biregular isomorphism $X \to Z$, defined over $\mathcal{L}$, with Z still defined over $\mathcal{K}$. Such an explicit method is given by considering the map

$$F : X \to Y^n \quad \text{defined by} \quad x \mapsto (R(x), R^{\sigma_2}(x), \ldots, R^{\sigma_n}(x)),$$

where $\Gamma = \{\sigma_1 = e, \sigma_2, \ldots, \sigma_n\}$ and then, as there is a natural permutation action of Γ, to consider the classical invariant theory to construct a regular map $\Psi : Y^n \to Z$, defined over $\mathcal{L}$, so that Z is defined over $\mathcal{K}$ and $\Psi \circ F : X \to Z$ is a biregular isomorphism.

In this article, we follow similar ideas as above to construct explicitly a rational map $R : X \to \mathcal{C}^m$, defined over $\mathcal{L}$, such that $Y = R(X)$ is defined over $\mathcal{K}$ and $R : X \to Y$ is a birational isomorphism. This explicit construction is done in terms of equations for X and of a Galois descent datum for X with respect to $\mathcal{L}/\mathcal{K}$. Since R is explicitly given, the algorithm can be used to compute explicit equations for Y over $\mathcal{K}$. Indeed, for the sake of completeness, in the last section we will work out an example where X is a complex algebraic curve of genus five defined over $\mathbb{Q}(i)$. This curve admits a group of conformal automorphisms isomorphic to $\mathbb{Z}_2^4$ and it is also endowed with an anticonformal involution; so, it is definable over $\mathbb{Q}$.

2. Preliminaries

Let $\mathcal{K}$ be a perfect field, let $\mathcal{C}$ be an algebraic closure of it, and let $\mathcal{L}$ be a subfield of $\mathcal{C}$ which is a finite Galois extension of $\mathcal{K}$. We denote by

$$\hat{\Gamma} = \text{Gal}(\mathcal{C}/\mathcal{K}) \quad \text{and} \quad \Gamma = \text{Gal}(\mathcal{L}/\mathcal{K})$$

the Galois group associated to the extensions $\mathcal{C}/\mathcal{K}$ and $\mathcal{L}/\mathcal{K}$ respectively.

Each $\eta \in \hat{\Gamma}$ induces a natural bijection

$$\hat{\eta} : \mathcal{C}^n \to \mathcal{C}^n \ \text{ given by } (y_1, \dots, y_n) \mapsto (\eta(y_1), \dots, \eta(y_n)),$$

and if $P \in \mathcal{C}[z_1, \dots, z_n]$ then we denote by P^η the polynomial obtained after applying η to the coefficients of P. In other words, the following diagram commutes.

$$\begin{array}{ccc} \mathcal{C}^n & \xrightarrow{\quad P \quad} & \mathcal{C} \\ \hat{\eta}\downarrow & & \downarrow\eta \\ \mathcal{C}^n & \xrightarrow{\quad P^\eta \quad} & \mathcal{C} \end{array}$$

Let

$$X = \{(y_1, \dots, y_n) \in \mathcal{C}^n : P_j(y_1, \dots, y_n) = 0,\ 1 \leq j \leq r\}$$

be an affine algebraic variety where each $P_j \in \mathcal{L}[z_1, \dots, z_n]$. If $\eta \in \hat{\Gamma}$ then $P_j^\eta \in \mathcal{L}[z_1, \dots, z_n]$ and

$$\hat{\eta}(X) = \{(y_1, \dots, y_n) \in \mathcal{C}^n : P_j^\eta(y_1, \dots, y_n) = 0,\ 1 \leq j \leq r\}.$$

Let us denote by $\rho : \hat{\Gamma} \to \Gamma$ the canonical epimorphism defined by restriction. Note that:

(1) if $\rho(\eta) = \sigma$ then $P_j^\eta = P_j^\sigma$, and
(2) if $\rho(\eta_1) = \rho(\eta_2)$ then $\hat{\eta_1}(X) = \hat{\eta_2}(X)$.

Then, if $\rho(\eta) = \sigma$ then we denote $\hat{\eta}(X)$ by X^σ.

Let $\{e_1, \dots, e_m\}$ be a basis of $\mathcal{L}$ as a $\mathcal{K}$–vector space. Then the matrix

$$A = \begin{bmatrix} e_1 & e_2 & \cdots & e_m \\ \sigma_2(e_1) & \sigma_2(e_2) & \cdots & \sigma_2(e_m) \\ \vdots & \vdots & \ddots & \vdots \\ \sigma_m(e_1) & \sigma_m(e_2) & \cdots & \sigma_m(e_m) \end{bmatrix} \in \mathrm{M}(m \times m, \mathcal{L})$$

is non-singular (see, for example, [**13**]).

The trace map

$$\mathrm{Tr} : \mathcal{L} \to \mathcal{K} \ \text{ given by } \ a \mapsto \Sigma_{j=1}^m \sigma_j(a)$$

extends naturally to polynomial rings

$$\mathrm{Tr} : \mathcal{L}[x_1, \dots, x_n] \to \mathcal{K}[x_1, \dots, x_n] \ \text{ given by } \ P \mapsto \Sigma_{j=1}^m P^{\sigma_j}.$$

LEMMA 2.1. *Under the above notations, we have the following.*

(a) *If* $P \in \mathcal{L}[x_1, \dots, x_n]$ *then* $P \in \mathrm{Span}_{\mathcal{L}}(\mathrm{Tr}(e_1 P), \dots, \mathrm{Tr}(e_m P))$.
(b) *If* $I < \mathcal{L}[x_1, \dots, x_n]$ *is an ideal so that* $P^\sigma \in I$ *for every* $\sigma \in \Gamma$ *and every* $P \in I$, *then* I *can be generated as ideal by polynomials in* $I \cap \mathcal{K}[x_1, \dots, x_n]$.

PROOF. For each $j = 1, \dots, m$ we define

$$Q_j := \mathrm{Tr}(e_j P) \in \mathcal{L}[x_1, \dots, x_n],$$

and notice that $Q_j \in \mathcal{K}[x_1, \dots, x_n]$ because $Q_j^\sigma = Q_j$ for each $\sigma \in \Gamma$.

As the matrix A is non-singular, there are values $\lambda_1, \dots, \lambda_m \in \mathcal{L}$ so that $A\lambda = E$, where $\lambda = {}^t[\lambda_1\ \lambda_2\ \dots\ \lambda_m]$ and $E = {}^t[1\ 0\ \dots\ 0]$. In other words, we have

$$\Sigma_{j=1}^m \lambda_j e_j = 1 \ \text{ and } \ \Sigma_{j=1}^m \lambda_j \sigma_k(e_j) = 0, \text{ for each } \ k = 2, \dots, m;$$

hence P can be written as

$$\lambda_1 Q_1 + \cdots + \lambda_m Q_m = \left(\Sigma_{j=1}^{m} \lambda_j e_j\right) P + \Sigma_{k=2}^{m} \left(P^{\sigma_k} \Sigma_{j=1}^{m} \lambda_j \sigma_k(e_j)\right)$$

and the first statement is proved. The second statement follows from the first one together with the fact that, if $P^{\sigma} \in I$ for every $\sigma \in \Gamma$ then $Q_j \in I$. □

LEMMA 2.2. *Let $Y \subset \mathcal{C}^n$ be an affine algebraic variety defined over $\mathcal{L}$. If $Y^{\sigma} = Y$ for every $\sigma \in \Gamma$ then Y is defined over $\mathcal{K}$. More precisely, if Y is defined by the polynomials $P_1, \ldots, P_r \in \mathcal{L}[x_1, ..., x_n]$ then Y is also defined by the polynomials $\mathrm{Tr}(e_j P_i) \in \mathcal{K}[x_1, \ldots, x_n]$, where $i \in \{1, \ldots, r\}$ and $j \in \{1, \ldots, m\}$.*

PROOF. As Y is defined over $\mathcal{L}$, its associated ideal of polynomials $\mathbb{I} < \mathcal{C}[x_1, \ldots, x_n]$ is generated by a finite collection of polynomials with coefficients in $\mathcal{L}$. Let $P \in \mathbb{I} \cap \mathcal{L}[x_1, \ldots, x_n]$. Let $\sigma \in \Gamma$ and let $\eta \in \mathrm{Gal}(\mathcal{C}/\mathcal{K})$ be so that $\rho(\eta) = \sigma$. Then, for $(b_1, \ldots, b_n) \in Y$, it holds that

$$0 = \sigma(P(b_1, \ldots, b_n)) = \eta(P(b_1, \ldots, b_n)) = P^{\eta}(\eta(b_1), \ldots, \eta(b_n)) = P^{\sigma} \circ \hat{\eta}(b_1, \ldots, b_n).$$

As $Y^{\sigma} = \hat{\eta}(Y)$ and we are assuming $Y^{\sigma} = Y$, we have that $\hat{\eta} : Y \to Y$ is a bijection. So, the above asserts that $P^{\sigma}(c_1, \ldots, c_n) = 0$ for each $(c_1, \ldots, c_n) \in Y$. This ensures that $P^{\sigma} \in \mathbb{I}$ and the desired result follows directly from Lemma 2.1. □

3. Constructive proof of Weil's Galois descent theorem

Let us assume that the algebraic variety $X \subset \mathcal{C}^n$ is defined by the polynomials $P_1, \ldots, P_r \in \mathcal{L}[x_1, \ldots, x_n]$ and that we have a Galois descent datum

$$\{f_{\sigma_j} : X \to X^{\sigma_j}\}_{j=1}^{m}$$

for X with respect to the Galois extension $\mathcal{L}/\mathcal{K}$, where $\Gamma = \mathrm{Gal}(\mathcal{L}/\mathcal{K}) = \{\sigma_1 = e, \ldots, \sigma_m\}$.

By Lemma 2.2, if $X^{\sigma} = X$ for every $\sigma \in \Gamma$ then X is already defined over $\mathcal{K}$. Thus, from now on, we assume this is not the case.

LEMMA 3.1. *We can assume that $X^{\sigma_i} \cap X^{\sigma_j} = \emptyset$ for $i \neq j$.*

PROOF. For each $\sigma_j \neq e$ we may find some $a_j \in \mathcal{L}$ so that $\sigma_j(a_j) \neq a_j$. Then we may consider the algebraic variety $\hat{X} \subset \mathcal{C}^{n+m-1}$ defined by the polynomials

$$P_1(x_1, \ldots, x_n) = \cdots = P_r(x_1, \ldots, x_n) = 0, x_{n+1} = a_2, \ldots, x_{n+m-1} = a_m.$$

Clearly, the map

$$Q : X \to \hat{X} \text{ given by } (x_1, \ldots, x_n) \mapsto (x_1, \ldots, x_n, a_2, \ldots, a_m)$$

defines a biregular isomorphism whose inverse is provided by the projection

$$Q^{-1} : \hat{X} \to X \text{ given by } (x_1, \ldots, x_n, a_2, \ldots, a_m) \to (x_1, \ldots, x_n).$$

By the construction, we see that $\hat{X}^{\sigma_i} \cap \hat{X}^{\sigma_j} = \emptyset$ for every $i \neq j$. It is easy to check that a Galois descent datum for $\hat{X}$ is given by $\{g_{\sigma} = Q^{\sigma} \circ f_{\sigma} \circ Q^{-1} : \sigma \in \Gamma\}$. □

3.1. A first isomorphism. Let us consider the rational map

$$\Phi : X \to \Pi_{\sigma\in\Gamma}\mathcal{C}^n \quad \text{defined by} \quad x \mapsto (f_\sigma(x))_{\sigma\in\Gamma}\,.$$

If $f_\sigma(x) = (\frac{r_{1,\sigma}(x)}{s_{1,\sigma}(x)}, \ldots, \frac{r_{n,\sigma}(x)}{s_{n,\sigma}(x)})$, where $s_{j,\sigma}$ and $r_{j,\sigma}$ are relatively prime polynomials (each one defined over $\mathcal{L}$), then the equality

$$f_\sigma(y_e) = y_\sigma = (y_{1,\sigma}, \ldots, y_{n,\sigma})$$

provides n polynomial equations

$$y_{j,\sigma}s_{j,\sigma}(y_e) = r_{j,\sigma}(y_e).$$

We see that $\Phi(X)$ is the affine algebraic variety defined over $\mathcal{L}$ given by

$$\begin{aligned}\Phi(X) = \{(y_\sigma)_{\sigma\in\Gamma} : P_1^\sigma(y_\sigma) = \cdots = P_r^\sigma(y_\sigma) = 0,\\ y_{j,\sigma}s_{j,\sigma}(y_e) = r_{j,\sigma}(y_e), \sigma \in \Gamma,\ j = 1, \ldots, n\}.\end{aligned}$$

LEMMA 3.2. *The map $\Phi : X \to \Phi(X)$ is a birational isomorphism. Furthermore, Φ is biregular provided that each f_σ is polynomial.*

PROOF. As each f_σ is a birational isomorphism, Φ induces a birational isomorphism between X and $\Phi(X)$. The inverse map is given by projection in the first coordinate (so it is regular)

$$\pi : \Pi_{\sigma\in\Gamma}\mathcal{C}^n \to \mathcal{C}^n \quad \text{given by} \quad (y_\sigma)_{\sigma\in\Gamma} \mapsto y_e.$$

If each f_σ is a polynomial, then Φ is a regular map and the lemma follows. □

3.2. A linear permutation action on $\Pi_{\sigma\in\Gamma}\mathcal{C}^n$ induced by Γ. Let us consider the following natural permutation action of Γ on the $\sigma-$coordinates:

$$\Theta : \Gamma \times \Pi_{\sigma\in\Gamma}\mathcal{C}^n \to \Pi_{\sigma\in\Gamma}\mathcal{C}^n \quad \text{given by} \quad \left(\tau, (y_\sigma)_{\sigma\in\Gamma}\right) \mapsto (y_{\tau\sigma})_{\sigma\in\Gamma}$$

LEMMA 3.3. *If $\tau \neq e$ then $\Theta(\tau)(\Phi(X)) \cap \Phi(X) = \emptyset$.*

PROOF. Let $(y_\sigma)_{\sigma\in\Gamma} \in \Theta(\tau)(\Phi(X)) \cap \Phi(X)$. The condition $(y_\sigma)_{\sigma\in\Gamma} \in \Phi(X)$ ensures that $y_\sigma \in X^\sigma$ and, in particular, $y_{\tau^{-1}} \in X^{\tau^{-1}}$. Moreover, the condition $(y_\sigma)_{\sigma\in\Gamma} \in \Theta(\tau)(\Phi(X))$ ensures that $\Theta(\tau^{-1})(y_\sigma)_{\sigma\in\Gamma} = (y_{\tau^{-1}\sigma})_{\sigma\in\Gamma} \in \Phi(X)$, that is, $y_{\tau^{-1}} \in X$. The above implies that $X \cap X^{\tau^{-1}} \neq \emptyset$, which is not possible by Lemma 3.1. □

3.3. A linearization of the Galois descent datum. Let $\tau \in \Gamma$ and let $\eta \in \mathrm{Gal}(\mathcal{C}/\mathcal{K})$ be so that $\rho(\eta) = \tau$. As $f_{\tau\sigma} = f_\sigma^\tau \circ f_\tau$, we see that

$$\left(\Phi^\tau \circ f_\tau\right)(x) = \left(f_\sigma^\tau(f_\tau(x))\right)_{\sigma\in\Gamma} = (f_{\tau\sigma}(x))_{\sigma\in\Gamma} = \left(\Theta(\tau) \circ \Phi\right)(x)$$

and, as $f_\sigma^\tau = f_\sigma^\eta = \hat{\eta} \circ f_\sigma \circ \hat{\eta}^{\,-1}$, where $\hat{\eta}$ acts on $\Pi_{\sigma\in\Gamma}\mathcal{C}^n$, we obtain the following equality

$$\left(\Phi \circ \hat{\eta}^{\,-1} \circ f_\tau\right)(x)) = \left(f_\sigma(\hat{\eta}^{\,-1}(f_\tau(x)))\right)_{\sigma\in\Gamma} = \left(\hat{\eta}^{\,-1}(f_\sigma^\tau(f_\tau(x)))\right)_{\sigma\in\Gamma} =$$

$$= \left(\hat{\eta}^{\,-1}(f_{\tau\sigma}(x))\right)_{\sigma\in\Gamma} = \left(\hat{\eta}^{\,-1} \circ \Theta(\tau) \circ \Phi\right)(x).$$

The above can be summarized in the following commutative diagram.

$$
\begin{array}{ccc}
X & \xrightarrow{\quad\Phi\quad} & \Phi(X) \\
{\scriptstyle f_\tau}\downarrow & & \downarrow{\scriptstyle \Theta(\tau)} \\
X^\tau & \xrightarrow{\quad\Phi^\tau\quad} & \Theta(\tau)(\Phi(X)) = \Phi^\tau(X^\tau) = \Phi(X)^\tau \\
\downarrow{\scriptstyle \hat{\eta}^{-1}} & & \downarrow{\scriptstyle \hat{\eta}^{-1}} \\
X & \xrightarrow{\quad\Phi\quad} & \Phi(X)
\end{array}
\tag{3.1}
$$

Similarly, it is not difficult to see that, for every $\tau \in \Gamma$ and every $\eta \in \mathrm{Gal}(\mathcal{C}/\mathcal{K})$, we have that

$$\Theta(\tau) \circ \hat{\eta} = \hat{\eta} \circ \Theta(\tau).$$

REMARK 3.4. Note that, from the commutative diagram (3.1), each isomorphism $f_\sigma : X \to X^\sigma$ is transformed into the linear (permutation) isomorphism $\Theta(\sigma) : \Phi(X) \to \Phi(X)^\sigma$. So, the Galois descent datum $\{f_\sigma\}_{\sigma\in\Gamma}$ for X is now transformed into the (linear) Galois descent datum $\{\Theta(\sigma)\}_{\sigma\in\Gamma}$. In other words, the above method says that we are able to change our model of X (in an explicit manner) to $W = \Phi(X)$ for which the Galois descent datum is given by permutation linear transformations.

3.4. A second isomorphism: Invariant theory. As $\Theta(\Gamma) \cong \Gamma$ is a finite group of permutations, it follows from Hilbert-Noether's theorem (see [**16, 17**] and also [**19**, Ch. 14]) that the algebra $\mathcal{C}[\Pi_{\sigma\in\Gamma}\mathcal{C}^n]^{\Theta(\Gamma)}$ of $\Theta(\Gamma)$-invariant polynomials with coefficients in $\mathcal{C}$, is finitely generated.

Let us consider a finite set of generators of such an $\mathcal{C}$-algebra, say

$$E_1((y_\sigma)_{\sigma\in\Gamma}), \ldots, E_N((y_\sigma)_{\sigma\in\Gamma}) \in \mathcal{C}[(y_\sigma)_{\sigma\in\Gamma}]^{\Theta(\Gamma)}.$$

At this point it is important to note that the finite permutation of coordinates produced by the action of $\Theta(\Gamma)$ does not depend on the field $\mathcal{C}$, that is, we may consider this permutation action on the product space $\Pi_{\sigma\in\Gamma}\mathcal{B}$, where $\mathcal{B}$ is the basic field of $\mathcal{C}$. It follows, in particular, that

$$E_1((y_\sigma)_{\sigma\in\Gamma}), \ldots, E_N((y_\sigma)_{\sigma\in\Gamma}) \in \mathcal{K}[(y_\sigma)_{\sigma\in\Gamma}].$$

For the constructiveness part we need to have explicitly computed such polynomials E_j (this can be done, for instance, with MAGMA [**2**] or Macauley2 [**7**]).

Consider the regular map

$$\Psi : \Pi_{\sigma\in\Gamma}\mathcal{C}^n \to \mathcal{C}^N \text{ given by } (y_\sigma)_{\sigma\in\Gamma} \mapsto (E_1((y_\sigma)_{\sigma\in\Gamma}), \ldots, E_N((y_\sigma)_{\sigma\in\Gamma})).$$

LEMMA 3.5. *The regular map Ψ satisfies the following properties:*

(a) $\Psi^\sigma = \Psi$ *for every* $\sigma \in \Gamma$.
(b) $\Psi \circ \Theta(\sigma) = \Psi$ *for every* $\sigma \in \Gamma$.
(c) *if* $\Psi(w) = \Psi(z)$ *then there is some* $\sigma \in \Gamma$ *so that* $w = \Theta(\sigma)(z)$.
(d) $Y = \Psi(\Phi(X))$ *and* $\Psi(\Pi_{\sigma\in\Gamma}\mathcal{C}^n)$ *are algebraic subvarieties of* $\mathcal{C}^N$.

PROOF. Properties (a) and (b) are easy to check. Property (c) is consequence of the fact that (i) a finite group is a reductive group [**11, 18**] and (ii) for a reductive group G, say acting linearly over $\mathcal{C}^d$, and a set of generators of the G-invariant polynomials, say $I_1, \ldots, I_m \in \mathcal{C}[z_1, \ldots, z_d]$, the map $A = (I_1, \ldots, I_m) : \mathcal{C}^d \to \mathcal{C}^m$

turns out to be a regular branched cover map with G as its deck group. In other words, $A(p) = A(q)$ if and only if there exists some $\alpha \in G$ so that $\alpha(p) = q$; the branch values of A agree with the $a-$images of those points with non-trivial stabilizer (for details see, for instance, [**11, 18**]). Property (d) follows from the previous ones. □

REMARK 3.6. The algebra of regular maps on $\Psi(\Pi_{\sigma\in\Gamma}\mathcal{C}^n)$ is known to be isomorphic to the algebra $\mathcal{C}[(y_\sigma)_{\sigma\in\Gamma}]^{\Theta(\Gamma)}$ of symmetric polynomials with respect to the linear group $\Theta(\Gamma)$; see [**4**]. This can be seen by considering the surjective homomorphism $\xi : \mathcal{C}[t_1, \dots, t_N] \to \mathcal{C}[(y_\sigma)_{\sigma\in\Gamma}]^{\Theta(\Gamma)}$ defined by $\xi(p) = p(E_1, \dots, E_N)$.

Properties (b) and (c) of Lemma 3.5 assert that Ψ is a finite regular (branched) cover with the finite algebraic group $\Theta(\Gamma)$ as its deck group.

LEMMA 3.7. *The map $\Psi : \Phi(X) \to Y$ is a biregular isomorphism. In particular, $R = \Psi \circ \Phi : X \to Y$ is a birational isomorphism; it is biregular provided that each f_σ is polynomial.*

PROOF. Set $\Phi(X) = W$. Since for each $\tau \in \Gamma - \{e\}$ one has that $\Theta(\tau)(W) \cap W = \emptyset$ (see Lemma 3.3), the polynomial map $\Psi : W \to Y$ is bijective. The set

$$\hat{W} := \cup_{\sigma\in\Gamma}\Theta(\sigma)(W)$$

is a reducible affine variety whose (pairwise disjoint) irreducible components are $\Theta(\sigma)(W)$, for $\sigma \in \Gamma$. Since these irreducible components are pairwise disjoint, we may see that the algebra of regular maps on $\hat{W}$, say $\mathcal{C}[\hat{W}]$, is the product of the algebras of regular maps of the components, that is,

$$\mathcal{C}[\hat{W}] = \Pi_{\sigma\in\Gamma}\mathcal{C}[\Theta(\sigma)(W)].$$

The above isomorphism is given by the restriction of each regular map of $\hat{W}$ to each of its irreducible components. Note that there is natural isomorphism

$$\rho : \mathcal{C}[W] \to \mathcal{C}[\hat{W}]^{\Theta(\Gamma)},$$

where $\mathcal{C}[\hat{W}]^{\Theta(\Gamma)}$ denotes the sub-algebra of $\Theta(\Gamma)$-invariant regular maps on $\hat{W}$. This isomorphism is given as follows. If $p \in \mathcal{C}[W]$ then for each $\sigma \in \Gamma$ we may consider the regular map $\rho_\sigma(p) = p \circ \Theta(\sigma)^{-1} \in \mathcal{C}[\Theta(\sigma)(W)]$. Then $\rho(p) := (\rho_\sigma(p))_{\sigma\in\Gamma}$ turns out to be an injective homomorphism. It is clear that every $\Theta(\Gamma)-$invariant regular map of $\hat{W}$ is obtained in that way (so ρ is surjective).

On the other hand, $\mathcal{C}[Y]$ is isomorphic to $\mathcal{C}[\hat{W}]^{\Theta(\Gamma)}$. To see this, one may consider the injective homomorphism $\chi : \mathcal{C}[Y] \to \mathcal{C}[\hat{W}]^{\Theta(\Gamma)}$ defined by $\chi(p) = p \circ \Psi$. Now, to see that χ is onto we only need to note that $\rho^{-1}(\chi(\mathcal{C}[Y]))$ is a sub-algebra of $\mathcal{C}[W]$, that W is irreducible and that Y has the same dimension as W.

In this way, $\chi^{-1} \circ \rho$ produces an isomorphism between $\mathcal{C}[W]$ and $\mathcal{C}[Y]$, that is, $\Psi : \Phi(X) \to Y$ is a biregular isomorphism and, in particular, $R = \Psi \circ \Phi : X \to Y$ is a birational isomorphism.

As $\Psi : W \to Y$ is biregular isomorphism and $\Phi : X \to W$ is biregular if each f_σ is polynomial, then $R : X \to Y$ turns out to be biregular provided that each f_σ is polynomial. □

LEMMA 3.8. *Y is defined over $\mathcal{K}$.*

PROOF. Let $\tau \in \Gamma$ and let $\eta \in \mathrm{Gal}(\mathcal{C}/\mathcal{K})$ be so that $\rho(\eta) = \tau$. It follows from part (a) of Lemma 3.5 that

$$\Psi \circ \hat{\eta} = \hat{\eta} \circ \Psi.$$

Now one sees that the bijection $\hat{\eta} : \Pi_{\sigma \in \Gamma} \mathcal{C}^n \to \Pi_{\sigma \in \Gamma} \mathcal{C}^n$ descends to the bijection $\hat{\eta} : \mathcal{C}^N \to \mathcal{C}^N$. Then it follows from diagram (3.1) that $\hat{\eta}(Y) = Y$, that is, $Y^\tau = Y$. As this holds for every $\tau \in \Gamma$, it follows from Lemma 2.2 that Y is defined over $\mathcal{K}$. □

REMARK 3.9. Above we have constructed an explicit birational isomorphism $R : X \to Y$. We should be able to construct explicit equations for Y, say given by polynomials $Q_1, \ldots, Q_m \in \mathcal{L}[t_1, \ldots, t_N]$ (again, this can be done by using MAGMA). As we already know that Y is defined over $\mathcal{K}$, so Y is also defined by some polynomials over $\mathcal{K}$. To obtain these polynomials over $\mathcal{K}$, we proceed to replace each of the polynomials Q_j, which is not already defined over $\mathcal{K}$, by the set of polynomials $\mathrm{Tr}(e_1 Q_j), \ldots, \mathrm{Tr}(e_m Q_j) \in \mathcal{K}[x_1, \ldots, x_n]$, where $\{e_1, \ldots, e_m\}$ is a basis of $\mathcal{L}$ as $\mathcal{K}$−vector space (see the proof of Lemma 2.1).

4. Applications to complex algebraic varieties and field of moduli

Weil's Galois descent theorem has also been used in the study of complex algebraic varieties and their fields of moduli; in particular, in the case of Belyi curves and dessin d'enfants (see, for example [**23**]).

Let us denote by $\mathrm{Gal}(\mathbb{C})$ the group of field automorphisms of the field of complex numbers and let X be a complex algebraic variety with a finite group $\mathrm{Aut}(X)$ of automorphisms. Let Γ_X be the subgroup of $\mathrm{Gal}(\mathbb{C})$ consisting of those elements σ with the property that X^σ and X are $\mathbb{C}$-isomorphic. The *field of moduli* of X, denoted by $\mathrm{M}(X)$, is the fixed field of Γ_X. In general, the determination of whether the field of moduli is a field of definition is a difficult task, even in the case of algebraic curves; see, for example [**1, 5, 9, 10, 12, 20, 21, 23**].

A *Galois descent datum* for X is a collection of isomorphisms $f_\sigma : X \to X^\sigma$ defined over $\mathbb{C}$ such that for every pair $\eta, \tau \in \Gamma_X$ it holds that $f_{\eta\tau} = f_\tau^\eta \circ f_\eta$. We shall suppose that X admits a Galois descent datum and that X is defined over a finite Galois extension $\mathcal{L}$ of $\mathrm{M}(X)$ (we remark that the second assumption is vacuous in the case of complex algebraic curves; see [**8**] and [**14**]). Let $\bar{\mathcal{L}}$ be the algebraic closure of $\mathcal{L}$ in $\mathbb{C}$ (which is also algebraic closure of $\mathrm{M}(X)$). As $\sigma \in \Gamma_X$ acts as the identity on $\mathrm{M}(X)$, it defines an automorphism of $\bar{\mathcal{L}}$. Galois theory implies that $\sigma \in \Gamma_X$ acts a field automorphism of $\mathcal{L}$. Now, as $\mathrm{Aut}(X)$ is finite, it can be seen that every automorphism of X is defined over $\bar{\mathcal{L}}$. If $\tau \in \mathrm{Aut}(\mathbb{C}/\bar{\mathcal{L}})$ then $X^\tau = X$ and, in particular, $\tau \in \Gamma_X$ and $f_\tau \in \mathrm{Aut}(X)$.

Let $\sigma \in \Gamma_X$. If $\tau \in \mathrm{Aut}(\mathbb{C}/\bar{\mathcal{L}})$ then $f_\sigma^\tau : X \to X^\sigma$. It follows that there is some $h_\tau \in \mathrm{Aut}(X)$ so that $f_\sigma^\tau = f_\sigma \circ h_\tau$. This ensures that f_σ is defined over $\bar{\mathcal{L}}$.

Let us consider the map

$$\Theta : \mathrm{Aut}(\mathbb{C}/\bar{\mathcal{L}}) \to \mathrm{Aut}(X) \;\; \text{given by} \;\; \sigma \mapsto f_\sigma^{-1}.$$

Since, for each $\sigma \in \Gamma_X$ the isomorphism f_σ is defined over $\bar{\mathcal{L}}$, the condition $f_{\eta\tau} = f_\tau^\eta \circ f_\eta$ ensures that the above is a homomorphism of groups. The kernel of Θ is a finite index subgroup of $\mathrm{Aut}(\mathbb{C}/\bar{\mathcal{L}})$ and its fixed field is then a finite extension of $\bar{\mathcal{L}}$; so equals to $\bar{\mathcal{L}}$. This implies that the kernel is the whole group $\mathrm{Aut}(\mathbb{C}/\bar{\mathcal{L}})$. In particular, if $\tau, \sigma \in \Gamma_X$ have the same restriction to $\bar{\mathcal{L}}$ then $f_\tau = f_\sigma$.

Now, since $\mathcal{L}$ is a finite Galois extension of $\mathrm{M}(X)$, we may see that there are only a finite number of algebraic varieties of the form X^{τ} which are isomorphic to X, where $\tau \in \mathrm{Gal}(\mathbb{C})$. In addition, we have a finite number of possible isomorphisms f_σ for $\sigma \in \Gamma_X$. Thereby, we might assume that X and all these isomorphisms are defined over $\mathcal{L}$. In this way, Weil's Galois descent theorem asserts the following.

COROLLARY 4.1. *Let X be a complex algebraic variety. Assume that* $\mathrm{Aut}(X)$ *is finite and that X is defined over a finite extension of its field of moduli* $\mathrm{M}(X)$. *Then*

(a) *X admits a Galois descent datum, say $\{f_\sigma\}_{\sigma\in\Gamma_X}$, if and only if there is an algebraic variety Y defined over* $\mathrm{M}(X)$ *and a birational isomorphism $R : X \to Y$ defined over the algebraic closure of* $\mathrm{M}(X)$, *such that $R = R^{\sigma} \circ f_\sigma$ for every $\sigma \in \Gamma_X$. If, moreover, all the isomorphisms f_σ are biregular, then R can be chosen to be biregular.*

(b) *If there is another isomorphism $\hat{R} : X \to \hat{Y}$, where $\hat{Y}$ is defined over* $\mathrm{M}(X)$ *and such that $\hat{R} = \hat{R}^{\sigma} \circ f_\sigma$, for every $\sigma \in \Gamma_X$ then there exists an isomorphism $J : Y \to \hat{Y}$ defined over* $\mathrm{M}(X)$ *so that $\hat{R} = J \circ R$.*

5. An Explicit Example: A curve of genus 5

A closed Riemann surface S of genus 5 admitting a group $H \cong \mathbb{Z}_2^4$ of conformal automorphisms is called a classical Humbert's curve. We may identify S/H with the Riemann sphere together with five cone points, which, up to conjugation by a Möbius transformation, can be supposed to be $\{\infty, 0, 1, \lambda_1, \lambda_2\}$. In [**3**] it was proved that S can be represented by the following irreducible and non-singular complex projective curve

$$\left\{\begin{array}{ccccccc} u_1^2 & + & u_2^2 & + & u_3^2 & = & 0 \\ \lambda_1 u_1^2 & + & u_2^2 & + & u_4^2 & = & 0 \\ \lambda_2 u_1^2 & + & u_2^2 & + & u_5^2 & = & 0 \end{array}\right\} \subset \mathbb{P}^4_{\mathbb{C}}.$$

We may consider an affine model by taking $u_1 = 1$. For our example we consider $\lambda_1 = -1$ and $\lambda_2 = i$. If we set $u_2 = x_1$, $u_3 = x_2$, $u_4 = x_3$ and $u_5 = x_4$, then the affine model (defined over $\mathbb{Q}(i)$) is given by

$$X : \left\{\begin{array}{ccccccc} 1 & + & x_1^2 & + & x_2^2 & = & 0 \\ -1 & + & x_1^2 & + & x_3^2 & = & 0 \\ i & + & x_1^2 & + & x_4^2 & = & 0 \end{array}\right\} \subset \mathbb{C}^4.$$

Notice that the involution

$$L : \mathbb{C}^4 \to \mathbb{C}^4 \quad \text{given by} \quad (x_1, x_2, x_3, x_4) \mapsto (-i\,\overline{x_1}, -i\,\overline{x_3}, -i\,\overline{x_2}, -i\,\overline{x_4})$$

keeps invariant X, so it defines a symmetry of X. It follows from Weil's Galois descent theorem that X is in fact definable over $\mathcal{K} = \mathbb{Q} = \mathbb{Q}(i) \cap \mathbb{R}$. Let us consider $\mathrm{Gal}(\mathbb{Q}(i)/\mathbb{Q}) = \{\sigma_1(z) = z, \sigma_2(z) = \overline{z}\}$ and set

$$f_{\sigma_2} : X \to \overline{X} = X^{\sigma_2} \quad \text{given by} \quad (x_1, x_2, x_3, x_4) \mapsto (ix_1, ix_3, ix_2, ix_4).$$

In this example,

$$\Phi : X \subset \mathbb{C}^4 \to \Phi(X) \subset \mathbb{C}^8 : (x_1, x_2, x_3, x_4) \mapsto (x_1, x_2, x_3, x_4, ix_1, ix_3, ix_2, ix_4).$$

Therefore, the equations defining $\Phi(X)$ are given by

$$\left\{\begin{array}{llll} z_1 - ix_1 = 0, & z_2 - ix_3 = 0 & z_3 - ix_2 = 0, & z_4 - ix_4 = 0 \\ 1 + x_1^2 + x_2^2 = 0, & -1 + x_1^2 + x_3^2 = 0, & i + x_1^2 + x_4^2 = 0 & \end{array}\right\}.$$

We have that $R : X \to Y = R(X) : (x_1, x_2, x_3, x_4) \mapsto (t_1, \dots, t_{14})$, where

$$t_1 = (1+i)x_1,\ t_2 = x_2 + ix_3,\ t_3 = x_3 + ix_2,\ t_4 = (1+i)x_4,\ t_5 = ix_1^2,$$
$$t_6 = x_2 + ix_3,\ t_7 = ix_2x_3,\ t_8 = ix_4^2,\ t_9 = x_1x_2 - x_1x_3,\ t_{10} = x_1x_3 - x_1x_2,$$
$$t_{11} = 0,\ t_{12} = 0,\ t_{13} = x_2x_4 - x_3x_4,\ t_{14} = x_3x_4 - x_2x_4.$$

The inverse map $R^{-1} : Y \to X$ is given as

$$(t_1, \dots, t_{14}) \mapsto (x_1, x_2, x_3, x_4) = \left(\frac{t_1}{1+i}, \frac{t_2 - it_3}{2}, \frac{t_3 - it_2}{2}, \frac{t_4}{1+i} \right).$$

Explicit equations for Y are given by

$$Y ::= \left\{ \begin{array}{c} 4 + t_2^2 - t_3^2 = 0, \\ t_1^2 + t_2t_3 = 0, \\ t_1^2 + t_4^2 - 2 = 0, \\ t_{14} = -t_{13} = t_4(t_3 - t_2)/2,\ t_{12} = t_{11} = 0, \\ t_{10} = -t_9 = -t_1(t_2 - t_3)/2,\ t_8 = t_4^2/2, \\ t_7 = (t_2^2 + t_3^2)/4,\ t_6 = t_2,\ t_5 = t_1^2/2. \end{array} \right\}$$

The above, in particular, asserts that (by forgetting the coordinates t_j, for $j \geq 5$) that the algebraic curve

$$\hat{Y} = \left\{ \begin{array}{c} 4 + w_2^2 - w_3^2 = 0 \\ w_1^2 + w_2w_3 = 0 \\ w_1^2 + w_4^2 - 2 = 0 \end{array} \right\} \subset \mathbb{C}^4$$

is isomorphic to X by the isomorphism

$$\hat{R} : X \to \hat{Y} \quad \text{given by} \quad (x_1, x_2, x_3, x_4) \mapsto (t_1, t_2, t_3, t_4) = (w_1, w_2, w_3, w_4).$$

REMARK 5.1. In this example, the MAGMA routine we can use is the following.

$>$ Q:=Rationals( );
$>$ $P < t >$:=PolynomialRing(Q);
$>$ $q := t^2 + 1$;
$>$ $K < r >$:=SplittingField(q);
$>$ $A < x_1, x_2, x_3, x_4 >$:=AffineSpace(K,4);
$>$ $B < t_1, \dots, t_{14} >$:=AffineSpace(K, 14);
$>$ X:=Scheme(A,$[1 + x_1^2 + x_2^2, -1 + x_1^2 + x_3^2, i + x_1^2 + x_4^2]$);
$>$ R:=map$< A- > B||[x_1 + z_1, \dots, x_3 * x_4 + z_3 * z_4] >$;
$>$ Image(R);
$>$ $R(X)$;

With the above routine, MAGMA provides equations for Z over $\mathbb{Q}$:

$$\left\{ \begin{array}{c} t_{14} = -t_{13} = t_4(t_3 - t_2)/2,\ t_{12} = t_{11} = 0\ t_9 + t_{10} = 0,\ t_6 - t_7 = 0\ t_5 + t_8 - 1 = 0, \\ t_4^2 - 2t_8 = 0,\ t_3^2 - 2t_7 - 2 = 0,\ t_2^2 - 2t_7 + 2 = 0, \\ t_8^2t_{10}^2 + t_8^2 - 2t_8t_{10}^2 - 2t_8 - \frac{1}{4}t_{10}^4 + t_{10}^2 + 1 = 0, \\ t_7t_{10}^2 + t_8t_{10}^2 + 2t_8 - t_{10}^2 - 2 = 0, \\ t_7t_8 - t_7 - t_8^2 + 2t_8 + \frac{1}{2}t_{10}^2 - 1 = 0, \\ t_7^2 - t_8^2 + 2t_8 - 2 = 0,\ t_2t_7 + t_2 - t_3t_8 + t_3 = 0, \\ t_2t_3 - 2t_8 + 2 = 0,\ t_1 - 1/2t_2t_{10} - \frac{1}{2}t_3t_{10} = 0, \\ t_2t_{10}^2 - 2t_3t_7 + 2t_3t_8 + t_3t_{10}^2 = 0, \\ t_2t_8 - t_2 - t_3t_7 + t_3 = 0. \end{array} \right\}.$$

References

[1] Michela Artebani and Saül Quispe, *Fields of moduli and fields of definition of odd signature curves*, Arch. Math. (Basel) **99** (2012), no. 4, 333–344, DOI 10.1007/s00013-012-0427-6. MR2990152

[2] Wieb Bosma, John Cannon, and Catherine Playoust, *The Magma algebra system. I. The user language*, J. Symbolic Comput. **24** (1997), no. 3-4, 235–265, DOI 10.1006/jsco.1996.0125. Computational algebra and number theory (London, 1993). MR1484478

[3] Angel Carocca, Víctor González-Aguilera, Rubén A. Hidalgo, and Rubí E. Rodríguez, *Generalized Humbert curves*, Israel J. Math. **164** (2008), 165–192, DOI 10.1007/s11856-008-0025-2. MR2391145

[4] David A. Cox, John Little, and Donal O'Shea, *Using algebraic geometry*, 2nd ed., Graduate Texts in Mathematics, vol. 185, Springer, New York, 2005. MR2122859

[5] Clifford J. Earle, *On the moduli of closed Riemann surfaces with symmetries*, Advances in the theory of riemann surfaces (Proc. Conf., Stony Brook, N.Y., 1969), Princeton Univ. Press, Princeton, N.J., 1971, pp. 119–130. Ann. of Math. Studies, No. 66. MR0296282

[6] Clifford J. Earle, *Diffeomorphisms and automorphisms of compact hyperbolic 2-orbifolds*, Geometry of Riemann surfaces, London Math. Soc. Lecture Note Ser., vol. 368, Cambridge Univ. Press, Cambridge, 2010, pp. 139–155. MR2665008

[7] D. R. Grayson and M. E. Stillman, *Macaulay2, a software system for research in algebraic geometry.* Available at http://www.math.uiuc.edu/Macaulay2/

[8] H. Hammer and F. Herrlich, *A remark on the moduli field of a curve*, Arch. Math. (Basel) **81** (2003), no. 1, 5–10, DOI 10.1007/s00013-003-4649-5. MR2002711

[9] Rubén A. Hidalgo, *Erratum to: Non-hyperelliptic Riemann surfaces with real field of moduli but not definable over the reals [MR2540787]*, Arch. Math. (Basel) **98** (2012), no. 5, 449–451, DOI 10.1007/s00013-012-0378-y. MR2922688

[10] Rubén A. Hidalgo and Sebastian Reyes, *Fields of moduli of classical Humbert curves*, Q. J. Math. **63** (2012), no. 4, 919–930, DOI 10.1093/qmath/har017. MR2999991

[11] David Hilbert, *Ueber die Theorie der algebraischen Formen* (German), Math. Ann. **36** (1890), no. 4, 473–534, DOI 10.1007/BF01208503. MR1510634

[12] Bonnie Huggins, *Fields of moduli of hyperelliptic curves*, Math. Res. Lett. **14** (2007), no. 2, 249–262, DOI 10.4310/MRL.2007.v14.n2.a8. MR2318623

[13] Thomas W. Hungerford, *Algebra*, Graduate Texts in Mathematics, vol. 73, Springer-Verlag, New York-Berlin, 1980. Reprint of the 1974 original. MR600654

[14] Shoji Koizumi, *The fields of moduli for polarized abelian varieties and for curves*, Nagoya Math. J. **48** (1972), 37–55. MR352095

[15] Aristides Kontogeorgis, *Field of moduli versus field of definition for cyclic covers of the projective line* (English, with English and French summaries), J. Théor. Nombres Bordeaux **21** (2009), no. 3, 679–692. MR2605539

[16] Emmy Noether, *Der Endlichkeitssatz der Invarianten endlicher Gruppen* (German), Math. Ann. **77** (1915), no. 1, 89–92, DOI 10.1007/BF01456821. MR1511848

[17] E. Noether, *Der Endlichkeitssatz der Invarianten endlicher linear Gruppen der Characteristik p,* Nachr. Akad. Wiss. Göttingen (1926), 28–35.

[18] D. Mumford, J. Fogarty, and F. Kirwan, *Geometric invariant theory*, 3rd ed., Ergebnisse der Mathematik und ihrer Grenzgebiete (2) [Results in Mathematics and Related Areas (2)], vol. 34, Springer-Verlag, Berlin, 1994. MR1304906

[19] Claudio Procesi, *Lie groups*, Universitext, Springer, New York, 2007. An approach through invariants and representations. MR2265844

[20] Sebastián Reyes-Carocca, *Field of moduli of generalized Fermat curves*, Q. J. Math. **63** (2012), no. 2, 467–475, DOI 10.1093/qmath/haq049. MR2925300

[21] Goro Shimura, *On the field of rationality for an abelian variety*, Nagoya Math. J. **45** (1972), 167–178. MR306215

[22] André Weil, *The field of definition of a variety*, Amer. J. Math. **78** (1956), 509–524, DOI 10.2307/2372670. MR82726

[23] J. Wolfart, *ABC for polynomials, dessins d'enfants and uniformization—a survey*, Elementare und analytische Zahlentheorie, Schr. Wiss. Ges. Johann Wolfgang Goethe Univ. Frankfurt am Main, vol. 20, Franz Steiner Verlag Stuttgart, Stuttgart, 2006, pp. 313–345. MR2310190

Departamento de Matemáticas y Estadística, Universidad de La Frontera. Temuco, Chile.
Email address: ruben.hidalgo@ufrontera.cl

Departamento de Matemáticas y Estadística, Universidad de La Frontera. Temuco, Chile.
Email address: sebastian.reyes@ufrontera.cl

Contemporary Mathematics
Volume **766**, 2021
https://doi.org/10.1090/conm/766/15384

Subcovers of curves and moduli spaces

Ernst Kani

To Herbert Lange on his 75th birthday

ABSTRACT. This article studies certain questions pertaining to the moduli space $M_g(g', n)$ which was introduced and studied by H. Lange in the 1970's. This space classifies the isomorphism classes of curves X of genus g which admit a subcover $f : X \to Y$ of degree n to a curve Y of genus g'.

Here we focus on the case that $g = 2$ and $g' = 1$, in which case the irreducible components of $M_2(1, n)$ are certain *Humbert surfaces*. It is shown here that $M_2(1, n)$ is always connected, and that it has $d(n) - 1$ irreducible components.

One of the questions discussed here in some detail concerns the intersection $M_2(1, n) \cap M_2(1, m)$ of two such moduli spaces. It turns out that the irreducible components of such an intersection are images of modular curves, and these can be described explicitly by using the *generalized Humbert schemes* $H(q)$ and the theory of binary quadratic forms.

1. Introduction

It is a pleasure to dedicate this paper[1] to Herbert Lange for his 75th birthday, not only because we have known each other for a long time, but also (and more importantly) because the results which are presented here are closely connected with some of his early work.

Starting with his thesis, much of Herbert Lange's early work (cf. Lange[**L1**] – [**L4**]) concerns the following topic:

PROBLEM 1. Describe the set of curves $X/\mathbb{C}$ which admit a non-rational subcover.

Thus, he was interested in studying the set $M_g(g', n)$ of (isomorphism classes of) curves X of genus g which have a subcover $f : X \to Y$ of degree n to some curve Y of genus $g' \geq 1$ inside the *moduli space* M_g of curves genus g, i.e.,

$$M_g(\mathbb{C}) = \{\text{isomorphism classes } \langle X \rangle \text{ of curves } X/\mathbb{C} \text{ of genus } g\}.$$

This was the subject both of his *dissertation* (the case $g = 2$) and of his *habilitation* (for arbitrary $g \geq 2$), and of at least 4 publications.

The following result is (a part of) Satz I of Lange[**L4**].

THEOREM 1.1 (Lange, 1977). (a) *If $g > g' \geq 1$ and $n \geq 2$, then the set $M_g(g', n)$ is a closed subset of M_g.*

[1]This is an expanded version of my lecture at the *Langefest*.

(b) *The subscheme $M_g(g',n)$ is equidimensional of dimension*

$$\dim M_g(g',n) = g-(n-2)(g'-1),$$

provided that either

$$g'\geq 2 \text{ and } \frac{g+1}{g'+1}\leq n\leq \frac{g+1}{g'-1} \quad \text{or that} \quad g'=1 \text{ and } n>\frac{g+1}{2}.$$

Moreover, $M_g(g',n)=\emptyset$ in all other cases, except possibly in the case that $n=\frac{g+1}{2}$ and $g'=1$.

These beautiful results of Lange naturally lead to further questions about the *geometric structure* of the subschemes $M_g(g',n)$. For example:

QUESTION 1. How many irreducible components does $M_g(g',n)$ have? When is $M_g(g',n)$ irreducible?

QUESTION 2. What is the "geometric type" of each irreducible component? When are they all rational or of *general type*?

QUESTION 3. Is $M_g(g',n)$ connected? *(This is a question of Accola/Previato* [**AP**] *in the case that $g=2$.)*

QUESTION 4. What can be said about the *intersection* of $M_g(g',n)$ with one or more $M_g(g'',n')$'s?

These and other related questions will be investigated in the case that $g=2$.

2. Humbert surfaces

When $g=2$, the irreducible components of $M_2(1,n)$ turn out to be open subsets of certain *Humbert surfaces;* these naturally live in the *moduli space* A_2 of *principally polarized* abelian surfaces. Thus,

$$A_2(\mathbb{C})=\{\text{isomorphism classes } \langle A,\lambda\rangle \text{ of principally polarized abelian surfaces } (A,\lambda)\},$$

where the pair (A,λ) consists of an abelian surface $A/\mathbb{C}$ together with a *principal polarization* $\lambda=\phi_\theta : A \xrightarrow{\sim} \hat{A}$; cf. [**LB**] or [**M2**].

REMARK 2.1. Via the *Torelli map* $\langle X\rangle\mapsto\langle J_X,\lambda_X\rangle$ we can (and will) view the moduli space M_2 as a subset of A_2, i.e. $M_2\subset A_2$. Here, J_X is the Jacobian surface of X, and

$$\lambda_X=\phi_{\theta_X} : J_X \xrightarrow{\sim} \hat{J}_X$$

is the polarization induced by the *theta-divisor* $\theta_X\simeq X$ on J_X.

The Humbert surfaces H_Δ are defined via the *Humbert invariant* Δ, which is a number which Humbert[**Hu**] attached in 1899 to a so-called "singular relation" (Humbert's terminology). Such a relation is one which is satisfied by the entries of a (normalized) period matrix of a principally polarized abelian surface (A,λ) and only exists for special abelian surfaces. (See §6 below for an algebraic description of this invariant.) Thus, the Humbert surface with Humbert invariant Δ is defined by

$$H_\Delta=\{(A,\lambda)\in A_2 : (A,\lambda) \text{ has a singular relation with Humbert invariant } \Delta\}$$

In van der Geer[**vdG**], Ch. IX, one finds a modern treatment of Humbert surfaces. Humbert's results are summarized in Kap. I, §1, of Hecke's dissertion [**He**].

The following result follows (more or less) from what Humbert[**Hu**] proved about his invariant.

THEOREM 2.2 (Humbert). *For each positive integer* $n \equiv 0, 1 \pmod 4$, *there exists an irreducible (analytic) surface* $H_n \subset A_2$ *(now called a* Humbert surface*) such that:*

(a) *If* $\langle A, \lambda\rangle \in A_2(\mathbb{C})$, *then* $\mathrm{End}(A) \neq \mathbb{Z} \Leftrightarrow \langle A, \lambda\rangle \in H_n(\mathbb{C})$, *for some* n*;*

(b) $M_2 = A_2 \setminus H_1$*;*

(c) *If* $\langle X\rangle \in M_2(\mathbb{C})$, *then* $\exists$ *a surjection* $f : X \to E$, *where* $E/\mathbb{C}$ *is some elliptic curve* $\Leftrightarrow \langle J_X, \lambda_X\rangle \in H_{N^2}(\mathbb{C})$, *for some* $N \geq 2$.

PROOF. For convenience of the reader,[2] here is a sketch of how these results follow from the results of Humbert[**Hu**]. (See also Remark 6.1 below.)

First of all, Humbert[**Hu**] explains in §3 (p. 237) that if one normalized period matrix of (A, λ) has a singular relation with invariant Δ, then every normalized period matrix of (A, λ) also has a singular relation with invariant Δ. Thus, this property is an invariant of the isomorphism class of (A, λ).

In §11 (p. 245) he shows that his invariant satisfies $\Delta \equiv 0, 1 \pmod 4$ and that $\Delta > 0$ (Théorème 14, p. 246).

In Théorème 12 (p. 245) he shows that one can choose the period matrix in such a way that the singular relation assumes a simple form. This means (in modern language) that for each $\Delta \equiv 0, 1 \pmod 4$ there is a linear relation $f_\Delta = 0$ in the Siegel half-space $\mathfrak{H}_2$ such that its image in $A_2 = \mathrm{Sp}_2(\mathbb{Z})\backslash\mathfrak{H}_2$ is H_Δ. Thus, H_Δ is an irreducible analytic surface in A_2; cf. also [**vdG**], Theorem (IX.2.4) on p. 212.

(a) In §138 Humbert shows (on p. 285) that if $(A, \lambda) \in H_\Delta$ for some Δ, the A is "singular" in his (and Hurwitz's) terminology, i.e., $\mathrm{End}(A) \neq \mathbb{Z}$.

I wasn't able to find that the converse implication was stated (or proved) in [**Hu**], except in special cases; cf. Théorème I in §104 and Théorème II in §105. However, Hecke[**He**] states at the end of §I.1 (without proof or reference) that one can deduce that the converse holds.

Note that part (a) follows directly from what was proven in [**K1**], Corollary 5.5.

(b) In Remarque II in §17 on p. 248, Humbert points out that $\Delta = 1$ if and only if $(A, \lambda) \simeq (E_1 \times E_2, \lambda_1 \otimes \lambda_2)$, where (E_i, λ_i) is a (principally polarized) elliptic curve, for $i = 1, 2$, and $\lambda_1 \otimes \lambda_2$ is the product polarization. This, together with Satz 2 of Weil[**We**], proves the assertion.

Note that in view of [**K1**], §5, this also follows Proposition 6 in [**K3**].

(c) This is Théorème 15 in §15 on p. 247. □

REMARK 2.3. (a) Each Humbert surface H_n is a Zariski-closed subset of A_2. This does not follow directly from Humbert[**Hu**].

(b) Part (c) of Theorem 2.2 had already been stated and proved by Biermann in 1883; cf. Krazer[**Kr**], Satz V on p. 485. Perhaps Humbert did not know this.

(c) The result of Theorem 2.2(c) was refined in Theorem 1.9 of [**K1**] in the following way:

$$(2.1) \qquad \langle J_X, \lambda_X\rangle \in H_{N^2} \;\Leftrightarrow\; \exists \text{ a minimal } f : X \to E \text{ with } \deg(f) = N.$$

[2]This follows a suggestion of the referee, who asked me to explain where these results are proved in Humbert's paper

Here and below, a cover $f : X \to E$ is called *minimal* if it does not factor over an isogeny of E of degree > 1.

COROLLARY 2.4. *For any $n \geq 2$ we have*

$$M_2(1,n) \;=\; \bigcup_{1<N|n} H^*_{N^2}, \tag{2.2}$$

where $H^*_{N^2} = H_{N^2} \cap M_2$. Thus, $M_2(1,n)$ is equidimensional of dimension 2, and has $d(n) - 1$ irreducible components, where $d(n)$ is the number of divisors of n. In particular, $M_2(1,n)$ is irreducible if and only if n is a prime number.

PROOF. Each subcover $f : X \to E$ factors as $f = h \circ f_{min}$, where $f_{min} : X \to E'$ is minimal, so the formula (2.2) follows from (2.1). The other assertions follow from (2.2) and Theorem 2.2, together with the fact that $H^*_{N^2} = H_{N^2} \cap M_2 \neq \emptyset$ when $N > 1$ (see Theorem 4.1 below). □

Note that this answers Question 1 of the introduction. We next turn to the study of Question 2.

3. The geometry of Humbert surfaces

To construct the moduli space $M_g(g', n)$, Lange[**L4**] adapts and generalizes Mumford's construction[**M1**] of the moduli space M_g.

Thus, he first constructs a suitable subvariety $H = H_{g,g',n}$ of some Hilbert scheme, and then shows that $M_g(g', n)$ is the quotient scheme of H under the action of the group $\mathrm{PGL}(m)$, where $m \geq 3$ is suitable.

Unfortunately, this construction sheds little light on the geometry of $M_g(g', n)$.

Since it suffices to construct the irreducible components of these schemes, and since for $M_2(1,n)$ these components are known by Corollary 2.4 to be Humbert surfaces, it suffices to construct these surfaces. For them, there is another construction method available, which is based on what G. Frey and I call the *the basic construction*; cf. [**FK1**], [**K2**], and [**FK2**].

The basic construction. Fix an integer $N \geq 2$, and let (E, E', ψ) be a triple consisting of two elliptic curves E and E' and an isomorphism

$$\psi : E[N] \xrightarrow{\sim} E'[N]$$

which is an *anti-isometry* with respect to the e_N-pairings (called *Weil-pairings* in [**Si**], p. 96) on the groups of N-torsion points of E and of E', respectively. This condition means that we have

$$e_N(\psi(P), \psi(Q)) \;=\; e_N(P,Q)^{-1}, \quad \forall P, Q \in E[N].$$

Given such a triple (E, E', ψ), construct the *abelian surface*

$$J_\psi := (E \times E')/\,\mathrm{Graph}(-\psi).$$

This surface has a "canonical" *principal polarization* λ_ψ, so $\langle J_\psi, \lambda_\psi \rangle \in A_2(\mathbb{C})$.

This principal polarization λ_ψ is "canonical" in the following sense. If we let $\pi_\psi : E \times E' \to J_\psi$ denote the quotient morphism, then λ_ψ is uniquely determined by the condition that

$$\hat{\pi}_\psi \circ \lambda_\psi \circ \pi_\psi = N(\lambda \otimes \lambda'),$$

where $\hat{\pi}_\psi : \hat{J}_\psi \to (E \times E')^\wedge$ is the dual isogeny and $\lambda \otimes \lambda'$ denotes the product polarization on $E \times E'$.

Moreover, if the theta-divisor $X_\psi \in \mathrm{Div}^+(J_\psi)$ of λ_ψ is irreducible, then we have a (minimal) cover $f_\psi : X_\psi \to E$ of degree N, and every (minimal) cover $f : X \to E$ of degree N arises this way.

Since this construction also works for families of elliptic curves over a base (cf. [**K2**]), it has a *modular interpretation*, i.e., it induces a map between the associated functors. To explain this in more detail, we first note the following result from [**FK2**]; cf. [**FK2**], Proposition 4.5.

THEOREM 3.1. *Let $\mathcal{Z}_N$ denote the functor which classifies isomorphism classes of triples (E, E', ψ), where $\psi : E[N] \to E'[N]$ is an anti-isometry. If $N \geq 3$, then $\mathcal{Z}_N$ is coarsely represented by the* modular diagonal quotient surface

$$Z_N \;:=\; \Delta_N^* \backslash (X(N) \times X(N)),$$

where $X(N) = \Gamma(N)\backslash\mathfrak{H}$ is the usual modular curve of level N on which the group $G_N := \Gamma(1)/(\pm\Gamma(N)) \simeq \mathrm{SL}_2(\mathbb{Z}/N\mathbb{Z})/(\pm 1)$ acts, and

$$\Delta_N^* \;=\; \{(g, \beta g \beta^{-1}) : g \in G_N\} \;\leq\; G_N \times G_N, \quad \textit{where } \beta = \begin{pmatrix} 1 & 0 \\ 0 & -1 \end{pmatrix}.$$

REMARK 3.2. (a) By construction, the modular diagonal quotient surface Z_N is a quotient of the product surface $Y(N) := X(N) \times X(N)$ of the modular curve $X(N)$. We thus have a finite surjective morphism

$$\Phi_N : Y(N) = X(N) \times X(N) \;\to\; \Delta_N^* \backslash Y(N) = Z_N.$$

(b) Since the curve $X(N)$ has a natural compactification (by adding cusps), the surfaces $Y(N)$ and Z_N also admit natural compactifications $\bar{Y}(N)$ and $\bar{Z}_N$, respectively. The geometry of the (desingularization of the) modular diagonal quotient surfaces $\bar{Z}_N$ was determined by Hermann[**H1**] and also by Kani and Schanz[**KS**].

The "basic construction" leads to the following result.

THEOREM 3.3. *The rule $(E, E', \psi) \mapsto (J_\psi, \lambda_\psi)$ defines a finite morphism*

$$\beta_N : Z_N \;\to\; A_2$$

whose image is the Humbert surface H_{N^2}. Thus, H_{N^2} is an irreducible surface which is a Zariski-closed subset of A_2. Moreover, the normalization of H_{N^2} is the symmetric modular diagonal quotient surface

$$Z_N^{sym} := \langle w_N \rangle \backslash Z_N,$$

where $w_N \in \mathrm{Aut}(Z_N)$ is the involution which is induced via Φ_N from the involution $\tau_N \in \mathrm{Aut}(Y(N))$ which interchanges the factors of $Y(N) = X(N) \times X(N)$. In particular, $\deg(\beta_N : Z_N \to H_{N^2}) = 2$.

PROOF. As is shown in [**K8**], the first assertions can be deduced from Theorem 15 of [**K7**], which is a refinement of [**K1**], Corollary 1.8 and of [**FK2**], Proposition 4.11. The assertion about the normalization of H_{N^2} was stated (without proof) in [**FK2**], loc. cit., and is proved in Theorem 7.3 of [**K8**]. □

The geometric structure of the desingularization of the natural compactification of Z_N^{sym} was determined by Hermann[**H2**] in the prime case.

THEOREM 3.4 (Hermann). *If $N \geq 3$ is prime, then the surface Z_N^{sym} is:*

- *rational, if $N \leq 13$,*
- *an elliptic K3-surface, if $N = 17$,*

- *an honest elliptic surface, if* $N = 19$,
- *of general type, if* $N \geq 23$.

REMARK 3.5. Thus, by Theorems 3.3 and 3.4 one has a precise description of the geometric type of the components H_{N^2} of $M_2(1,n)$, at least when N is prime. In particular, this gives the birational description of $M_2(1,n)$, when n is prime. It is to be expected that a similar description exists for all N's, but this has not been worked out (as far as I know).

On the other hand, the geometric description of Z_N, which is a double cover of Z_N^{sym}, has been given for all N by Hermann[**H1**] and by Kani and Schanz[**KS**], as was mentioned above.

4. Connectedness properties

We now turn to discuss Question 3, which was (implicitly) raised by Accola and Previato[**AP**]. Recall that this question asks whether $M_2(1,n)$ is *connected* for all $n \geq 2$. That this is indeed the case follows from the following much stronger assertion (which also partially answers Question 4).

THEOREM 4.1. *If* $n > m \geq 2$, *then* $H_{n^2} \cap H_{m^2} \cap M_2 \neq \emptyset$. *Moreover,* $H_{n^2} \cap H_{m^2}$ *is a finite union of (images of)* modular curves.

REMARK 4.2. The first part of Theorem 4.1 is a special case of Corollary 8.2 below, which will be proven in §8. The second part follows from the more precise assertion (10.3) in Theorem 10.5.

The first part of this theorem was also proven by Franciosi/Pardini/Rollenske [**FPR**] by another method.

COROLLARY 4.3. *The moduli space* $M_2(1,n)$ *is connected for any* $n \geq 2$.

PROOF. By Corollary 2.4 (and Theorem 3.3) we know that the Humbert surfaces $H_{N^2}^* = H_{N^2} \cap M_2$ with $N|n$ and $N \geq 2$ are the irreducible components of $M_2(1,n)$. Thus, by Theorem 4.1 we see that any two irreducible components of $M_2(1,n)$ have a non-empty intersection, so $M_2(1,n)$ is connected. □

5. Intersections of Humbert surfaces

We now come to the discussion of Question 4, which concerns the intersection of different moduli spaces. In our situation ($g = 2$) this question reduces to the following.

Question 4* *What are the irreducible components of the intersection* $M_2(1,n) \cap M_2(1,m)$, *when* $\gcd(n,m) = 1$?

In view of the above structure theorem (Corollary 2.4) for the $M_2(1,n)$'s, it suffices to answer the following question.

QUESTION 5. How can we describe the irreducible components of the intersection of two Humbert surfaces?

REMARK 5.1. The intersection $H_{n^2} \cap H_{m^2} \cap M_2$ classifies curves X of genus 2 with two minimal morphisms $f_1 : X \to E_1$ and $f_2 : X \to E_2$ of degrees n and m.

To understand the components of this intersection, we will generalize Humbert's construction of the Humbert surfaces H_n in A_2 in order to construct certain one-dimensional subschemes of A_2.

These generalized Humbert schemes will be defined by considering a *refinement* of the Humbert invariant, which Humbert introduced in his work.

However, while Humbert's invariant is a number, the *refined Humbert invariant* is a *quadratic form*, as will be explained presently.

The basic idea. As will be explained below, each integral, positive definite quadratic form q defines a subset

$$H(q) \subset A_2,$$

called a *generalized Humbert scheme.* Its definition relies on the notion of the refined Humbert invariant.

As a preview, we state some of the basic properties of the generalized Humbert schemes. These will be verified in §7.

PROPOSITION 5.2. (a) *The set $H(q)$ depends only on the* $\mathrm{GL}_r(\mathbb{Z})$*-equivalence class of the quadratic form $q = q(x_1, \ldots, x_r)$. Moreover,*

$$q_1 \to q_2 \quad \Rightarrow \quad H(q_1) \subset H(q_2). \tag{5.1}$$

where the symbol $q_1 \to q_2$ means that q_1 primitively represents the form q_2; cf. §7.

(b) *The usual Humbert surface is $H_n = H(nx^2)$.*

(c) *We have that $H(q) \neq A_2$, but $H(q)$ may be empty. Indeed, $H(q) = \emptyset$, if q is not positive-definite or if $r > 3$.*

(d) *If $n \neq m$, then*

$$H_n \cap H_m = \bigcup_{q \to n,m} H(q), \tag{5.2}$$

where the union is over all integral, positive definite binary *quadratic forms q which* represent *both n and m* primitively*; cf.* §8.

REMARK 5.3. If $m \neq n$, then up to $\mathrm{GL}_2(\mathbb{Z})$-equivalence there are only finitely many positive binary forms q satisfying the condition $q \to n, m$ because this condition implies that $|\operatorname{disc}(q)| \leq 4mn$. Thus, by Proposition 5.2(a), the union on the right hand side of (5.2) consists only of finitely many $H(q)$'s.

The *intersection formula* (5.2) constitutes a *first step* towards understanding the components of the intersection of two Humbert surfaces.

However, further work is necessary to obtain a better picture, and this leads to the following questions.

QUESTION 6. For which integral *binary* quadratic forms q is $H(q) \neq \emptyset$?

QUESTION 7. What is the geometric structure of $H(q)$? Is $H(q)$ irreducible?

QUESTION 8. Can (the components of) $H(q)$ be described by images of modular curves? If so, which modular curves map to $H(q)$?

These questions will be addressed below.

6. The refined Humbert invariant

The *refined Humbert invariant* $q_{(A,\lambda)}$ of a principally polarized abelian surface $(A,\lambda) \in A_2(\mathbb{C})$ is defined as follows.

DEFINITION. Let A be an abelian surface with a principal polarization $\lambda : A \xrightarrow{\sim} \hat{A}$, and let $\theta = \theta_\lambda \in \mathrm{Div}(A)$ be an associated theta-divisor, i.e., $\lambda = \phi_\theta$ in the notation of [**M2**]. Let $\mathrm{NS}(A)$ denote the Néron-Severi group of A, i.e., $\mathrm{NS}(A) := \mathrm{Div}(A)/\equiv$, where $\equiv$ denotes numerical equivalence of divisors, and put

$$\mathrm{NS}(A,\lambda) \ := \ \mathrm{NS}(A)/\mathbb{Z}\theta_\lambda.$$

Consider the quadratic form on $\mathrm{NS}(A)$ defined by

$$q_{(A,\lambda)}(D) = (D.\theta_\lambda)^2 - 2(D.D), \quad \forall D \in \mathrm{NS}(A),$$

where $(\, . \,)$ denotes the intersection pairing on the Néron-Severi group $\mathrm{NS}(A)$. It then follows from the definition and the Hodge index theorem that $q_{(A,\lambda)}$ defines a *positive-definite quadratic form* on the module $\mathrm{NS}(A,\lambda)$; cf. [**K1**]. This induced quadratic form on $\mathrm{NS}(A,\lambda)$ is called *refined Humbert invariant* of (A,λ) and is also denoted by $q_{(A,\lambda)}$.

REMARK 6.1. If $\bar{D} \in \mathrm{NS}(A,\lambda)$ is *primitive*, i.e., if $\mathrm{NS}(A,\lambda)/\mathbb{Z}\bar{D}$ is torsionfree (and $\bar{D} \neq 0$), then $\overline{D}$ gives rise to a "singular relation" of A (in the sense of Humbert), and conversely every such relation arises in this way, as was shown in [**K1**], §5. Moreover, $\Delta := q_{(A,\lambda)}(\bar{D})$ is equal to the *Humbert invariant* Δ attached to the singular relation; cf. [**K1**], Corollary 5.5.

7. Generalized Humbert schemes

Using the *refined Humbert invariant* $q_{(A,\lambda)}$, which was defined in the previous section, we can generalize the notion of a Humbert surface as follows.

NOTATION. If $q : \mathbb{Z}^r \to \mathbb{Z}$ is an integral, positive-definite quadratic form in r variables, then we put

$$H(q) := \{\langle A,\lambda\rangle \in A_2(\mathbb{C}) : q_{(A,\lambda)} \to q\}.$$

Here the symbol $q_{(A,\lambda)} \to q$ indicates that the form $q_{(A,\lambda)}$ *primitively represents* the form q. This means that there exists an injective homomorphism

$$h : \mathbb{Z}^r \hookrightarrow \mathrm{NS}(A,\lambda)$$

such that $\mathrm{NS}(A,\lambda)/h(\mathbb{Z}^r)$ is torsionfree and such that the h-restriction of $q_{(A,\lambda)}$ to $\mathbb{Z}^r$ is q, i.e., $q_{(A,\lambda)} \circ h = q$.

If $q_{(A,\lambda)} \to q$ and h is surjective, then we say that $q_{(A,\lambda)}$ is *equivalent* to q and write $q_{(A,\lambda)} \sim q$.

PROOF OF PROPOSITION 5.2. (a) This is clear from the definition. Indeed, if q_1 is $\mathrm{GL}_r(\mathbb{Z})$-equivalent to q_2, then $\forall \langle A,\lambda\rangle \in A_2(\mathbb{C})$ we have that $q_{(A,\lambda)} \to q_1 \Leftrightarrow q_{(A,\lambda)} \to q_2$, and so $H(q_1) = H(q_2)$.

Similarly, if $\langle A,\lambda\rangle \in H(q_1)$, then $q_{(A,\lambda)} \to q_1$, and if $q_1 \to q_2$, then $q_{(A,\lambda)} \to q_2$, so $\langle A,\lambda\rangle \in H(q_2)$. Thus $H(q_1) \subset H(q_2)$. This proves (5.1).

(b) This follows from Remark 6.1.

(c) If q is any (non-zero) quadratic form, then $q \to nx^2$, for some $n \neq 0$, and then by parts (a) and (b) we have that $H(q) \subset H(nx^2) = H_n$. By Humbert

(Theorem 2.2), H_n is either empty or a surface, so $H_n \neq A_2$ because $\dim A_2 = 3$. This proves the first statement.

Since $q_{(A,\lambda)}$ is positive-definite, then the same is true for any q with $q_{(A,\lambda)} \to q$, and so it follows that $H(q) = \emptyset$, if q is not positive-definite.

Moreover, since $\mathrm{NS}(A) \simeq \mathbb{Z}^\rho$, where $\rho \leq h^{1,1}(A) = 4$ (cf. [**vdG**], p. 207), we see that $\mathrm{NS}(A,\lambda) \simeq \mathbb{Z}^r$, with $r \leq 3$. Thus, the same is true for any q with $q_{(A,\lambda)} \to q$, and so it follows that $H(q) = \emptyset$, if q has rank > 3.

(d) The fact that the right hand side of (5.2) is contained in the left hand side follows immediately from (5.1).

Conversely, suppose that $\langle A, \lambda\rangle \in H_m \cap H_n$. Then there exist primitive vectors $v, w \in M := \mathrm{NS}(A,\lambda)$ such that $q_{(A,\lambda)}(v) = m$ and $q_{(A,\lambda)}(w) = n$. If v and w were linearly dependent, then $v = \pm w$ and hence $q_{(A,\lambda)}(v) = q_{(A,\lambda)}(w)$, contrary to the hypothesis. Thus, v and w are linearly independent and hence $M_0 := \mathbb{Z}v + \mathbb{Z}w$ has rank 2. Let M_1 be the saturation of M_0 in M. Then the restriction q of $q_{(A,\lambda)}$ to M_1 is a positive definite, binary quadratic form which is primitively represented by $q_{(A,\lambda)}$, and so $\langle A, \lambda\rangle \in H(q)$. Moreover, $m = q(v)$ is primitively represented by q (because v is primitive in M, hence also in M_1. Similarly, $n = q(w)$ is primitively represented by q. Thus q is one of the forms of the right side of (5.2), so $\langle A, \lambda\rangle \in \cup H(q)$. □

THEOREM 7.1. *If q is an integral quadratic form with $q \to N^2$, for some integer $N \geq 1$, then $H(q)$ is a closed subset of A_2. Moreover, if $H(q) \neq \emptyset$ then $H(q)$ is equidimensional of dimension* $\dim H(q) = 3 - r$, *where r is the number of variables of q.*

REMARK 7.2. This theorem follows easily from the results below; cf. §10, where this is proved. It is to be expected that the same result holds for all integral quadratic forms, but the methods below do not prove this.

DEFINITION. A (non-empty) subset of A_2 of the form $H(q)$, where q is a quadratic form, will be called a *generalized Humbert scheme.*

8. Special quadratic forms

We now want to classify those binary quadratic forms q for which $H(q) \neq \emptyset$, and thus answer Question 6. Here we will only consider those forms which primitively represent a square because only those lead to elliptic subcovers of genus 2 curves; cf. Theorem 2.2(c) and Proposition 5.2(b).

NOTATION. Write $q = [a, b, c]$ for the binary quadratic form

$$q(x, y) = ax^2 + bxy + cy^2$$

on $\mathbb{Z}^2$. Moreover, let Q denote the set of integral binary quadratic forms q which satisfy the following two conditions:

(i) q is positive-definite;

(ii) $q(x, y) \equiv 0, 1 \pmod 4$, $\forall x, y \in \mathbb{Z}$.

In addition, for $n \in \mathbb{N}$, let

$$Q(n) = \{q \in Q : q \to n\}$$

denote the set of forms $q \in Q$ which *primitively represent* n, i.e.,

$$q(x, y) = n, \quad \text{for some } x, y \in \mathbb{Z} \text{ with } \gcd(x, y) = 1.$$

We can now classify the binary q's for which $H(q) \neq \emptyset$ (and such that $H(q)$ lies on some Humbert surface H_{N^2}) in the following way.

THEOREM 8.1. *Let q be an integral binary quadratic form which primitively represents a square, and let $N \geq 1$. Then:*

$$H(q) \neq \emptyset \ \text{ and } \ H(q) \subset H_{N^2} \quad \Leftrightarrow \quad q \in Q(N^2). \tag{8.1}$$

If this is the case, then there exists $\langle A, \lambda \rangle \in H(q)$ such that $q_{(A,\lambda)}$ is equivalent to q.

PROOF. If $q \in Q(N^2)$, then the Existence Theorem 32 of [**K5**] shows that that there exists $\langle A, \lambda \rangle \in H(q)$ with $q_{(A,\lambda)} \sim q$, so the last assertion holds. Since $q \to N^2x^2$, we have by (5.1) and Proposition 5.2(b) that $H(q) \subset H_{N^2}$.

Conversely, suppose that $\langle A, \lambda \rangle \in H(q)$. Then $q_{(A,\lambda)} \to q$, so $q \in Q$ because properties (i) and (ii) hold for $q_{(A,\lambda)}$ and hence are inherited by q. By hypothesis, $q \to M^2$, for some $M \geq 1$. Thus $q \in Q(M^2)$ and so by the Existence Theorem 32 of [**K5**] there exists $\langle A', \lambda' \rangle \in H(q)$ such that $q_{(A',\lambda')} \sim q$. Since $\langle A', \lambda' \rangle \in H(q)$ and since $H(q) \subset H_{N^2}$ by our hypothesis, it follows that $q_{(A',\lambda')} \to N^2$ (cf. Proposition 5.2(b)), and hence also $q \to N^2$ because $q \sim q_{(A',\lambda')}$. Thus $q \in Q(N^2)$, as desired. □

Note that Theorem 8.1 answers Question 6 for the $H(q)$'s that appear in the intersection $H_{N^2} \cap H_{m^2}$. Furthermore, this theorem allows us to give a simple proof of the first part of Theorem 4.1. Indeed, one can prove just as easily the following more general result.

COROLLARY 8.2. *If $m \equiv \varepsilon \pmod 4$, where $\varepsilon \in \{0, 1\}$, and if $m, N > 1$, then*

$$H_m \cap H_{N^2} \cap M_2 \neq \emptyset.$$

PROOF. Note that $q := [N^2, 2\varepsilon N, m] \in Q(N^2) \cap Q(m)$, so $H(q) \neq \emptyset$ by Theorem 8.1. Since $q \to N^2$ and $q \to m$, we see that $H(q) \subset H_m \cap H_{N^2}$ by (5.2).

Moreover, since $q(x, y) = (Nx + \varepsilon y)^2 + (m - \varepsilon^2)y^2 > 1$, we see that $q \not\to 1$. Thus $H(q) \not\subset H_1$ by (8.1), so $H(q) \not\subset H_1 = A_2 \setminus M_2$ by Theorem 2.2(b), and hence $H(q) \cap M_2 \neq \emptyset$. Thus also $H_m \cap H_{N^2} \cap M_2 \neq \emptyset$. □

9. The components of $H(q)$

We now consider Question 7 which in part asks about the irreducibility of $H(q)$. The following two results give a partial answer to this question.

THEOREM 9.1. *If $q = [a, b, c] \in Q(N^2)$ and if $8 \nmid c(q) := \gcd(a, b, c)$, then $H(q)$ has at most $2^{\omega(c(q))}$ irreducible components, where $\omega(c) = |\{p \mid c\}|$ denotes the number of distinct prime divisors of $c = c(q)$.*

In particular, if q is primitive, *i.e., if $c(q) = 1$, then $H(q)$ is an irreducible curve.*

THEOREM 9.2. *Let $q \in Q(N^2)$. If $|\operatorname{disc}(q)| > 4N^4$ and if N is odd, then $H(q)$ has precisely $2^{\omega(c(q))}$ irreducible components, except when $q \sim [N^2, 0, 4d]$, for some integer $d \geq 1$.*

The proofs of these two theorems will be discussed below in §11.

REMARK 9.3. (a) If $8|c(q)$, then $H(q)$ has at most $2^{\omega(c(q))+1}$ irreducible components, and an analogue of Theorem 9.2 holds (but there are more exceptions). Moreover, the number of components can also be determined in the exceptional cases.

(b) If $q \in Q(N^2)$ has discriminant $\text{disc}(q) = -16d$ with $\gcd(d, N) = 1$, then $H(q)$ is irreducible, as was proved in Theorem 4 of [**K3**]. In fact, that theorem also gives the *birational structure* of $H(q)$ in this case. The result is that $H(q)$ is birational to the modular curve $X_0^+(d) = X_0(d)/\langle w_d\rangle$, where w_d is the *Fricke involution* on $X_0(d) = \Gamma_0(d)\backslash\mathfrak{H}$, except when q is an ambiguous form (see below). In latter case $H(q)$ is birational to a quotient of $X_0^+(d)$, which is given explicitly in Theorem 57 of [**K3**].

When N is an odd prime, then the number of components of the $H(q)$'s can be completely determined. However, this number depends on whether $N \equiv 1 \pmod 4$ or not, as the following result shows. To state it, recall (from Gauss) that a binary integral form q is said to be *ambiguous* if $q \sim [a, ak, c]$, for some integers $a, k, c \in \mathbb{Z}$.

THEOREM 9.4. *Let $N > 2$ be a prime and let $q \in Q(N^2)$. Then $H(q)$ has at most 2 irreducible components. Moreover, if $N \equiv 1\,(mod 4)$, then we have that*

$$H(q) \text{ is irreducible} \quad \Leftrightarrow \quad q \text{ is primitive, i.e., } c(q) = 1,$$

whereas if $N \equiv 3\,(mod 4)$, then we have that

$$H(q) \text{ is irreducible} \quad \Leftrightarrow \quad q \text{ is primitive or ambiguous.}$$

9.1. Numerical examples. By using the *reduction theory* of binary quadratic forms and the above results (and others from [**K7**]), one can work out the intersections of Humbert surfaces with small invariants, and hence also the intersections of Lange's moduli spaces $M_2(1, n)$.

PROPOSITION 9.5. *The intersection of some the Humbert surfaces are as follows:*

$$\begin{aligned}
H_1 \cap H_4 &= H[1,0,4],\\
H_1 \cap H_5 &= H[1,0,4],\\
H_4 \cap H_9 &= H[4,0,5] \cup H[4,0,9] \cup H[4,4,9],\\
H_4 \cap H_{16} &= H[4,0,12] \cup H[4,0,16] \cup H[4,4,5] \cup H[4,4,8] \cup H[4,4,16],\\
H_4 \cap H_{25} &= H[1,0,4] \cup H[4,0,9] \cup H[4,0,21] \cup H[4,0,25] \cup H[4,4,17] \cup H[4,4,25].
\end{aligned}$$

Moreover, all the $H(q)$'s listed above are irreducible. Thus, if we put as before $H^*(q) = H(q) \cap M_2$, then the intersection of some of Lange's moduli spaces are as follows:

$$\begin{aligned}
M_2(1,2) \cap M_2(1,3) &= H_4 \cap H_9 \cap M_2\\
&= H^*[4,0,5] \cup H^*[4,0,9] \cup H^*[4,4,9],\\
M_2(1,2) \cap M_2(1,5) &= H_4 \cap H_{25} \cap M_2\\
&= H^*[4,0,9] \cup H^*[4,0,21] \cup H^*[4,0,25] \cup H^*[4,4,17] \cup H^*[4,4,25].
\end{aligned}$$

The above proposition suggests that the number of irreducible components of such intersections becomes quite large as we increase the parameters. This is also borne out by the following table which gives the number of irreducible components of $H_{N^2} \cap H_{M^2}$ for small values of N and M:

$N \backslash M$	2	3	4	5
2	$*$	3	5	6
3	3	$*$	6	9
4	5	6	$*$	12
5	6	9	12	$*$

REMARK 9.6. In the above table of intersections, each $H(q)$ is irreducible. But this will not be true in general, as Theorems 9.2 and 9.4 and the following table show. The latter gives the number of irreducible components of $H_{N^2} \cap H_m$, when $m \equiv 0, 1 \mod 4$:

$N^2 \backslash m$	1	4	5	8	9	12	13	16	17	20	21	24	25
1	$*$	1	1	2	1	2	2	2	3	3	2	3	3
4	1	$*$	3	4	3	4	5	5	5	6	5	6	6
9	1	3	3	5	$*$	6	5	6	8	7	8^*	10^*	9
16	2	5	5	6	6	9	9	$*$	9	12	10	11	12
25	3	6	7^*	8	9	9	10	12	15	16^*	11	13	$*$

Here the starred entries 8^*, 10^*, 7^* and 16^* are those for which the intersection $H_{N^2} \cap H_m$ contains reducible $H(q)$'s. This table is discussed in more detail in [**K7**].

9.2. Application to curves with automorphisms. Let G be a finite group and $0 \leq g' < g$. Then it is known that the set

$$M_g(G, g') := \{\langle X \rangle \in M_g(\mathbb{C}) : G \leq \mathrm{Aut}(X) \text{ and } g_{X/G} = g'\}$$

is a closed subset of the moduli space M_g; cf. Baily[**Ba**] and Kuribayashi[**Ku**].

In the case that $g = 2$, these subschemes turn out to be generalized Humbert schemes $H^*(q) = H(q) \cap M_2$, except possibly when $\dim M_2(G, g') = 0$.

THEOREM 9.7. *The G-loci $M_2(G, g')$ in M_2 of dimension ≥ 1 are as follows:*

$$\begin{aligned} M_2(C_2, 0) &= M_2 \\ M_2(C_2, 1) &= H_4^* \\ M_2(V_4, 0) &= H_4^* \\ M_2(D_4, 0) &= H^*[4, 0, 4] \\ M_2(D_6, 0) &= H^*[4, 4, 4] \end{aligned}$$

where $C_2 = \mathbb{Z}/2\mathbb{Z}, V_4 = C_2 \times C_2$, and D_n is the dihedral group of order $2n$.

PROOF. The first equality is obvious since every curve $\langle X \rangle \in M_2(\mathbb{C})$ is hyperelliptic, and the second is clear from (2.1). The uniqueness of the hyperelliptic involution shows that $M_2(C_2, 1) = M_2(C_4, 0)$, so the third equality holds. The last 2 equalities follow directly from Theorem 4(a) of [**K5**].

Moreover, it follows from the discussion on p. 141 of [**AP**] that $M_2(G, 0)$ is either empty or consists of single point, if G is not one of the above cases. Furthermore, it easy to see that $M_2(G, 1) = \emptyset$ if $G \neq C_2$. Thus, the above list is complete. $\square$

REMARK 9.8. The curves belonging to H_4^*, $H^*[4, 0, 4]$ and to $H^*[4, 4, 4]$ have the following explicit equations:

(a) $y^2 = x(x-1)(x-\alpha)(x-\beta)(x-\alpha\beta)$ (Jacobi, 1832)

(b) $y^2 = x(1-x^2)(1-\kappa^2 x^2)$ (Legendre, 1832)

(c) $y^2 = x^6 + ax^3 + 1$ (Bolza, 1888)

Note that Accola and Previato [**AP**] give on p. 142 the associated period matrices for these curves (and others).

10. The structure of $H(q)$

The aim of this section is to describe all the irreducible components of the generalized Humbert schemes $H(q)$ in the case that $q \in Q(N^2)$ and thus answer Question 8. As an application, we can then prove the second part of Theorem 4.1 and also Theorem 7.1.

For this, we'll use the (modified) "basic construction map"

$$\tilde{\beta}_N = \beta_N \circ \Phi_N : Y(N) = X(N) \times X(N) \overset{\Phi_N}{\to} Z_N \overset{\beta_N}{\to} H_{N^2},$$

where, as above, $\Phi_N : X(N) \times X(N) \to \Delta_N^*\backslash(Y(N) = Z_N$ is the quotient map and β_N is the basic construction map of §3.

Thus, since $H(q) \subset H_{N^2}$, we can expect that $H(q)$ can be described by the images of suitable curves on the product surface $X(N) \times X(N)$. These "suitable curves" are the *modular correspondences* on $Y(N) = X(N) \times X(N)$.

10.1. The modular correspondences T_α^N. We begin by recalling the classical definition of the modular correspondences on the product surface $X(N) \times X(N)$ which were introduction by Klein[**K**] in 1880.

NOTATION. For $d \geq 1$, let $\mathcal{M}_d$ denote the set of *primitive* (integral) matrices of determinant d. Thus,

$$\mathcal{M}_d \;=\; \Gamma(1)\alpha_d\Gamma(1), \quad \text{where } \Gamma(1) = \mathrm{SL}_2(\mathbb{Z}), \text{ and } \alpha_d = \left(\begin{smallmatrix} 1 & 0 \\ 0 & d \end{smallmatrix}\right).$$

PROPOSITION 10.1 (Klein). *For each $\alpha \in \mathcal{M}_d$ there is an irreducible curve*

$$T_\alpha^N \;\subset\; X(N) \times X(N)$$

which depends only on the double coset $\pm\Gamma(N)\alpha\Gamma(N)$.

PROOF. Klein/Fricke[**KF**], p. 597, or Shimura[**Sh**], p. 170. □

REMARK 10.2. Analytically T_α^N is constructed as follows. Recall first that

$$X(N) = \Gamma(N)\backslash\mathfrak{H}, \quad \text{where } \Gamma(N) = \mathrm{Ker}(\mathrm{SL}_2(\mathbb{Z}) \to \mathrm{SL}_2(\mathbb{Z}/N\mathbb{Z}))$$

is the principal congruence subgroup of level N. Now if $\alpha \in \mathcal{M}_d$, and if $\Gamma_\alpha \subset \mathfrak{H}\times\mathfrak{H}$ denotes the graph of α, where we view α as a fractional linear transformation on the upper half-plane $\mathfrak{H}$, then the modular correspondence T_α^N is the image of Γ_α with respect to the quotient map

$$\mathcal{H} \times \mathcal{H} \to (\Gamma(N)\backslash\mathfrak{H}) \times (\Gamma(N)\backslash\mathfrak{H}) = X(N) \times X(N).$$

10.2. The quadratic form of a modular correspondence. In order to connect the modular correspondences T_α^N with the $H(q)$'s, it is useful to associate to each $\alpha \in \mathcal{M}_d$ and $N \geq 1$ a binary quadratic form q_α^N in the following way.

NOTATION. For $\alpha \in \mathcal{M}_d$ and $N \geq 1$ put

$$q_\alpha^N := [N^2, 2mt, m^2(t^2 + 4d)/N^2].$$

Here, $t = \mathrm{trace}(\beta\alpha)$, where $\beta = \left(\begin{smallmatrix} 1 & 0 \\ 0 & -1 \end{smallmatrix}\right)$ and $m|N$ is given by

$$N/m = \gcd(x - w, y, z, N), \quad \text{if } \beta\alpha = \left(\begin{smallmatrix} x & y \\ z & w \end{smallmatrix}\right). \tag{10.1}$$

PROPOSITION 10.3 (Key Lemma). (a) *If* $\alpha \in \mathcal{M}_d$, *and* $N \geq 1$, *then* $q_\alpha^N \in Q(N^2)$. *Moreover, if* m *is defined by* (10.1), *then*

$$\text{disc}(q_\alpha^N) = -16m^2d \quad \textit{and} \quad \gcd(N/m, d) = 1, \tag{10.2}$$

(b) *If* $q \in Q(N^2)$, *then there are unique positive integers* $m|N$ *and* d *such that* (10.2) *holds, and there is a (primitive) matrix* $\alpha \in \mathcal{M}_d$ *such that* $q \sim q_\alpha^N$.

PROOF. (a) This is Corollary 12 of [**K5**]. (Note that the statement of that corollary has a typo: "N-primitive" should be a replaced by "primitive".)

(b) The existence and uniqueness of m and d satisfying (10.2) is proven in Proposition 7 of [**K5**], and the existence of $\alpha \in \mathcal{M}_d$ with $q \sim q_\alpha^N$ follows from Proposition 15 (or from Theorem 16) of [**K5**]. □

DEFINITION. The forms $q \in Q(N^2)$ which satisfy (10.2) will be called *forms of type* (N, m, d).

10.3. The structure theorem. We now come to the main structure theorem for the $H(q)$'s in the case that $q \in Q(N^2)$. This states that $H(q)$ is a union of certain *modular curves* which are obtained as the images of modular correspondences with respect to the "basic construction map" $\tilde{\beta}_N$. To justify the term "modular curve", we first observe:

PROPOSITION 10.4. *For each* $\alpha \in \mathcal{M}_d$ *and* $N \geq 1$, *the image*

$$\overline{T}_\alpha^N := \tilde{\beta}_N(T_\alpha^N) \subset H_{N^2} \subset A_2$$

of the modular correspondence T_α^N *is a closed irreducible curve on* H_{N^2} *and on* A_2.

PROOF. By Proposition 10.1 we know that T_α^N is a closed irreducible curve on $Y(N) = X(N) \times X(N)$, hence $\tilde{\beta}_N(T_\alpha^N)$ is also a closed and irreducible curve on H_{N^2} and on A_2 because β_N and hence $\tilde{\beta}_N$ is a finite morphism; cf. Theorem 3.3. □

THEOREM 10.5 (Structure Theorem). *If* q *is a binary form of type* (N, m, d), *then*

$$H(q) = \bigcup_\alpha \overline{T}_\alpha^N, \tag{10.3}$$

where the union is over all $\alpha \in \mathcal{M}_d$ *such that* $q_\alpha^N \sim q$. *This is a finite union because for* $\alpha_1, \alpha_2 \in \mathcal{M}_d$ *we have that*

$$\gamma\beta\alpha_1\gamma^{-1} \equiv \pm\beta\alpha_2 \ (\textit{mod}\, N), \textit{ for some } \gamma \in \Gamma(1) \quad \Rightarrow \quad \overline{T}_{\alpha_1}^N = \overline{T}_{\alpha_2}^N. \tag{10.4}$$

The proof of the Structure Theorem 10.5 will be discussed in more detail in the next section. Here we observe that we can use it to complete the proofs of Theorems 4.1 and 7.1. For this, we first note the following consequence of Theorem 10.5.

COROLLARY 10.6. *Let* $q_i \in Q(N_i^2)$, *for* $i = 1, 2$. *Then*

$$|H(q_1) \cap H(q_2)| = \infty \quad \Leftrightarrow \quad H(q_1) = H(q_2) \quad \Leftrightarrow \quad q_1 \sim q_2. \tag{10.5}$$

PROOF. If $q_1 \sim q_2$, then $H(q_1) = H(q_2)$, and if $H(q_1) = H(q_2)$, then $H(q_1) \cap H(q_2) = H(q_1)$ is infinite, since $H(q_1)$ is a (non-empty) union of irreducible curves by Theorems 8.1 and 10.5. (Note that q_i has type (N_i, m_i, d_i) for some $m_i|N_i$ by

Proposition 10.3(b).) Thus, it suffices to show that $|H(q_1) \cap H(q_2)| = \infty \Rightarrow q_1 \sim q_2$.

Now since $H(q_i)$ is a union of irreducible curves by Theorem 10.5, it follows that $|H(q_1) \cap H(q_2)| < \infty$ except when $H(q_1)$ and $H(q_2)$ have a common component. By Theorem 10.5, such a component has the form $\overline{T}_\alpha^N$ with $q_\alpha^N \sim q_i$, for $i = 1, 2$. But then $q_1 \sim q_2$, which proves (10.5). □

PROOF OF THEOREM 4.1. The first assertion follows from Corollary 8.2, and the second follows from (10.3). □

PROOF OF THEOREM 7.1. Since $H(q) = \emptyset$, if $r > 3$ (cf. Proposition 5.2(c)), we may assume that $r \leq 3$.

If $r = 1$, then $q = N^2x^2$ because $q \to N^2$. Thus $H(q) = H_{N^2}$ by Proposition 5.2(b), and so $H(q)$ is a closed subset of A_2 by Theorem 3.3 because β_N is a finite and hence a proper morphism. Moreover, H_{N^2} is irreducible (hence equidimensional) of dimension $2 = 3 - r$ by Theorem 3.3 (or by Theorem 2.2).

If $r = 2$, i.e., if q is a binary quadratic form, and if $H(q) \neq \emptyset$, then $q \in Q(N^2)$, and then by Theorem 10.5 we know that $H(q)$ is a finite union of modular curves of the form $\overline{T}_\alpha^N = \tilde{\beta}_N(T_\alpha^N)$. By Proposition 10.4, these are all closed and irreducible curves on H_{N^2}, so $H(q)$ is a closed subset of H_{N^2} which is equidimensional of dimension $1 = 3 - r$.

Finally, if $r = 3$, then by Lemma 10.7 below we have that $H(q)$ is a finite set, so $H(q)$ is equidimensional of dimension $0 = 3 - r$, if $H(q) \neq \emptyset$. □

In the above proof we had used the following fact which is interesting in itself since it does not require the hypothesis that q primitively represents a square.

LEMMA 10.7. *If q is a ternary form, then there exists an integer $N \geq 1$ and binary quadratic forms $q_i \in Q(N^2)$ with $q_1 \not\sim q_2$ such that $H(q) \subset H(q_1) \cap H(q_2)$. In particular, $H(q)$ is a finite set.*

PROOF. The last assertion follows from (10.5), so it is enough to prove the first. Moreover, if $H(q) = \emptyset$, then this assertion holds trivially by taking any two forms $q_i \in Q(N^2)$ with $q_1 \not\sim q_2$; for example, we can take $q_i = [N^2, 0, 4d_i]$, with $d_1 \neq d_2$.

Thus, assume that there exists $(A, \lambda) \in H(q)$. Then $q_{(A,\lambda)} \sim q$ (cf. the proof of Proposition 5.2(c)), so $\rho(A) = r + 1 = 4$. By the structure theorems for $\mathrm{End}(A)$, this implies that $A \sim E \times E$, where $E/\mathbb{C}$ is a CM elliptic curve; cf. [**vdG**], p. 207.

Thus, A has an elliptic subgroup $E_1 \sim E$; put $N = (\theta_\lambda.E_1)$ as in §6. Then by Proposition 10(a) of [**K4**] there exists an elliptic curve $E_2/\mathbb{C}$ and an anti-isometry $\psi : E_1[N] \to E_2[N]$ such that $(A, \lambda) \simeq (J_\psi, \lambda_\psi)$ in the notation of Theorem 3.3. This means that (E_1, E_2, ψ) is an N-presentation of (A, λ).

To construct q_1 and q_2, let d_i be two values which are primitively represented by the degree form q_{E_1,E_2} of $\mathrm{Hom}(E_1, E_2)$, which is a positive binary quadratic form because $E_1 \sim E_2 \sim E$ are CM elliptic curves. By elementary quadratic form theory (cf. Cox[**Co**], Lemma 2.25), we can choose d_1 and d_2 such there exists a prime $p|d_2$ with $p \nmid 2Nd_1$. Then by Theorem 25 of [**K5**], there exists $m_i|N$ such that $q_{(A,\psi)} \to q_i$, where q_i is a form of type (N, m_i, d_i), so in particular, $q_i \in Q(N^2)$. Since $\mathrm{disc}(q_i) = -16m_i^2d_i$, we see that $\mathrm{disc}(q_1) \neq \mathrm{disc}(q_2)$ because $p|\,\mathrm{disc}(q_2)$ but $p \nmid \mathrm{disc}(q_1)$. Thus $q_1 \not\sim q_2$. Since $q \sim q_{(A,\lambda)} \to q_i$, we have by Proposition 5.2(a) that $H(q) \subset H(q_1) \cap H(q_2)$, as desired. □

11. Method of proof

The first main technique for studying the points lying on the Humbert surface H_{N^2} is closely related to the basic construction (cf. §3). This gives rise to the following terminology which was introduced in [**K4**]:

DEFINITION. An *N-presentation* of a principally polarized abelian surface (A,λ) is a 4-tuple (E_1,E_2,ψ,π) where $E_i/\mathbb{C}$ are elliptic curves, $\psi : E_1[N]\to E_2[N]$ is an anti-isometry, and

$$\pi : E_1\times E_2\to A$$

is an isogeny such that $\mathrm{Ker}(\pi)=\mathrm{Graph}(-\psi)$ and such that

$$\pi^*\theta_\lambda\equiv N(\theta_1+\theta_2),$$

where θ_λ is the theta-divisor of (A,λ) and $\theta_i=pr_i^*(0_{E_i})$, where $pr_i : E_1\times E_2\to E_i$ is the ith projection for $i=1,2$.

REMARK 11.1. It follows from the basic construction (cf. §3 or [**K4**], Proposition 10) that

$$(A,\lambda)\text{ has an }N\text{-presentation}\quad\Leftrightarrow\quad\langle A,\lambda\rangle\in H_{N^2}.$$

A second ingredient for studying points on H_{N^2} is a useful *modular description* of the modular correspondences T_α^N and hence of their images $\overline{T}_\alpha^N$. This allows us to find an N-presentation of the elements $\langle A,\lambda\rangle\in\overline{T}_\alpha^N(\mathbb{C})$.

Such a modular description was determined in [**K6**]. Specializing this to our situation, we obtain:

PROPOSITION 11.2. *Let $\alpha\in\mathcal{M}_d$, and let $z:=\langle E,E',\psi\rangle\in Z_N(\mathbb{C})$, where $N\geq 2$. Then $z\in\Phi_N(T_\alpha^N)$ if and only if there exists a cyclic isogeny $f : E\to E'$ of degree d and a (symplectic) level N-structure $\phi : (\mathbb{Z}/N\mathbb{Z})\times(Z/N\mathbb{Z})\xrightarrow{\sim} E[N]$ such that*

$$\psi^{-1}\circ f_{|E[N]} \;=\; \phi\circ[\beta\alpha]_N\circ\phi^{-1}, \tag{11.1}$$

where $[\beta\alpha]_N\in\mathrm{End}((\mathbb{Z}/N\mathbb{Z})^2)$ denotes the endomorphism of $(\mathbb{Z}/N\mathbb{Z})^2$ defined by the matrix $\beta\alpha$ mod N. Here, as before, $\beta=\left(\begin{smallmatrix}1&0\\0&-1\end{smallmatrix}\right)$.

PROOF. This follows easily from the modular description given in Theorem 10 of [**K6**]. □

PROOF OF THEOREM 10.5. (Sketch; full details in [**K7**]). Let $z=\langle E,E',\psi\rangle\in\Phi_N(T_\alpha^N)$, where $\alpha\in\mathcal{M}_d$, so $\beta_N(z)=\langle J_\psi,\lambda_\psi\rangle\in\overline{T}_\alpha^N$. Thus, , if $\pi : E\times E'\to J_\psi$ is the quotient map, then (E,E',ψ,π) is an N-presentation of (J_ψ,λ_ψ). Moreover, by Proposition 11.2 we know there exists a cyclic isogeny $f : E\to E'$ of degree d such that (11.1) holds. It thus follows from Proposition 28 and Proposition 11 of [**K5**] that $q_{(J_\psi,\lambda_\psi)}\to q_\alpha^N$, and so $\overline{T}_\alpha^N\subset H(q_\alpha^N)=H(q)$, for any $q\sim q_\alpha^N$; cf. Proposition 5.2(a). This shows that the right hand side of (10.3) is contained in the left hand side.

Conversely, let $\langle A,\lambda\rangle\in H(q)$, so $q_{(A,\lambda)}\to q$. By using Theorem 31 of [**K5**] (together with [**K5**], Lemma 30) we can find an N-presentation (E,E',ψ,π) of (A,λ) and a cyclic isogeny $f : E\to E'$ of degree d such that (11.1) holds for some $\alpha\in\mathcal{M}_d$ and some level N-structure $\phi : (\mathbb{Z}/N\mathbb{Z})^2\to E[N]$. Thus $z=\langle E,E',\psi\rangle\in\Phi_N(T_\alpha^N)$ by Proposition 11.2 and so $\langle A,\lambda\rangle=\beta_N(z)\in\overline{T}_\alpha^N$. This proves (10.3).

To prove the assertion (10.4), note that if $\alpha_1, \alpha_2 \in \mathcal{M}_d$, then by [**K6**], Proposition 15 and Lemma 11, we have that

$$T^N_{\alpha_1} = T^N_{\alpha_2} \quad \Leftrightarrow \quad \alpha_2 \equiv \pm\alpha_2 \pmod N.$$

Using this and the modular description of Φ_N, we see that

(11.2)

$$\Phi_N(T^N_{\alpha_1}) = \Phi_N(T^N_{\alpha_2}) \Leftrightarrow \gamma\beta\alpha_1\gamma^{-1} \equiv \pm\beta\alpha_2 \pmod N, \text{ for some } \gamma \in \Gamma(1),$$

and from this (10.4) follows because $\overline{T}^N_{\alpha_i} = \beta_N(\Phi_N(T^N_{\alpha_i}))$, for $i = 1, 2$. □

In order to deduce Theorems 9.1 and 9.2 from the Structure Theorem 10.5, further work is necessary. In view of (10.3) and (10.4), this leads to the study the following three problems:

(1) Determine the $\mathrm{SL}_2(\mathbb{Z}/N\mathbb{Z})$-conjugacy classes of the matrices $\alpha \pmod N$.
(2) Study the $\pm$-action on the conjugacy classes.
(3) Examine the converse of implication (10.4).

The solutions of these problems will now be discussed in turn.

1. Conjugacy classes of matrices. By an easy extension of the work of Nobs[**No**], who treated the case of a prime power $N = p^r$, it is possible to determine explicit representatives for the $\mathrm{SL}_2(\mathbb{Z}/N\mathbb{Z})$-conjugacy classes of 2×2 matrices mod N.

NOTATION. For an integral 2×2 matrix $\alpha = \left(\begin{smallmatrix} x & y \\ z & w \end{smallmatrix}\right) \in M_2(\mathbb{Z})$ and integer $N \geq 2$ put

$$g_{\alpha,N} = \gcd(x - w, y, z, N).$$

Note that $g_{\alpha,N}$ depends only on $\alpha \pmod N$ and that $g_{\alpha,N}$ is invariant under conjugation by $\mathrm{SL}_2(\mathbb{Z})$; cf. [**K5**], Corollary 13. Furthermore, for $t, d \in \mathbb{Z}$ put

$$S_N(t, d) = \{a \in (Z/N\mathbb{Z})^\times : \exists(\xi, \eta) \in \mathbb{Z}^2 \text{ with } \xi^2 + t\xi\eta + d\eta^2 \equiv a \pmod N\}.$$

As in [**No**], it is easy to see that $S_N(t, d)$ is a subgroup of $(\mathbb{Z}/N\mathbb{Z})^\times$.

The following result gives an overview of the $\mathrm{SL}(\mathbb{Z}/N\mathbb{Z})$-conjugacy classes of matrices $\alpha \pmod N$.

PROPOSITION 11.3. (a) *Let $\alpha \in M_2(\mathbb{Z})$ and $N \geq 2$, and put $g = g_{\alpha,N}$ and $m = \frac{N}{g}$. Then there exist integers $x, t, d, z, z' \in \mathbb{Z}$ with $0 \leq x < g$ and $zz' \equiv 1 \pmod m$ such that*

$$\gamma\alpha\gamma^{-1} \equiv xI + g\begin{pmatrix} 0 & -dz' \\ z & t \end{pmatrix} \pmod N, \quad \text{for some } \gamma \in \mathrm{SL}_2(\mathbb{Z}). \tag{11.3}$$

(b) *Let $g|N$ and put $m = \frac{N}{g}$. If $\alpha_1, \alpha_2 \in M_2(\mathbb{Z})$ have the form*

$$\alpha_i = x_i + g\begin{pmatrix} 0 & -d_i z'_i \\ z_i & t_i \end{pmatrix},$$

where $0 \leq x_i < g$ and $z_i z'_i \equiv 1 \pmod m$, for $i = 1, 2$, then the following conditions are equivalent:

(i) *There exists $\gamma \in \mathrm{SL}_2(\mathbb{Z})$ such that $\gamma\alpha_1\gamma^{-1} \equiv \alpha_2 \pmod N$.*

(ii) *We have that $x_1 = x_2$, $t_1 \equiv t_2 \pmod m$, $d_1 \equiv d_2 \pmod m$, and $a \in S_m(t_1, d_1)$, where $a \equiv z_1 z'_2 \pmod m$.*

This leads to the following result which refines the Structure Theorem 10.5.

THEOREM 11.4. *Let q be a form of type (N, m, d) and put $g = \frac{N}{m}$. Then there exist integers x, t and δ with $0 \leq x < g$ such that if $z_1, \ldots, z_s$ is a system of representatives of $(\mathbb{Z}/m\mathbb{Z})^\times/S_m(t,\delta)$, and if $\alpha_1, \ldots, \alpha_s \in \mathcal{M}_d$ are primitive matrices such that*

$$\beta\alpha_i \equiv xI + g\begin{pmatrix} 0 & -\delta z_i' \\ z_i & t \end{pmatrix} \quad (mod\, N),$$

where $z_i z_i' \equiv 1\,(mod\, m)$, for $1 \leq i \leq s$, then

$$H(q) = \bigcup_{i=1}^{s} \overline{T}_{\alpha_i}^N. \tag{11.4}$$

In particular, the number of irreducible components of $H(q)$ satisfies the estimate

$$|\operatorname{Irr}(H(q))| \leq i_m(t,\delta) := |(\mathbb{Z}/m\mathbb{Z})^\times : S_m(t,\delta)|. \tag{11.5}$$

REMARK 11.5. The above is a shortened version of Theorem 40 of [**K7**], which also gives the precise recipe of how x, t, and δ are determined from q.

Thus, to find an upper bound for the number of irreducible components of $H(q)$, it suffices by (11.5) to bound the index $i_m(t,\delta)$. For prime powers m, Nobs[**No**] computed $S_m(t,d)$ and hence also $i_m(t,d)$. Using this, we obtain:

PROPOSITION 11.6. *Let $t, \delta \in \mathbb{Z}$ be integers and put $\Delta_m(t,\delta) = \gcd(t^2 - 4\delta, m)$. Then*

$$i_m(t,\delta) \leq 2^{\omega(\Delta_m(t,\delta))}, \tag{11.6}$$

except when $8|\Delta_m(t,\delta)$, in which case $i_m(t,\delta) \leq 2^{\omega(\Delta_m(t,\delta))+1}$. Furthermore, equality holds in (11.6) *if $2 \nmid \Delta_m(t,\delta)$.*

REMARK 11.7. In [**K7**] it is shown that in the situation of Theorem 11.4 we have that

$$\Delta_m(t,\delta) = \gcd(c(q), m), \tag{11.7}$$

where, as before, $c(q)$ is the content of the form q. Using this, we see that Theorem 9.1 follows immediately from (11.5) and (11.6).

2. The $\pm$-action on conjugacy classes. In view of the explicit representatives of the $\mathrm{SL}_2(\mathbb{Z}/N\mathbb{Z})$-conjugacy classes of matrices given in Proposition 11.3, it is an easy exercise to determine when the $\pm$-action identifies some of the $\mathrm{SL}_2(\mathbb{Z}/N\mathbb{Z})$-conjugacy classes or not. For example, the following result follows easily from Proposition 11.3.

LEMMA 11.8. *Let $\alpha_1, \ldots, \alpha_s \in \mathcal{M}_d$ be as in* Theorem 11.4, *and suppose that N is odd. If $x \neq 0$ or if $t \not\equiv 0\,(\mathrm{mod}\, m)$, then we have that*

$$-\beta\alpha_i \not\equiv \gamma\beta\alpha_j\gamma^{-1}\,(\mathrm{mod}\, N), \quad \text{for } 1 \leq i,j \leq s \text{ and all } \gamma \in \mathrm{SL}_2(\mathbb{Z}). \tag{11.8}$$

REMARK 11.9. (a) If x and t do not satisfy the conditions of Lemma 11.8, then $q \sim [N^2, 0, 4d]$, so q has type (N, N, d). This explains why this form plays a special role in Theorem 9.2.

(b) If N is even, then there is an analogue of Lemma 11.8, but then there are more exceptional cases. In this case the exceptional q's are those of the form $[N^2, 0, 4d]$, $[N^2, N^2, (\frac{N}{2})^2 + 4d]$ and $[N^2, \varepsilon N^2, (\frac{\varepsilon N}{2})^2 + d]$, where $\varepsilon \in \{0, 1\}$ and $d \equiv 1 + \varepsilon N\,(\mathrm{mod}\, 4)$ in the last case; cf. [**K7**].

3. The converse of the implication (10.4). This task is much more difficult than the previous two, and complete results are not yet available, except when N is an odd prime.

To understand the difficulties, recall from the proof of Theorem 10.5 that the converse of (10.4) does hold for the modular curves on Z_N; cf. (11.2). However, since the basic construction map $\beta_N : Z_N \to H_{N^2}$ is generically $2:1$ (cf. Theorem 3.3), we can expect that distinct modular curves on Z_N are mapped to the same modular curve $\overline{T}_\alpha^N$ on H_{N^2}. However, we have the following result.

PROPOSITION 11.10. *Every modular curve $\Phi_N(T_\alpha^N)$ on Z_N is stable under the involution w_N of* Theorem 3.3. *Thus, if*

$$\pi_N : Z_N \;\to\; Z_N^{sym} \;=\; \langle w_N\rangle\backslash Z_N$$

denotes the quotient morphism, then we have for $\alpha_1, \alpha_2 \in \mathcal{M}_d$ that

(11.9)

$$\pi_N\beta_N(T_{\alpha_1}^N) = \pi_N\beta_N(T_{\alpha_2}^N) \;\Leftrightarrow\; \gamma\beta\alpha_1\gamma^{-1} \equiv \pm\beta\alpha_2 \pmod N, \textit{ for some } \gamma \in \mathrm{SL}_2(\mathbb{Z}).$$

PROOF. (Sketch; details in [**K7**]). It is not difficult to see (cf. [**K6**]) that $\tau_N(T_N^\alpha) = T_{\alpha^*}^N$, where τ denotes the involution of $X(N)\times X(N)$ which interchanges the factors (cf. Theorem 3.3) and $\alpha^* = \det(\alpha)\alpha^{-1}$ denotes the adjoint of α. Thus

$$w_N(\Phi_N(T_\alpha^N)) \;=\; \Phi_N(T_{\alpha^*}^N).$$

Using Proposition 11.3 we see that $\beta\alpha^*$ is conjugate mod N to $\beta\alpha$, and so we obtain that $w_N(\Phi_N(T_\alpha^N)) = \Phi_N(T_\alpha^N)$ by (11.2). Thus $\Phi_N(T_\alpha^N)$ is w_N-stable, as claimed.

From this it follows that the map π_N maps the modular curves on Z_N bijectively to those of Z_N^{sym}, and so (11.9) follows from (11.2). □

REMARK 11.11. Since $\beta_N : Z_N \to H_{N^2}$ is w_N-invariant, it factors over π_N as

$$\beta_N \;=\; \nu_N \circ \pi_N,$$

where $\nu_N : Z_N^{sym} \to H_{N^2}$ is finite and birational; cf. Theorem 3.3. Thus, ν_N is generically an isomorphism, and so only finitely many of the modular curves on Z_N^{sym} are not mapped bijectively to H_{N^2}; these then lie in the inverse image of the *conductor locus* of H_{N^2}. The following result and Remark 11.13 below make this precise.

THEOREM 11.12. *Let q be a form of type (N,m,d) which satisfies the condition*

$$|\{(x,y)\in\mathbb{Z}^2 : q(x,y) = N^2, \gcd(x,y) = 1\}| \;=\; 2. \tag{11.10}$$

Then the converse of (10.4) *holds for the matrices $\alpha_i \in \mathcal{M}_d$ with $q_{\alpha_i}^N \sim q$.*

REMARK 11.13. If $d > N^4/(4m^2)$, then the reduction theory of binary quadratic forms shows that (11.10) holds. Thus, there are only finitely many $H(q)$'s on H_{N^2} which fail to satisfy the condition (11.10) because for them the absolute value of the discriminant of q is bounded by $4N^4$, and so there are only finitely equivalence classes of such q's.

PROOF OF THEOREM 11.12. (Sketch; details in [**K7**]). Suppose that $\overline{T}_{\alpha_1}^N = \overline{T}_{\alpha_2}^N$, and let $\langle A,\lambda\rangle \in \overline{T}_{\alpha_i}^N \subset H(q)$ be a non-CM point, so $q_{(A,\lambda)} \sim q$. Then the hypothesis $\overline{T}_{\alpha_1}^N = \overline{T}_{\alpha_2}^N$ implies that there exist $z_i \in \Phi_N(T_{\alpha_i}^N)$ such that $\beta_(z_1) = \beta_N(z_2) = \langle A,\lambda\rangle$. Using the hypothesis (11.10), one then shows that in fact $z_1 = z_2$.

Since $T^N_{\alpha_i}$ has infinitely many non-CM points, it follows that $\Phi_N(T^N_{\alpha_1}) = \Phi_N(T^N_{\alpha_2})$, and so the converse of (10.4) follows by using (11.2). $\square$

PROOF OF THEOREM 9.2. If $q \in Q(N^2)$, then q has type (N, m, d) for some unique m, d by Proposition 10.3(b), so by (10.2) and the hypothesis we have $16m^2d = |\operatorname{disc}(q)| > 4N^4$, or $d > N^4/(4m^2)$. Thus, by Remark 11.13 we have that q satisfies (11.10). Thus, by Theorem 11.12 we know that the converse of (10.4) holds for the $\alpha_i \in \mathcal{M}_d$'s such that $q^N_{\alpha_i} \sim q$.

Moreover, since by hypothesis $q \not\sim [N^2, 0, 4d]$, we have by Remark 11.9(a) and Lemma 11.8 and the converse of (10.4) that the α_i's of Theorem 11.4 give distinct modular curves $\overline{T}^N_{\alpha_i}$ on H_{N^2}, and so by Theorem 11.4 we see that $|\operatorname{Irr}(H(q))| = i_m(t, \delta)$, and so the desired formula follows from Proposition 11.6 and Remark 11.7. $\square$

In the case that q does not satisfy condition (11.10), the analysis of the converse of (10.4) is much more delicate. In the case of prime level $N > 2$, we have the following technical result which (almost) suffices to prove Theorem 9.4.

PROPOSITION 11.14. *Let $q \in Q(N, N, d)$, where $N > 2$ is prime. If condition* (11.10) *does not hold for q, then $N \nmid d$ and $N \mid c(q)$, so $\beta_N^{-1}(H(q)) \subset Z_N$ consists of two irreducible components. On the other hand, we have that*

$$H(q) \text{ is irreducible} \quad \Leftrightarrow \quad N \equiv 3 \pmod 4. \tag{11.11}$$

PROOF. (Sketch; details in [**K8**]). This is rather technical. The first two assertions are easily verified. Thus $\omega(\gcd(c(q), N)) = 1$, so we have by (11.2) and a refinement of Theorem 11.4 and Proposition 11.6 that $\beta_N^{-1}(H(q))$ consists of two distinct components $\Phi_N(T^N_{\alpha_i})$, $i = 1, 2$. Thus, the matrices $\alpha_i \in \mathcal{M}_d$ are such that $q \sim q^N_{\alpha_1} \sim q^N_{\alpha_2}$, but $\beta\alpha_1$ is not mod N-conjugate to $\pm\beta\alpha_2$.

To prove (11.11), suppose that $z_i \in \Phi_N(T^N_{\alpha_i})$ are non-CM points such that $\langle A, \lambda\rangle = \beta(z_1) = \beta(z_2)$, so (A, λ) has two distinct N-presentations. A long and delicate computation using these two N-presentations shows that $\beta\alpha_1$ is mod N-conjugate to $\pm\beta\alpha_2$ if and only $p \equiv 1 \pmod 4$. Thus, if $p \equiv 1 \pmod 4$, then we have a contradiction, so in particular $\beta_N(T^N_{\alpha_1})$ and $\beta_N(T^N_{\alpha_2})$ are distinct (and intersect only in CM-points). On the other hand, if $p \equiv 3 \pmod 4$, then reversing the above argument shows that each $\langle A, \lambda\rangle \in H(q)$ (which is not a CM point) is the image of a point $z_i \in \Phi_N(T^N_{\alpha_i})$ on each of the two components, so $\overline{T}^N_{\alpha_1} = \overline{T}^N_{\alpha_2} = H(q)$. $\square$

To connect the the above hypotheses with those of Theorem 9.4, we observe the following fact which is proved in [**K8**].

LEMMA 11.15. *If q be an imprimitive form of type (N, m, d), where $N > 2$ is prime, then $m = N$. If in addition $q \not\sim [N^2, 0, 4d]$, then condition* (11.10) *does not hold for q if and only if q is ambiguous.*

EXAMPLE 11.16. The generalized Humbert scheme $H[9, 6, 9] \subset H_9$.

The form $q = [9, 6, 9]$ has type $(3, 3, 2)$ and condition (11.10) fails for q because $\pm(1, 0)$ and $\pm(0, 1)$ are primitive solutions of $q(x, y) = 9$. (Note also that $q \sim [24, 24, 9]$, so q is an ambiguous form.) Thus, Proposition 11.14 shows that the converse of (10.4) does not hold for q, and that $H(q)$ is an irreducible curve; cf. also Theorem 9.4.

Note that (the proof of) Proposition 11.14 implies that $H(q)$ lies in the *conductor locus* of H_9,i.e., in the (closed) subset of non-normal points of H_9, so in particular $Z_3^{sym} \not\subset H_9$, and so H_9 is *not normal*. In fact, it turns out that H_{N^2} is not normal for any $N \geq 3$; cf. [**K8**].

Acknowledgment

I would like to thank the organizers of the *Langefest* and of the conference *Geometry at the Frontier III* for inviting me and for providing travel funds and for their kind hospitality at the conference. In addition, I would like to thank the referee for his/her useful comments and suggestions which improved this paper. Moreover, I gratefully acknowledge receipt of funding from the Natural Sciences and Research Council of Canada (NSERC).

References

[AP] R. D. M. Accola and E. Previato, *Covers of tori: genus two*, Lett. Math. Phys. **76** (2006), no. 2-3, 135–161, DOI 10.1007/s11005-006-0067-5. MR2235401

[Ba] W. L. Baily Jr., *On the automorphism group of a generic curve of genus* > 2, J. Math. Kyoto Univ. **1** (1961/62), 101–108; correction, 325, DOI 10.1215/kjm/1250525063. MR142552

[Co] D. A. Cox, *Primes of the form* $x^2 + ny^2$, A Wiley-Interscience Publication, John Wiley & Sons, Inc., New York, 1989. Fermat, class field theory and complex multiplication. MR1028322

[FPR] M. Franciosi, R. Pardini, and S. Rollenske, (d, d')*-elliptic curves of genus two*, Ark. Mat. **56** (2018), no. 2, 299–317, DOI 10.4310/ARKIV.2018.v56.n2.a6. MR3893776

[FK1] G. Frey and E. Kani, *Curves of genus* 2 *covering elliptic curves and an arithmetical application*, Arithmetic algebraic geometry (Texel, 1989), Progr. Math., vol. 89, Birkhäuser Boston, Boston, MA, 1991, pp. 153–176. MR1085258

[FK2] G. Frey and E. Kani, *Curves of genus 2 with elliptic differentials and associated Hurwitz spaces*, Arithmetic, geometry, cryptography and coding theory, Contemp. Math., vol. 487, Amer. Math. Soc., Providence, RI, 2009, pp. 33–81, DOI 10.1090/conm/487/09524. MR2555986

[He] E. Hecke, *Höhere Modulfunktionen und ihre Anwendung auf die Zahlentheorie* (German), Math. Ann. **71** (1911), no. 1, 1–37, DOI 10.1007/BF01456926. MR1511639

[H1] C. F. Hermann, *Modulflächen quadratischer Diskriminante* (German), Manuscripta Math. **72** (1991), no. 1, 95–110, DOI 10.1007/BF02568268. MR1107455

[H2] C. F. Hermann, *Symmetrische Modulflächen der Diskriminante* p^2 (German, with German summary), Manuscripta Math. **76** (1992), no. 2, 169–179, DOI 10.1007/BF02567753. MR1174909

[Hu] G. Humbert, *Sur les fonctions abéliennes singulières. I–III.* J. de Math. (ser. 5) **5** (1899), 233–350; **6** (1900), 279–386; **7** (1901), 97–123.

[K1] E. Kani, *Elliptic curves on abelian surfaces*, Manuscripta Math. **84** (1994), no. 2, 199–223, DOI 10.1007/BF02567454. MR1285957

[K2] E. Kani, *Hurwitz spaces of genus 2 covers of an elliptic curve*, Collect. Math. **54** (2003), no. 1, 1–51. MR1962943

[K3] E. Kani, *The moduli spaces of Jacobians isomorphic to a product of two elliptic curves*, Collect. Math. **67** (2016), no. 1, 21–54, DOI 10.1007/s13348-015-0148-9. MR3439838

[K4] E. Kani, *Elliptic subcovers of a curve of genus 2. I. The isogeny defect* (English, with English and French summaries), Ann. Math. Qué. **43** (2019), no. 2, 281–303, DOI 10.1007/s40316-018-0105-6. MR3996071

[K5] E. Kani, *Elliptic subcovers of a curve of genus 2 II. The refined Humbert invariant*, J. Number Theory **193** (2018), 302–335, DOI 10.1016/j.jnt.2018.05.011. MR3846811

[K6] E. Kani, Modular correspondences on $X(N)$. Preprint, 27 pages.

[K7] E. Kani, Generalized Humbert schemes and intersections of Humbert surfaces. Preprint, 35pp.

[K8] E. Kani, Modular curves on Humbert Surfaces. Preprint, 68 pages.

[KS] E. Kani and W. Schanz, *Modular diagonal quotient surfaces*, Math. Z. **227** (1998), no. 2, 337–366, DOI 10.1007/PL00004379. MR1609061

[K] F. Klein, *Zur Theorie der elliptischen Modulfunctionen* (German), Math. Ann. **17** (1880), no. 1, 62–70, DOI 10.1007/BF01444120. MR1510051

[KF] F. Klein, *Vorlesungen über die Theorie der elliptischen Modulfunktionen. Band I: Grundlegung der Theorie* (German), Ausgearbeitet und vervollständigt von Robert Fricke. Nachdruck der ersten Auflage. Bibliotheca Mathematica Teubneriana, Band 10, Johnson Reprint Corp., New York; B. G. Teubner Verlagsgesellschaft, Stuttgart, 1966. MR0247996

[Kr] A. Krazer, *Lehrbuch der Thetafunktionen.* Teubner, Leipzig, 1903; Chelsea Reprint, New York, 1970.

[Ku] A. Kuribayashi, *On analytic families of compact Riemann surfaces with non-trivial automorphisms*, Nagoya Math. J. **28** (1966), 119–165. MR217280

[L1] H. Lange, *Über die Modulschemata der Kurven vom Geschlecht* 2 *mit* 1, 2 *oder* 3 *Weierstrasspunkten*, J. Reine Angew. Math. **277** (1975), 27–36, DOI 10.1515/crll.1975.277.27. MR407028

[L2] H. Lange, *Moduli spaces of algebraic curves with rational maps*, Math. Proc. Cambridge Philos. Soc. **78** (1975), no. 2, 283–292, DOI 10.1017/S0305004100051689. MR407031

[L3] H. Lange, *Über die Modulvarietät der Kurven vom Geschlecht* 2, J. Reine Angew. Math. **281** (1976), 80–96, DOI 10.1515/crll.1976.281.80. MR407029

[L4] H. Lange, *Kurven mit rationaler Abbildung* (German), J. Reine Angew. Math. **295** (1977), 80–115, DOI 10.1515/crll.1977.295.80. MR476754

[LB] H. Lange and C. Birkenhake, *Complex abelian varieties*, Grundlehren der Mathematischen Wissenschaften [Fundamental Principles of Mathematical Sciences], vol. 302, Springer-Verlag, Berlin, 1992. MR1217487

[M1] D. Mumford, *Geometric invariant theory*, Ergebnisse der Mathematik und ihrer Grenzgebiete, Neue Folge, Band 34, Springer-Verlag, Berlin-New York, 1965. MR0214602

[M2] D. Mumford, *Abelian varieties*, Tata Institute of Fundamental Research Studies in Mathematics, No. 5, Published for the Tata Institute of Fundamental Research, Bombay; Oxford University Press, London, 1970. MR0282985

[No] A. Nobs, *Die irreduziblen Darstellungen der Gruppen $SL_2(Z_p)$, insbesondere $SL_2(Z_2)$. I* (German), Comment. Math. Helv. **51** (1976), no. 4, 465–489, DOI 10.1007/BF02568170. MR444787

[Sh] G. Shimura, *Introduction to the arithmetic theory of automorphic functions*, Publications of the Mathematical Society of Japan, No. 11. Iwanami Shoten, Publishers, Tokyo; Princeton University Press, Princeton, N.J., 1971. Kanô Memorial Lectures, No. 1. MR0314766

[Si] J. H. Silverman, *The arithmetic of elliptic curves*, Graduate Texts in Mathematics, vol. 106, Springer-Verlag, New York, 1986. MR817210

[vdG] G. van der Geer, *Hilbert modular surfaces*, Ergebnisse der Mathematik und ihrer Grenzgebiete (3) [Results in Mathematics and Related Areas (3)], vol. 16, Springer-Verlag, Berlin, 1988. MR930101

[We] A. Weil, *Zum Beweis des Torellischen Satzes* (German), Nachr. Akad. Wiss. Göttingen. Math.-Phys. Kl. IIa. **1957** (1957), 33–53. MR0089483

Department of Mathematics and Statistics, Queen's University, Kingston, Ontario, Canada K7L 3N6

Contemporary Mathematics
Volume **766**, 2021
https://doi.org/10.1090/conm/766/15385

Algebraic geometry and convolutional codes

Ángel Luis Muñoz Castañeda and José I. Iglesias-Curto

This paper is dedicated to the memory of José M. Muñoz Porras

ABSTRACT. We make an overview of the application of algebraic geometry in the study of the fundamentals of convolutional coding theory. We will make a special emphasis on the ways we have to see a convolutional code as an algebraic-geometric object as well as their classification, and how the free distance comes into play in this framework.

1. Introduction

Convolutional codes were introduced by P. Elias ([**3**]) and they have been widely used and studied since then. Particularly important is the study of the distance properties of these codes. The Hamming distance on $\mathbb{F}_q^n$ can be extended to $\mathbb{F}_q[z]^n$ and the minimum distance between code-words of the convolutional code is called the free distance of the code $d_f(\mathcal{C})$. The larger the free distance the better. In [**22**], it was shown that the free distance of convolutional codes of rate k/n and complexity δ is bounded from above, and those codes that attain that bound are called maximum separable distance codes or just MDS codes. In [**23**], it was given a method for constructing an MDS convolutional code for any rate k/n and complexity δ, provided the base field $\mathbb{F}_q$ is large enough. However, there is still not a general method for constructing such codes and neither a characterization of them.

Algebraic-geometric tools have been used in coding theory for long time. Remarkable examples are [**6, 7**], where V. Goppa defined the so-called Goppa linear codes from the geometric data (X, D, G), where X is a smooth projective curve over $\mathbb{F}_q$, G is an effective divisor and D is the sum of distinct rational points not belonging to the support of G. The algebraic-geometric nature of Goppa linear codes allow to implement easy algebraic decoding algorithms with good error-correction capability (see [**16**] for instance). Regarding convolutional codes, V. Lomazde gave in [**12**] an algebraic-geometric description of them in terms of sheaves on the projective line. For that, he used the link between convolutional codes and linear dynamical systems [**21**]. The linear dynamical systems point of view can be very

2020 *Mathematics Subject Classification.* Primary 94B10, 94B27; Secondary 14D20, 14H99, 14N10.

Key words and phrases. Convolutional codes, Goppa codes, moduli spaces, algebraic curves, free distance.

Research supported by grant MTM2015-66760-P of MINECO.

valuable, since it establishes a close relation between reachable and observable input/state/output representations and observable convolutional codes [**21**], the most desirable codes in practice. However, this link is not as useful for the general classification problem as one would desire. Although there exists a projective moduli space for linear dynamical systems with fixed numerical parameters ([**8, 11, 20**]), the above connection to convolutional codes is not one-to-one for those that are not observable. Furthermore, there is no such connection between families.

In this paper, we will make an overview of the applications of algebraic geometry in the study of the classification of convolutional codes and the characterization of those that are MDS. We will also focus on convolutional Goppa codes, recalling that every observable convolutional code is a Goppa code over the projective line $\mathbb{P}^1$, and showing that these algebraic-geometric codes can be used to construct MDS convolutional codes, particularly of those with rate $1/n$.

2. Convolutional codes. Classification and free distance

We will give some basic definitions regarding convolutional codes. For a deep exposition of the algebraic theory of convolutional codes, we refer the reader to [**4, 13**].

DEFINITION 2.1. Let p be a prime number and $q = p^l$ for some natural number l. A convolutional code of dimension k and length n (or just rate k/n) is a submodule $\mathcal{C} \subset \mathbb{F}_q[z]^n$ of rank k. A convolutional encoder for $\mathcal{C}$ is an injective polynomial matrix $\mathcal{G}(z) \in \mathrm{Mat}_{k\times n}(\mathbb{F}_q[z])$ whose row space is equal to $\mathcal{C}$.

Let $\mathcal{C} \subset \mathbb{F}_q[z]^n$ be a convolutional code of dimension k, and let $\mathcal{G}(z)$ be a convolutional encoder for $\mathcal{C}$. If $\mathcal{G}_j(z)$ denotes the j-th row of $\mathcal{G}(z)$, $\deg(\mathcal{G}_j(z))$ is defined as the maximum degree among its coefficients and it is denoted by ν_j. After permuting the rows of $\mathcal{G}(z)$ we may assume that $\nu_1 \geq \ldots \geq \nu_k$. The memory of $\mathcal{G}(z)$ is defined as ν_1 and the complexity is defined as $\delta_{\mathcal{G}(z)} = \sum_{i=1}^{k} \nu_i$. Likewise, we may define the column-degrees of $\mathcal{G}(z)$ in a similar way as the row-degrees, and we denote them by $m_1, \ldots, m_n$.

Given an encoder $\mathcal{G}(z)$ of a convolutional code $\mathcal{C}$ of dimension k and length n, any other encoder can be obtained by multiplying by a $k \times k$ unimodular matrix, $U(z) \cdot \mathcal{G}(z)$. Multiplying by unimodular matrices do not vary the highest degree of the full size minors of $\mathcal{G}(z)$, thus this is a parameter of the code instead of the encoder which is called complexity of the code and it is denoted by $\delta_{\mathcal{C}}$. If we define $\delta := \min_{\mathcal{G}(z)}\{\delta_{\mathcal{G}(z)}\}$, we say that $\mathcal{G}(z)$ is a minimal encoder of $\mathcal{C}$ if its complexity attaches the minimum. Clearly, $\mathcal{G}(z)$ is minimal if and only if $\delta_{\mathcal{G}(z)} = \delta = \delta_{\mathcal{C}}$.

DEFINITION 2.2. Let $\mathcal{C} \subset \mathbb{F}_q[z]^n$ be a convolutional code of dimension k and length n. It is said to be observable if there is an integer N large enough such that if $v(z), v'(z) \in \mathbb{F}_q[z]^n$ satisfy $d(v(z), v'(z)) \geq N$ and $v(z) + v'(z) \in \mathcal{C}$ then both belong to the code.

Observable codes are characterized as follows: a convolutional code $\mathcal{C} \subset \mathbb{F}_q[z]^n$ of dimension k and length n is observable if and only if it is the kernel of a surjective morphism of modules, $\mathbb{F}_q[z]^n \twoheadrightarrow \mathbb{F}_q[z]^{n-k}$, called syndrome formers of the code [**21**]. Then, if $\mathcal{C} \subset \mathbb{F}_q[z]^n$ is an observable convolutional code of dimension k, $E := \mathbb{F}_q[z]^n/\mathcal{C}$ is a free module of rank $n-k$ and it induces a morphism of schemes $\mathbb{A}^1 \to \mathrm{Grass}(k, n)$, which extends uniquely to $\mathbb{P}^1$ since the Grassmannian is a proper scheme. From this observation, it follows that there is a one-to-one correspondence between observable convolutional codes of dimension k and length n and rational points of the scheme $\mathrm{Hom}_{\mathbb{F}_q}(\mathbb{P}^1, \mathrm{Grass}(k,n))$. Although this space can be thought as a moduli space for observable convolutional codes, a general classification must be given.

2.1. General classification. In order to improve the knowledge on their structure and properties, a classification of convolutional codes has been given as points of a certain algebraic variety [**15**]. The start point is the convolutional encoder, that can be extended to an exact sequence

$$0 \to \mathbb{F}_q[z]^k \to \mathbb{F}_q[z]^n \to \overline{L} \to 0 \,, \tag{2.1}$$

with $\overline{L}$ a free $(n-k)$ rank module. After homogenezing and shifting by column degrees of $\mathcal{G}(z)$, sequence (2.1) gives rise to an exact sequence of $\mathbb{F}_q[x_0, x_1]$-graded modules

$$0 \to \mathbb{F}_q[x_0, x_1]^k \overset{\phi_G}{\to} \bigoplus_{i=1}^{n} \mathbb{F}_q[x_0, x_1][m_i] \to \mathrm{Coker}\phi_G \to 0 \tag{2.2}$$

Further, since the categories of coherent sheaves on $\mathbb{P}^1_{\mathbb{F}_q}$ and that of equivalence classes of graded $\mathbb{F}_q[x_0, x_1]$-modules are equivalent, sequence (2.2) leads to an exact sequence of sheaves,

$$0 \longrightarrow \mathcal{O}^k_{\mathbb{P}^1} \longrightarrow \bigoplus_{i=1}^{n} \mathcal{O}_{\mathbb{P}^1}(m_i) \longrightarrow Q \to 0 \,, \tag{2.3}$$

with Q a sheaf with Hilbert polynomial $P(r) = (n-k)(r+1) + \mathrm{gr}\, Q$. Hence, Q represents the encoder and is itself represented by a point in the quotient scheme

$$Q \in \mathrm{Quot}^{P(r)}_{\oplus \mathcal{O}_{\mathbb{P}}(m_i)}$$

which is in turn a subscheme of the Grassmannian $\mathrm{Grass}\big(k, H^0(\oplus_{i=1}^n \mathcal{O}_{\mathbb{P}^1}(m_i))\big)$. Notice however that we are representing just an encoder as a point of a certain Grassmannian. In order to represent a code let us consider two facts. First, that if $\mu_i = \max_{\mathcal{G}(z)}\{i\text{-th column degree}\}$ then for any encoder we have

$$\bigoplus_{i=1}^{n} \mathcal{O}_{\mathbb{P}^1}(m_i) \hookrightarrow \bigoplus_{i=1}^{n} \mathcal{O}_{\mathbb{P}^1}(\mu_i)$$

Second, that if Forney indices are taken in increasing order, $\nu_k \leq \ldots \leq \nu_1$ then the above map allows us to rewrite (2.3) as

$$0 \longrightarrow \bigoplus_{i=1}^{k} \mathcal{O}(\nu_1 - \nu_i) \longrightarrow \bigoplus_{i=1}^{n} \mathcal{O}(\mu_i) \longrightarrow \mathcal{Q} \longrightarrow 0 \tag{2.4}$$

and consequently the sheaf $\mathcal{Q}$ represents the code rather than a particular encoder. $\mathcal{Q}$ has Hilbert polynomial $P(r) = (n-k)(r+1) + \sum_{i=1}^n \mu_i + \delta - k\nu_1$. which depends on the code parameters: length n, dimension k, complexity δ, memory $m = \nu_1$ and column indices $\{\mu_i\}$. Hence, a convolutional code, with its associated sheaf $\mathcal{Q}$ can be represented as a point in the scheme $\mathrm{Quot}^{P(r)}_{\oplus \mathcal{O}_{\mathbb{P}}(\mu_i)}$, which is in turn a subscheme of the Grassmannian $\mathrm{Grass}(k(\nu_1+1) - \delta, H^0(\bigoplus_{i=1}^{n} \mathcal{O}_{\mathbb{P}^1}(\mu_i)))$.

Once we know in which variety we can represent convolutional codes we wish to determine which points of that variety do in fact correspond to a convolutional code. Notice that the way in which we obtained the exact sequence (2.4) can be taken backwards to recover the encoding map of a convoplutional code. Hence, a quotient sheaf $\mathcal{Q}$ of $\oplus\mathcal{O}_{\mathbb{P}}(\mu_i)$ corresponds to a convolutional code, and therefore our task is to determine which points of the Grassmannian represent such quotient sheaves.

Let M be a matrix whose row span is a vector subspace of $H^0(\bigoplus_{i=1}^{n}\mathcal{O}_{\mathbb{P}^1}(\mu_i))$. Let us define the matrix $\widehat{M}$ as follows: given a partition of M

$$M = (M^{(1)}|M^{(2)}|\ldots|M^{(n)})$$

where the block $M^{(i)}$ has $\mu_i + 1$ columns, denoted by $m_0^{(i)}, \ldots, m_{\mu_i}^{(i)}$, then $\widehat{M}$ is defined as

$$\left(\begin{array}{cccccc|cc|cccccc} m_0^{(1)} & m_1^{(1)} & m_2^{(1)} & \ldots & m_{\mu_1}^{(1)} & 0 & \ldots & \ldots & m_0^{(n)} & m_1^{(n)} & m_2^{(n)} & \ldots & m_{\mu_n}^{(n)} & 0 \\ 0 & m_0^{(1)} & m_1^{(1)} & \ldots & m_{\mu_1-1}^{(1)} & m_{\mu_1}^{(1)} & \ldots & \ldots & 0 & m_0^{(n)} & m_1^{(n)} & \ldots & m_{\mu_n-1}^{(n)} & m_{\mu_n}^{(n)} \end{array}\right)$$

LEMMA 2.3 ([**15**]). *A point in* $\mathrm{Grass}(\lambda, H^0(\mathbb{P}^1, \bigoplus_{i=1}^{n}\mathcal{O}(\mu_i)))$ *which as a vector subspace is the row span of a matrix* M*, represents a quotient sheaf with rank* $n-k$ *of the sheaf* $\bigoplus_{i=1}^{n}\mathcal{O}(\mu_i)$*, with* $k \leq \lambda$*, if and only if the rank of* $\widehat{M}$ *is* $\lambda + k$*.*

COROLLARY 2.4. *If the Forney indices of the represented convolutional codes are distinct (if and only if* $\delta < k\nu_k$*, i.e.* $k < \lambda$*) the set of points satisfying the previous lemma is a closed subset. If the Forney indices are equal (if and only if* $\delta = k\nu_k$*, i.e.* $k = \lambda$*) then the set of points satisfying the previous lemma is an open subset.*

2.2. Free distance and generalized Singleton bound. In order to measure the error-correction capability of a convolutional code, we have to introduce a metric in the space $\mathbb{F}_q[z]^n$ to be able to measure the distances between a received message and the codewords and between the codewords themselves. To do so, we extend the Hamming distance in $\mathbb{F}_q^n$ to the polynomial framework by declaring $d(c(z), c'(z)) := \sum_{i=0}^{\infty} d_H(c_i, c'_i)$, where c_i and c'_i are the vectors of degree i in $c(z)$ and $c'(z)$ respectively.

The free distance, d_f, of a convolutional code $\mathcal{C} \subset \mathbb{F}_q[z]^n$ is the greatest natural number d such that $d \leq d(c(z), 0)$ for every codeword $c(z) \in \mathcal{C}$. A central problem in convolutional coding theory is to find an explicit expression of the free distance d_f in terms of the other parameters of the codes, as the dimension, the length and the complexity. Although this problem is still open, there is a well-known bound for the free distance in terms of these parameters, called *generalized Singleton bound* [**22**],

$$d_f \leq (n-k)(\lfloor \delta/k \rfloor + 1) + \delta + 1. \tag{2.5}$$

The convolutional codes which attain the generalized Singleton bound are called maximum distance separable or just MDS.

As we have seen before, there is a scheme $\mathcal{C}(k, n, \delta, m, \underline{\mu})$ parametrizing convolutional codes of length n, dimension k, complexity δ, memory m and column-degrees $\underline{\mu} = \{\mu_1, \ldots, \mu_n\}$, which is a subscheme of the Grassmannian $\mathrm{Grass}(\kappa, \theta)$, where $\kappa = k(m+1) - \delta$ and $\theta = \sum_{i=1}^{n}(\mu_i + 1)$. A major question is whether the locus parametrizing MDS convolutional codes has an algebraic-geometric structure and, if so, what geometric properties does it have. The next theorem gives some insight on this question.

THEOREM 2.5. [**18**] *The free distance function* $d_f : \mathcal{C}(k, n, \delta, m, \underline{\mu}) \to \mathbb{Z}$ *is lower semicontinuous. In particular, the locus parametrizing* MDS *convolutional codes,* $\mathrm{MDS}(k, n, \delta, m, \underline{\mu})$*, is an open subscheme.*

Since $\mathrm{MDS}(k, n, \delta, m, \underline{\mu})$ is Zariski open inside $\mathcal{C}(k, n, \delta, m, \underline{\mu})$ one can ask for the system of algebraic equations defining $\mathcal{C}(k, n, \delta, m, \underline{\mu}) \setminus \mathrm{MDS}(k, n, \delta, m, \underline{\mu})$. Observe that finding such system of equations would lead to an algebraic characterization of MDS convolutional codes. In the next Section we will describe the system of algebraic equations characterizing the complement of the MDS locus in a family of 1-dimensional convolutional codes.

3. Convolutional Goppa codes

The subclass of observable convolutional codes is of great importance in the theory of convolutional codes. Observe that there is a one-to-one correspondence

$$\left\{ \begin{array}{l} \text{Submodules } \mathcal{C} \subset \mathbb{F}_q[z]^n \text{ of} \\ \text{rank } k \text{ with free cokernel} \end{array} \right\} \longleftrightarrow \left\{ \begin{array}{l} \text{Subspaces } \mathcal{C} \subset \mathbb{F}_q(z)^n \\ \text{of dimension } k \end{array} \right\},$$

where the left-to-the-right arrow is given by localizing at the generic point , while the right-to-the-left arrow is given by the kernel of the composition of maps $\mathbb{F}_q[z]^n \hookrightarrow \mathbb{F}_q(z)^n \to \mathbb{F}_q(z)^n/\mathcal{C}$. Note that this let us define (observable) convolutional Goppa codes in the same way as linear Goppa codes are defined but over the function field $\mathbb{F}_q(z)$.

3.1. The definition. We will give the definition of convolutional Goppa code as well as an easy but important example that will appear latter. For a complete exposition of convolutional Goppa codes, we refer the reader to [**1, 14**].

Let X be a smooth projective irreducible curve over $\mathbb{F}_q(z)$ of genus g. Let G be an effective divisor, and let $p_1, \dots, p_n \in X$ be n distinct rational points of X such that $p_i \notin \mathrm{Supp}(G)$. Denote by $D = p_1 + \ldots + p_n$ the corresponding divisor. There is an exact sequence of invertible sheaves

$$0 \longrightarrow \mathcal{O}_X(G - D) \longrightarrow \mathcal{O}_X(G) \xrightarrow{ev} \mathcal{O}_D \longrightarrow 0,$$

which leads to the following long exact sequence in cohomology

$$0 \longrightarrow H^0(X, \mathcal{O}_X(G - D)) \longrightarrow H^0(X, \mathcal{O}_X(G)) \xrightarrow{ev} H^0(X, \mathcal{O}_D) \simeq \mathbb{F}_q(z)^n \dots$$

Given a linear subspace, $\Gamma \subset H^0(X, \mathcal{O}_X(G))$, the convolutional Goppa code associated to the geometric data X, Γ, D, and denoted by $\mathcal{C}_X(\Gamma, D)$, is defined as $\mathcal{C}_X(\Gamma, D) := \mathrm{Im}(ev|_\Gamma) \subset \mathbb{F}_q(z)^n$. Observe that $ev|_\Gamma : \Gamma \to \mathbb{F}_q(z)^n$ is given by

$$ev|_\Gamma(s) := (s(p_1), \dots, s(p_n)).$$

If $\Gamma = H^0(X, \mathcal{O}_X(G))$ we say that the code is defined by the complete linear series and we denote it just by $\mathcal{C}_X(G, D)$.

EXAMPLE 3.1. Consider the projective line $\mathbb{P}^1$ over the field $\mathbb{F}_q(z)$ and fix homogeneous coordinates x_0, x_1. Let $p_\infty = [1; 0]$ be the point at infinity and let $t = x_0/x_1$ be the affine coordinate. Consider n distinct rational points in the affine part $p_i = [\alpha_i; 1]$. Set $G = (n-1)p_\infty$ and $D = p_1 + \ldots + p_n$. Then the evaluation map is given by

$$\begin{aligned} ev : H^0(\mathbb{P}^1, \mathcal{O}_{\mathbb{P}^1}(G)) \simeq \mathbb{F}_q(z)[t]_{<n} &\to \mathbb{F}_q(z)^n \\ P(t) &\mapsto (P(\alpha_1), \dots, P(\alpha_n)) \end{aligned} \tag{3.1}$$

Consider the k-dimensional subspace $\Gamma =< 1, t, \ldots, t^{k-1} >\subset H^0(X, \mathcal{O}_X(G))$. Then, $\mathcal{C}_X(\Gamma, D)$ is given by the subspace spanned by the rows of the matrix

$$\mathcal{G}(z) = \begin{pmatrix} 1 & 1 & \ldots & 1 \\ \alpha_1 & \alpha_2 & \ldots & \alpha_n \\ \vdots & \vdots & \ddots & \vdots \\ \alpha_1^{k-1} & \alpha_2^{k-1} & \ldots & \alpha_n^{k-1} \end{pmatrix}$$

3.2. Every observable convolutional code is Goppa over $\mathbb{P}^1$. Let $X \subset \mathbb{P}^{n-1}$ be a twisted smooth projective curve over $\mathbb{F}_q(z)$. Note that if we want to define a convolutional Goppa code over X, it is mandatory for the curve to have enough rational points. Since the base field is the function field $\mathbb{F}_q(z)$ this is equivalent to assuming that there exists one rational point. To show this, suppose $p \in X$ is a rational point. We can assume that the point is general (after applying a projectivity on $\mathbb{P}^{n-1}$ if necessary) and its affine coordinates $(\alpha_1, \ldots, \alpha_{n-1})$ do not lie in $\overline{\mathbb{F}_q} \setminus \mathbb{F}_q$. Therefore, for every natural number i, the i-th power of the Frobenius morphism applied to p gives $F^i(p) = (\alpha_1^i, \ldots, \alpha_{n-1}^i)$ with $\alpha_j^i \neq \alpha_j^l$. Thus, $p, F(p), \ldots, F^{n-1}(p)$ are n distinct rational points. Furthermore, since $X \subset \mathbb{P}^{n-1}$ is twisted, it follows that p can be chosen so that $p, F(p), \ldots, F^{n-1}(p)$ are in general position in $\mathbb{P}^{n-1}$.

At this point a natural question is how large is the sub-class of convolutional Goppa codes inside the class of observable convolutional codes. This question has also been addressed in the case of linear Goppa codes in [**17**]. However, the techniques used there are not applicable in convolutional coding theory since they make use of the fact that the base field has finitely many elements.

The main result is this direction is the following theorem.

THEOREM 3.2. [**10**] *Every observable convolutional code of length n can be obtained as a convolutional Goppa code over any smooth twisted curve $X \subset \mathbb{P}^{n-1}$ over $\mathbb{F}_q(z)$ with at least one rational point.*

PROOF. Since the proof shows how to find the geometric data that define the Goppa structure we present a sketch of it. Fix homogeneous coordinates $x_0, \ldots, x_n$ in $\mathbb{P}^{n-1}$. Given a twisted immersion $i : X \hookrightarrow \mathbb{P}^{n-1}$, we define $\mathcal{L} := i^*\mathcal{O}_{\mathbb{P}^1(1)}$ and we fix an effective divisor G with $\mathcal{L} \simeq \mathcal{O}_{\mathbb{P}^1}(G)$. Since X is twisted, we have an injective linear map $i^* : H^0(\mathbb{P}^1, \mathcal{O}_{\mathbb{P}^1}(1)) \hookrightarrow H^0(X, \mathcal{L})$ and we define $E := \operatorname{Im}(i^*)$. For every rational point $p \in X$ there is an evaluation map $\phi_p : H^0(X, \mathcal{L}) \to \mathbb{F}_q(z)$ ($\mathcal{L} \otimes_{\mathcal{O}_X} \mathbb{F}_q(z) \simeq \mathbb{F}_q(z)$ is canonical since the divisor G has been fixed) whose restriction to E is denoted by $\overline{\phi}_p$. Let $p_1, \ldots, p_n \in X$ be rational points in general position and $D = p_1 + \ldots + p_n$ the corresponding divisor. Then, $\overline{\phi}_{p_1}, \ldots, \overline{\phi}_{p_n}$ is a basis for E^*. In order to give the linear subspace $\Gamma \subset H^0(X, \mathcal{L})$ that defines the Goppa structure on the code $\mathcal{C}$ we have to consider an encoder $\mathcal{G}(z) = (a_{ij})$. Let $\sigma_1, \ldots, \sigma_n$ be the dual basis of $\overline{\phi}_{p_1}, \ldots, \overline{\phi}_{p_n}$. Now, we define k elements in $H^0(X, \mathcal{L})$ by $s_i = a_{i1}\sigma_1 + \ldots + a_{in}\sigma_n$. Then, $\Gamma =< s_1, \ldots, s_k >$ makes it possible to define $\mathcal{C}$ as $\mathcal{C}_X(\Gamma, D)$. □

It is worth to note that the projective line satisfies the conditions of the theorem and, therefore, every observable convolutional code is a Goppa code over $\mathbb{P}^1$. To understand better this result, let us look at Example 3.1, where the invertible sheaf $\mathcal{L}((n-1)p_\infty)$ has been chosen, as well as n distinct rational points $p_i = [\alpha_i; 1]$. Obviously, the evaluation map is injective and, therefore, a linear isomorphism.

Thus, given a convolutional code $\mathcal{C}$ of length n and dimension k, it follows that $\mathcal{C} = \mathcal{C}_X(\Gamma, D)$, where $\Gamma = ev^{-1}(\mathcal{C})$. Furthermore, a basis $s_1(t), \ldots, s_k(t)$ of the k-dimensional vector space Γ can be computed explicitly in terms of the basic symmetric polynomials S_r of degree $r \leq n$ in $n-1$ variables, the affine coordinates α_i and the entries a_{ij} of an encoder of $\mathcal{C}$ **[10]**,

$$s_i(t) = \sum_{r=1}^{n}(-1)^r t^{n-r} \sum_{j=1}^{n} a_{ij}\lambda_j S_r(\alpha_1, \ldots, \widehat{\alpha}_j, \ldots, \alpha_n), \ \lambda_j = \frac{1}{\prod_{l=1, j\neq l}^{n}(\alpha_j - \alpha_l)}$$

3.3. Convolutional Goppa codes defined by complete linear series. As we have seen before, every observable convolutional code has a Goppa structure defined over the projective line. However, this is not the case if we impose the condition for the Goppa structure to be defined by a complete linear series. More precisely: let $\mathcal{C} \subset \mathbb{F}_q(z)^n$ be a convolutional code of dimension k and length n, and let $\mathcal{G}(z)$ be an encoder for $\mathcal{C}$. Consider the projective space $\mathbb{P}^{m-1}$ and fix homogeneous coordinates as well as a tautological divisor H, that is, $\mathcal{O}_{\mathbb{P}^{m-1}}(1) = \mathcal{O}_{\mathbb{P}^{m-1}}(H)$. Consider the set $\mathcal{M}_{g,n,d}(\mathbb{P}^{m-1})$ of triples (X, D, i) where X is a smooth projective curve of genus g with at least one rational point, $i : X \hookrightarrow \mathbb{P}^{m-1}$ is a twisted embedding of degree d, and $D = p_1 + \ldots + p_n$ is an effective divisor of degree n with $i(p_j) \neq H$ for every $j = 1, \ldots, n$. To each of those curves we can associate an invertible sheaf $\mathcal{L} = i^*\mathcal{O}_{\mathbb{P}^{m-1}}(1)$ and an associated effective divisor $G = X \cap H$ with $\mathrm{Supp}(G) \cap \mathrm{Supp}(D) = \emptyset$. Thus, any triple $(X, D, i) \in \mathcal{M}_{g,n,d}(\mathbb{P}^{m-1})$ gives a convolutional Goppa code $\mathcal{C}_X(G, D)$ defined by a complete linear series. Then, we can ask ourselves the following natural question.

PROBLEM 3.3. *How many complete Goppa structures among those obtained from $\mathcal{M}_{g,n,d}(\mathbb{P}^{m-1})$ can be given to an observable convolutional code?*

Before presenting the main result regarding this problem, let us show that the input data imposes some constrains on the degree $d = \deg(\mathcal{L})$. On one hand, X being twisted implies that $\dim H^0(X, \mathcal{L}) \geq m$. On the other hand, $\mathcal{L}$ is generated by its global sections, thus $H^1(X, \mathcal{L}(-x)) = 0$ for every rational point $x \in X$. Therefore, $\dim H^0(X, \mathcal{L}) = 1 - g + d$. We also want the code to be defined by the complete linear series, so it must holds $\dim H^0(X, \mathcal{L}) = k$ and $H^0(\mathcal{L}(-D)) = 0$. The first condition implies that $d = k + g - 1$, while the second condition can be ensured by declaring $d < n$. We will be considering the special situation $k = m$.

Let $\mathcal{C} \subset \mathbb{F}_q(z)^n$ be a convolutional code of dimension k and length n, and $\mathcal{G}(z)$ be an encoder for $\mathcal{C}$. Each of its n columns can be considered as a point in the $(k-1)$-dimensional projective space, $p_i = [a_{i1}; \ldots; a_{ik}] \in \mathbb{P}^{k-1}$, and we call them the column-points of the encoder. Let $i : X \hookrightarrow \mathbb{P}^{k-1}$ be a twisted curve of degree $d = k - 1 + g < n$ with at least one rational point, $\mathcal{L} = i^*\mathcal{O}_{\mathbb{P}^{k-1}}(1)$ and let G be the associated effective divisor.

THEOREM 3.4. **[10]** *$\mathcal{C}$ is a convolutional Goppa code over the twisted curve $X \subset \mathbb{P}^{k-1}$ defined by the complete linear series $H^0(X, \mathcal{L})$ if and only X passes through the n column-points of an encoder of $\mathcal{C}$.*

Note that Theorem 3.4 reduces Problem 3.3 to a problem in enumerative algebraic geometry, closely related to those solved in Gromov-Witten theory, now over the function field $\mathbb{F}_q(z)$.

COROLLARY 3.5. **[10]** *In case $g = 0$ or $g = 1$ we have:*

(1) *Every observable convolutional code of dimension up to* 2 *is a convolutional Goppa code defined by a complete linear series over* $\mathbb{P}^1$.
(2) *Each observable convolutional code of dimension* 3 *and length equal or less than* 8 *is a convolutional code defined by a complete linear series over an elliptic curve.*

3.4. 1-dimensional MDS convolutional codes. Let $\mathcal{C} \subset \mathbb{F}_q[z]^n$ be an observable convolutional code of dimension one. It follows from the beginning of Section 3 and Section 3.2 that it admits a description as a convolutional Goppa code over the projective line. We can consider the divisor $G = (n-1)p_\infty$ and the pairwise distinct rational points $p_i = (a_i z + b)$, with $a_i, b \in \mathbb{F}_q$ for $i = 1, \ldots, n$ in order to define such structure. Then, there is a global section $s(t) = \lambda_0 + \lambda_1 t + \ldots + \lambda_\delta t^\delta$, with $\delta < n$, such that $ev|_{<s(t)>}$ is an encoder for it.

Let us consider the family of 1-dimensional observable convolutional codes generated by the global sections $s_\lambda(t) = \lambda_0 + \lambda_1 t + \ldots + \lambda_\delta t^\delta \in H^0(\mathbb{P}^1, \mathcal{O}(G))$, where we let $\lambda = (\lambda_0, \ldots, \lambda_\delta)$ to vary along $\mathbb{F}_q^\delta \setminus \{(0, \ldots, 0)\}$. Observe that this family $\mathcal{C}_\lambda$ is parametrized by $\mathbb{P}^\delta_{\mathbb{F}_q}$. Members of this family have already been considered to generate MDS 1-dimensional convolutional codes (see [**5**] for instance, where $b = 0$, $a_i = \alpha^i$ with $\alpha \in \mathbb{F}_q$ an element of order greater than $n-1$ and $\lambda_i = 1$). As remarked in Section 2.2, the MDS locus in this family forms a Zariski open subset (indeed it is dense in this case) and we may consider to find the algebraic equations characterizing the complement inside $\mathbb{P}^\delta_{\mathbb{F}_q}$. To do so, define the functions

$$s_\lambda^{(j)}(t) = \sum_{r=j}^{\delta} \binom{r}{j} \lambda_r t^{r-j}$$

with $0 \leq j \leq \delta$. Then we have

THEOREM 3.6. [**18**] *The* MDS *locus in* $\mathbb{P}^\delta_{\mathbb{F}_q}$ *is non empty and*

$$\{\lambda \in \mathbb{P}^\delta_{\mathbb{F}_q} | \mathcal{C}_\lambda \text{is MDS}\} = \{\lambda \in \mathbb{P}^\delta_{\mathbb{F}_q} | s_\lambda^{(j)}(b) \neq 0,\ \forall\ 0 \leq j \leq \lambda\}$$

4. Further applications and future work

The theory of convolutional Goppa codes can be generalized in distinct ways, in order to get further constructions. As a matter of fact, by considering fibrations one may obtain families of codes depending on certain parameters instead of a single one. By varying the parameters we can check when the codes are optimal.

It is also of great interest to make use of the underlying structure of these codes in order to design decoding algorithms. This has been done for classical block Goppa codes. In the case of convolutional Goppa codes this is a subject of ongoing work.

In addition, there is also a class of codes which is optimal in a certain sense, named Maximum Distance Profile codes, specially suited for shift decoding. The characterization of such codes in terms of their algebraic geometric structure is a promising research topic.

References

[1] J. A. Domínguez Pérez, J. M. Muñoz Porras, and G. Serrano Sotelo, *Convolutional codes of Goppa type*, Appl. Algebra Engrg. Comm. Comput. **15** (2004), no. 1, 51–61, DOI 10.1007/s00200-004-0151-y. MR2142430

[2] J. A. Domínguez Pérez, J. M. Muñoz Porras, and G. Serrano Sotelo, One dimensional convolutional Goppa codes over the projective line 2011 [Online]. Available: arXiv:1107.2059.
[3] P. Elias, Coding for noisy channels, IRE Convention Record, Part 4 (1955) 37–46
[4] G. David Forney Jr., *Convolutional codes. I. Algebraic structure*, IEEE Trans. Inform. Theory **IT-16** (1970), 720–738, DOI 10.1109/tit.1970.1054541. MR275988
[5] Heide Gluesing-Luerssen and Barbara Langfeld, *A class of one-dimensional MDS convolutional codes*, J. Algebra Appl. **5** (2006), no. 4, 505–520, DOI 10.1142/S0219498806001855. MR2253601
[6] V. D. Goppa, *Codes that are associated with divisors* (Russian), Problemy Peredači Informacii **13** (1977), no. 1, 33–39. MR0497293
[7] V. D. Goppa, *Codes on algebraic curves* (Russian), Dokl. Akad. Nauk SSSR **259** (1981), no. 6, 1289–1290. MR628795
[8] Uwe Helmke, *A compactification of the space of rational transfer functions by singular systems*, J. Math. Systems Estim. Control **3** (1993), no. 4, 459–472. MR1318607
[9] J. I. Iglesias Curto, J. M. Muñoz Porras, F. J. Plaza Martín, and G. Serrano Sotelo, *Convolutional Goppa codes defined on fibrations*, Appl. Algebra Engrg. Comm. Comput. **23** (2012), no. 3-4, 165–178, DOI 10.1007/s00200-012-0173-9. MR3000507
[10] José Ignacio Iglesias Curto, Ángel Luis Muñoz Castañeda, José María Muñoz Porras, and Gloria Serrano Sotelo, *Every convolutional code is a Goppa code*, IEEE Trans. Inform. Theory **59** (2013), no. 10, 6628–6641, DOI 10.1109/TIT.2013.2271033. MR3106850
[11] V. G. Lomadze, *Finite-dimensional time-invariant linear dynamical systems: algebraic theory*, Acta Appl. Math. **19** (1990), no. 2, 149–201. MR1074223
[12] Vakhtang Lomadze, *Convolutional codes and coherent sheaves*, Appl. Algebra Engrg. Comm. Comput. **12** (2001), no. 4, 273–326, DOI 10.1007/s002000000048. MR1855040
[13] Robert J. McEliece, *The algebraic theory of convolutional codes*, Handbook of coding theory, Vol. I, II, North-Holland, Amsterdam, 1998, pp. 1065–1138. MR1667948
[14] J. M. Muñoz Porras, J. A. Domínguez Pérez, J. I. Iglesias Curto, and G. Serrano Sotelo, *Convolutional Goppa codes*, IEEE Trans. Inform. Theory **52** (2006), no. 1, 340–344, DOI 10.1109/TIT.2005.860447. MR2238161
[15] José María Muñoz Porras and José Ignacio Iglesias Curto, *Classification of convolutional codes*, Linear Algebra Appl. **432** (2010), no. 10, 2701–2725, DOI 10.1016/j.laa.2009.12.011. MR2608186
[16] N. J. Patterson, *The algebraic decoding of Goppa codes*, IEEE Trans. Inform. Theory **IT-21** (1975), 203–207, DOI 10.1109/tit.1975.1055350. MR379009
[17] R. Pellikaan, B.-Z. Shen, and G. J. M. van Wee, *Which linear codes are algebraic-geometric?*, IEEE Trans. Inform. Theory **37** (1991), no. 3, 583–602, DOI 10.1109/18.79915. MR1145820
[18] Francisco J. Plaza-Martín, José I. Iglesias-Curto, and Gloria Serrano-Sotelo, *On the construction of 1-D MDS convolutional Goppa codes*, IEEE Trans. Inform. Theory **59** (2013), no. 7, 4615–4625, DOI 10.1109/TIT.2013.2251926. MR3071348
[19] M. S. Ravi and J. Rosenthal, *A smooth compactification of the space of transfer functions with fixed McMillan degree*, Acta Appl. Math. **34** (1994), no. 3, 329–352, DOI 10.1007/BF00998684. MR1273616
[20] Joachim Rosenthal, *A compactification of the space of multivariable linear systems using geometric invariant theory*, J. Math. Systems Estim. Control **2** (1992), no. 1, 111–121. MR1149908
[21] Joachim Rosenthal, J. M. Schumacher, and Eric V. York, *On behaviors and convolutional codes*, IEEE Trans. Inform. Theory **42** (1996), no. 6, 1881–1891, DOI 10.1109/18.556682. Codes and complexity. MR1465746
[22] Joachim Rosenthal and Roxana Smarandache, *Maximum distance separable convolutional codes*, Appl. Algebra Engrg. Comm. Comput. **10** (1999), no. 1, 15–32, DOI 10.1007/s002000050120. MR1713642
[23] Roxana Smarandache, Heide Gluesing-Luerssen, and Joachim Rosenthal, *Constructions of MDS-convolutional codes*, IEEE Trans. Inform. Theory **47** (2001), no. 5, 2045–2049, DOI 10.1109/18.930938. MR1842538

Departamento de Matemáticas & RIASC, Universidad de León, León 24007, Spain
Email address: amunc@unileon.es

Departamento de Matemáticas & IUFFYM, Universidad de Salamanca, Salamanca 37008, Spain
Email address: joseig@usal.es

Contemporary Mathematics
Volume **766**, 2021
https://doi.org/10.1090/conm/766/15386

Adelic and idelic pairings related to Weil reciprocity on algebraic curves

Luis Manuel Navas Vicente and Francisco J. Plaza Martín

Abstract. In this paper we present an overview of some recent results concerning the properties of a function field embedded in the adeles. Specifically, for an algebraic curve over a perfect field, Weil reciprocity allows us to define a subgroup of the idele class group whose topology encodes arithmetic properties of the base field and geometric properties of the curve. In particular we establish a relationship with the Picard group. The construction is inspired by Tate's results on the character group of the adeles over a global field.

1. Introduction

Our aim is to begin a study of the function field Σ_X of an algebraic curve X over an arbitrary perfect field k, using the machinery of the geometric adeles and ideles. This means studying the topological embedding of the field in the additive and multiplicative settings and how it reflects certain arithmetic and geometric data. With this in mind, let us give, by way of introduction, a list of examples that serve as inspiration and motivation.

1.1. Tate's thesis. In the idelic formulation of Class Field Theory over a global field Σ, one considers the idele class group $\mathbb{I}/\Sigma^*$ and more specifically $\mathbb{I}^1/\Sigma^* N$ where $\mathbb{I}^1$ are the ideles of norm 1 and N the norm group. This is one of the standard approaches, whereby one studies local class field theory via the local reciprocity map and then puts together these local data to construct the global versions.

In his well-known thesis ([**5**, Chap. XV, Tate's Thesis], [**23**]), Tate reformulated Hecke's proof of the analytic continuation and functional equation of the zeta and L-functions of an algebraic number field using the Fourier transform and the associated Pontryagin duality at each place, archimedean and non-archimedean. This approach necessitates studying characters on the adeles and ideles, highlighting some interesting relationships between the additive and multiplicative structures, for example the appearance of gamma factors, which combine additive and multiplicative characters (over $\mathbb{R} = \mathbb{Q}_\infty$ this is clear in the well-known definition of the gamma function as an integral).

2020 *Mathematics Subject Classification.* Primary 14H05; Secondary 19F15, 11R37.

Key words and phrases. Weil reciprocity, class field theory, algebraic curves, function fields, residues.

Research of both authors supported by grants PGC2018-099599-B-I00, MTM2015-66760-P, J416/463AC03 and of the second author also by PGC2018-096504-B-C32.

In particular, self-duality at the local level results in self-duality of the adeles $\mathbb{A}_\Sigma$, in the following sense: on the rational adeles $\mathbb{A}_\mathbb{Q}$ there is a standard additive character $\chi : \mathbb{A}_\mathbb{Q} \to S^1$ projecting onto the exponential map $e^{-2\pi i x}$ at the archimedean place. We then obtain the standard additive character $\chi_\Sigma(\alpha) = \chi_\mathbb{Q}(\mathrm{tr}_{\Sigma/\mathbb{Q}}(\alpha))$ on $\mathbb{A}_\Sigma$, and the pairing $(\alpha, \beta) \mapsto \chi(\alpha\beta)$ induces an isomorphism of $\mathbb{A}_\Sigma$ with its continuous character group $\widehat{\mathbb{A}_\Sigma} = \mathrm{Hom}(\mathbb{A}_\Sigma, S^1)$.

In addition, since χ is trivial on Σ, the field Σ is a totally isotropic subspace. The previous result, along with Pontryagin duality and standard facts about the adelic topology, show that $\Sigma^\perp/\Sigma$ is discrete and compact, hence trivial, so that the pairing above also induces an isomorphism of the character group of the quotient $\mathbb{A}_\Sigma/\Sigma$ with the ground field Σ.

An exposition of these results in the context of the Langlands Program, of which this can be said to be the Gl(1) case, can be found in [**10**].

In the geometric context of a curve over a perfect field, we will no longer have the machinery of integration on locally compact topological groups which allowed Tate, and later Weil in the distributional sense, to use Pontryagin duality. However, focusing on the isomorphism between the character group of $\mathbb{A}_\Sigma$ and Σ, we can look for other pairings which may lead to similar results in the geometric context, ultimately finding a suitable analog in Weil reciprocity.

1.2. The cross-ratio. As an illustration of how Weil Reciprocity characterizes the adeles corresponding to a function, let us consider the case when X is the projective line $X = \mathbb{P}^1_k$. Fix a function z having a unique zero and pole, so that the function field of the curve is $\Sigma_X = k(z)$. Assuming that k is algebraically closed, z may be thought of as an affine coordinate. Recall that the cross-ratio of four distinct points in affine coordinates is given by

$$(z_1, z_2, z_3, z_4) = \frac{(z_3 - z_1)(z_4 - z_2)}{(z_3 - z_2)(z_4 - z_1)}.$$

The cross-ratio satisfies the symmetry relation $(z_1, z_2, z_3, z_4) = (z_3, z_4, z_1, z_2)$ and in general its properties can be used to prove many classical theorems in Projective Geometry, such as those of Menealus, Ceva, Carnot, etc.

Now, note that the rational functions

$$F(z) = \frac{z - z_1}{z - z_2}, \qquad G(z) = \frac{z - z_3}{z - z_4}$$

have divisors of zeros and poles equal to $z_1 - z_2$ and $z_3 - z_4$ respectively. Hence the symmetry of the cross-ratio can be written as

$$F(z_3)F(z_4)^{-1} = G(z_1)G(z_2)^{-1}.$$

If $F, G \in \Sigma^*_{\mathbb{P}^1_k}$ have disjoint supports, then this can be generalized to

$$\prod F(x)^{v_x(G)} = \prod G(x)^{v_x(F)} \qquad \forall F, G \in \Sigma^*_{\mathbb{P}^1_k},$$

where the product is over closed points. It can be easily seen that the restriction on the supports can be avoided by writing the above relation instead as

$$\prod_{x \in \mathbb{P}^1_k} (-1)^{v_x(F)v_x(G)} \frac{F^{v_x(G)}}{G^{v_x(F)}}(x) = 1 \qquad \forall F, G \in \Sigma^*_{\mathbb{P}^1_k}. \tag{1.1}$$

This leads to the insight that for $\mathbb{P}^1$, the symmetry of the cross-ratio is equivalent to the Weil Reciprocity Law (we have seen one direction; the converse can be proved using resultants, among other methods). By using that any 0 degree divisor on $\mathbb{P}^1$ (over an algebraically closed field) is the divisor of a rational function, one finds easily that in this case reciprocity characterizes the function field inside the ideles:

PROPOSITION. *For $X = \mathbb{P}^1$ over k algebraically closed, given an idele $(\alpha_x) \in \mathbb{I}_X$ such that over closed points*

$$\prod_{x \in X} (-1)^{v_x(\alpha_x)v_x(G)} \left(\frac{\alpha_x(x)^{v_x(G)}}{G(x)^{v_x(\alpha_x)}} \right) = 1$$

for all $G \in \Sigma_X^*$, there exists $F \in \Sigma_X^*$ such that $F(x) = \alpha_x(x)$ for all $x \in X$.

1.3. Residues. From another point of view, namely, by looking at the infinitesimal level, one finds a suitable pairing in terms of residues. Let X be a complete, non-singular, irreducible algebraic curve over $k = \mathbb{C}$ and x a closed point of X. For integers $m < n$, consider the exact sequence of sheaves on X

$$0 \to \mathcal{O}_X(mx) \to \mathcal{O}_X(nx) \to \mathcal{O}_X(nx)/\mathcal{O}_X(mx) \to 0,$$

inducing the long exact sequence in cohomology:

$$\cdots \to H^0(X, \mathcal{O}_X(nx)) \to H^0(X, \mathcal{O}_X(nx)/\mathcal{O}_X(mx)) \to H^1(X, \mathcal{O}_X(mx)) \to \cdots.$$

Taking direct and inverse limits in n and m respectively (noting $H^0(X, \mathcal{O}_X(mx)) = H^1(X, \mathcal{O}_X(nx)) = 0$ for $m \ll 0$ and $n \gg 0$), we obtain the exact sequence

$$0 \to H^0(X \setminus \{x\}, \mathcal{O}_X) \to K_x = (\widehat{\mathcal{O}}_{X,x})_{(0)} \to \varprojlim_m H^1(X, \mathcal{O}_X(mx)) \to 0.$$

Denote the canonical bundle by ω_X. Then by the Riemann-Roch Theorem and Serre duality. we find that

$$\varprojlim_m H^1(X, \mathcal{O}_X(mx)) = \left(\varinjlim_m H^0(X, \omega_X(-mx)) \right)^* = H^0(X \setminus \{x\}, \omega_X)^*$$

and the map $K_x \to H^0(X \setminus \{x\}, \omega_X)^*$ sends α to the linear form $\omega \mapsto \operatorname{Res}_x \alpha\omega$. Hence the subspace $H^0(X \setminus \{x\}, \mathcal{O}_X)$ is totally isotropic with respect to the residue pairing on K_x:

$$(f, g) \mapsto \operatorname{Res}_x(f \, \mathrm{d}g) : K_x \times K_x \longrightarrow k.$$

We may repeat this argument in the ring of adeles $\mathbb{A}_X$ over Σ_X, which contains Σ_X via the diagonal embedding. With x replaced by an effective divisor D and taking the limit, we find that Σ_X is totally isotropic with respect to the residue pairing on adeles

$$(\alpha, \beta) \mapsto \operatorname{Res}_x(\alpha \, \mathrm{d}\beta) : \mathbb{A}_X \times \mathbb{A}_X \longrightarrow k. \tag{1.2}$$

Furthermore, this will allow us to characterize the function field in the adeles in the sense that for a fixed $\alpha \in \mathbb{A}_X$, if $\sum_x \operatorname{Res}_x(\alpha \, \mathrm{d}g) = 0$ for all $g \in \Sigma_X$, then in fact $\alpha \in \Sigma_X + \prod_{x \in X} k$ (see Theorem 6.4). The key point is the following. The Weil Reciprocity Law allows us to introduce a bilinear pairing on the group of ideles $\mathbb{I}_X = \mathbb{A}_X^*$ and the pairing (1.2) is exactly the pairing induced on the level of Lie algebras (this approach is carried out in [**13**]). Thus, once more, we see how the Weil Reciprocity Law might be useful in characterizing algebraic functions, differentials, etc. inside the adele ring.

1.4. Goals and the main results. In this paper we aim at giving a unified approach to the facts pointed out in §1.2 and §1.3. This goal will be attained by introducing certain subgroups and quotients of the ideles which reflect some properties of the curve X (such as the genus) and of the base field k. At the end of the paper, we will present some evidence on how these objects may be useful for studying algebraic extensions of Σ_X and, hopefully, in furthering the geometric Langlands program. A second line of future research consists of reinterpreting our results in terms of K-theory, along the lines established in [**4**, **20**] for local and global fields.

Let us now briefly describe the main construction and principal results of this paper.

For a smooth, complete and connected curve X over a perfect field k, the multiplicative group of its function field Σ_X can be embedded as a subgroup of the norm-1 idele group $\mathbb{I}_X^1$. The pairing which we will use to study duality will be the global symbol arising from the local multiplicative symbols constructed in ([**12**,**17**]). This pairing leads us to consider the corresponding "orthogonal" which we will denote $(\Sigma_X^*)^{\perp\!\!\!\perp}$. The Weil Reciprocity Law is equivalent to the existence of an induced inclusion $\Sigma_X^* \subseteq (\Sigma_X^*)^{\perp\!\!\!\perp}$.

Based on the ideas of Tate and Weil, it is natural to study the topology of the quotient $(\Sigma_X^*)^{\perp\!\!\!\perp}/\Sigma_X^*$ which, as can be expected, is related to the arithmetic properties of the base field (e.g. we have results like Proposition 4.1). Geometrically, the relation between this quotient and the curve is summarized in one of our main results, Theorem 5.2, which proves the exactness of

$$0 \to \operatorname{Hom}_{\mathrm{gr}}\big(\operatorname{Pic}^0(X), k^*\big) \to (\Sigma_X^*)^{\perp\!\!\!\perp}/\Sigma_X^* \to \operatorname{Pic}^0(X) \tag{1.3}$$

when k is algebraically closed k, as well as an analog for finite fields. As a corollary we have that $(\Sigma_X^*)^{\perp\!\!\!\perp}/\Sigma_X^* = 0$ for $X = \mathbb{P}_1$ over an algebraically closed field, but note that the quotient need not be trivial for other curves or other base fields.

1.5. Notation and set-up. All through this article we will be working with a smooth, complete and connected curve X over a perfect field k. To simplify the technical details, we also assume that k is algebraically closed in the function field of X, as well as the existence of a k-rational point. We use the following notation:

- Σ_X is the function field of X.
- When we simply write $x \in X$, it is implied that x is a closed point of X, with associated valuation v_x, local ring $\mathcal{O}_{X,x}$, and maximal ideal $\mathfrak{m}_x$.
- Let $\widehat{\mathcal{O}}_{X,x}$ be the v_x-completion of $\mathcal{O}_{X,x}$, $\widehat{\mathfrak{m}}_x$ its maximal ideal, and K_x its quotient field.
- Denote the residue field at $x \in X$ by $k(x) = \widehat{\mathcal{O}}_{X,x}/\widehat{\mathfrak{m}}_x$. Since k is assumed perfect, $k(x)/k$ is a finite separable extension of degree $\deg(x) = \dim_k k(x)$.
- In general, for a ring R, we let R^* denote the group of invertible elements.
- For $x \in X$, let $N_{k(x)/k}$ denote the norm map $k(x)^* \to k^*$ and let N_x be its kernel; i.e. the subgroup of $k(x)^*$ consisting of elements of norm 1.

REMARK 1.1. In fact, the hypothesis that k is algebraically closed in the function field of X is redundant, as it follows from the other assumptions (smoothness, completeness, connectedness, perfect base field) and the existence of a k-rational point. Under these assumptions, X is geometrically integral by [**18**, Lemma 0CDW], and therefore geometrically irreducible. From [**8**, Proposition 4.5.9], it follows that the separable algebraic closure of k in Σ_X is k and, since k is perfect, this is the algebraic closure.

2. Adeles, ideles, and the function field

Cassels, Serre and Weil ([**5**, Chp. II] and [**17, 21**]) established the list of basic topological properties of the adeles and ideles over global fields. In the geometric case, however, we have not found a similar comprehensive account in the literature. Here we will give a summary of the main results needed for our objectives.

REMARK 2.1. The details regarding the properties of the adelic and idelic topologies are rather technical and in some cases straightforward, so we will omit the proofs. The abstract theory of topological groups and vector spaces is enough for deducing most of the claims of this section. See, for example [**6, 9**].

2.1. Adeles. The ring of adeles $\mathbb{A}_X$ of Σ_X/k is the restricted direct product

$$\mathbb{A}_X = \prod_{x\in X}{}' K_x = \Big\{(\alpha_x)_{x\in X} \mid \alpha_x \in \widehat{\mathcal{O}}_{X,x} \text{ for almost all } x \in X\Big\}$$

over closed points. Let $\mathbb{A}_X^+$ be the subring $\mathbb{A}_X^+ = \prod_{x\in X}\widehat{\mathcal{O}}_{X,x}$. The *idele group* $\mathbb{I}_X$ is the group $\mathbb{A}_X^*$ of invertible elements of $\mathbb{A}_X$. It is the restricted product of K_x^* with respect to the unit groups $\widehat{\mathcal{O}}_{X,x}^*$. For $x \in X$ and an adele $\alpha = (\alpha_x)_{x\in X}$ wih $v_x(\alpha_x) \geq 0$, let $\alpha(x)$ be the image in the residue field $k(x)$ of the x-component α_x.

The adelic topology is the restriction of the direct product topology induced by the valuation topology at each (closed) point x. A neighborhood basis at (0) for this adelic topology is given by either of the following collections:

(1) The k-vector subspaces $V \subseteq \mathbb{A}_X$ commensurable with $\mathbb{A}_X^+$, i.e. such that $\dim_k(V + \mathbb{A}_X^+)/(V\cap\mathbb{A}_X^+) < \infty$.
(2) For a divisor $D = \sum_x n_x x$, which may be assumed to be effective, the sets $U_D = \big\{\alpha \in \mathbb{A}_X \mid v_x(\alpha_x) \geq n_x \ \forall x \in X\big\}$, equal to the products $\prod_x \widehat{\mathfrak{m}}_x^{n_x}$ with $n_x \geq 0$ for all x and $n_x = 0$ for almost all x. Note $\mathbb{A}_X^+ = U_0$.

It is easy to check that the U_D are clopen subgroups, satisfying $\mathbb{A}_X = \bigcup_D U_D$ and $(0) = \bigcap_D U_D$ where D runs over $\operatorname{Div}(X)$. These facts are frequently referred to.

2.2. Ideles. Recall that for a topological ring R, the subspace topology on the group of units R^* does not in general make R^* into a topological group. Instead one uses the topology induced by the map $x \mapsto (x, x^{-1}) : R^* \to R\times R$. This is also the case for the ideles $\mathbb{I}_X = \mathbb{A}_X^*$. A neighborhood basis at 1 is given by the sets

$$V_D = \prod_{x\in\operatorname{supp} D}(1+\widehat{\mathfrak{m}}_x^{n_x}) \times \prod_{x\notin\operatorname{supp} D}\widehat{\mathcal{O}}_{X,x}^*, \tag{2.1}$$

where $D = \sum_x n_x x$ is an effective divisor on X. As in the additive context, the sets V_D are clopen subgroups,

$$\mathbb{I}_X \cap \mathbb{A}_X^+ = V_0 = \bigcup_D V_D,$$

$\bigcap_D V_D = (1)$, where D runs over effective divisors, and $V_D \subseteq V_{D'}$ if $D \leq D'$. In addition,

$$V_D \subseteq (1 + U_D) \cap \mathbb{I}_X, \tag{2.2}$$

for an effective divisor D, which implies that the inclusion $\mathbb{I}_X \hookrightarrow \mathbb{A}_X$ is continuous.

There is a natural notion of degree on the ideles, which is the additive version of the content or norm commonly used over global fields. Given an idele α, since $v_x(\alpha) = 0$ for almost all x, there is a homomorphism

$$\alpha \mapsto D(\alpha) = \sum_{x \in X} v_x(\alpha_x)\, x : \quad \mathbb{I}_X \to \mathrm{Div}(X) \tag{2.3}$$

from ideles to divisors. Its kernel is the neighborhood $V_0 = \prod_x \widehat{\mathcal{O}}^*_{X,x}$. The subgroup $\mathbb{I}^1_X$ consisting of ideles α with $\deg D(\alpha) = \sum_x \deg(x) v_x(\alpha_x) = 0$ corresponds to the norm one ideles for global fields. It plays a central role throughout this discussion.

As also happens over global fields, one can check that the adelic and idelic topologies induce the same topology on $\mathbb{I}^1_X$. In particular, since $V_0 = \mathbb{I}^1_X \cap \mathbb{A}^+_X$, we see that $\mathbb{I}^1_X$ is an open subgroup of $\mathbb{I}_X$. Its basic properties are summarized in the following list.

THEOREM 2.2.

(1) $\mathbb{I}^1_X$ *is a Hausdorff topological group.*
(2) $\mathbb{I}^1_X$ *is locally compact if and only if* k *is finite.*
(3) $\mathbb{I}^1_X$ *is not locally connected.*

PROOF. See [**11**, Theorem 2.1]. □

2.3. The function field. Assigning to a rational function its germ at a point induces an injective morphism $\Sigma_X \hookrightarrow K_x$, leading to a diagonal embedding $\Sigma_X \hookrightarrow \mathbb{A}_X$. We of course want to know how Σ_X "sits inside" $\mathbb{A}_X$. From the general theory of linear topological vector spaces (see [**9**]), a subspace V inherits the discrete topology if and only if its intersection with an open neighborhood is finite-dimensional, in which case V is also closed.

PROPOSITION 2.3. *The function field* Σ_X*, embedded diagonally in* $\mathbb{A}_X$*, is discrete and closed in the adelic topology.*

PROOF. An argument similar to that given in §1.3 can be used to prove that the kernel and cokernel of $\Sigma_X \to \mathbb{A}_X/\mathbb{A}^+_X$ are $H^0(X, \mathcal{O})$ and $H^1(X, \mathcal{O})$ respectively (this means that $\Sigma_X \in \mathrm{Gr}(\mathbb{A}_X, \mathbb{A}^+_X)$, the infinite Grassmannian studied in [**12, 16**]). In particular, recalling that $\mathbb{A}^+_X = U_0$ is open, it follows that $\mathbb{A}^+_X \cap \Sigma_X = H^0(X, \mathcal{O}_X)$ is finite-dimensional (by the hypotheses on the curve) and open. By the remark above, the conclusion follows. □

On the other hand, removing a point from the restricted direct product yields an embedding of Σ_X with dense image. This is a well-known result for global fields, known as the Strong Approximation Theorem [**5**]. The geometric version is also a known result, but necessitates a different proof since Σ_X is not a global field if k is infinite.

THEOREM 2.4 (Geometric Strong Approximation Theorem). *For a closed point* $x_0 \in X$*, the diagonal embedding* $\Sigma_X \to \prod'_{x \neq x_0} K_x$ *of the function field* Σ_X *in the restricted product of* $(K_x, \widehat{\mathcal{O}}_{X,x})$ *is injective with dense image in the adelic topology.*

This means that, having fixed principal parts $\alpha_x \in K_x/\widehat{\mathcal{O}}_{X,x}$ for $x \neq x_0$, we can find a function F with principal part α_x at $x \neq x_0$.

PROOF. A proof using the Riemann-Roch Theorem is given in [**11**]. $\square$

Turning to the multiplicative context, recall that the idele class group $\mathbb{I}_X^1/\Sigma_X^*$ is one of the central objects in Class Field Theory over a global field, and its compactness in that case is in itself an important result. In our case, however, compactness is going to be the exception rather than the rule.

THEOREM 2.5.

(1) *Σ_X^* is closed and discrete in $\mathbb{I}_X^1$.*
(2) *The quotient $\mathbb{I}_X^1/\Sigma_X^*$ is compact if and only if k is finite.*

PROOF. The first part follows from the corresponding additive result and the equality of the adelic and idelic topologies on $\mathbb{I}_X^1$. Compactness for finite fields is a known result. The converse may be proved using Theorem 2.2. See [**11**, Theorem 2.4]. $\square$

In particular, for algebraically closed base fields k we lose the local compactness of $\mathbb{I}_X^1$ and with it the compact/discrete relationship provided by Pontryagin duality, which is central in Tate's approach via adelic and idelic character groups. There is no longer a Haar measure and consequently no local or global analytic Fourier transforms.

3. Reciprocity pairing on ideles

In order to find a suitable analog to the duality results of Tate which were discussed in §1.1, we need a pairing on the ideles with respect to which Σ_X^* is totally isotropic. This requirement leads naturally to the local and global symbols that arise from Weil Reciprocity, as we now see.

3.1. The local and global symbols. Recall the discussion of the case $X = \mathbb{P}^1$ in §1.2 and (1.1). This is a special case of a fundamental result [**22**].

THEOREM 3.1 (The Weil Reciprocity Law). *Under the stated hypotheses on the curve X, we have:*

$$\prod_{x\in X} (-1)^{\deg(x)v_x(F)v_x(G)} N_{k(x)/k}\left(\frac{F^{v_x(G)}}{G^{v_x(F)}}(x)\right) = 1 \qquad (F, G \in \Sigma_X^*).$$

To work with ideles, we need to generalize Serre's definiton of the local symbol in [**17**], as a pairing $\Sigma_X^* \times \Sigma_X^* \to k^*$ for algebraically closed k. Under our hypotheses, we may define the local symbol at a closed point $x \in X$ as the pairing $K_x^* \times K_x^* \to k^*$ given by

$$(3.1) \qquad (f, g)_x = (-1)^{\deg(x)\cdot v_x(f)\cdot v_x(g)} N_{k(x)/k}\left(\frac{f^{v_x(g)}}{g^{v_x(f)}}(x)\right) \qquad (f, g \in K_x^*).$$

It is immediate that this reduces to Serre's definition in [**17**]. Next, we define the global symbol $\langle\cdot,\cdot\rangle : \mathbb{I}_X \times \mathbb{I}_X \to k^*$ as

$$(3.2)\ \langle\alpha, \beta\rangle = \prod_{x\in X} (\alpha_x, \beta_x)_x = \prod_{x\in X} (-1)^{\deg(x)\cdot v_x(\alpha_x)\cdot v_x(\beta_x)} N_{k(x)/k}\left(\frac{\alpha^{v_x(\beta_x)}}{\beta^{v_x(\alpha_x)}}(x)\right).$$

As detailed in [12,15], these symbols and Weil Reciprocity can be derived from the study of the infinite Grassmannian $\mathrm{Gr}(\mathbb{A}_X, \mathbb{A}_X^+)$ given by subspaces $U \subseteq \mathbb{A}_X$ such that the kernel and cokernel of $U \to \mathbb{A}_X/\mathbb{A}_X^+$ are finite-dimensional, which in particular contains the point Σ_X. The action of $\mathbb{I}_X^1$ on $\mathrm{Gr}(\mathbb{A}_X, \mathbb{A}_X^+)$ yields a central extension with an associated commutator pairing which yields precisely the global symbol (3.2). The corresponding action of K_x^* on $\mathrm{Gr}(K_x, \widehat{\mathcal{O}}_{X,x})$ yields the local symbol (3.1).

This approach also gives an alternative proof of the Weil Reciprocity Law. Indeed, since the above mentioned central extension becomes trivial when restricted to the subgroup $\Sigma_X^* \subset \mathbb{I}_X^1$, the global pairing on Σ_X^* is accordingly trivial also; that is,

$$\langle f, g\rangle = \prod_{x\in X} (f,g)_x = 1 \qquad (f, g \in \Sigma_X^*)\,, \tag{3.3}$$

which, after substituting (3.1), takes on the classical form of the Weil Reciprocity Law.

As could be expected, the global pairing is locally constant, which amounts to continuity in the idelic topology. Indeed, every pair of ideles (α, β) has a clopen neighborhood on which $\langle\cdot,\cdot\rangle$ is constant.

The interpretation of the global symbol as a commutator pairing allows us to easily deduce that it is a symbol in the sense of algebraic K-theory as in [20], namely, it satisfies

$$\begin{aligned} \langle\alpha_1\alpha_2, \beta\rangle &= \langle\alpha_1, \beta\rangle \cdot \langle\alpha_2, \beta\rangle, & \langle\alpha, \beta\rangle \cdot \langle\beta, \alpha\rangle &= 1, \\ \langle\alpha, -\alpha\rangle &= 1, & \langle\alpha, \alpha\rangle &= \langle\alpha, -1\rangle, \\ \langle\alpha, 1-\alpha\rangle &= 1, \quad \text{if } \alpha, 1-\alpha \in \mathbb{I}_X, & \langle\alpha, \alpha\rangle &= 1, \quad \text{if } \alpha \in \mathbb{I}_X^1. \end{aligned}$$

REMARK 3.2. In a manner similar to the method Tate ([19]) used to deduce the residue theorem from the properties of a cocycle, we now rely on the properties of local symbols to obtain our results. The consequences of local symbols have been widely studied ([1, 2]) and, among them, the above-mentioned proof of the Weil reciprocity law given in [12]. Indeed, the ideas of [12] together with the techniques of [16] might be used to study analogues for arithmetic curves. Furthermore, Artin and Whaples ([3]) characterized certain fields in terms of a *product formula* for valuations; however, they need archimedean valuations in the case of function fields, while we have only non-archimedean valuations. Finally, the relation of our results to the classical approach of class field theory in terms of K-theory (e.g. [20]) deserves further research.

3.2. The radical and orthogonal complements. The radicals of the global symbol on $\mathbb{I}_X$ and $\mathbb{I}_X^1$ will be denoted respectively by

$$\begin{aligned} R &= \{\alpha \in \mathbb{I}_X \mid \langle\alpha, \beta\rangle = 1 \;\forall\beta \in \mathbb{I}_X\} \\ R^1 &= \{\alpha \in \mathbb{I}_X^1 \mid \langle\alpha, \beta\rangle = 1 \;\forall\beta \in \mathbb{I}_X^1\}. \end{aligned}$$

We have the following characterizations (recall the definition of the neighborhoods V_D in (2.1)):

THEOREM 3.3. *The radical of the global reciprocity pairing on the group of norm one ideles $\mathbb{I}_X^1$ satisfies:*

(1) $k^* \cdot \prod_{x\in X}\big((1+\widehat{\mathfrak{m}}_x)N_x\big) = R^1 \cap V_0$, *where* N_x *is the kernel of the norm,*
$$N_x = \{\lambda \in k(x) \mid N_{k(x)/k}(\lambda) = 1\}.$$
(2) *If* k *is infinite, then* $R^1 \subseteq V_0$.
(3) k *is algebraically closed if and only if* $R^1 = k^* \prod_{x\in X}(1+\widehat{\mathfrak{m}}_x)$.

The radical on the total group of ideles $\mathbb{I}_X$ *similarly satisfies:*

(1) $\prod_{x\in X}\big((1+\widehat{\mathfrak{m}}_x)N_x\big) = R \cap V_0$.
(2) *If* k *is infinite, then* $R \subseteq V_0$.
(3) k *is algebraically closed if and only if* $R = \prod_{x\in X}(1+\widehat{\mathfrak{m}}_x)$.

PROOF. The reasoning for $\mathbb{I}_X$ is similar to that for $\mathbb{I}^1_X$. The inclusion in R^1 is easily checked, but one needs a k-rational point (stated in our main hypotheses) for equality. For the reverse implication in the third part we use the general characterization of the kernel of the norm maps given in [**14**]. See [**11**, Theorem 3.1] for details. □

Note that for a finite field k, the complete radical R may contain elements of non-zero degree.

Just as one takes the quotient of the idele class group by the norm group in Class Field Theory, it is natural to consider
$$\bar{\mathbb{I}}^1_X = \mathbb{I}^1_X / \prod_{x\in X}(1+\widehat{\mathfrak{m}}_x)$$
with the quotient topology. The quotient map is open but not necessarily closed since the above product need not be compact.

We list its basic topological properties.

THEOREM 3.4. *Consider the quotient group* $\bar{\mathbb{I}}^1_X = \mathbb{I}^1_X/\prod_{x\in X}(1+\widehat{\mathfrak{m}}_x)$. *Then:*

(1) $\bar{\mathbb{I}}^1_X$ *is a Hausdorff and totally disconnected topological group.*
(2) $\bar{\mathbb{I}}^1_X$ *is profinite if and only if* k *is finite.*

PROOF. The argument relies on showing that a neighborhood basis of $\bar{\mathbb{I}}^1_X$ at 1 is given by the subgroups $\bar{V}_S = \prod_{x\in S}\{1\} \times \prod_{x\notin S} k(x)^*$ where $S \subset X$ runs over finite sets of closed points. The profinite topological groups are precisely those which are Hausdorff, compact and totally disconnected. See for example [**6**] for these aspects of topological groups. □

With this result in mind, we proceed to study the pairing induced on $\bar{\mathbb{I}}^1_X$ by the global pairing on $\mathbb{I}^1_X$. We will use the same symbol for both pairings. We will take "orthogonal complements" of subsets $G \subset \mathbb{I}^1_X$ with respect to the global symbol, denoted by
$$G^\perp = \{\alpha \in \mathbb{I}^1_X \mid \langle \alpha, g\rangle = 1 \ \forall g \in G\},$$
as well as complements of subsets $G \subseteq \bar{\mathbb{I}}^1_X$ with respect to the induced pairing on the quotient, denoted by
$$G^{\perp\!\!\perp} = \{\alpha \in \bar{\mathbb{I}}^1_X \mid \langle \alpha, g\rangle = 1 \ \forall g \in G\}.$$

PROPOSITION 3.5. *The orthogonal complement satisfies the following properties:*

(1) *For any subset* $G \subseteq \mathbb{I}^1_X$, *the orthogonal complement* $G^\perp$ *is a closed subgroup of* $\mathbb{I}^1_X$ *containing the radical* R^1.

(2) $\mathbb{I}^1_X \subseteq (k^*)^\perp$, *with equality when* k *is infinite. In the latter case, if* $k^* \subseteq G \subseteq \mathbb{I}_X$ *and* $\alpha \in \mathbb{I}_X$ *satisfies* $\langle \alpha, g\rangle = 1\ \forall g \in G$, *then actually* $\alpha \in \mathbb{I}^1_X$.

(3) *If* $p : \mathbb{I}^1_X \to \bar{\mathbb{I}}^1_X$ *is the projection map, then for any subset* $G \subseteq \bar{\mathbb{I}}^1_X$ *we have* $p^{-1}(G^{\perp\!\!\perp}) = p^{-1}(G)^\perp$, *and for any subset* $G \subseteq \mathbb{I}^1_X$, *we have* $p(G)^{\perp\!\!\perp} = p(G^\perp)$.

(4) *For any subset* $G \subseteq \bar{\mathbb{I}}^1_X$, *the orthogonal complement* $G^{\perp\!\!\perp}$ *is a closed and totally disconnected subgroup of* $\bar{\mathbb{I}}^1_X$.

PROOF. This is not hard to check. See [**11**, Proposition 3.2]. □

4. The Main Quotient $(\Sigma^*_X)^{\perp\!\!\perp}/\Sigma^*_X$ – Topology

The global symbol and its associated complement allows us to finally study an analogy of Tate duality as was discussed in the introduction (§1.1), namely, the quotients $\bar{\mathbb{I}}^1_X$ and, foremost, $(\Sigma^*_X)^{\perp\!\!\perp}$. Here we will see how the latter encodes properties of the base field and, in the next section, how it depends on the curve.

Since $\Sigma^*_X \cap \prod(1 + \widehat{\mathfrak{m}}_x) = \{1\}$, we may consider Σ^*_X as a subset $\Sigma^*_X \subseteq \bar{\mathbb{I}}^1_X$. Furthermore, since Σ^*_X is totally isotropic with respect to the global reciprocity symbol (3.3), we have in fact

$$\Sigma^*_X \subseteq (\Sigma^*_X)^{\perp\!\!\perp} \subseteq \bar{\mathbb{I}}^1_X.$$

By Proposition 3.5 we can then conclude that

$$(\Sigma^*_X)^{\perp\!\!\perp} = p((\Sigma^*_X)^\perp) \simeq (\Sigma^*_X)^\perp / \prod_{x\in X}(1 + \widehat{\mathfrak{m}}_x). \tag{4.1}$$

As usual, we summarize the main topological properties of these embeddings and the corresponding quotients.

THEOREM 4.1.

(1) Σ^*_X *is closed in* $\bar{\mathbb{I}}^1_X$.

(2) $(\Sigma^*_X)^{\perp\!\!\perp}$ *is closed, Hausdorff, totally disconnected and with empty interior in* $\bar{\mathbb{I}}^1_X$.

(3) $(\Sigma^*_X)^{\perp\!\!\perp}/\Sigma^*_X$ *is closed in* $\bar{\mathbb{I}}^1_X/\Sigma^*_X$. *Both are Hausdorff totally disconnected topological groups.*

(4) $\bar{\mathbb{I}}^1_X/\Sigma^*_X$ *is profinite if and only if* k *is finite.*

(5) *If* k *is finite, then* $(\Sigma^*_X)^{\perp\!\!\perp}/\Sigma^*_X$ *is profinite.*

$(\Sigma^*_X)^{\perp\!\!\perp}$ is not necessarily a discrete subgroup of $\bar{\mathbb{I}}^1_X$. In fact, the next result implies that this is can only happen for algebraically closed k (Proposition 4.2).

PROPOSITION 4.2. *Consider the subgroup* $\prod_{x\in X} N_x \subset \bar{\mathbb{I}}^1_X$. *Then*

$$\prod_{x\in X} N_x \subseteq (\Sigma^*_X)^{\perp\!\!\perp}/\Sigma^*_X. \tag{4.2}$$

In addition, if $(\Sigma^*_X)^{\perp\!\!\perp}/\Sigma^*_X$ *is discrete, then* k *must be algebraically closed.*

PROOF. The characterization of the radical given in Theorem 3.3 implies the inclusion. If k is not algebraically closed then there are infinitely many closed points x over $\bar{k}$ which are not k-rational. Using the basic facts about the norm map in general separable extensions given in [**14**], we see that $N_x \neq \{1\}$ for these points. This implies that $\left(\prod_{x\in X} N_x\right) \cap \bar{V}_S \neq \{\bar{1}\}$ in $\bar{\mathbb{I}}^1_X$ for any finite $S \subset X$ and also that $p\left(\prod_{x\in X} N_x\right) \cap p(\bar{V}_S) \neq \{\bar{1}\}$ where $p : \bar{\mathbb{I}}^1_X \to \bar{\mathbb{I}}^1_X/\Sigma^*_X$ is the projection. Thus

$p\big(\prod_{x\in X} N_x\big)$ is not discrete, and the inclusion (4.2) implies that $(\Sigma_X^*)^{\perp\!\!\perp}/\Sigma_X^*$ is not discrete either. $\square$

COROLLARY 4.3. *Let $X = \mathbb{P}_1$. If k is infinite, then*

$$\prod_{x\in X} N_x = (\Sigma_X^*)^{\perp\!\!\perp}/\Sigma_X^*,$$

and hence $(\Sigma_X^*)^{\perp\!\!\perp}/\Sigma_X^* = 0$ if and only if k is algebraically closed.

PROOF. When $X = \mathbb{P}_1$, given $\alpha \in (\Sigma_X^*)^{\perp\!\!\perp}$, then $D(\alpha) = D(f)$ for some $f \in \Sigma_X^*$. It is enough to show that $f^{-1}\cdot\alpha \in \prod_{x\in X} N_x$. Fix a k-rational point x_0 and let x_1 be any other closed point. Let $\beta \in \Sigma_X^*$ have divisor $\deg(x_1)x_0 - x_1$. Then the relation $\langle\alpha,\beta\rangle = 1$ implies that $N_{k(x_1)/k}(f^{-1}\alpha) = 1$. $\square$

5. The Main Quotient $(\Sigma_X^*)^{\perp\!\!\perp}/\Sigma_X^*$ – Geometry

In this final section we explore some relationships between the quotient $(\Sigma_X^*)^{\perp\!\!\perp}/\Sigma_X^*$ and the Picard group of the curve X. In particular we have non-triviality when X is an elliptic curve over $\mathbb{C}$. First, note that the map from degree 0 ideles $\mathbb{I}_X^1$ to $\mathrm{Div}^0(X)$ defined by (2.3) induces a map

$$(\Sigma_X^*)^{\perp\!\!\perp}/\Sigma_X^* \longrightarrow \mathrm{Pic}^0(X).$$

The study of this map sheds light on the structure of $(\Sigma_X^*)^{\perp\!\!\perp}/\Sigma_X^*$.

The global reciprocity pairing on $\mathbb{I}_X^1$ defines a map $\mathbb{I}_X^1 \to \mathrm{Hom}_{\mathrm{gr}}(\mathbb{I}_X^1, k^*)$ which in turn, by (3.3) and (4.1), factors through the quotient, yielding a map

$$\begin{aligned}\Phi : (\Sigma_X^*)^{\perp\!\!\perp} &\longrightarrow \mathrm{Hom}_{\mathrm{gr}}(\bar{\mathbb{I}}_X^1/\Sigma_X^*, k^*)\\ \alpha &\longmapsto \big(\beta \mapsto \langle\alpha,\beta\rangle\big).\end{aligned}$$

Note that here we are considering only group homomorphisms $\mathrm{Hom}_{\mathrm{gr}}$, discarding the previous topological structure or equivalently, giving all groups the discrete topology. In addition, the global pairing (3.2) also induces a map

$$\begin{aligned}\Psi : \mathrm{Div}^0(X) &\longrightarrow \mathrm{Hom}_{\mathrm{gr}}(\bar{V}_0/k^*, k^*)\\ \sum_{x\in X} n_x x &\longmapsto \Big(\alpha \mapsto \prod_{x\in X} N_{k(x)/k}(\alpha(x))^{n_x}\Big)\end{aligned} \tag{5.1}$$

where $\bar{V}_0 = \prod_x \widehat{\mathcal{O}}_{X,x}^* / \prod_x (1+\widehat{\mathfrak{m}}_x) \subseteq \bar{I}_X^1$. The maps Φ and Ψ are injective if k is algebraically closed (see Theorem 3.3).

Now, there is an exact sequence

$$0 \to \bar{V}_0/k^* \to \bar{\mathbb{I}}_X^1/\Sigma_X^* \to \bar{\mathbb{I}}_X^1/\bar{V}_0\cdot\Sigma_X^* \to 0$$

and we can identify

$$\mathbb{I}_X^1/V_0 \simeq \bar{\mathbb{I}}_X^1/\bar{V}_0 \simeq \mathrm{Div}^0(X)\,, \qquad \mathbb{I}_X^1/\Sigma_X^* V_0 \simeq \bar{\mathbb{I}}_X^1/\Sigma_X^*\bar{V}_0 \simeq \mathrm{Pic}^0(X)\,.$$

It is then not hard to check that the following diagram is commutative:

$$\begin{array}{ccccccc}
0 & \longrightarrow & (\Sigma_X^*)^{\perp} \cap \mathbb{A}_X^{+} & \longrightarrow & (\Sigma_X^*)^{\perp} & \longrightarrow & \mathrm{Div}^0(X)\\
 & & \Big\downarrow & & \Big\downarrow{\scriptstyle\Phi} & & \Big\downarrow{\scriptstyle\Psi}\\
0 & \longrightarrow & \mathrm{Hom}_{\mathrm{gr}}\big(\mathrm{Pic}^0(X), k^*\big) & \longrightarrow & \mathrm{Hom}_{\mathrm{gr}}\big(\bar{\mathbb{I}}_X^1/\Sigma_X^*, k^*\big) & \longrightarrow & \mathrm{Hom}_{\mathrm{gr}}\big(\bar{V}_0/k^*, k^*\big).
\end{array}$$

The Picard group appears when we consider the corresponding map defined by the following diagram:

$$\begin{array}{ccc} (\Sigma_X^*)^\perp & \longrightarrow & \mathrm{Div}^0(X) \\ \downarrow & & \downarrow \\ (\Sigma_X^*)^{\perp\!\!\perp}/\Sigma_X^* & \longrightarrow & \mathrm{Pic}^0(X). \end{array}$$

PROPOSITION 5.1. *There are maps ι, π such that the following sequence is exact:*

$$0 \to \prod_{x\in X}(1+\widehat{\mathfrak{m}}_x)N_x \overset{\iota}{\to} (\Sigma_X^*)^\perp \cap \mathbb{A}_X^+ \overset{\pi}{\to} \mathrm{Hom}_{\mathrm{gr}}\big(\mathrm{Pic}(X), k^*\big).$$

As a consequence, there is an induced exact sequence

$$0 \to k^* \prod_{x\in X} N_x \to \big((\Sigma_X^*)^{\perp\!\!\perp} \cap \bar{V}_0\big) \overset{\bar{\pi}}{\to} \mathrm{Hom}_{gr}\big(\mathrm{Pic}^0(X), k^*\big).$$

The maps π and $\bar{\pi}$ are surjective if k is either finite or algebraically closed.

PROOF. ι is the inclusion since by Theorem 3.3, we have

$$\prod_{x\in X}(1+\widehat{\mathfrak{m}}_x)N_x \subseteq \mathbb{A}_X^+ \cap (\Sigma_X^*)^\perp = \ker\big((\Sigma_X^*)^\perp \to \mathrm{Div}(X)\big).$$

To define π on an idele λ, consider the standard action of ideles on divisors,

$$(5.2) \qquad D = \sum_{x\in X} n_x x \longmapsto \lambda(D) = \prod_{x\in X} N_{k(x)/k}(\lambda(x))^{n_x} \; : \; \mathrm{Div}(X) \to k^*.$$

One checks easily that $\lambda(D(f)) = 1$ for $\lambda \in \ker\big((\Sigma_X^*)^\perp \to \mathrm{Div}(X)\big)$ and $f \in \Sigma_X^*$, hence λ induces a map $\mathrm{Pic}(X) = \mathrm{Div}(X)/\Sigma_X^* \to k^*$, which is $\pi(\lambda)$. The details regarding the maps in the second sequence and exactness of both are given in [**11**]. □

We are finally in a position to state the main result.

THEOREM 5.2. *Let k be either finite or algebraically closed. There is an exact sequence*

$$(5.3) \qquad 0 \to \mathrm{Hom}_{gr}\big(\mathrm{Pic}^0(X), k^*\big) \to (\Sigma_X^*)^{\perp\!\!\perp}/(\Sigma_X^* \prod_{x\in X} N_x) \to \mathrm{Pic}^0(X).$$

Moreover, for $k = \mathbb{C}$, the last arrow is surjective.

PROOF. We give a brief sketch of the proof. First, apply the Snake Lemma to

$$\begin{array}{ccccccccc} 0 & \longrightarrow & \Sigma_X^* & \longrightarrow & (\Sigma_X^*)^{\perp\!\!\perp} & \longrightarrow & (\Sigma_X^*)^{\perp\!\!\perp}/\Sigma_X^* & \longrightarrow & 0 \\ & & \downarrow & & \downarrow & & \downarrow & & \\ 0 & \longrightarrow & \Sigma_X^*/k^* & \longrightarrow & \mathrm{Div}^0(X) & \longrightarrow & \mathrm{Pic}^0(X) & \longrightarrow & 0. \end{array}$$

Bearing in mind the other results in this section, we obtain the exactness of the sequence

$$0 \to k^* \to \ker\big((\Sigma_X^*)^{\perp\!\!\perp} \to \mathrm{Div}^0(X)\big) \to (\Sigma_X^*)^{\perp\!\!\perp}/\Sigma_X^* \to \mathrm{Pic}^0(X)$$

and we apply Proposition 5.1.

To prove the surjectivity for $k = \mathbb{C}$, we have relied on the following facts: (1) any rational function is a quotient of products of prime forms [**7**], (2) the divisors of the form $a - b$ generate $\mathrm{Div}^0(X)$ and (3) the explicit form of the reciprocity symbols which we discussed in §3. □

COROLLARY 5.3. *If $k = \mathbb{C}$ and X is an elliptic curve, then*

$$(\Sigma_X^*)^{\perp\!\!\perp}/\Sigma_X^* \neq 0.$$

PROOF. We may write X as a Tate curve $X \simeq \mathbb{C}^*/q^{\mathbb{Z}}$ with $|q| < 1$. Then the map

$$z \longmapsto e^{2\pi i \frac{\log|z|}{\log|q|}} : \mathbb{C}^*/q^{\mathbb{Z}} \to \mathbb{C}^*$$

is a non-trivial group homomorphism

$$\mathrm{Pic}^0(X) \simeq X \simeq \mathbb{C}^*/q^{\mathbb{Z}} \to \mathbb{C}^*$$

from which one deduces that $\mathrm{Hom}_{\mathrm{gr}}(\mathrm{Pic}^0(X), \mathbb{C}^*) \neq 0$. □

6. Applications

6.1. Study of field extensions. One application of these results is to study field extensions of Σ_X in the adeles. On the one hand, we have first an example of the kind of results one may expect to obtain via topological characterizations.

THEOREM 6.1. *For a field extension Ω/Σ_X inside the adeles, $\Sigma_X \subset \Omega \subset \mathbb{A}_X$, the following conditions are equivalent:*

(1) $\dim_k \Omega \cap \mathbb{A}_X^+ < \infty$.
(2) *Ω is discrete (and therefore closed).*
(3) $\Omega^* \subset \mathbb{I}_X^1$.
(4) $\Omega \cap \mathbb{A}_X^+ = k$.
(5) $\Omega^* \cap \mathbb{A}_X^+ = k^*$.

PROOF. See [**11**, Theorem 2.5]. □

REMARK 6.2. If any of the conditions in the theorem hold, then it also follows that Ω^* is discrete in $\mathbb{I}_X$. However, the converse is false and is incorrectly stated in the above reference.

On the other hand, the following result shows how the group $(\Sigma_X^*)^{\perp}$, defined by orthogonality with respect to the global reciprocity symbol, is also relevant for such a study.

THEOREM 6.3. *Let k be algebraically closed and let Ω be a field with $\Sigma_X^* \subset \Omega^* \subset (\Sigma_X^*)^{\perp}$. Then*

(1) *Ω^* is discrete in $\mathbb{I}_X$.*
(2) *The natural map $\Omega^* \to (\Sigma_X^*)^{\perp\!\!\perp}$ is injective and its image is discrete.*
(3) *There is a natural injection*

$$\Omega^*/\Sigma_X^* \hookrightarrow \mathrm{Pic}^0(X).$$

PROOF. See [**11**, Theorem 5.2]. □

6.2. Residues. Let us finish by sketching how the discussion in §1.3 is connected to our general approach. We refer the reader to [**13**] for the details.

Recalling that the group $\mathbb{A}_X$ is defined as a restricted product, we may set:

$$\operatorname{Lie}\mathbb{I}_X = \{1+\epsilon\alpha_x \in k[\epsilon]/(\epsilon^2)\otimes_k K_x \text{ s.t. } 1+\epsilon\alpha_x \in k[\epsilon]/(\epsilon^2)\otimes_k \widehat{\mathcal{O}}_{X,x} \text{ for almost all } x\}$$

and there is thus a canonical isomorphism $\mathbb{A}_X \simeq \operatorname{Lie}\mathbb{I}_X$. Accordingly, the global symbol (3.2) yields a pairing

$$\langle\cdot,\cdot\rangle_{\text{Lie}} : \mathbb{A}_X \times \mathbb{A}_X \longrightarrow k$$

where $\langle\cdot,\cdot\rangle_{\text{Lie}}$ is defined as follows. For $\alpha,\beta \in \mathbb{A}_X$, $\langle\alpha,\beta\rangle_{\text{Lie}}$ is defined by the relation

$$1+\epsilon_1\epsilon_2\langle\alpha,\beta\rangle_{\text{Lie}} = [1+\epsilon_1\alpha, 1+\epsilon_2\beta],$$

where the second bracket is the commutator. The explicit computation in [**13**, Thm 4.10] yields

$$\langle\alpha,\beta\rangle_{\text{Lie}} = \sum_{x\in X} \operatorname{tr}_{k(x)/k} \operatorname{Res}_x \alpha\, \mathrm{d}\beta \qquad (\alpha,\beta\in\mathbb{A}_X)$$

where $\operatorname{tr}_{k(x)/k}$ is the trace and $\operatorname{Res}_x : K_x \simeq k(x)((z_x)) \to k(x)$ maps $\alpha_x = \sum_i a_i z_x^i \in k(x)((z_x))$ to a_{-1} (this does not depend on the choice of a formal parameter z_x at x).

In particular, via this mechanism, the Weil Reciprocity Law now implies the Residue Theorem, which is the statement that Σ_X is totally isotropic for $\langle\cdot,\cdot\rangle_{\text{Lie}}$.

We finish with the following result, which can be thought of as an algebraicity criterion for adeles based on the Residue Theorem.

THEOREM 6.4. *Under the hypotheses on the curve X and the field k in §1.5 and* $\operatorname{char} k = 0$, *we have*

$$\Sigma_X^{\perp} = \Sigma_X + \prod_{x\in X} k(x).$$

PROOF. The Residue Theorem implies the reverse inclusion $\Sigma_X + \prod_{x\in X} k(x) \subseteq \Sigma_X^{\perp}$. Now, let $\alpha\in\Sigma_X^{\perp}$. Then the linear map

$$\bar\alpha : \mathbb{A}_X \longrightarrow k \quad : \quad \beta \longmapsto \langle\alpha,\beta\rangle_{\text{Lie}} = \sum_{x\in X} \operatorname{tr}_{k(x)/k} \operatorname{Res}_x \alpha\,\mathrm{d}\beta$$

factors through the quotient to give a map $\mathbb{A}_X/\Sigma_X \to k$. It is also straightforward that it factors through

$$\mathbb{A}_X/(\Sigma_X + U_D) \longrightarrow k$$

where D is an effective divisor on X such that $D + D(\alpha) = D + \sum_{x\in X} v_x(\alpha)x \geq 0$. The following diagram provides a geometrical interpretation of $\mathbb{A}_X/(\Sigma_X + U_D)$:

$$\begin{array}{ccccccccc}
0 \to & H^0(X,\mathcal{O}_X(-D)) & \to & \Sigma_X & \to & H^0(X,\Sigma_X/\mathcal{O}_X(-D)) & \to & H^1(X,\mathcal{O}_X(-D)) & \to 0 \\
 & \| & & \| & & \| & & \| & \\
0 \to & U_D \cap \Sigma_X & \to & \Sigma_X & \to & \mathbb{A}_X/U_D & \to & \mathbb{A}_X/(U_D+\Sigma_X) & \to 0.
\end{array}$$

Since X is smooth, the dualizing sheaf is given by the canonical sheaf ω_X. Thus, the linear map $\bar\alpha$ is given by a meromorphic differential $\omega \in H^0(X,\omega_X(D)) \simeq H^1(X,\mathcal{O}_X(-D))^*$ via the relation $\bar\alpha(\beta) = -\sum_{x\in X}\operatorname{Res}_x \beta\omega$, that is to say, $\mathrm{d}\alpha = \omega$.

Now, let us consider a non-empty open subset $U \subset X$ and a rational function $f\in\Sigma_X$ such that $\mathrm{d}f = \omega$. Since $f\in\Sigma_X^{\perp}$, it follows that also $\alpha - f \in \Sigma_X^{\perp}$, which is tantamount to assuming that there is a non-empty open subset U such that

$d(\alpha - f)|_U = 0$. Since $\prod_{x\in X} k(x) \in \Sigma_X^\perp$, we may assume without loss of generality that $\alpha|_U = 0$. Accordingly, for such an α, the condition $\alpha \in \Sigma_X^\perp$ means that

$$\sum_{x\in X\setminus U} \operatorname{tr}_{k(x)/k} \operatorname{Res}_x \alpha \, dg = 0 \qquad \forall g \in \Sigma_X,$$

but this implies that $\alpha_x \in k(x)$ for all $x \in X \setminus U$ and the conclusion follows. □

Acknowledgments

We would like to dedicate this paper to José María Muñoz Porras, Professor at the University of Salamanca, who passed away on December 31st, 2018. He was our friend and a brilliant mathematician, who actively participated in several meetings of "Geometry at the Frontier" of which he had very pleasant memories. He had in mind the study of algebraic extensions in terms of adeles, and many ideas of this paper have been inspired by him.

In particular, the results presented here are a detailed version of the talk titled *Algebraic Curves, Residues and Weil Reciprocity Law* given at the *Geometry at the Frontier III* conference in November 2018.

References

[1] G. W. Anderson and F. Pablos Romo, *Simple proofs of classical explicit reciprocity laws on curves using determinant groupoids over an Artinian local ring*, Comm. Algebra **32** (2004), no. 1, 79–102, DOI 10.1081/AGB-120027853. MR2036223

[2] E. Arbarello, C. De Concini, and V. G. Kac, *The infinite wedge representation and the reciprocity law for algebraic curves*, Theta functions—Bowdoin 1987, Part 1 (Brunswick, ME, 1987), Proc. Sympos. Pure Math., vol. 49, Amer. Math. Soc., Providence, RI, 1989, pp. 171–190, DOI 10.1007/bf01228409. MR1013132

[3] E. Artin and G. Whaples, *Axiomatic characterization of fields by the product formula for valuations*, Bull. Amer. Math. Soc. **51** (1945), 469–492, DOI 10.1090/S0002-9904-1945-08383-9. MR13145

[4] J.-L. Brylinski and P. Deligne, *Central extensions of reductive groups by* $\mathbf{K}_2$, Publ. Math. Inst. Hautes Études Sci. **94** (2001), 5–85, DOI 10.1007/s10240-001-8192-2. MR1896177

[5] *Algebraic number theory*, Proceedings of an instructional conference organized by the London Mathematical Society (a NATO Advanced Study Institute) with the support of the International Mathematical Union. Edited by J. W. S. Cassels and A. Fröhlich, Academic Press, London; Thompson Book Co., Inc., Washington, D.C., 1967. MR0215665

[6] D. Dikranjan, *Introduction to Topological Groups*, `https://users.dimi.uniud.it/~dikran.dikranjan/ITG.pdf`.

[7] J. D. Fay, *Theta functions on Riemann surfaces*, Lecture Notes in Mathematics, Vol. 352, Springer-Verlag, Berlin-New York, 1973. MR0335789

[8] A. Grothendieck, *Éléments de géométrie algébrique. IV. Étude locale des schémas et des morphismes de schémas IV* (French), Inst. Hautes Études Sci. Publ. Math. **32** (1967), 361. MR238860

[9] G. Köthe, *Topological Vector Spaces*, Grundlehren der mathematischen Wissenschaften, Springer-Verlag Berlin Heidelberg, 1983.

[10] S. S. Kudla, *Tate's thesis*, An introduction to the Langlands program (Jerusalem, 2001), Birkhäuser Boston, Boston, MA, 2003, pp. 109–131. MR1990377

[11] J. M. Muñoz Porras, L. M. Navas Vicente, F. Pablos Romo, and F. J. Plaza Martín, *An idelic quotient related to Weil reciprocity and the Picard group*, Collect. Math. **71** (2020), no. 1, 151–171, DOI 10.1007/s13348-019-00252-7. MR4047704

[12] J. M. Muñoz Porras and F. Pablos Romo, *Generalized reciprocity laws*, Trans. Amer. Math. Soc. **360** (2008), no. 7, 3473–3492, DOI 10.1090/S0002-9947-08-04554-6. MR2386233

[13] J. M. Muñoz Porras and F. J. Plaza Martín, *Weil Reciprocity Law and the Theorem of Residues*, `http://arxiv.org/abs/1905.11774`

[14] T. Ono, *On algebraic groups defined by norm forms of separable extensions*, Nagoya Math. J. **11** (1957), 125–130. MR85229
[15] F. Pablos Romo, *On the tame symbol of an algebraic curve*, Comm. Algebra **30** (2002), no. 9, 4349–4368, DOI 10.1081/AGB-120013323. MR1936475
[16] F. J. Plaza Martín, *Arithmetic infinite Grassmannians and the induced central extensions*, Collect. Math. **61** (2010), no. 1, 107–129, DOI 10.1007/BF03191229. MR2604862
[17] J.-P. Serre, *Groupes algébriques et corps de classes* (French), Publications de l'institut de mathématique de l'université de Nancago, VII. Hermann, Paris, 1959. MR0103191
[18] *The Stacks project*, `https://stacks.math.columbia.edu` (2019).
[19] J. Tate, *Residues of differentials on curves*, Ann. Sci. École Norm. Sup. (4) **1** (1968), 149–159. MR227171
[20] J. Tate, *Symbols in arithmetic*, Actes du Congrès International des Mathématiciens (Nice, 1970), Gauthier-Villars, Paris, 1971, pp. 201–211. MR0422212
[21] A. Weil, *Basic number theory*, Classics in Mathematics, Springer-Verlag, Berlin, 1995. Reprint of the second (1973) edition. MR1344916
[22] A. Weil, *Généralisation des fonctions abéliennes*, J. Math. Pures et Appl. **17** (1938), pp. 47–87.
[23] A. Weil, *Sur la théorie du corps de classes* (French), J. Math. Soc. Japan **3** (1951), 1–35, DOI 10.2969/jmsj/00310001. MR44569

Departamento de Matemáticas and IUFFYM, Universidad de Salamanca, Plaza de la Merced 1-4, 37008 Salamanca, Spain
Email address: `navas@usal.es`

Departamento de Matemáticas and IUFFYM, Universidad de Salamanca, Plaza de la Merced 1-4, 37008 Salamanca, Spain
Email address: `fplaza@usal.es`

Contemporary Mathematics
Volume **766**, 2021
https://doi.org/10.1090/conm/766/15387

Remarks on the size of the Tamagawa product

Hector Pasten

ABSTRACT. The Tamagawa product of an elliptic curve is a quantity that naturally appears in the Birch and Swinnerton-Dyer conjecture, and it is often conjectured to be small with respect to the conductor. In this note we propose a precise version of such a conjectural bound, we give some evidence for it, and we investigate arithmetic consequences beyond the context of the Birch and Swinnerton-Dyer conjecture.

1. Tamagawa factors

Let E be an elliptic curve over $\mathbb{Q}$. The Tamagawa factor of E at a prime p is denoted by $c_p(E)$. We recall that $c_p(E) = [\mathscr{E}(\mathbb{F}_p) : \mathscr{E}_0(\mathbb{F}_p)]$, where $\mathscr{E}$ is the Neron model of E over $\mathbb{Z}$ and $\mathscr{E}_0$ is the identity component of $\mathscr{E}$. Equivalently, $c_p(E)$ is the number of irreducible components of $\mathscr{E} \otimes \mathbb{F}_p$ over $\mathbb{F}_p$. Thus, $c_p(E) = 1$ if E has good reduction at p.

The Tamagawa product $\tau(E)$ is defined by

$$\tau(E) = \prod_p c_p(E)$$

where the product is over all primes p. Tate [**Ta75**] proved that $\tau(E)$ is precisely the "fudge factor" appearing in the Birch and Swinnerton-Dyer conjectural formula for the first non-zero Taylor coefficient of $L(E,s)$ at $s = 1$. Due to this, the Tamagawa product plays an important role in various questions involving other quantities appearing in the Birch and Swinnerton-Dyer conjecture, such as the study of the conjectural size of Tate-Shafarevich groups [**dW98, Ni00, Hi07**], statistics of ranks of elliptic curves [**Wa08**], or heuristics for uniform boundedness of ranks [**PPVW16**].

The main point in this note is to clarify the role of $\tau(E)$ in Diophantine problems beyond the context of the Birch and Swinnerton-Dyer conjecture. We will be interested in the size of $\tau(E)$, first with respect to the minimal discriminant of E, and then with respect to the conductor N_E. It turns out that showing that $\tau(E)$ is small with respect to the conductor is a difficult open problem closely related to the *abc*-conjecture. To me, it seems plausible that the following holds

CONJECTURE 1.1. *Let $\epsilon > 0$. For all elliptic curves E over $\mathbb{Q}$ we have*

$$\log \tau(E) \le \left(\frac{7\log 3}{3} + \epsilon\right)\frac{\log N_E}{\log\log N_E} + O_\epsilon(1).$$

Supported by FONDECYT Regular grant number 1190442.

Presumably, the constant $7(\log 3)/3 = 2.563...$ is not optimal, but is its fairly small and it is not too difficult to give some evidence for it. Among the Diophantine applications, we will explain how this conjecture relates to sub-exponential versions of the *abc*-conjecture.

2. Notation and preliminaries

We will use the following standard arithmetic functions for positive integers n:

- For a prime p the p-adic valuation of n is $v_p(n)$.
- $d(n)$ is the number of divisors of n. It satisfies $d(n) = \prod_{p|n}(1 + v_p(n))$.
- $\omega(n)$ is the number of different prime factors of n, without counting repetitions.
- $\mathrm{rad}(n)$ is the radical of n, i.e. it is the product of the different prime factors of n, without counting repetitions.

For any given $\epsilon > 0$, one has the following classical estimates:

$$\omega(n) \le (1+\epsilon)\frac{\log n}{\log\log n} + O_\epsilon(1), \quad \log d(n) \le (\log 2 + \epsilon)\frac{\log n}{\log\log n} + O_\epsilon(1).$$

In addition, we will consider de Weger's function (cf. [**dW98**])

$$W(n) = \prod_{p|n} v_p(n).$$

Erdös conjectured and de Weger proved that for any given $\epsilon > 0$

$$W(n) \le \left(\frac{\log 3}{3} + \epsilon\right)\frac{\log n}{\log\log n} + O_\epsilon(1).$$

For an elliptic curve E over $\mathbb{Q}$ we denote the absolute value of the minimal discriminant of E by Δ_E, and the conductor by N_E. The Ogg's formula and the Neron-Ogg-Shafarevich theorem gives that $N_E|\Delta_E$ and these two integers have the same prime factors.

We already observed that the Tamagawa factor of E at a prime p of good reduction satisfies $c_p(E) = 1$. Tate's algorithm [**Ta75**] provides further information:

THEOREM 2.1. *Let E be an elliptic curve over $\mathbb{Q}$ and let p be a prime. The Tamagawa factor $c_p(E)$ satisfies:*

(i) *$c_p(E) \le 4$ if E has additive reduction at p*
(ii) *$c_p(E) \le 2$ if E has non-split multiplicative reduction at p*
(iii) *$c_p(E) = v_p(\Delta_E)$ if E has split multiplicative reduction at p.*

The *abc*-conjecture and Szpiro's conjecture will be relevant in our discussion, so let us recall them here.

CONJECTURE 2.2 (*abc*-conjecture). *Let $\epsilon > 0$. There is a number κ_ϵ depending only on ϵ, such that for all coprime positive integers a, b, c with $a + b = c$ we have*

$$c \le \kappa_\epsilon \cdot \mathrm{rad}(abc)^{1+\epsilon}.$$

CONJECTURE 2.3 (Szpiro's conjecture). *Let $\epsilon > 0$. There is a number K_ϵ depending only on ϵ, such that for all elliptic curves E over $\mathbb{Q}$ we have $\Delta_E \le K_\epsilon \cdot N_E^{6+\epsilon}$.*

It is a classical result that, up to the precise values of the exponents, these two conjectures are essentially equivalent [**Fr89**].

3. Smallness of the Tamagawa factor

From Theorem 2.1 we get $\tau(E) \leq 4^{\omega(\Delta_E)} d(\Delta_E)$. Thus, the estimates in Section 2 give the following well-known lemma (cf. [**dW98, Hi07, PPVW16**])

LEMMA 3.1. *For all elliptic curves E over $\mathbb{Q}$ we have*

$$\log \tau(E) = O\left(\frac{\log \Delta_E}{\log\log \Delta_E}\right).$$

In particular, $\log \tau(E) = o(\log \Delta_E)$.

In fact, a more careful analysis gives

PROPOSITION 3.2. *Let $\epsilon > 0$. For all elliptic curves E over $\mathbb{Q}$ we have*

$$\log \tau(E) \leq \left(\frac{\log 3}{3} + \epsilon\right) \frac{\log(\Delta_E N_E)}{\log\log \Delta_E} + O_\epsilon(1).$$

PROOF. Let us factor $\Delta_E = AM$ where the primes of additive reduction appear in A, and those of multiplicative reduction appear in M. If $p|A$ then $v_p(N_E) \geq 2$, and since $N_E|\Delta_E$ we deduce

$$v_p(\Delta_E N_E) \geq \begin{cases} v_p(\Delta_E) & \text{if } p|M \\ 4 & \text{if } p|A. \end{cases}$$

Thus, $\tau(E) \leq W(\Delta_E N_E)$. We conclude by de Weger's bound on $W(n)$, using $N_E|\Delta_E$. ☐

Lemma 3.1 and Proposition 3.2 show that, in a sense, $\tau(E)$ is small. It is natural to expect bounds of this sort where Δ_E is replaced by the conductor N_E.

CONJECTURE 3.3 (Folklore). *For all elliptic curves E over $\mathbb{Q}$ we have* $\log \tau(E) = o(\log N_E)$.

Lemma 3.1 gives the following well-known fact (cf. [**Hi07, dW98**])

PROPOSITION 3.4. *Szpiro's conjecture implies Conjecture 3.3. In fact, Szpiro's conjecture implies the stronger estimate*

$$\log \tau(E) = O\left(\frac{\log N_E}{\log\log N_E}\right).$$

From our Proposition 3.2 we can in fact deduce more:

PROPOSITION 3.5. *Szpiro's conjecture implies Conjecture* 1.1.

PROOF. Szpiro's conjecture gives $\log(\Delta_E N_E) \leq (7+\epsilon)\log N_E + O_\epsilon(1)$. ☐

Thus, Conjecture 1.1 seems plausible, as it follows from Szpiro's conjecture. Furthermore, the following result is easy to show, and it provides unconditional evidence for Conjecture 1.1:

PROPOSITION 3.6. *Let $\epsilon > 0$. For all elliptic curves E over $\mathbb{Q}$ without primes of split multiplicative reduction, we have*

$$\log \tau(E) \leq (\log 4 + \epsilon)\frac{\log N_E}{\log\log N_E} + O_\epsilon(1).$$

PROOF. From the results in Section 2 we get

$$\log \tau(E) \leq \omega(N_E) \log 4 \leq (\log 4 + \epsilon) \frac{\log N_E}{\log \log N_E} + O_\epsilon(1).$$

□

We observe that there are plenty of elliptic curves as in Proposition 3.6. For instance, all elliptic curves with integral j-invariant satisfy the requirements, because they cannot have multiplicative reduction at any prime.

The hard case in conjectures 3.3 and 1.1 is, of course, when E has primes of split multiplicative reduction. In [**Pa17**] I proved the following result, which provides further unconditional evidence for Conjecture 3.3 including the cases of split multiplicative reduction:

THEOREM 3.7. *Let $\epsilon > 0$. For all semi-stable elliptic curves E over $\mathbb{Q}$, we have*

$$\log \tau(E) \leq \left(\frac{11}{2} + \epsilon\right) \log N_E + O_\epsilon(1).$$

In particular, for these elliptic curves we have $\log \tau(E) = O(\log N_E)$.

See [**Pa17**] for a more general theorem allowing controlled additive reduction. Unlike the previous results discussed so far, the proof of Theorem 3.7 is far from elementary —it is based on the theory of automorphic forms, Arakelov theory on Shimura curves, proved cases of Colmez's conjecture, and zero density estimates for L-functions, among other tools.

Improving from the estimate $\log \tau(E) = O(\log N_E)$ provided by Theorem 3.7, to the bound $\log \tau(E) = O(\log N_E / \log \log N)$ predicted by 1.1 (possibly with other numerical coefficient) might seem reachable —after all, it is just a $\log \log$ factor! However, I suspect that this is a deep problem. As an indication of this, in the next section we will see that such an improvement would lead to some remarkable Diophantine consequences.

4. Diophantine consequences

Along with Theorem 3.7, in [**Pa17**] I also proved the following unconditional result towards the abc-conjecture

THEOREM 4.1 ($d(abc)$-theorem). *Let $\epsilon > 0$. There is a number K_ϵ depending only on ϵ such that for all coprime positive integers a, b, c with $a + b = c$ we have*

$$d(abc) \leq K_\epsilon \cdot \mathrm{rad}(abc)^{8/3+\epsilon}.$$

In particular, $\log d(abc) = O(\log \mathrm{rad}(abc))$.

Here, I would like to propose a conjectural refinement of the $d(abc)$-theorem:

CONJECTURE 4.2. *There are constants K, M such that for all coprime positive integers a, b, c with $a + b = c$ we have*

$$\log d(abc) \leq K \cdot \frac{\log \mathrm{rad}(abc)}{\log \log \mathrm{rad}(abc)} + M.$$

It is easy to see that this refined $d(abc)$ estimate follows from the abc-conjecture

PROPOSITION 4.3. *The abc-conjecture implies that Conjecture 4.2 holds and that K can be taken as any real number with $K > \log 8$.*

PROOF. Since $abc \le c^3$, we have

$$\log d(abc) \le (\log 2 + \epsilon/3)\frac{\log(abc)}{\log\log(abc)} + O_\epsilon(1) \le (\log 8 + \epsilon)\frac{\log c}{\log\log c} + O_\epsilon(1).$$

The *abc*-conjecture states $\log c \le (1+\epsilon)\log \operatorname{rad}(abc) + O_\epsilon(1)$, hence the result. □

It turns out that the bound for $\tau(E)$ proposed in Conjecture 1.1 also implies Conjecture 4.2 without invoking the *abc*-conjecture, but the proof is less direct.

THEOREM 4.4. *Conjecture* 1.1 *implies that Conjecture* 4.2 *holds and that* K *can be taken as any real number with* $K > 7\log 3$.

PROOF. Let a, b, c be coprime positive integers with $a + b = c$. Consider the Frey elliptic curve

$$E: \quad y^2 = x(x-a)(x+b).$$

We have that E is semi-stable away from $p = 2$. Furthermore, $\Delta_E = 2^r(abc)^2$ and $N_E = 2^s\operatorname{rad}(abc)$, where $|s|, |t| \le 8$ (cf. p.256-257 [**Si09**]),.

Each $p > 2$ dividing N_E is a prime of multiplicative reduction for E, and $p = 2$ also is, unless $v_2(abc) \le 4$ (cf. [**DK95**]).

By the Chinese remainder theorem, we can choose an integer $D \le 8N_E$ coprime to $2N_E$ with the following properties regarding E^D (the quadratic twist of E by D):

(1) Each odd $p|N_E$ is a prime of split multiplicative reduction for E^D.
(2) If E has multiplicative reduction at 2, then E^D has split multiplictive reduction at 2.

If E has good or multiplicative reduction at $p = 2$ then

$$\tau(E^D) \ge \prod_p v_p(\Delta_E) \ge 2^{-8}\prod_p (1 + v_p(abc)) = 2^{-8}d(abc).$$

If E has additive reduction at $p = 2$ then $v_2(abc) \le 4$ and we have

$$\tau(E^D) \ge \prod_{p>2} v_p(\Delta_E) \ge \frac{1}{5}\prod_p (1 + v_p(abc)) = \frac{1}{5}d(abc).$$

Note that $N_{E^D} \le N_E D^2 = O(N_E^3)$. Conjecture 1.1 applied to E^D gives

$$\log \tau(E^D) \le \left(\frac{7\log 3}{3} + \epsilon\right)\frac{3\log N_E}{\log\log N_E} + O_\epsilon(1)$$

which proves the result. □

Finally, we show that Conjecture 1.1 implies an *abc*-inequality which, although is not as strong as the full *abc*-conjecture, it is certainly stronger than anything we can prove at present —e.g. the exponential bounds $\log c = O_\epsilon(\operatorname{rad}(abc)^{\alpha+\epsilon})$ for $\alpha = 15$ [**ST89**], $\alpha = 2/3$ [**SY91**], and $\alpha = 1/3$ [**SY01**].

COROLLARY 4.5. *Conjecture* 1.1 *implies the following sub-exponential version of the abc-conjecture:*

Let $\epsilon > 0$. *For all coprime positive integers* a, b, c *with* $a + b = c$ *we have*

$$\log c \le \exp\left((7\log 3 + \epsilon)\frac{\log \operatorname{rad}(abc)}{\log\log \operatorname{rad}(abc)}\right) + O_\epsilon(1).$$

PROOF. In view of Theorem 4.4, the crude bound

$$\log c \leq \left(\max_{p|abc} v_p(abc)\right) \log \operatorname{rad}(c) \leq d(abc) \log \operatorname{rad}(abc)$$

suffices. □

In particular, a bound for the Tamagawa product of elliptic curves as the one proposed in Conjecture 1.1 would allow one to get bounds for the height of solutions of various Diophantine equations, such as S-unit equations and Mordell equations, in a form that is much stronger than what is available today —these applications for abc-estimates are standard, see for instance [**MP13, vK14, vKM16**].

References

[DK95] Fred Diamond and Kenneth Kramer, *Modularity of a family of elliptic curves*, Math. Res. Lett. **2** (1995), no. 3, 299–304, DOI 10.4310/MRL.1995.v2.n3.a6. MR1338788

[Fr89] Gerhard Frey, *Links between solutions of $A - B = C$ and elliptic curves*, Number theory (Ulm, 1987), Lecture Notes in Math., vol. 1380, Springer, New York, 1989, pp. 31–62, DOI 10.1007/BFb0086544. MR1009792

[Hi07] Marc Hindry, *Why is it difficult to compute the Mordell-Weil group?*, Diophantine geometry, CRM Series, vol. 4, Ed. Norm., Pisa, 2007, pp. 197–219. MR2349656

[vK14] Rafael von Känel, *Integral points on moduli schemes of elliptic curves*, Trans. London Math. Soc. **1** (2014), no. 1, 85–115, DOI 10.1112/tlms/tlu003. MR3296485

[vKM16] R. von Känel, B. Matschke. *Solving S-unit, Mordell, Thue, Thue-Mahler and generalized Ramanujan-Nagell equations via Shimura-Taniyama conjecture.* arXiv preprint arXiv:1605.06079 (2016).

[MP13] M. Ram Murty and Hector Pasten, *Modular forms and effective Diophantine approximation*, J. Number Theory **133** (2013), no. 11, 3739–3754, DOI 10.1016/j.jnt.2013.05.006. MR3084298

[Ni00] Abderrahmane Nitaj, *Invariants des courbes de Frey-Hellegouarch et grands groupes de Tate-Shafarevich* (French), Acta Arith. **93** (2000), no. 4, 303–327, DOI 10.4064/aa-93-4-303-327. MR1759478

[PPVW16] Jennifer Park, Bjorn Poonen, John Voight, and Melanie Matchett Wood, *A heuristic for boundedness of ranks of elliptic curves*, J. Eur. Math. Soc. (JEMS) **21** (2019), no. 9, 2859–2903, DOI 10.4171/JEMS/893. MR3985613

[Pa17] H. Pasten, *Shimura curves and the abc conjecture.* (2017) arXiv:1705.09251

[Si09] Joseph H. Silverman, *The arithmetic of elliptic curves*, 2nd ed., Graduate Texts in Mathematics, vol. 106, Springer, Dordrecht, 2009. MR2514094

[ST89] C. L. Stewart and R. Tijdeman, *On the Oesterlé-Masser conjecture*, Monatsh. Math. **102** (1986), no. 3, 251–257, DOI 10.1007/BF01294603. MR863221

[SY91] C. L. Stewart and Kun Rui Yu, *On the abc conjecture*, Math. Ann. **291** (1991), no. 2, 225–230, DOI 10.1007/BF01445201. MR1129361

[SY01] C. L. Stewart and Kunrui Yu, *On the abc conjecture. II*, Duke Math. J. **108** (2001), no. 1, 169–181, DOI 10.1215/S0012-7094-01-10815-6. MR1831823

[Ta75] J. Tate, *Algorithm for determining the type of a singular fiber in an elliptic pencil*, Modular functions of one variable, IV (Proc. Internat. Summer School, Univ. Antwerp, Antwerp, 1972), Springer, Berlin, 1975, pp. 33–52. Lecture Notes in Math., Vol. 476. MR0393039

[Wa08] Mark Watkins, *Some heuristics about elliptic curves*, Experiment. Math. **17** (2008), no. 1, 105–125. MR2410120

[dW98] Benjamin M. M. de Weger, *$A + B = C$ and big Sh's*, Quart. J. Math. Oxford Ser. (2) **49** (1998), no. 193, 105–128, DOI 10.1093/qjmath/49.193.105. MR1617347

DEPARTAMENTO DE MATEMÁTICAS, PONTIFICIA UNIVERSIDAD CATÓLICA DE CHILE, FACULTAD DE MATEMÁTICAS, 4860 AV. VICUÑA MACKENNA, MACUL, RM, CHILE

Email address: `hpasten@gmail.com`

Selected Published Titles in This Series

766 **Paola Comparin, Eduardo Esteves, Herbert Lange, Sebastián Reyes-Carocca, and Rubí E. Rodríguez, Editors,** Geometry at the Frontier, 2021

765 **Michael Aschbacher,** Quaternion Fusion Packets, 2021

764 **Gabriel Cunningham, Mark Mixer, and Egon Schulte, Editors,** Polytopes and Discrete Geometry, 2021

763 **Tyler J. Jarvis and Nathan Priddis, Editors,** Singularities, Mirror Symmetry, and the Gauged Linear Sigma Model, 2021

762 **Atsushi Ichino and Kartik Prasanna,** Periods of Quaternionic Shimura Varieties. I., 2021

761 **Ibrahim Assem, Christof Geiß, and Sonia Trepode, Editors,** Advances in Representation Theory of Algebras, 2021

760 **Olivier Collin, Stefan Friedl, Cameron Gordon, Stephan Tillmann, and Liam Watson, Editors,** Characters in Low-Dimensional Topology, 2020

759 **Omayra Ortega, Emille Davie Lawrence, and Edray Herber Goins, Editors,** The Golden Anniversary Celebration of the National Association of Mathematicians, 2020

758 **Jan Šťovíček and Jan Trlifaj, Editors,** Representation Theory and Beyond, 2020

757 **Kaïs Ammari and Stéphane Gerbi, Editors,** Identification and Control: Some New Challenges, 2020

756 **Joeri Van der Veken, Alfonso Carriazo, Ivko Dimitrić, Yun Myung Oh, Bogdan D. Suceavă, and Luc Vrancken, Editors,** Geometry of Submanifolds, 2020

755 **Marion Scheepers and Ondřej Zindulka, Editors,** Centenary of the Borel Conjecture, 2020

754 **Susanne C. Brenner, Igor Shparlinski, Chi-Wang Shu, and Daniel B. Szyld, Editors,** 75 Years of Mathematics of Computation, 2020

753 **Matthew Krauel, Michael Tuite, and Gaywalee Yamskulna, Editors,** Vertex Operator Algebras, Number Theory and Related Topics, 2020

752 **Samuel Coskey and Grigor Sargsyan, Editors,** Trends in Set Theory, 2020

751 **Ashish K. Srivastava, André Leroy, Ivo Herzog, and Pedro A. Guil Asensio, Editors,** Categorical, Homological and Combinatorial Methods in Algebra, 2020

750 **A. Bourhim, J. Mashreghi, L. Oubbi, and Z. Abdelali, Editors,** Linear and Multilinear Algebra and Function Spaces, 2020

749 **Guillermo Cortiñas and Charles A. Weibel, Editors,** K-theory in Algebra, Analysis and Topology, 2020

748 **Donatella Danielli and Irina Mitrea, Editors,** Advances in Harmonic Analysis and Partial Differential Equations, 2020

747 **Paul Bruillard, Carlos Ortiz Marrero, and Julia Plavnik, Editors,** Topological Phases of Matter and Quantum Computation, 2020

746 **Erica Flapan and Helen Wong, Editors,** Topology and Geometry of Biopolymers, 2020

745 **Federico Binda, Marc Levine, Manh Toan Nguyen, and Oliver Röndigs, Editors,** Motivic Homotopy Theory and Refined Enumerative Geometry, 2020

744 **Pieter Moree, Anke Pohl, L'ubomír Snoha, and Tom Ward, Editors,** Dynamics: Topology and Numbers, 2020

743 **H. Garth Dales, Dmitry Khavinson, and Javad Mashreghi, Editors,** Complex Analysis and Spectral Theory, 2020

742 **Francisco-Jesús Castro-Jiménez, David Bradley Massey, Bernard Teissier, and Meral Tosun, Editors,** A Panorama of Singularities, 2020

For a complete list of titles in this series, visit the
AMS Bookstore at **www.ams.org/bookstore/conmseries/**.